AF389140

Mining Technologies Innovative Development

Mining Technologies Innovative Development

Editors

Sergey Zhironkin
Dawid Szurgacz

MDPI • Basel • Beijing • Wuhan • Barcelona • Belgrade • Manchester • Tokyo • Cluj • Tianjin

Editors
Sergey Zhironkin
Surface Mining Department
T.F. Gorbachev Kuzbass State
Technical University
Kemerovo
Russia

Dawid Szurgacz
DOH Hydraulics Center
Polska Grupa Górnicza S.A.
Katowice
Poland

Editorial Office
MDPI
St. Alban-Anlage 66
4052 Basel, Switzerland

This is a reprint of articles from the Special Issue published online in the open access journal *Energies* (ISSN 1996-1073) (available at: www.mdpi.com/journal/energies/special_issues/mining_technologies_innovative_development).

For citation purposes, cite each article independently as indicated on the article page online and as indicated below:

LastName, A.A.; LastName, B.B.; LastName, C.C. Article Title. *Journal Name* **Year**, *Volume Number*, Page Range.

ISBN 978-3-0365-3225-7 (Hbk)
ISBN 978-3-0365-3224-0 (PDF)

Contents

Preface to "Mining Technologies Innovative Development"

Modern challenges of mining industry innovative development require a technological breakthrough in the extraction of minerals based on a wide range of fundamental and applied research. The purpose of this book is to highlight the advanced opinions of leading scientists on the ways and prospects of the progress of mining technologies, to widen the polemic regarding the innovative development of mining technologiesamong leading research schools of the world mining science. Safe, "smart"and environmentally friendly mining is the core value promoted by this book. All its chapters address a broad range of mining technologies innovative development, pushing the boundaries of mining science and highlighting the complex nature of securing a prosperous future for the mining sector.

Sergey Zhironkin, Dawid Szurgacz
Editors

 *energies*

Editorial

Mining Technologies Innovative Development: Economic and Sustainable Outlook

Sergey Zhironkin [1,2,3,*] and **Dawid Szurgacz** [4]

1   Institute of Trade and Economy, Siberian Federal University, 79 Svobodny Avenue, 660041 Krasnoyarsk, Russia

2   School of Core Engineering Education, National Research Tomsk Polytechnic University, 30 Lenina Street, 634050 Tomsk, Russia

3   Department of Engineering and Economy, Mezhdurechensk Branch, T.F. Gorbachev Kuzbass State Technical University, 36 Stroiteley Street, 652881 Mezhdurechensk, Kemerovo Region, Russia

4   Center of Hydraulics DOH Ltd., 41-906 Bytom, Poland; dawidszurgacz@doh.com.pl

*   Correspondence: zhironkinsa@kuzstu.ru

check for
updates

Citation: Zhironkin, S.; Szurgacz, D. Mining Technologies Innovative Development: Economic and Sustainable Outlook. *Energies* **2021**, *14*, 8590. https://doi.org/ 10.3390/en14248590

Received: 24 November 2021
Accepted: 11 December 2021
Published: 20 December 2021

Publisher's Note: MDPI stays neutral with regard to jurisdictional claims in published maps and institutional affiliations.

1. Introduction

Today science faces the task of ensuring the innovative development of the mineral extractive sector of the economy in resource-rich countries, in the context of unfolding two opposite trends. The first trend is the growing volatility of commodity prices caused by increased competition both between the largest mining companies and between interchangeable types of raw materials, traditional and alternative energy. The second trend is the growing long-term demand for mineral resources, which is unabated due to the world population growth and the convergence of developed and developing countries.

The volatility inherent to the commodity market in the past two years has been exacerbated by an unprecedented exogenous factor—the COVID-19 pandemic, which caused a reduction in global industrial output and labor hours equivalent to the loss of 195 million jobs in the global labor market [1]. As a result of the pandemic in 2020, prices for such raw materials as nickel, copper and coal showed a double-digit decline. Most of the forty largest mining companies have lowered their production forecasts by 7–9%, while their net profit and total market capitalization have decreased by almost two times [2]. In such conditions, innovative development of mining technologies is the only way to ensure cost reduction and, thereby, the long-term stability of mining companies and to avoid resource crises.

The innovative development of mining technologies is also driven by the need for saturation the long-term demand for raw materials, the growth of which is attributed by experts to the "new industrialization" in BRICS countries, North Africa, South America, as well as to the large-scale digitalization and development of electric transport [3]. Moreover, the two aforementioned trends in the development of the global mining sector can be described as converging, due to the stimulating effect of climatic and pandemic factors on the long-term demand for raw materials. Consequently, we can talk about a consensus of views regarding the long-term prospects for demand for minerals, both in developed and developing countries, which is a positive factor in the development of technological modernization of the mining industry. This is of particular importance for the economies of the countries leading the extraction of raw materials on their continents [4].

At the same time, in the last decade, there has been a decrease in labor productivity in the mineral extractive sector, among the reasons for which one can single out the achievement of the limit of technologies and the specific capacity of equipment, the "closure" of the chains using raw materials within the framework of existing industries. In such conditions, an increase in the specific capacity of equipment for the extraction of minerals inevitably turns into an increase in capital and operating costs, which reduces the interest of investors in the modernization of the industry as a whole. On the contrary, there is a great need to improve technologies for automating mining processes, ensuring labor safety in the complex extraction of minerals. This is dictated by the peculiarities of mining in the 21st century, such as the depletion of rich deposit reserves, the need to develop poor and hard-to-reach deposits, the limited mineral resources for enrichment, the need to solve environmental problems and as a result, the increase in costs.

The extraction of certain types of mineral raw materials directly determines the possibilities of promoting in society the values created with the help of Industry 4.0 technologies. For example, the growth rate of demand for copper, without which it is impossible to develop high-tech industries and energy, has doubled over the past 20 years—from 12 to 24 million tons, with the prospect of growth by 2035 by another 50% [5]. In general, humanity cannot stop extracting mineral resources, but it must use the possibilities of their complex extraction and waste-free processing. To solve this problem, it is necessary to integrate the efforts of the scientific community, mining companies, national states and the public in a "triple" and "quadruple" helix, in which innovative technologies will be introduced in a timely manner for the benefit of not only mining companies, but also the whole society. This, in turn, should emphasize the desire of the mining scientific community to expand the range of decision-making centers in the process of developing discussions regarding the future innovative development of the mining industry.

Thus, the development of innovative mining technologies should provide an answer to a number of challenges facing the mining industry.

The first challenge is the need for a radical increase in labor productivity in the mineral extractive sector in the process of improving mining equipment and technologies. Over the past two decades, mining companies have been actively investing in the implementation of digital technologies of Industry 4.0 (automated and unmanned equipment, integration of exploration and geophysical surveys, design of mining enterprises, environmental modeling, etc.). However, for a full-scale technological transition throughout the mining industry, it is necessary to modernize traditional extractive technologies "inherited" from the 20th century. In other words, despite the amazing opportunities that digital technologies offer for reducing costs, increasing productivity and improving labor safety and technogenic security, these benefits cannot be realized without the innovative development of mining technologies and processes themselves. Innovative modernization of the mining industry means profound changes in its technological core, including the transition to integrated extraction of reserves and waste-free technologies, comprehensive replacement of obsolete equipment incompatible with digitalization and minimization of the risks associated with failures of production systems.

The second challenge facing the innovative development of the mining industry is the transition to geotechnology consistent with Mining 4.0. The current stage of geotechnology development is based on the return of mining into the category of high-tech industries, thanks to the expansion of the use of unmanned robotic complexes in underground and surface mines, digital telemetry, machine-to-machine Internet communications and artificial intelligence in the analysis of mining technology and design information. Such "smart mining" is associated with a more accurate assessment of mineral reserves based on current prices and costs, optimization of material flows, labor costs, equipment performance, advanced forecasting of equipment failures, virtual digital training of mining personnel and the creation of fully unmanned sites. The research of innovative geotechnology (underground, opencast, construction) aimed at improving drilling and blasting, excavation and loading, transport operations, planning and operational management of processes must go side by side with Mining 4.0 so that mining enterprises of the future can completely exclude people from mineral extraction processes. The future of geotechnology development lies in the highest level of coordination of equipment complexes and flexibility in managing technological processes in mines, which should make mining profitable at any level of prices and demand for raw materials.

The third challenge is "green mining", which means the diffusion of innovations originating in related industries into the mining industry. Today, the prevailing opinion is that the extraction of minerals as a natural-technological interaction is a divergent process that alienates a person from nature, despite the expanding range of nature-saving technologies [6]. At the same time, governments in many countries are tightening environmental requirements for mining companies, which negatively affects their profitability. Therefore, the innovative development of "green" mining should be a convergent process that brings together the production and consumption of minerals on the one hand, and the preservation of the environment on the other. Such opportunities for mining-and-environmental convergence are provided by a complete transition to land and water-saving technologies, recycling of water and other resources, and complete extraction of useful components in several cycles. This is the only way to solve the problem of reducing the technogenic impact on the environment from mining operations and to save used resources by reducing the costs of companies. This mainstream implementation of environmental innovations in the mining industry will allow it to develop in the context of transitioning to decarbonization—the reduction of greenhouse gases by mining enterprises in order to reduce the risks associated with climate change.

The fourth challenge facing the innovative development of mining is improving labor safety and avoiding social consequences and economic losses caused by industrial accidents. Despite the fact that mining is associated with a large number of risk factors for accidents that threaten the lives of many people, and not completely predictable processes in the Earth's crust, there is a consensus that the root cause of emergencies (gas explosions, landslides, rock bumps, rock bursts, flooding, etc.) is the human factor. At the same time, today, the role of people in the extraction of minerals is limited to the operating and maintenance of equipment with a highly specific capacity, which, on the one hand, can harm employees, and on the other, cause economic losses in the case of breakdowns due to the fault of personnel. Therefore, the development of technologies for ensuring labor safety at mining enterprises is closely related both to innovations in mining engineering and geotechnology and in the organization of mining operations. At a fundamental level, the creation of innovations in labor safety in the mining industry is associated with predicting the action of threats and the factors that form them, using "Smart Sensors" and "Cloud Analysis" of data on the state of rock arrays, equipment and work processes.

All of the listed challenges of the mining industry's innovative development require a technological breakthrough in the extraction of minerals based on a wide range of fundamental and applied research. The purpose of this Special Issue is to highlight the advanced opinions of leading scientists on the ways and prospects of the progress of mining technologies. As Guest Editors, we got a chance to contribute to the polemic regarding the innovative development of mining technologies, and to make sure of the necessity to involve the leading research schools of the world of mining science.

We welcome a positive response from the world mining science community, accompanied by nineteen high-quality articles in quantity, that confirm a sustained interest in the innovative development of mining technologies, as well as in reducing the man-made impact on the environment and improving occupational safety. Safe, "smart" and environmentally-friendly mining is the core value promoted by this Special Edition. All articles published in this Special Issue address a broad range of mining technologies innovative development, pushing the boundaries of mining science and highlighting the complex nature of securing a prosperous future for the mining sector.

The next section is devoted to a review of the research papers published in the 'Energies' Special Issue "Mining Technologies Innovative Development", reflecting the above-mentioned challenges.

2. Special Issue Articles' Short Review

Articles published in this Special Issue as a part of Energies can be grouped in accordance with their thematic areas, which correspond to the aforementioned challenges for the innovative development of mining technologies.

With regard to the first challenge—the improvement of mining equipment in order to radically increase labor productivity in the mineral resource sector—eight articles address innovations in the development of tumbling and ball mills, powered roof supports, conveyor belts and drives maintenance, in-engine oil systems for heavy-duty equipment, drilling tools and equipment selection for surface mining. Their provisions are connected with increasing the efficiency of mining equipment as a key factor of the productivity of mining enterprises.

The article by Góralczyk et al. [7] provides an overview of innovations in the operating of the comminution process in tumbling mills from the point of view of reducing energy consumption. The authors note that despite a significant number of studies carried out, in real industrial practice, the energy efficiency of the comminution process in tumbling mills remains low in relation to the optimal level of ore particle size for beneficiation, the efficiency of which depends on reducing the amount of waste. The relevance of the presented study lies in the fact that the modernization of heavy equipment is high-cost; process optimization using special models and control systems is the most preferred solution to reduce energy consumption. Therefore, research is currently focused on new methods of performance analysis, as well as control of internal dynamics.

Szurgacz et al. [8] presented a methodology for tests and assessment of powered roof support, which determines the possibility of automatic control during coal mining from deep seams. The powered roof support system has three main parts: a shearer, a scraper conveyor and a powered roof support. The article reflects a comprehensive methodology for estimation of the prospects of a powered roof support in real conditions, combining four directions, and presents the results of tests, certification and executive documentation, allowing determination of the prospects for the potential use of the system in the future. The conclusions drawn from the results of the presented study are a valuable source of information for the design of mining equipment systems in terms of improving the efficiency and safety of labor.

Bajda et al. [9] presented new results from tests of the strength parameters of conveyor belt splices carried out in laboratory and industrial conditions, which indicated the reasons for the decrease in strength. A conveyor belt, which is critical to the reliable and efficient operation of equipment, is the most expensive and least durable element, so repairing belt damages is important to ensure the smooth transport of bulk materials. A conveyor belt consists of a core, covers and edges, and cutting of the core causes concentrated stresses of the belt splices. During a splicing procedure, parts of a belt are joined with each other. In view of the above, the analysis performed by the authors made it possible to establish that the strength of conveyor belt splicing decreases mainly due to improper preparation of the spliced surfaces and various mechanical properties of the belts. The value of the conclusions obtained by the authors lies in the ability to increase the strength of conveyor belt splicing without the use of additional equipment and materials.

Bortnowski et al. [10] investigated models of copper ore ball mill drive systems in order to come up with a solution to reduce the energy intensity of starting a filled mill. The authors took as a basis two designs of drive systems with relatively high efficiency—low-speed synchronous motor with permanent magnets without a gearbox, and asynchronous high-efficiency motor with a gearbox and a fluid coupling. The average energy consumption per unit mass of crushed material and the hourly electricity consumption were taken as criteria for evaluating the energy consumption of copper ore ball mill drive systems. To analyze the data obtained, the Monte Carlo method was applied, using which it was revealed that the least energy-intensive solution for starting a copper ore ball mill is to use a drive with a low-speed synchronous motor with permanent magnets.

Patyk et al. [11] propose a new methodology for the selection of mining equipment for surface mining, based on the adoption of multi-criteria decisions in the design of sections, allowing optimization of the excavation and loading transport processes in open pits, taking into account operating costs. The authors developed a universal set of criteria for the selection of mining equipment, including those taking into account technical parameters and the experience in mining operations accumulated by the mining company. The study used the ELECTRE III ranking method to select alternative options for choosing mining equipment. The equipment selection methodology obtained by the authors can be applied in surface mining sections with similar mining conditions. The authors associate further prospects of this study with the formation of universal criteria for choosing mining equipment, taking into account data on the state of the environment and the quality of minerals.

The article by Szurgacz et al. [12] presents the results of a thermal imaging study of the state of the conveyors' power units of the main system for transporting coal from a longwall. The study used a non-contact method for measuring infrared radiation; subsequently, the obtained thermograms were analyzed to compare the temperature distribution, taking into account the operating time of the conveyors. As a result of the application of the proposed method, changes in the thermal state of the operating conveyor were identified, indicating a possible malfunction of the probability of fire, which allows early maintenance and minimizes the risk of their occurrence. The obtained data on the thermal state of the conveyor belt drive will help to improve the maintenance of the mine.

The article by Grzesiek et al. [13] presents an innovative method of data analysis for structure breakpoint detection in engine oil pressure data, which makes it possible to quickly predict equipment breakdowns. The presented method is based on the authors' findings showing that when a degradation process occurs, there is a change in engine oil pressure. As a result, the authors proposed an original method of statistical processing and analyzing data, which are transformed into a nearly monotonic function that can describe the process of changing the regime within the process. The use of the proprietary technique will allow for removing ambiguity in oil pressure analysis and improve its visualization, which, in turn, will simplify the procedure and increase the accuracy of diagnostics of mining equipment failures.

Bazaluk et al. [14] analyzed the constructive, technological and operational methods of increasing the productivity of the drilling tool, which is necessary to enlarge the volume of drilling and increase the production of hydrocarbons. The article presents a new analytical model of the interaction of a drill bit with a well, which takes into account the effect of imperfections in the manufacture of a bit on its strength. As a result of the study, a significant lateral force was identified that arises due to manufacturing errors and presses the drill bit against the borehole wall, which also depends on the material from which the drilling tool is made. The authors found that a geometrical imperfection of the drill bit has a minimal effect on the drill pipe system based on carbon fiber composite material, and this effect is maximized in steel drill pipes. The research conducted allows us to determine the permissible errors in the manufacturing of a drill bit to ensure its safe operation.

With regard to the second challenge—the innovative development of geotechnology up to the level of Mining 4.0—six articles were published devoted to the innovative technology of liquid hydrocarbons production, modeling the ventilation process of mines, studying the stability of mine workings and introducing digital technologies in open pit mining. The provisions of these articles are connected with increasing the efficiency of mining operations and the implementation of Mining 4.0.

Bazaluk et al. [15] presented the results of their study of innovative technology for the production of liquid hydrocarbons, which makes it possible to intensify this process by increasing the productivity of the borehole with an increase in the area of its filtering surface. The innovation presented in the author's study consists of the formation of circular cuts of a large diameter perpendicular to the borehole axis for the destruction of the rock mass between them as a result of the application of static or dynamic loads by the mass of the drilling tool. The main advantage of the proposed innovative technology for the intensification of hydrocarbon fluid production is low energy consumption for expanding the diameter of the well. The article presents a special tool developed by the authors for drilling annular notches and gives the results of a numerical analysis of the expansion of the borehole diameter in reservoir rocks, which increases in twenty times.

Janus et al. [16] published in their article the results of numerical modeling of the airflow in mine workings, innovative due to the rejection of the widespread simplification—the adoption of a constant geometry of the working model along its entire length. Instead, the authors used the exact reproduction of the sections of the modeled mine working. The author's method is based on laser scanning of mine workings with multi-point velocity measurements in selected cross-sections, and on the modeling of SAS turbulence. The methodology proposed by the authors made it possible to simplify the flow analysis and modeling process significantly, while its results are quite consistent with measurements on the site.

In an article by Adach-Pawelus et al. [17], the problem of stability in a group of headings in the copper ore mines was considered (by the example of the Legnica-Glogow Copper Belt), and a new method of application of a roof bolting system was proposed to protect headings driven in unfavorable conditions in a high horizontal stress field. An analysis of stability was performed by authors for a group of four headings in the Polkowice-Sieroszowice mine using the finite element method. The Hoek-Brown classification was used to determine rocks parameters (RocLab 1.0 software); the measuring device was a CSIRO HI probe, and in-situ measurements were used to identify stress field parameters. The authors concluded that the stability of headings depends on the direction of the maximum horizontal stress component, the shape, depth and cross-section surface area of the heading, as well as on the surrounding rock's stress and strain parameters.

Wajs et al. [18] presented an innovative method for obtaining and integrating digital data for the accurate inventorying of surface mines, including the physical and technical parameters of overburden and mineral, mining and geometric data on its occurrence and mine workings, the boundaries of support pillars and buffer zones, the success of mining operations in relation to established boundaries, the impact of mining operations on the land, documentation of the hazards of landslides and the environmental impact of mining operations, etc. The implementation of the author's method is presented in the form of the results of a measurement experiment carried out at the Mikoszów granite surface mine (Poland) using mobile LiDAR systems, combining sensors with automatic and global navigation on a mobile platform, which generates a precise 3D cloud of geospatial points. In the future, the implementation of the method presented by the authors should contribute to using the technological level of Industry 4.0 in surface mining.

Wodecki et al. [19] proposed a simple, reliable system for the automatic identification of drilling cycles in the process of monitoring the work of heavy-duty drilling rigs. Despite the fact that monitoring of drilling processes is widely used for analyzing the physical and technical properties of rocks and assessing bit wear, modern on-board monitoring and automatic data analysis systems are not highly reliable or versatile. In order to overcome these shortcomings, the article discusses innovative hardware and software monitoring based on electro-acoustic measurements, thereby identifying the stage of preliminary drilling with an intermediate amplitude, which masks real drilling cycles. The tests of the proposed monitoring system installed on the drilling rig in the landfill conditions consisted of comparing the data with that of the on-board monitoring system installed by the equipment manufacturer. The test results indicate the high efficiency and reliability of the authors' system.

Borkowski [20] presented a new vision of the problem of grinding copper ore with the use of a high-pressure water jet. The results of the published study indicate that the effect of the grinding method proposed by the author allows increasing the specific surface area of copper ore particles significantly, which, in turn, simplifies its further processing. The analysis of the efficiency of grinding copper ore by the water jet method in comparison with mechanical grinding (in a planetary ball mill) according to the criterion of specific energy consumption and the efficiency of the grinding process showed that the use of a high-pressure water jet gives lower energy absorption. The results of the study were supplemented with a provision on the relationship between the efficiency of grinding copper ore and the strength characteristics of the host rock; in particular, the highest efficiency was found for sandstone and shale ores.

The third challenge facing the innovative development of mining technologies—"green mining"—is reflected in the article by Pactwa et al. [21], in which traditional and new approaches to the integrated development of subsoil are analyzed, and post-mining is highlighted as the most advanced method of reclamation of mine workings, taking into account environmental, social and economic results. The prospects for the environmental and energy policy of the European Union are associated with a reduction in the number of mines; therefore, the authors argue the need for advanced planning of their future reclamation. For this purpose, the article analyzes the strengths and weaknesses of the new reclamation methods, in comparison with the traditional method of filling or flooding the worked-out underground space at the end of mining. The authors took into account the views of a number of stakeholder groups regarding the feasibility of reusing underground structures and concluded in favor of preserving the mine infrastructure and using it in post-mining as being consistent with the imperatives of sustainable development.

The next four articles are devoted to the fourth challenge—improving labor safety at mining enterprises and reducing social-and-economic losses caused by accidents. These articles consider the development of innovative technologies to reduce the risk of sudden methane emissions from coal seams and improve mobile sensors, UAV application in search and rescue actions in the mines, and geomechanics. The provisions of these articles are connected with achieving the target of zero fatal accidents caused by technogenic factors in mines.

Tutak et al. [22] presented a new approach to the analysis of mine ventilation impact on the risk of methane release and coal combustion. The authors consider the selection of a ventilation system that is suitable for a given coal mining area. The article considers the case of reconstruction of the entire ventilation system of the active section of the longwall, with the transition from a U-type to Y-type ventilation. Following the research presented in the article, an important conclusion was made regarding the use of a Y-type ventilation system, ensuring safety and efficiency in areas with a high risk of methane emissions, while reducing safety due to the spontaneous combustion of coal (which is not typical for U-type ventilation). The research results presented in the article significantly expand knowledge in the field of increasing the efficiency of ventilation in underground workings and providing safety in underground coal mining.

Zietek et al. [23] consider the prospects of introducing an innovative air quality measurement system using portable personal devices to report hazardous gas concentrations in deep mines, which is the cornerstone of safe mining. This system is based on the use of inexpensive microcontrollers and gas analyzers, as well as on the use of common models of smartphones for the calculation and graphical display of results. The existing systems of air hazard evaluations, despite their availability, do not allow for storing the data received from each miner during the shift and require expensive equipment for their analysis. The system proposed by the authors is devoid of these shortcomings. The tests monitoring this system at a mine located in Poland showed very encouraging results.

Zimroz et al. [24] published the results of a study on an innovative way to conduct search and rescue operations in deep mines using unmanned aerial vehicles (UAVs) that are able to locate victims in an accident who are unable to move or give a signal. The authors developed and presented a method for the automatic processing of an acoustic signal to detect a specific sound—a human voice—taking into account noise, including that from the UAVs. The method proposed in the article is based on the time-frequency representation and measurement of the distance between the UAV noise and the obtained data. The article reflects the results of two laboratory experiments of the method proposed by the authors and one experiment carried out in the mine. During the experiments, satisfactory results were obtained, confirming the effectiveness of the innovative method of search and rescue actions in an underground mine proposed by the authors.

Dwornik et al. [25] proposed a new method for the automatic detection of disturbances in the Earth's surface, applicable in clusters of intensive mining, based on SAR interferograms. The authors describe the procedures and results of the application of coherence and entropy and the spatial distribution of the interferogram phase, which were applied within a field area in mining areas in southern Poland (based on SAR Sentinel-1 images). Comparison of the results of applying the proprietary method with the results obtained by the method based on circular Gabor filters showed that the detection rate for the proprietary method lies within 34–83%, compared to 30–53% for the method based on Gabor filters. This indicates a significant improvement in the methods of detecting disturbances of the Earth's surface, as carried out by the authors.

3. Conclusions

The articles published in the Special Issue "Mining Technologies Innovative Development" cover the main challenges important for future prospects of subsoil extraction as a public effective and profitable business, as well as technologically advanced industry. In the near future, the mining industry must overcome the problems of structural changes in raw material demand and raise the productivity up to the level of high-tech industries to maintain profits. This means the formation of a comprehensive and integral response to such challenges as the need for innovative modernization of mining equipment and an increase in its reliability, the widespread introduction of Industry 4.0 technologies in the activities of mining enterprises, the transition to "green mining" and the improvement of labor safety and avoidance of man-made accidents. The answer to these challenges is impossible without involving a wide range of the scientific community in the publication of research results and the exchange of views and ideas.

To solve the problem, this Special Issue has combined the works of researchers from the world's leading centers of mining science on the development of mining machines and mechanical systems, surface and underground geotechnology, mineral processing, digital systems in mining, mine ventilation and labor protection, in geo-ecology. A special place among the articles in the Special Issue is given to the post-mining technologies research.

We associate the further advancement of scientific thought in the field of innovative development of mining technologies, on the one hand, with deepening research in the areas of searching for answers to the key challenges facing the mining industry. On the other hand, we are confident that future Special Issue of Energies, devoted to innovative development in mining technologies, will help consolidate and popularize the ideas of scientists and research teams from the world's leading mining countries, including Poland, the Russian Federation, China and Australia.

Author Contributions: All authors contributed equally to this work. All authors have read and agreed to the published version of the manuscript.

Funding: This research received no external funding.

Institutional Review Board Statement: Not applicable.

Informed Consent Statement: Not applicable.

Conflicts of Interest: The authors declare no conflict of interest.

References

1. International Labour Organization. COVID-19 and the World of Work. Available online: https://www.ilo.org/global/topics/coronavirus/lang--en/index.htm (accessed on 15 November 2021).
2. Mine 2020. Resilient and Resourceful: PwC's 17th Annual Review of Global Trends in the Mining Industry. Available online: https://www.pwc.com/gx/en/industries/energy-utilities-resources/publications/mine-2020.html (accessed on 15 November 2021).
3. OECD. How Exports of Mineral Commodities Contribute to Economy-Wide Growth. Available online: https://www.oecd.org/trade/topics/trade-in-raw-materials/ (accessed on 15 November 2021).
4. Trenberth, K. Earth's Global Energy Budget. *Bull. Am. Meteorol. Soc.* **2009**, *90*, 311–323. [CrossRef]
5. Dong, D.; Tukker, A.; Van der Voet, E. Modeling copper demand in China up to 2050. *J. Ind. Ecol.* **2019**, *23*, 1363–1380. [CrossRef]

6.	Engebretsen, R.; Brugger, F. Divergent corporates: Explaining mining companies divergent performance in health impact assessments. *Resour. Policy* **2021**, *74*, 102355. [CrossRef]

7.	Góralczyk, M.; Krot, P.; Zimroz, R.; Ogonowski, S. Increasing Energy Efficiency and Productivity of the Comminution Process in Tumbling Mills by Indirect Measurements of Internal Dynamics—An Overview. *Energies* **2020**, *13*, 6735. [CrossRef]

8.	Szurgacz, D.; Zhironkin, S.; Cehlár, M.; Vöth, S.; Spearing, S.; Liqiang, M. A Step-by-Step Procedure for Tests and Assessment of the Automatic Operation of a Powered Roof Support. *Energies* **2021**, *14*, 697. [CrossRef]

9.	Bajda, M.; Hardygóra, M. Analysis of Reasons for Reduced Strength of Multiply Conveyor Belt Splices. *Energies* **2021**, *14*, 1512. [CrossRef]

10.	Bortnowski, P.; Gładysiewicz, L.; Król, R.; Ozdoba, M. Energy Efficiency Analysis of Copper Ore Ball Mill Drive Systems. *Energies* **2021**, *14*, 1786. [CrossRef]

11.	Patyk, M.; Bodziony, P.; Krysa, Z. A Multiple Criteria Decision Making Method to Weight the Sustainability Criteria of Equipment Selection for Surface Mining. *Energies* **2021**, *14*, 3066. [CrossRef]

12.	Szurgacz, D.; Zhironkin, S.; Vöth, S.; Pokorný, J.; Spearing, A.J.S.; Cehlár, M.; Stempniak, M.; Sobik, L. Thermal Imaging Study to Determine the Operational Condition of a Conveyor Belt Drive System Structure. *Energies* **2021**, *14*, 3258. [CrossRef]

13.	Grzesiek, A.; Zimroz, R.; Śliwiński, P.; Gomolla, N.; Wyłomańska, A. A Method for Structure Breaking Point Detection in Engine Oil Pressure Data. *Energies* **2021**, *14*, 5496. [CrossRef]

14.	Bazaluk, O.; Velychkovych, A.; Ropyak, L.; Pashechko, M.; Pryhorovska, T.; Lozynskyi, V. Influence of Heavy Weight Drill Pipe Material and Drill Bit Manufacturing Errors on Stress State of Steel Blades. *Energies* **2021**, *14*, 4198. [CrossRef]

15.	Bazaluk, O.; Slabyi, O.; Vekeryk, V.; Velychkovych, A.; Ropyak, L.; Lozynskyi, V. A Technology of Hydrocarbon Fluid Production Intensification by Productive Stratum Drainage Zone Reaming. *Energies* **2021**, *14*, 3514. [CrossRef]

16.	Janus, J.; Krawczyk, J. Measurement and Simulation of Flow in a Section of a Mine Gallery. *Energies* **2021**, *14*, 4894. [CrossRef]

17.	Adach-Pawelus, K.; Pawelus, D. Influence of Driving Direction on the Stability of a Group of Headings Located in a Field of High Horizontal Stresses in the Polish Underground Copper Mines. *Energies* **2021**, *14*, 5955. [CrossRef]

18.	Wajs, J.; Trybała, P.; Górniak-Zimroz, J.; Krupa-Kurzynowska, J.; Kasza, D. Modern Solution for Fast and Accurate Inventorization of Open-Pit Mines by the Active Remote Sensing Technique—Case Study of Mikoszów Granite Mine (Lower Silesia, SW Poland). *Energies* **2021**, *14*, 6853. [CrossRef]

19.	Wodecki, J.; Góralczyk, M.; Krot, P.; Ziętek, B.; Szrek, J.; Worsa-Kozak, M.; Zimroz, R.; Śliwiński, P.; Czajkowski, A. Process Monitoring in Heavy Duty Drilling Rigs—Data Acquisition System and Cycle Identification Algorithms. *Energies* **2020**, *13*, 6748. [CrossRef]

20.	Borkowski, P.J. Comminution of Copper Ores with the Use of a High-Pressure Water Jet. *Energies* **2020**, *13*, 6274. [CrossRef]

21.	Pactwa, K.; Konieczna-Fuławka, M.; Fuławka, K.; Aro, P.; Jaśkiewicz-Proć, I.; Kozłowska-Woszczycka, A. Second Life of Post-Mining Infrastructure in Light of the Circular Economy and Sustainable Development—Recent Advances and Perspectives. *Energies* **2021**, *14*, 7551. [CrossRef]

22.	Tutak, M.; Brodny, J.; Szurgacz, D.; Sobik, L.; Zhironkin, S. The Impact of the Ventilation System on the Methane Release Hazard and Spontaneous Combustion of Coal in the Area of Exploitation—A Case Study. *Energies* **2020**, *13*, 4891. [CrossRef]

23.	Ziętek, B.; Banasiewicz, A.; Zimroz, R.; Szrek, J.; Gola, S. A Portable Environmental Data-Monitoring System for Air Hazard Evaluation in Deep Underground Mines. *Energies* **2020**, *13*, 6331. [CrossRef]

24.	Zimroz, P.; Trybała, P.; Wróblewski, A.; Góralczyk, M.; Szrek, J.; Wójcik, A.; Zimroz, R. Application of UAV in Search and Rescue Actions in Underground Mine—A Specific Sound Detection in Noisy Acoustic Signal. *Energies* **2021**, *14*, 3725. [CrossRef]

25.	Dwornik, M.; Porzycka-Strzelczyk, S.; Strzelczyk, J.; Malik, H.; Murdzek, R.; Franczyk, A.; Bała, J. Automatic Detection of Subsidence Troughs in SAR Interferograms using Mathematical Morphology. *Energies* **2021**, *14*, 7785. [CrossRef]

Article

Automatic Detection of Subsidence Troughs in SAR Interferograms Using Mathematical Morphology

Maciej Dwornik [1], Stanisława Porzycka-Strzelczyk [1,2], Jacek Strzelczyk [2], Hubert Malik [2], Radosław Murdzek [2], Anna Franczyk [1,*] and Justyna Bała [1]

1 Department of Geoinformatics and Applied Computer Science, AGH University of Science and Technology, 30-059 Krakow, Poland; dwornik@agh.edu.pl (M.D.); kontakt@satim.pl (S.P.-S.); jbala@agh.edu.pl (J.B.)
2 SATIM, 30-046 Krakow, Poland; jacek.strzelczyk@satim.pl (J.S.); hubert.malik@satim.pl (H.M.); radoslaw.murdzek@satim.pl (R.M.)
* Correspondence: franczyk@agh.edu.pl

Abstract: In this paper, an automatic algorithm for the detection of subsidence areas in SAR interferograms is proposed. It is based on the analysis of spatial distribution of the interferogram phase, and its coherence and entropy. The developed method was tested for differential interferograms generated on the basis of Sentinel-1 SAR images covering mining areas in South Poland. The obtained results were compared with those achieved using a method based on circular Gabor filters. Performed analysis revealed that the detection rate for the proposed method varied from 34% to 83%. It is an improved method based on Gabor filters that achieved a detection rate from 30% to 53%.

Keywords: radar interferometry; subsidence trough; automatic detection; mathematical morphology

Citation: Dwornik, M.; Porzycka-Strzelczyk, S.; Strzelczyk, J.; Malik, H.; Murdzek, R.; Franczyk, A.; Bała, J. Automatic Detection of Subsidence Troughs in SAR Interferograms Using Mathematical Morphology. *Energies* **2021**, *14*, 7785. https://doi.org/10.3390/en14227785

Academic Editors: Sergey Zhironkin and Dawid Szurgacz

Received: 26 October 2021
Accepted: 17 November 2021
Published: 20 November 2021

Publisher's Note: MDPI stays neutral with regard to jurisdictional claims in published maps and institutional affiliations.

1. Introduction

Differential interferometric SAR (DInSAR) analysis is a significant source of information in crisis management dealing with the problem of ground deformations [1–3]. DInSAR results can be used for monitoring landslides, volcanic activity [4], earthquakes [5], and mining-induced subsidence [6]. Nowadays, a very large amount of SAR data acquired by Sentinel-1 is freely available, making differential radar interferometry more accessible. However, to take real advantage of DInSAR in crisis management, it is necessary to ensure its automatic processing, which is a subject of many works [7–9]. A system that allows for the continuous monitoring of endangered areas and can alert about critical values of ground deformations is needed. Such a solution requires a technique for the automatic identification of subsidence patterns in SAR interferograms. There are many scientific papers on this problem that use different methods. In [10], the automatic detection of subsidence troughs in SAR interferograms based on circular Gabor filter was proposed. Its detection rate varied from 30% to 53% with a relatively low number of false alarms. In [11], circlet transform was compared with Hough transform [12,13]. There are also works that are based on a convolutional neural network [14,15]. Each of those methods has its limitations, which is why their efficiency in trough detection in noisy satellite images is low. Methods [10–13] use circular or elliptical troughs in the detection process. In many cases, the shape of the troughs on interferograms is deformed, which significantly reduces the effectiveness of these solutions. Moreover, the proposed method is much less computationally expensive, which is quite important due to the large size of the processed SAR images. However, most of the solutions [10–15] can be used to support the detection of subsidence in larger SAR images, but they are not suitable for fully automated operations.

Here, a new algorithm for the automatic detection of subsidence areas in radar interferograms is proposed. It identifies elliptical interferometric fringes that usually correspond to subsidence troughs. The developed method is intended for monitoring mining and postmining ground deformation. When elliptical fringes are caused by other deformation phenomena or atmospheric effects, they are also recognized. Therefore, the interpretation of the obtained results should be connected with a priori information about the studied region.

In the proposed algorithm, mathematical morphology methods are used for analyzing the phase, coherence, and entropy of interferograms, and for the identification of subsidence troughs. To speed up detection, the proposed algorithm consists of two stages. In the first stage, potential areas of subsidence troughs occurrence are identified. Afterwards, those areas are verified. The proposed method was tested on interferograms generated on the basis of Sentinel-1 SAR data.

2. Methodology

The proposed algorithm consists of two main steps. The first step eliminates from further analysis parts of the interferogram where, there are most probably no subsidence troughs. A map of the interferogram phase is used with maps of its coherence and entropy (calculated for data converted into unsigned int 8-bit types with Equation (1) in a moving window). For the phase map, lines with a sudden phase change (approximately from $-\pi$ to $+\pi$) are identified. Thos is motivated by the fact that such phase changes are characteristic for interferometric fringes that occur in the case of subsidence troughs represented by a full phase cycle. For this reason, subsidence troughs without full phase cycle may not be included in further analysis. This needs to be improved in the future.

$$Entropy = -\sum_{n=0}^{255} p_n \cdot \log_2 p_n \tag{1}$$

where p_n—sum of pixels with intensivity n divided by the number all pixels in window [16].

The identified lines that are very short are removed from the next steps of analysis. The continuity of the remained lines is improved by the closing operation with a linear structural element [17]. Areas inside those lines constitute a set of potential regions, where the centers of subsidence troughs may occur. An example of the obtained results is presented in the Figure 1 (obtained on the basis of the generated interferogram using Sentinel-1 data).

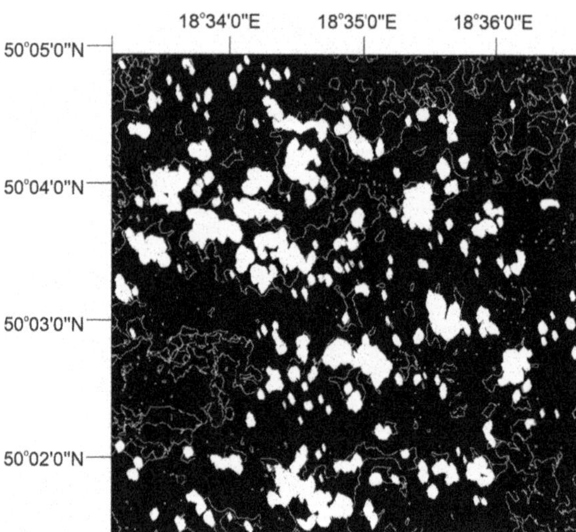

Figure 1. Identified lines with potential (white) regions where centers of subsidence troughs may occur.

From the map presented in Figure 1, all regions with low coherence (noisy regions) and low entropy (homogeneous regions) were removed by considering two logical images presenting coherence larger than a defined threshold ($\gamma_{th} = 0.6$) and entropy above a defined threshold ($H_{th} = 0.82$), respectively. To define the appropriate thresholds, tests were performed on several interferograms (different than those for which tests of the method were performed). Those interferograms, generated on the basis of Sentinel-1 images, differed from each other by time interval (6 and 12 days), and a number of subsidence troughs (from 1 to several). For the analyzed values of thresholds, the number of correctly detected troughs was divided by the number of total alarms. Thresholds that gave the highest value of this proportion were chosen as optimal. A procedure for the automatic selection of thresholds (γ_{th} and H_{th}) should be proposed in the future.

For a map with potential regions where the centers of subsidence troughs may occur, and logical maps of coherence and entropy, multiplication is applied. In Figure 2 ,the obtained results (areas surrounded by white lines) are presented on the phase image. Those areas represent more reliable regions (compared to those presented in Figure 1) where centers of the subsidence trough may occur (called perspective areas). The location and extent of perspective areas are an input for the second step of the detection procedure.

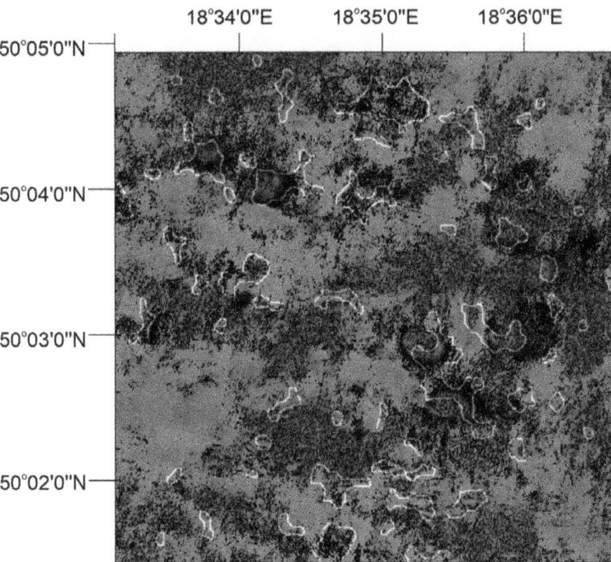

Figure 2. Location and extent of perspective areas presented in interferogram phase.

The second stage of the detection algorithm is much more time-consuming than the first one. For each pixel belonging to perspective areas, the center of the theoretical ellipse was set. The method of obtaining values the of theoretical ellipses radius is presented in the next section and Figure 3. After that, from this center, the rays in each direction (with 5° step) were created. The length of the rays was set up while taking into account the resolution of the analyzed images and the information about the probable maximal spatial extent of the subsidence trough (in our case, for Sentinel-1 data, this was set to 75 pixels). The correlation between the phase values along those rays and a saw function was calculated, and is presented in Figure 4 and described in Equation (2) .

$$\rho(r) = \pi - 2\pi \cdot \frac{\text{mod}\,(r, \omega)}{\omega} \tag{2}$$

where r—distance to the center of the theoretical ellipse, ω—a period of saw function, *mod*—modulo operation. The saw function represents the phase changes characteristic for the interferometric fringes. The shape and period of the saw function are based on a typical period in troughs in tested images.

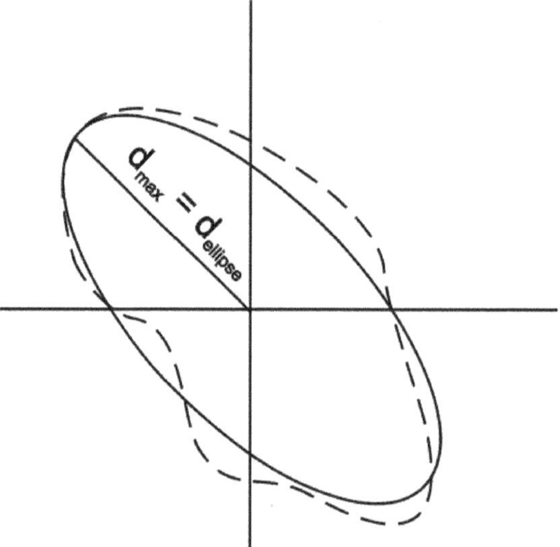

Figure 3. Theoretical ellipse and calculated shape of subsidence pattern.

After that, for each ray, distance d_{max} (from the center of the theoretical ellipse) to the maximal value of the calculated correlation was determined. The line connecting the d_{max} values for each ray determines the shape of the subsidence pattern (Figure 3—dotted line). The maximal and minimal values of dmax were used to define the minor and major axes, and orientation of the theoretical ellipse (Figure 3—solid line).

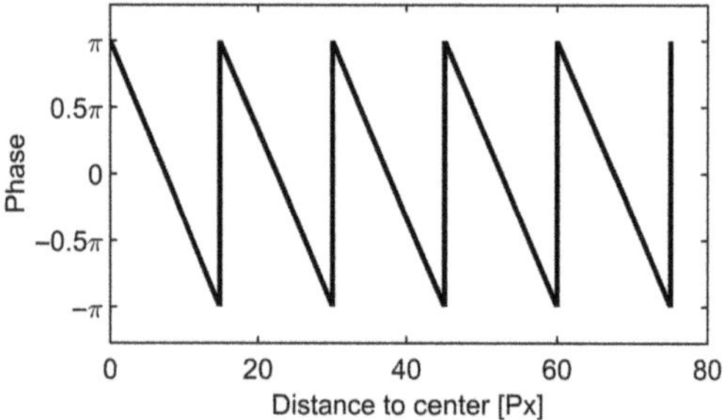

Figure 4. Example of saw function with period ω = 15 px.

In the last step of the algorithm, for each pixel from perspective areas, the fitting error ($d_{max}(k)$) between the distance of the maximal correlation function (d_{max}) for each radius and theoretical ellipse ($d_{ellipse}$) was calculated according to Equation (3).

$$d_{error} = \sqrt{\sum_{k=1}^{Number\ of\ radius} \left(d_{max}(k) - d_{ellipse}(k)\right)^2} \tag{3}$$

Values of d_{error} were normalized and inverted to the range of <0,1>, where 1 means $\forall_k : d_{max}(k) = d_{ellipse}(k)$. Lastly, only those pixels for which the error function was greater than a threshold (in our case, 0.85) were identified as subsidence areas. The lower the value of the threshold was, the greater the number of false alarms. The greater value of the threshold can suggest that some subsidence areas could be undetected.

3. Experimental Analysis

The developed method was applied to a dataset covering a selected region of the Upper Silesian Coal Basin (USCB) in South Poland. USCB is located in southern Poland and in the Ostrava-Karvina region in the Czech Republic. Mining activity there has been conducted for over 200 years. The USCB region is Poland's largest hard-coal basin and one of the largest hard-coal deposits in Europe. It is also a large metropolitan region with 37 towns and nearly 3 million residents. It is estimated that an area of around 600 km^2 already suffers from subsidence in USCB [18].

This is the same test site that was used in [10], where the detection method based on circular Gabor filters was presented and tested. For the selected region, the subsidence phenomenon is characteristic. The ground deformations are caused by extensive and long-lasting coal exploitation. The occurrence of subsidence troughs was confirmed many times using classical levelling or DInSAR analysis [19,20]. In the DInSAR method, a pair of SAR images that correspond to the same area and had been acquired at different times by the same SAR system were used. First, the analyzed SAR images need to be coregistrated, and the phase difference ($\Delta\phi$) is then calculated for each pixel of the image. In this way, the flattened interferogram is obtained. To remove the component associated with the topography of the studied region, the Digital Elevation Model (DEM) is used. DEM is converted into the synthetic interferogram, which is then subtracted from the flattened interferogram. As a result, the differential interferogram is obtained. Its phase values can be directly linked to ground deformations. The subsidence phenomenon, especially related to mining activity, usually manifests itself on the differential interferograms as elliptical interferometric fringes. The density of the fringes is correlated with the size of the deformations.

In the presented work, as in [10], the efficiency of the presented subsidence trough detection algorithm was tested using three differential interferograms generated on the basis of Sentinel-1 SAR data. The interferogram concerns the following time spans: 10–22 October 2016 (interferogram I), 10 October–3 November 2016 (interferogram II) and 22 October–3 November 2016 (interferogram III). All interferograms were prepared using SNAP. The interferograms were filtered twice, using the Goldstein and Lee filtering methods. Goldstein filtering was applied before the removal of the topographic phase, whereas the Lee filtering method was applied to differential interferograms. The parameters of both methods are shown in Table 1. The multilooking procedure was not applied during processing step. More information about the used datasets and their processing parameters can be found in [10]. The main stages of the subsidence trough detection algorithm were performed using the Mathworks MATLAB 2019b computing package.

Table 1. Parameters of filtering methods.

Goldstein Filtering	Lee Filtering
FFT size: 64 × 64	window size: 7 × 7
Window size: 3 × 3	target window size: 3 × 3
Coherence threshold : 0.2	sigma: 0.7
Adaptive filter exponent : 1.0	

The proposed algorithm of the automatic detection of subsidence areas was applied to both interferograms computed using VV and VH channels of SAR images. Results are presented in Table 2 and Figures 5–7. Table 2 $k1$ shows a number of undetected troughs that had a very irregular shape. $k2$ is a number of areas incorrectly classified as subsidence troughs (false alarms).

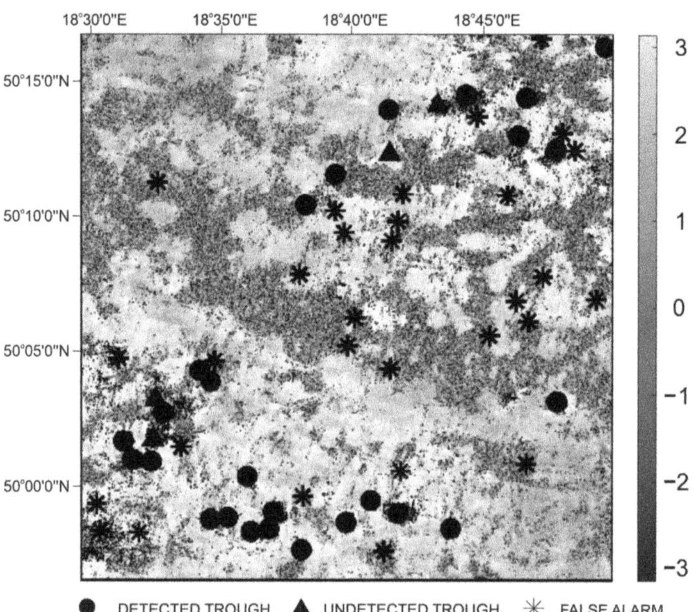

Figure 5. Results of subsidence area detection for interferogram I (time period: 10–22 October 2016; used VV polarimetric channel).

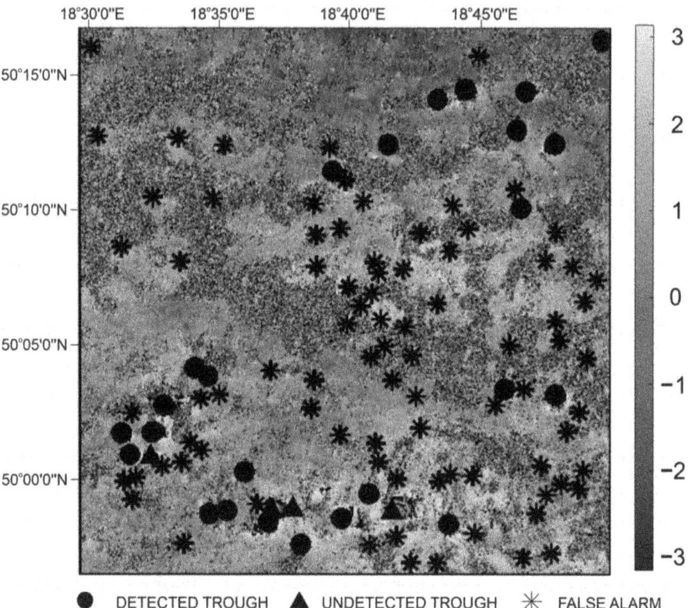

Figure 6. Results of subsidence area detection for interferogram II (time period: 10 October–3 November 2016; used VV polarimetric channel).

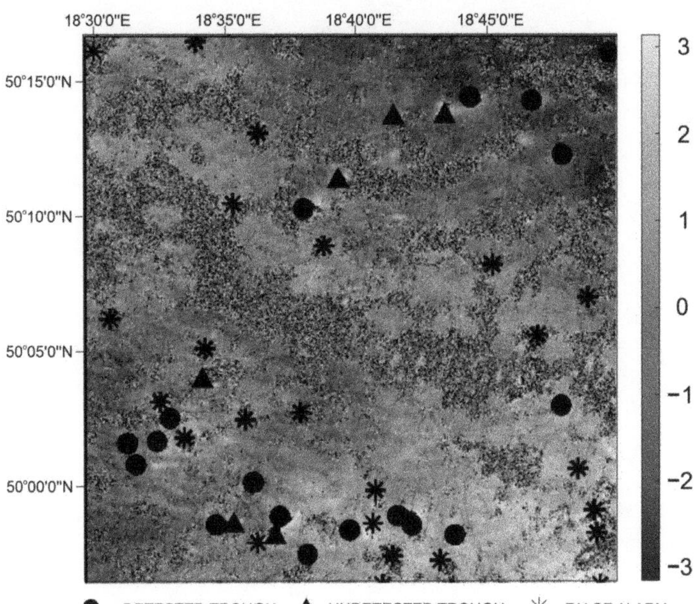

Figure 7. Results of subsidence area detection for interferogram III (time period: 22 October–3 November 2016; used VV polarimetric channel).

Table 2. Results of subsidence trough detection. $k1$—number of undetected troughs; $k2$—number of false alarms.

Interferogram I: 10–22 October 2016			
Polarimetric Channel	Detected Troughs	Undetected Troughs	False Alarms
VV	26	4	30
VH	25	5	35
VH and VV	25	5 ($k1 = 20\%$)	4 ($k2 = 0\%$)
Interferogram II: 10 October 2016–3 November 2016			
Polarimetric Channel	Detected Troughs	Undetected Troughs	False Alarms
VV	25	4	86
VH	10	19	102
VH and VV	10	19 ($k1 = 10\%$)	39 ($k2 = 7\%$)
Interferogram III: 22 October 2016–3 November 2016			
Polarimetric Channel	Detected Troughs	Undetected Troughs	False Alarms
VV	18	6	25
VH	17	7	34
VH and VV	17	7 ($k1 = 14\%$)	9 ($k2 = 11\%$)

Table 2 shows that, for all analyzed cases, the best results were obtained for VV channels. In the case of VH channels, fewer subsidence troughs were detected, and more false alarms were noticed. Taking into account the results of both polarimetric channels (VH and VV), the number of false alarms could be significantly reduced (even more than sevenfold for interferogram I). However, in such cases, the number of correctly detected troughs can also decrease more than twice the time (interferogram II). The best results of subsidence area detection were obtained for interferogram I (Figure 5). In this case, for VV channels, 26 subsidence troughs (86.7%) were correctly identified, and 30 false alarms were created. Table 3 shows that, for interferogram I, the results for VH channels are only slightly worse than for VV channels. Therefore, it is beneficial to take into account both channels. VH and VV results are characterized by a high number of correctly detected subsidence troughs (25, which accounts for 83.3% of all troughs) and a low number of false alarms (only 4 cases).

Table 3. Comparison of obtained results by the proposed algorithm and the method based on circular Gabor filters.

	Number of Troughs	Method Based on Circular Gabor Filters		Method Proposed in This Paper	
		Detected Troughs	False Alarms	Detected Troughs	False Alarms
Int. I	30	16	21	25	4
Int. II	29	12	5	10	39
Int. III	24	7	0	17	9

The detection results for interferogram II (VV channel) are presented in Figure 6. The number of correctly identified subsidence troughs was high and totalled 25 (86.2% of all troughs). However, for interferogram II, the number of false alarms was very high, equal to 86 cases. Additionally, results for the VH channel were significantly worse, and false alarms decreased by using a logical product of VV- and VH-based results, compound detection accuracy.

For the third interferogram (VV channel), the proposed method allowed for correctly detecting 18 of 24 (75%) subsidence troughs (Figure 7). The number of false alarms was 25, which may be significantly reduced to 9 cases when the results of VH and VV are taken into account. In such a case, the number of undetected troughs increased from 6 to 7.

Table 2 also shows that, for all analyzed interferograms, the number of undetected troughs with irregular shapes ($k1$) and a number of false alarms with a shape similar to subsidence pattern ($k2$) were very low, and in both cases did not exceed 20%. In the case of a method based on a circular Gabor filter [10], the $k1$ and value of $k2$ reached 60%. For the algorithm proposed in this paper, the irregularity of the subsidence pattern is thus not a crucial problem in the detection procedure. Figure 8 shows the example of two undetected subsidence patterns. The subsidence pattern presented in Figure 8 (left image) was not detected because of its strong irregularity in different directions. This irregularity produced very high error function values. Figure 8 (right) shows the subsidence pattern, where an incomplete interferometric fringe pattern occurred (due to the relatively low ground subsidence magnitude—less than half of the radar wavelength). Figure 9 shows examples of false alarms from interferogram I. They appeared in regions where there were no similar patterns to those characteristic for subsidence troughs. Such regions should be removed from analysis in the first step. However, probably due to a not restrictive enough threshold on coherence and entropy, they passed to the second step of the analysis.

Table 3 shows the comparison between obtained results using the proposed algorithm and those from a method based on circular Gabor filters [10]. .

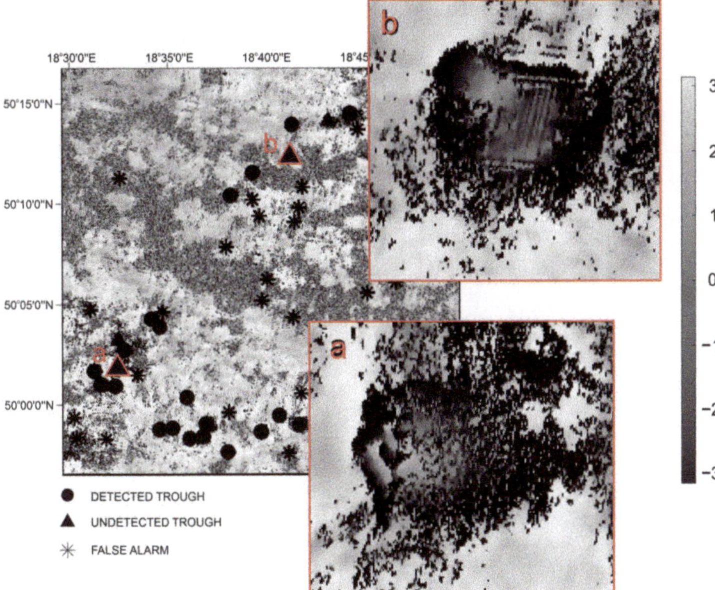

Figure 8. Example of two undetected troughs from interferogram I.

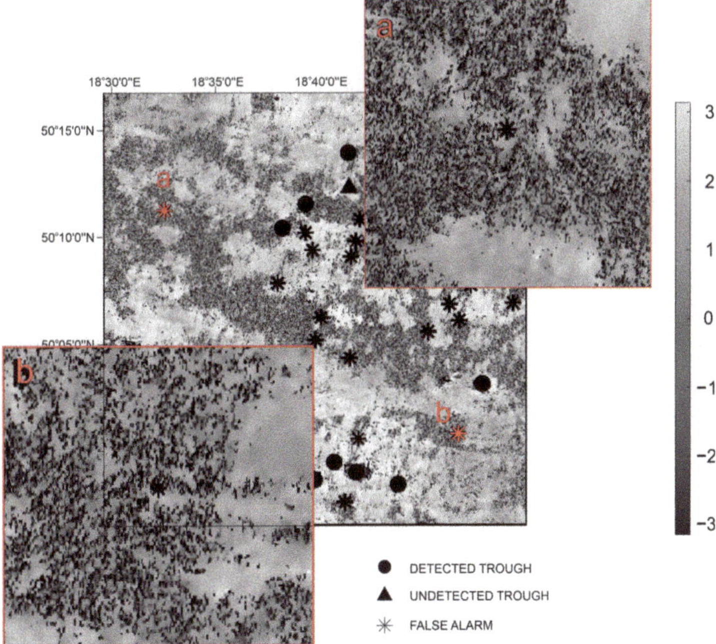

Figure 9. Example of two false alarms from interferogram I.

Table 3 shows that, for interferograms I and III, the number of correctly detected troughs by the proposed method was higher than that in the method based on circular Gabor filters. However, for interferogram III, the number of false alarms also increased. In the case of interferogram II, the proposed method gave worse results than those of the method based on circular Gabor filters. For this interferogram, the number of correctly detected troughs was similar for both methods, but the number of false alarms in the method proposed in this paper was significantly higher.

4. Conclusions

This work proposed a new algorithm for the automatic detection of subsidence areas in SAR interferograms using a method based on mathematical morphology. The proposed method was tested on three interferograms generated on the basis of Sentinel-1 SAR data. Detection results were compared with those obtained using a method based on circular Gabor filters. The detection rate for the proposed method varied from 34% to 83%. This is an improvement over the method based on circular Gabor filters that achieved detection rates from 30% to 53%. However, for both methods, different numbers of false alarms were reported. For the proposed method to be more operational, additional work should be devoted to the automatic selection of thresholds and dealing with subsidence patterns without a full phase cycle .

Author Contributions: Conceptualization, M.D., S.P.-S. and J.S.; methodology, M.D.; software, M.D., H.M. and R.M.; validation, S.P.-S., J.S., H.M. and R.M.; formal analysis, M.D., H.M. and R.M.; investigation, M.D., S.P.-S. and J.B.; resources, S.P.-S. and J.S.; data curation, M.D., A.F. and J.B.; writing—original draft preparation, S.P.-S. and M.D.; writing—review and editing, S.P.-S., M.D., A.F. and J.B.; visualization, M.D., H.M. and R.M.; supervision, A.F. and J.B.; project administration, A.F. and J.B.; funding acquisition, S.P.-S., M.D., A.F. and J.B. All authors have read and agreed to the published version of the manuscript.

Funding: This work was supported by the AGH—University of Science and Technology, Faculty of Geology, Geophysics and Environmental Protection, as a part of statutory project. The work is a part of MineSAR Project.

Institutional Review Board Statement: Not applicable.

Informed Consent Statement: Not applicable.

Data Availability Statement: The data accessed upon request from any of the author.

Conflicts of Interest: The authors declare no conflict of interest.

References

1. Gupta, R.P. *Remote Sensing Geology*, 3rd ed.; Springer: Berlin, Germany, 2017.
2. Sedighi, M.; Arabi, S.; Nankali, H.R.; Amighpey, M.; Tavakoli, F.; Soltanpour, A.; Motagh, M. Subsidence Detection Using InSAR and Geodetic Measurements in the North-West of Iran. In Proceedings of the 2009 Fringe Workshop, Frascati, Italy, 30 November–4 December 2009.
3. Porzycka, S.; Strzelczyk, J. Preliminary results of ground deformations monitoring within mining area of "Prosper-Haniel" coal mine. In Proceedings of the 12th International Multidisciplinary Scientific GeoConferences (SGEM), Balchik Municipality, Bulgaria, 17–22 June 2012; Volume 2, pp. 895–899.
4. Lu, Z.; Dzurisin, D. *InSAR Imaging of Aleutian Volcanoes: Monitoring a Volcanic Arc from Space*; Springer Science and Business Media: Chichester, UK, 2014.
5. Stramondo, S.; Chini, M.; Bignami, C.; Salvi, S.; Atzori, S. X-, C-, and L-band DInSAR investigation of the April 6, 2009, Abruzzi earthquake. *IEEE Geosci. Remote Sens. Lett.* **2010**, *8*, 49–53 [CrossRef]
6. Nádudvari, A. Using radar interferometry and SBAS technique to detect surface subsidence relating to coal mining in Upper Silesia from 1993–2000 and 2003–2010. *Environ. Socio Econ. Stud.* **2016**, *4*, 24–34. [CrossRef]
7. Deguchi, T. Automatic InSAR processing and introduction of its application studies. In Proceedings of the 26th Asian Conference on Remote Sensing, Hanoi, Vietnam, 7–11 November 2005.
8. d'Oreye, N.; Celli, G. Automatic InSAR systematic processing and web based tool for efficient data mining: Application to volcano monitoring in Africa. In Proceedings of the "Fringe 2009 Workshop", Frascati, Italy, 30 November–4 December 2009.
9. Albino, F.; Biggs, J.; Yu, C.; Li, Z. Automated Methods for Detecting Volcanic Deformation Using Sentinel-1 InSAR Time Series Illustrated by the 2017–2018 Unrest at Agung, Indonesia.*J. Geophys. Res. Solid Earth* **2020**, *125*, 017908. [CrossRef]
10. Porzycka-Strzelczyk, S.; Rotter, P.; Strzelczyk, J. Automatic detection of subsidence troughs in SAR interferograms based on circular Gabor filters. *IEEE Geosci. Remote Sens. Lett.* **2018**, *15*, 873–876. [CrossRef]
11. Bała, J.; Dwornik, M.; Franczyk, A. Automatic subsidence troughs detection in SAR interferograms using circlet transform. *Sensors* **2021**, *21*, 1706. [CrossRef] [PubMed]

12. Klimczak, M.; Bała, J. Application of the Hough transform for subsidence troughs detection in SAR images. *SGEM* **2017**, *17*, 819–826.
13. Yip, R.K.; Tam, P.K.; Leung, D.N. Modification of Hough transform for circles and ellipses detection using a 2-dimensional array. *Pattern Recognit.* **1992**, *25*, 1007–1022. [CrossRef]
14. Rotter, P.; Muron, W. Automatic Detection of Subsidence Troughs in SAR Interferograms Based on Convolutional Neural Networks. *IEEE Geosci. Remote Sens. Lett.* **2021**, *18*, 82–86. [CrossRef]
15. Pei, J.; Huang, Y.; Huo, W.; Zhang, Y.; Yang J.; Yeo, T. SAR Automatic Target Recognition Based on Multiview Deep Learning Framework. *IEEE Trans. Geosci. Remote Sens.* **2018**, *56*, 2196–2210. [CrossRef]
16. Gonzalez, R.C.; Woods, R.E.; Eddins, S.L. *Digital Image Processing Using MATLAB*; Prentice Hall: Hoboken, NJ, USA, 2003.
17. Dougherty, E.R. *Mathematical Morphology in Image Processing*; Marcel Dekker Inc.: New York, NY, USA, 1993.
18. Graniczny, M.; Kowalski, Z.; Tkowska, A.P.A.; Przyłucka, M. Observation of the Mining-Induced Surface Deformations Using C and L SAR Bands: The Upper Silesian Coal Basin (Poland) Case Study. In *Lecture Notes in Earth System Sciences*; Springer: Berlin/Heidelberg, Gemany, 2013; pp. 249–255.
19. Perski, Z. Applicability of ERS-1 and ERS-2 InSAR for land subsidence monitoring in the Silesian Coal mining region, Poland. *Int. Arch. Photogrametry Remote Sens.* **1998**, *32*, 555–558.
20. Mirek, K.; Isakow, Z. Preliminary analysis of InSAR data from south-west part of Upper Silesian Coal Basin.*Mineral Resour. Manag.* **2009**, *25*, 239–246.

 *energies*

Article

Second Life of Post-Mining Infrastructure in Light of the Circular Economy and Sustainable Development—Recent Advances and Perspectives

Katarzyna Pactwa [1,*], Martyna Konieczna-Fuławka [1], Krzysztof Fuławka [2], Päivi Aro [3], Izabela Jaśkiewicz-Proć [2] and Aleksandra Kozłowska-Woszczycka [1]

[1] Faculty of Geoengineering, Mining and Geology, Wrocław University of Science and Technology, 15 Na Grobli Street, 50-421 Wrocław, Poland; martyna.konieczna-fulawka@pwr.edu.pl (M.K.-F.); aleksandra.kozlowska@pwr.edu.pl (A.K.-W.)

[2] KGHM Cuprum Ltd. Research & Development Centre, 2-8 Sikorskiego Street, 53-659 Wrocław, Poland; kfulawka@cuprum.wroc.pl (K.F.); ijaskiewicz@cuprum.wroc.pl (I.J.-P.)

[3] School of Business and Information Management, Oulu University of Applied Sciences, Business, Yliopistonkatu 9, 90570 Oulu, Finland; paivi.aro@oamk.fi

* Correspondence: katarzyna.pactwa@pwr.edu.pl; Tel.: +48-71-320-68-51

Abstract: Current EU policy will force a significant reduction of hard coal mines in the near future due to environmental restrictions. There are also numerous non-coal underground mines that will be excavated in the next few years. Taking the above into consideration, it is worth starting to plan further steps in terms of reclamation of these facilities. Within this manuscript, both recently used and novel approaches to underground space reclamation have been reviewed. Selected methods of reclamation were analyzed in terms of their strengths and weaknesses, and the results were compared with the effect of a commonly used approaches (i.e., filling or flooding of underground space after mine termination). The analysis has been performed in the scope of sustainable development. Taking into account the opinion of many stakeholder groups and underground facilities, reuse was considered as an action aimed at fulfilling sustainable development goals and the circular economy concept. Based on numerous surveys, the challenges and opportunities have been determined as well. Finally, most perspectives concerning underground mine reclamation, including environmental impact, social acceptance, and profitability have been proposed and described.

Keywords: underground laboratory; new ways of mine reclamation; sustainability; circular economy

check for **updates**

Citation: Pactwa, K.; Konieczna-Fuławka, M.; Fuławka, K.; Aro, P.; Jaśkiewicz-Proć, I.; Kozłowska-Woszczycka, A. Second Life of Post-Mining Infrastructure in Light of the Circular Economy and Sustainable Development—Recent Advances and Perspectives. *Energies* **2021**, *14*, 7551. https://doi.org/10.3390/en14227551

Academic Editors: Sergey Zhironkin and Dawid Szurgacz

Received: 11 October 2021
Accepted: 10 November 2021
Published: 12 November 2021

Publisher's Note: MDPI stays neutral with regard to jurisdictional claims in published maps and institutional affiliations.

1. Introduction

The worldwide trend of the substitution of fossil energy by atomic and renewable energy sources has had a significant impact on the demand for raw mineral materials [1]. As a result, there are numerous mining projects and initiatives aimed at exploring and excavating new resources of mineral raw materials [2,3]. Growing needs in the field of renewable energy and electric mobility are related to the high importance of critical raw materials (CRMs) [4]. In turn, due to EU environmental policy and lack of competitiveness [5], most underground hard coal mines will be closed down over the next 15–20 years. Between 2014 and 2017 alone, 27 coal mines in Poland, Germany, the Czech Republic, Hungary, Romania, Slovakia, Slovenia, and the United Kingdom were closed [6]. Furthermore, considering the efficiency of underground coal mining it is expected that most coal regions will be subjected to termination by 2030, which poses a great risk of local economic crisis and numerous job losses in these areas. The map of potential job losses related to coal mine closure in the EU is presented in Figure 1.

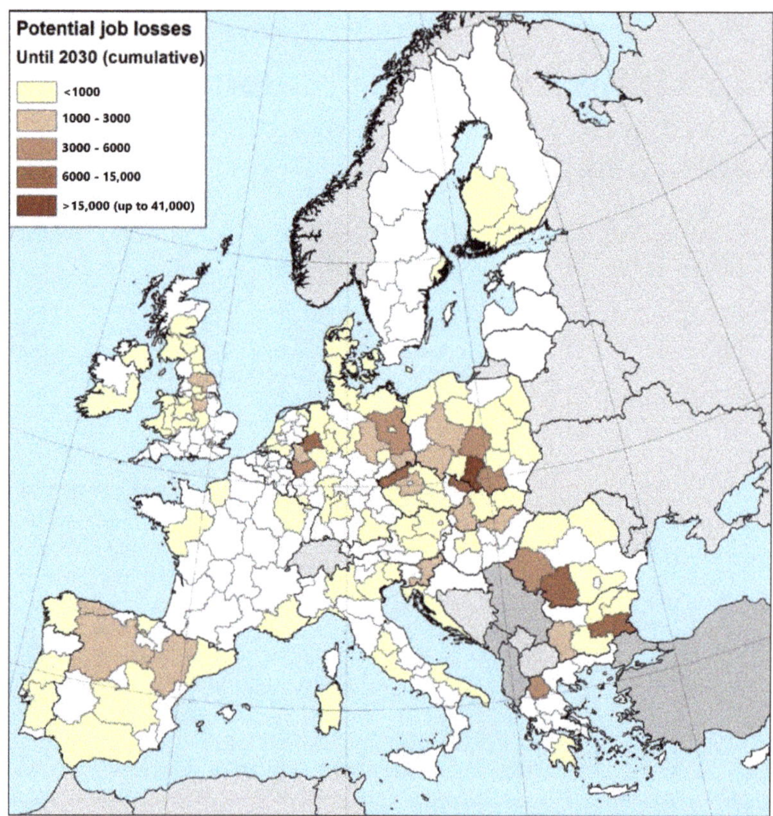

Figure 1. Potential job losses caused by the closure of coal mines in the EU [6].

The consequences of closing mines will also be felt by employees of mining-related enterprises (i.e., companies operating due to the existence of mines). This applies both to entities supplying products (mining machinery and equipment, specialized materials) and services, as well as those using minerals as the basic raw material for their production. As research [7] shows, some of them are not preparing for a mining shutdown scenario.

As past experiences showed, the unplanned and sudden transformation of mining areas has a negative impact on the economy [5,8] and environmental sustainability [9–14]. One of the inglorious examples of such an impact was the rapid termination of underground hard coal mining in the Lower Silesian Coal Basin (LSCB) in Poland, at the end of the 20th century. According to [15] after mines liquidation, the local unemployment increased from 15% in 1997 up to 32% in 2002. The consequences of the liquidation of industries are best illustrated by the data over the years. In the former Wałbrzych Voivodeship, employment in industry (in total) amounted to 148.5 thousand in 1980, 99.1 thousand in 1990 (the year when the decision to close the Wałbrzych mines was made), and 70.1 thousand in 1996 [16]. At the same time, in 1996, the last coal cart was excavated at the "Thorez" mine. Employment in the Wałbrzych Voivodeship for two decades is shown in Figure 2.

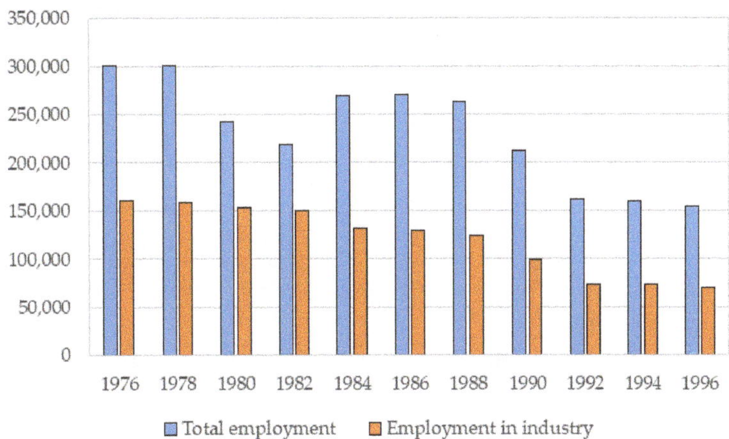

Figure 2. Employment in the Wałbrzych Voivodeship in the years between 1976 and 1996 (own study based on [16]).

Considering the numbers of closed mines and those which are planned to be set up, it may be concluded that even with the rapid development of new mining projects, most workers will stay unemployed. At the same time, post-mining underground space, in most cases, has not been considered as a suitable place for activities other than mining and are often filled with rocks and water to prevent its further effect on the local environment [17–19]. In general, widely considered methods of underground mines reclamation aim to utilize the most cost-effective and efficient way of backfilling to prevent further ground subsidence after mine closure [20].

A lack of other forms of underground space reclamation may be related to the harsh and dangerous environment [21,22] and limitations in the scope of national regulations [23]. Still, as it was pointed out in a paper [24], *"Each revitalization project should be performed with consideration of the social, economic, and environmental aspects"*. Therefore, it is important to develop solutions for post-mining underground workings reclamation which will at least partially ensure social and economic sustainability after the end of resources exploitation to achieve sustainable development (SD) and circular economy (CE) goals. Here, the circular economy is understood not in the classical sense [25], but in a broader sense—by planning the use of remnants (land, facilities, machines, etc.) after the completed mining activities. A second life for mines is, due to the variety of opportunities, a second life also for communities, landscapes, and/or science.

The authors [26] notice a growing need for post-mining space development, which is dictated, firstly, by the increased demand for space, and secondly by the need to use this space effectively in order to achieve the goals of sustainable development. It should be noted that the re-use of the exploited underground space does not have to be limited to one method of development.

Choosing how to reuse underground post-mining spaces becomes a challenge. There are few generally accepted practices or patterns for the redevelopment of closed mines' resources in the literature. The research conducted so far focuses mainly on individual case studies and does not give a diagnosis towards the correct choice of the method of reuse. This paper [27] presents an innovative solution and provides a universal framework for the development of abandoned mines. However, such an approach has still not been taken into consideration by most of the mining companies.

The aim of the article is to analyze alternative methods of underground working use and their possible impact on local and regional sustainability and the circular economy. The analysis was performed based on a literature review and surveys conducted with representatives of chosen underground facilities located in the EU.

2. Recently Utilized Alternative Ways of Mining Reclamation

When analyzing the potential ways of repurposing or revitalizing underground workings, one may conclude that the most popular and easy to implement method involves setting up underground tourist routes within the stable areas of old mines [28,29]. According to [30], in Poland there are over 200 underground touristic routes with over 1500 employees. As it was pointed out in research [31], such a method of mine reclamation provides numerous short-, medium-, and long-term advantages for the local society, environment, and regional economy after mine closure. Moreover, according to the World Travel and Tourism Council's estimations [32], tourism was the fastest growing industry worldwide until the COVID-19 pandemic. Examples of underground, profitable tourist routes in Europe are, among others, Wieliczka salt Mine in Poland [33], Zollverein Coal Mine Industrial Complex, Germany [34], Iwami Ginzan Silver Mine in Japan [35], and the Ruskeala Mining Park [17]. Still, considering the involvement of the mining industry in local budgets, there is no possibility to fill this gap only with a simple transformation into the tourism industry [35]. One of the infamous examples of "tourism after mining" may be the Hokkaido coal-mining region in Japan, where after a mine closure in Yubari City, even with strong support of local administration and high investment into tourism, the local community suffered bankruptcy and the population in the town shrank from 12,000 to 500 people [36]. Therefore, a sustainable strategy aimed at competitive advantages of each region is a key element during the process of transformation into "tourism cities".

The latest experiences during the pandemic have drawn attention to digital solutions that allow visitors to contact tourist facilities remotely [37]. Examples include the provision of materials by the Coal Mining Museum in Zabrze in the form of short virtual walks [38]. The materials available on Google Arts and Culture include, among others, photos from the Museum of Mines of Mercury Monte Amiata [39]. Digital technologies are used to provide information before visiting tourist sites. They can maintain the continuity of the tourism industry. They are useful for older people or with disabilities [37]. They can be described as consistent with the idea of SD.

On the other hand, post-mining underground spaces, due to their unique environment, offers several opportunities in the field of science, research, and education. As a result, continued growth in the use of underground space for research purposes can be observed within the past several years [40,41]. Rapid development is visible mainly in the scope of advanced physical measurements deep under the ground surface [42]. This is due to the presence of natural coverage that provides a significant reduction of cosmic ray flux or flux of neutrons when compared to the surface. As a result, numerous underground laboratories aimed strictly at astrophysical measurements have been set up during the last twenty years. Examples of such underground facilities may be Laboratory for Underground Nuclear Astrophysics (Luna), Italy [43,44], Jiangmen Underground Neutrino Observatory, (JUNO), China [45], Sudbury Neutrino Observatory (SNO), Canada [46], and Pyhasalmi mine, Finland [47,48].

Distinctive environmental conditions also affect the possibility of the development of new technologies for the mining industry. The simultaneous effect of such factors as high humidity, dust, lack of light, and problems with the ground control cause great difficulties in mapping the mine environment in laboratory conditions. Therefore, facilities located deep underground designed exclusively for the needs of technology development are strongly desired in the whole mining industry. Such a method of reclamation fits also into current EU policy and highlights the necessity of safe and more efficient mineral raw materials exploitation. Research performed in a real underground mining environments potentially allows one to develop and implement more reliable and economically effective products. Facilities of that type may be found in Sweden (e.g., Äspö Hard Rock Laboratory) where research on smart integrated test environments for the mining industry was performed [49]; Germany (e.g., Reiche Zeche) where, for example, methods of bioleaching are under development [50,51]; or Poland (e.g., Experimental Mine "Barbara"), where research on coal gasification [52], safety in the underground environment [53], and blasting technology [54,55] were performed recently.

An interesting and sustainable solution is the use of abandoned mines as energy storage. There are several technological solutions in this area [56]. They represent a way to deal with the problem of integrating the share of renewable energy sources with energy systems. An example is the project of storing excess heat from solar power plants using post-mining infrastructure in the Ruhr region [57]. Menendez et al. [58] and Li et al. [59] also wrote about solutions leading to the use of the potential of mines in a pro-ecological manner.

Furthermore, growing interest in underground farming and food production inside old post-mining areas can be observed in the last few years. As it was pointed out by GreenForges [60], underground spaces permit creating stable environmental conditions where food may be grown in a constant, predictable, and sustainable way. Such kinds of facilities have been set up in London [61] and in the Pyhasalmi mine, where a project related to industrial-scale insects and food production is under development [62].

It has to be highlighted that the underground space which may be repurposed after the termination of mining works in most cases is big enough to consider not only one way of underground site adaptation, but the setting up of a number of smaller facilities which may work independently. The example of a multipurpose experimental site, the ÄSPÖ HRL, has been presented in Figure 3.

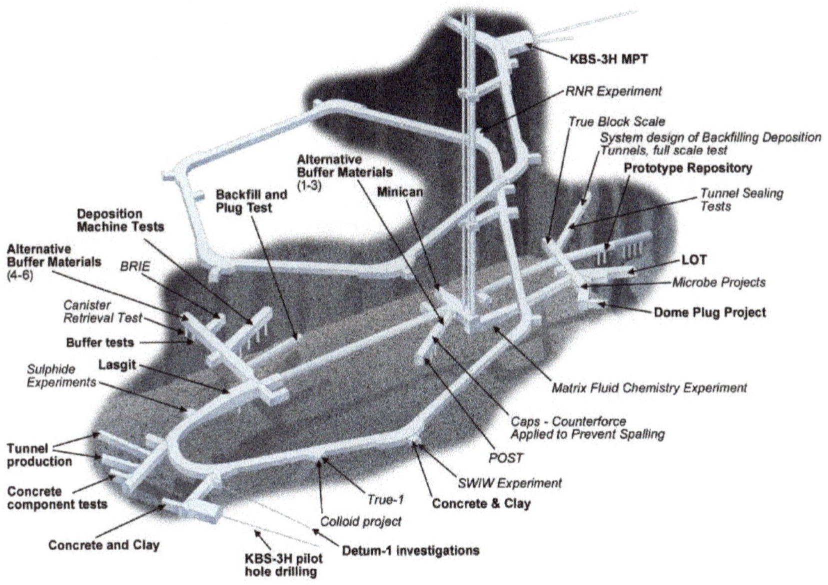

Figure 3. Allocation of experiments at Äspö HRL [57]. Reprint with permission of Äspö HRL, 2021, https://www.skbinternational.se.

However, it must be borne in mind that such a method of mine reclamation, in the case of deeply located underground workings, may not generate enough profits to maintain this type of facility by itself. Therefore, when analyzing the possibility of mine workings, a detailed business model should be developed.

3. Material and Methods

The possibility of effective underground space re-development is strongly dependent on the properly developed business model, quality of value proposition, method and effectiveness of service design, and the presence of stakeholders who are essential in the process of implementing and exploiting new ideas. The whole process of mining reclamation should be preceded by SWOT analyses which could highlight the most efficient way of further underground space development. All methods, surveys, and analyses have been performed within EU donated Baltic Sea Underground Innovation (BSUIN) and Empowering Underground Laboratories Network Usage (EUL) projects.

3.1. Development of the Business Model

A business model describes the logic of how a company intends to maintain its business lucrative, as well as the rationale of how a company or an organization creates, delivers, and captures value [63]. A business can be described with a business model canvas. Here, the business model canvas consists of nine building blocks: customer segments, value proposition, channels, customer relationships, revenue streams, key resources, key activities, key partnerships, and cost structure. One of its key benefits is bringing clarity to the core aims of an organization while identifying its strengths, weaknesses, and priorities. The example of the Canvas survey worksheet developed for conceptual laboratory which is considered to set up in one of KGHM mines in future is presented in Figure 4.

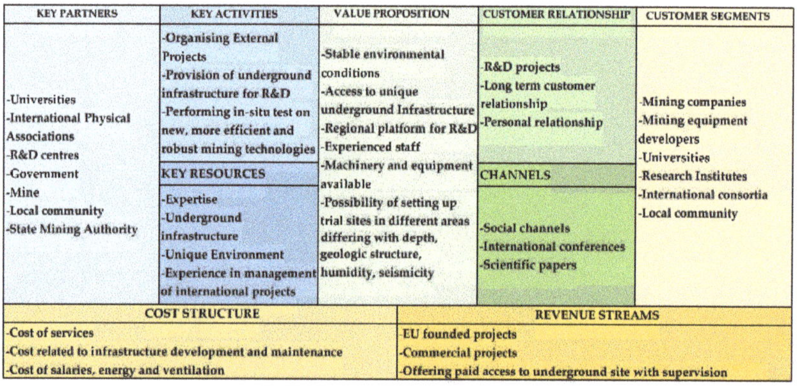

KEY PARTNERS	KEY ACTIVITIES	VALUE PROPOSITION	CUSTOMER RELATIONSHIP	CUSTOMER SEGMENTS
-Universities -International Physical Associations -R&D centres -Government -Mine -Local community -State Mining Authority	-Organising External Projects -Provision of underground infrastructure for R&D -Performing in-situ test on new, more efficient and robust mining technologies **KEY RESOURCES** -Expertise -Underground infrastructure -Unique Environment -Experience in management of international projects	-Stable environmental conditions -Access to unique underground Infrastructure -Regional platform for R&D -Experienced staff -Machinery and equipment available -Possibility of setting up trial sites in different areas differing with depth, geologic structure, humidity, seismicity	-R&D projects -Long term customer relationship -Personal relationship **CHANNELS** -Social channels -International conferences -Scientific papers	-Mining companies -Mining equipment developers -Universities -Research Institutes -International consortia -Local community
COST STRUCTURE			**REVENUE STREAMS**	
-Cost of services -Cost related to infrastructure development and maintenance -Cost of salaries, energy and ventilation			-EU founded projects -Commercial projects -Offering paid access to underground site with supervision	

Figure 4. The general structure of the canvas business model worksheet for facility set up for development of new mining technologies.

3.2. Process of Service Development

Service design is an approach that uses design thinking to create new or improve existing services so that they are more desirable and useful by and for customers, as well as efficient and effective for service providers [64]. The following are the main principles of service design [65] that should be considered during the entire service design process, which includes the following stages: exploration, creation, reflection and implementation [66]:

- *Human-centered*: denoting the need to fully understand the impressions and experiences of all people affected by the service;
- *Collaborative*: meaning the need to integrate different points of view by involving different stakeholders in the service design process;
- *Iterative*: explaining that service design is an iterative process consisting of: exploration, adaptation, experimentation and implementation;
- *Sequencing*: meaning that the entire service process should be viewed as a sequence of interrelated activities;
- *Real*: about the fact that the customer needs research and concept prototyping should be real;
- *Holistic in nature*: emphasizing the need to have a broad perspective and considering the entire service environment. The service development process for underground

workings also has been described based on interviews with representatives of BSR underground laboratories in Figure 5.

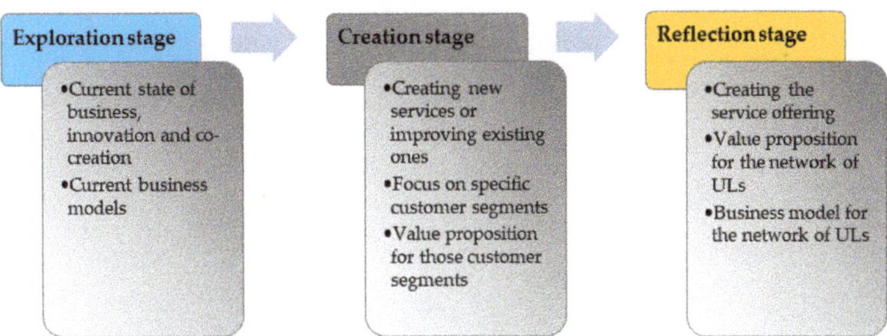

Figure 5. Stages of underground facility service development process.

3.3. Surveys with Stakeholders

The identification of stakeholders' opinions was based on interviews conducted with representatives of underground mines, state mining authorities, mining universities, research and development centers, and underground laboratories in BSR. The data collection questionnaires have been sent to the participant of the EUL project, from BSR countries. The structure of the questionnaire has been presented in Figure 6.

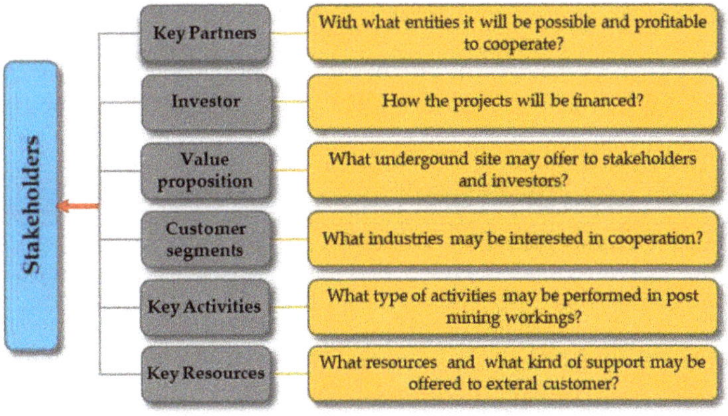

Figure 6. The topics which need to be solved during the identification of stakeholders.

3.4. The SWOT Analysis

Knowing the potential business model development, service design approach and stakeholders some, methods of underground mine reclamation may be determined. Still, in order to determine which idea is most feasible, a SWOT analysis has to be conducted. Such an approach is easy to implement and allows one to obtain all strengths, weaknesses, opportunities, and threats of each proposed solution. The details about SWOT analysis have been described in [67].

All abovementioned data have been collected among specialists and researchers representing underground facilities from Sweden, Germany and Switzerland, Poland, Finland and Russia who were involved in the process of data collection. Moreover, the analysis presented herein has been supported by representatives of KGHM copper mines, Poland, representatives of R&D centers, and universities from the European Underground Laboratories Association. As a result, a number of questionnaires have been collected. These data were used for determining what kind of repurpose are utilized in EU mines and what are the perspective ways concerning the business models, service design process, possible stakeholders, and SWOT.

4. Results and Discussion

Based on 29 questionnaires which were filled up by representatives of R&D centers, underground labs, universities, and representatives of state mining authorities the ways of underground workings readaptation were described and analyzed. The whole process of choosing the type of reclamation need to be supported by the development of the business model, service development, identification of stakeholders, and finally an analysis of strengths and weaknesses (Figure 7).

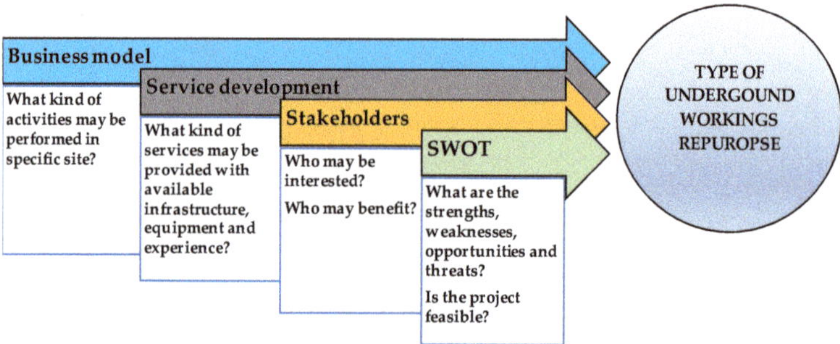

Figure 7. The topics which need to be solved during the identification of stakeholders.

Depending on the content analysis of present ways of underground laboratories functioning in BSR—on the basis of their business models, it may be concluded that the methods of underground space reclamation are in different stages of service business development due to the different nature of their contexts, the geomechanical conditions of surrounding rockmass, and location. Still, the actions aimed at repurposing of post-mining underground workings may be divided into three main groups. The first group gathers activities that lead to the permanent filling and liquidation of empty underground space, while the second is aimed at further exploitation of the unique environment. The last group is the multipurpose facility, combining different ways of workings adaptation. The classification of ways of underground mines reclamation is presented in Figure 8.

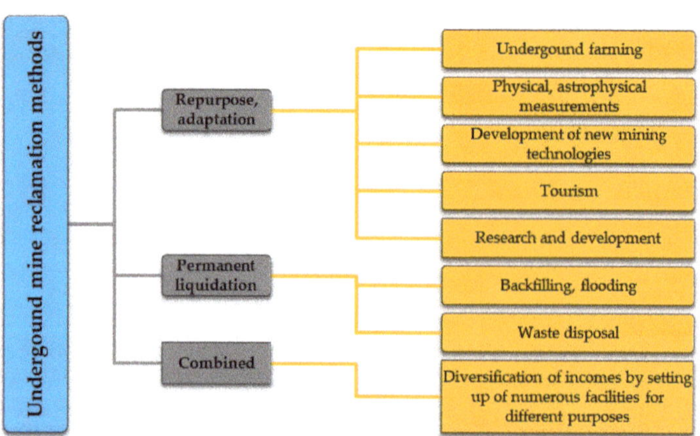

Figure 8. Prospective methods of underground workings repurposing according to surveys performed within the scope of EUL project.

All the above-presented methods, after detailed planning, may be economically justified and have a potential for further development. At the same time, all approaches have some strengths and weaknesses. A detailed analysis may be found in Supplementary Materials.

It is worth mentioning that according to SWOT analysis, the biggest strength of underground spaces is that they provide access to a unique environment with stable environmental conditions in terms of temperature, humidity, and natural background radiation. In turn, these unique conditions, are also related to unique hazards which are not observed in other facilities located on the surface. Therefore, numerous hazards and safety issues are mentioned as the biggest weakness in most of the analyses. Such a situation also affects the potential threats. Namely, hazardous environment, and high accident rates in the mining industry, turn into a lack of potential investors, which may be one of the biggest challenges during the running up of new venture. Subsidies by local governments or EU funds may be the solution, while according to current EU policy, the extended research and actions aimed into sustainable development, green technologies and development of new effective exploitation methods are strongly required. This may be a great opportunity.

In order to summarize the collected data, the quantitative results of SWOT analysis have been presented in Figures 9–11.

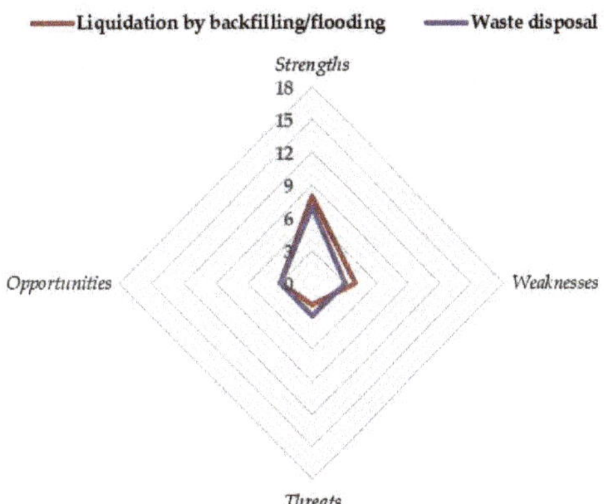

Figure 9. The quantitative results of SWOT analysis for reclamation methods aimed at permanent liquidation of underground workings.

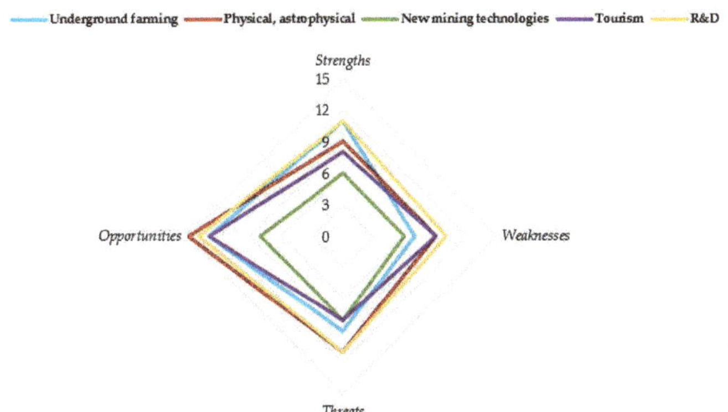

Figure 10. The quantitative results of SWOT analysis for reclamation methods aimed at adaptation and repurpose of underground workings.

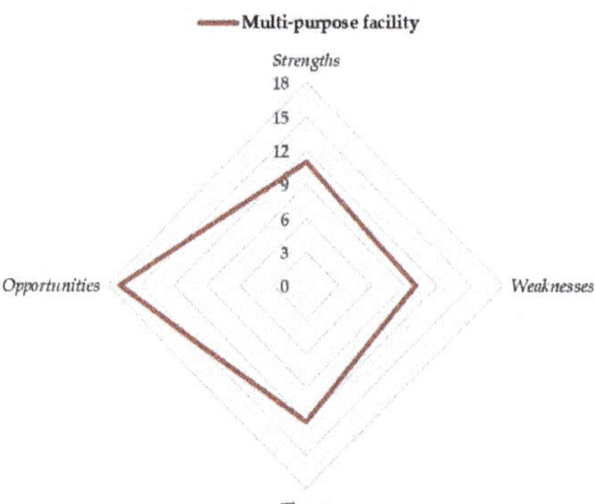

Figure 11. The quantitative results of SWOT analysis for reclamation methods aimed at combined/multidisciplinary repurposing of underground workings.

According to performed analyses, one may conclude that the most commonly used approach, based on permanent liquidation of underground workings in general, is the safest solution due to the low number of threats and weaknesses (Figure 9). Such an approach does not require further investments, and therefore is preferably utilized by most of the mining companies. However, more sophisticated solutions aimed at further exploitation of underground spaces may bring numerous advantages to local society as well as the science and mining industry. In fact, strengths and opportunities in analyzed cases are distributed parallel to weaknesses and threats related to each type of venture. As it may be noticed in analyzing Figure 10, the most promising facilities are those which are focused on astrophysical and physical measurements and those which are strictly related to performing R&D activities. At the same time, these approaches are burdened with the highest risk of project implementation. In turn, faculty dedicated solely for the development of new mining technologies is expected to provide the lowest incomes and is characterized by a low number of opportunities, and simultaneously is not characterized by many weaknesses and threats.

The best rate of opportunities and strengths in comparison to weaknesses and threats may be achieved by the setting up of numerous multidisciplinary facilities in the area of abandoned mines (Figure 11). Such a solution allows for the diversification of incomes. Moreover, as it may be concluded when analyzing the case study from Pyhäsalmi Mine, Finland, such approaches are proved to be economically efficient. However, what is worth mentioning is the fact that representatives of the mine and cooperating universities have started the marketing campaign a few years before mine closure. Such an approach would allow one to perform a smooth transition between an excavation-based business model to a research/science business model.

When analyzing the general challenges related to the transition from mining activities to science, tourism, and R&D, it must be borne in mind that most of the threats and weaknesses of such facilities are related to their hazardous environment and unfriendly conditions [21]. As a result, the way of perceiving this issue by potential investors is rather negative. Thus, much more attention has to be paid to marketing and the development of proper service design and business models.

After examining the service design and business models of currently working underground laboratories in Europe, it was concluded that in order to develop an economically justified service business, the managers of underground mines need to attract more investors and perform more proactive marketing communication. According to our analyses, the most important customer segments at the moment are universities and research institutes which gather funding mostly from international projects performed within large framework programs such as Horizon 2020, Horizon Europe, Interreg BSR, and EIT Raw

Materials. Still, these founding sources are temporary, and without commercialization of performed activities, the revenue streams are not substantial enough for the future. Moreover, the financial resources are generally scarce in many of the ULs, while their fixed costs are significant.

5. Conclusions

In this manuscript, the currently used and prospective methods of underground space reclamation have been reviewed and analyzed. The advantages and weaknesses of the selected methods of post-mining space repurposing have been examined in terms of their strengths and weaknesses. As our analysis showed, there a few methods of development of post-mining workings which will act as a base for achieving goals of sustainable development and circular economy, and the commonly used method of mine filling with water and wastes may be successfully replaced by more economically justified projects, which at the same time may support local communities in terms of both generated incomes, as well as in terms of maintenance of workplaces.

The detailed review of recent advances in that matter proved that use of post-mining underground space may be profitable not only in the scope of tourism but may also be a great accelerator of technology and science development.

Unfortunately, there are some challenges and further work aimed at broadening public awareness, and the awareness of investors is still needed. The heavy cost structure required is also a big challenge. A coherent vision and strategy for the future is a good starting point, as it would help to find new business opportunities and to attract new customer segments. Further, the organizational structure of the underground laboratories is quite complicated for many reasons. This can be tackled by good communication and organizational efforts by the underground laboratories, and they should also consider customer relationship management practices. On the other hand, there are unique strengths of such facilities lying with their human, intellectual, and environmental resources not present in other branches of industry.

Moreover, there is a high possibility that novel approaches to mine reclamation will gather social acceptance due to clearly visible profits for the local community. In an example, any R&D, science, or commercial activities will generate further income to the local budget. In turn, set up of the tourist route may also generate a great impact on the local tourism industry. At the same time in most cases, the environmental impact may be less than in the case of regular mining activities, which is in perfect accordance with the current EU policy and regulations.

Of course, it must be kept in mind that data presented in this manuscript do not represent universal solutions which may be implemented in every site. This is a general highlight for mine companies or potential investors of the methods of reclamation which are possible, and of which successful projects with a similar approach have been running at the moment. Therefore, each site, before implementation of any idea, should perform a detailed risk analysis, SWOT, develop a business model for the particular case, and design the offer of services which may lead a particular venture to successful finalization. Still, knowing the potential directions of development, managers of such facilities may look more broadly on their current activities, and hopefully will choose a way of further development.

Supplementary Materials: The following are available online at https://www.mdpi.com/article/10.3390/en14227551/s1, Table S1: Way of underground space repurposing: underground farming, Table S2: Way of underground space repurposing: physical, asrtophysical, measurements, Table S3: Way of underground space repurposing: development of new mining technologies, Table S4: Way of underground space repurposing: tourism, Table S5: Way of underground space repurposing: research and development, Table S6: Way of underground space repurposing: liquidation by backfilling/flooding, Table S7: Way of underground space repurposing: waste disposal, Table S8: Way of underground space repurposing: multi-purpose facility.

Author Contributions: Conceptualization, K.F., K.P., and M.K.-F.; methodology, K.F., K.P., M.K.-F., and P.A.; software, P.A.; validation, M.K.-F. and K.P.; formal analysis, K.P., M.K.-F., and I.J.-P.; investigation, K.P. and K.F.; resources, K.F., I.J.-P., and P.A.; data curation, K.P., K.F., I.J.-P., and P.A.; writing—original draft preparation, K.P., M.K.-F., K.F., P.A., I.J.-P., and A.K.-W.; writing—review and editing, K.P., M.K.-F., K.F., and A.K.-W.; visualization, M.K.-F. and P.A.; supervision, K.P.; project administration, K.F.; funding acquisition, K.F. and K.P. All authors have read and agreed to the published version of the manuscript.

Funding: This research was funded by European Regional Development Fund (ERDF), grant number Interreg Baltic Sea Region #X010 EUL project, and "The APC was funded by Polish Ministry of Science and Higher Education: Subsidy 2021 for WUST.

Institutional Review Board Statement: Not applicable.

Informed Consent Statement: Not applicable.

Data Availability Statement: The data presented in this study are available on request from the corresponding author.

Acknowledgments: We would like to thank all partners and contributors from EUL project which were respond to our questionnaires and took part in interviews and surveys.

Conflicts of Interest: The authors declare no conflict of interest.

References

1. Buchholz, P.; Brandenburg, T. Demand, Supply, and Price Trends for Mineral Raw Materials Relevant to the Renewable Energy Transition Wind Energy, Solar Photovoltaic Energy, and Energy Storage. *Chem. Ing. Tech.* **2018**, *90*, 141–153. [CrossRef]
2. Probst, L.; Frideres, L.; Cambier, B.; Hommel, S.; Luxembourg, P. *Sustainable Supply of Raw Materials—Innovative Mineral and Metallurgical Extraction and Processing*; European Commission: Luxembourg, 2016; Volume 17.
3. Sahu, H.B.; Prakash, N.; Jayanthu, S. Underground Mining for Meeting Environmental Concerns—A Strategic Approach for Sustainable Mining in Future. *Procedia Earth Planet. Sci.* **2015**, *11*, 232–241. [CrossRef]
4. Directorate-General for Internal Market, I.; Bobba, S.; Carrara, S.; Huisman, J.; Mathieux, F.; Pavel, C. *Critical Raw Materials for Strategic Technologies and Sectors in the EU: A Foresight Study*; Publications Office of the European Union: Luxembourg, 2020; ISBN 978-92-76-15336-8.
5. Kretschmann, J.; Efremenkov, A.B.; Khoreshok, A.A. From Mining to Post-Mining: The Sustainable Development Strategy of the German Hard Coal Mining Industry. In Proceedings of the IOP Conference Series: Earth and Environmental Science, Yurga, Russian, 17–19 November 2016; IOP Publishing: Bristol, UK, 2017; Volume 50. [CrossRef]
6. Dias, P.; Kanellopoulos, K.; Medarac, H.; Kapetaki, Z.; Miranda-Barbosa, E.; Shortall, R.; Veronica, C.; Telsnig, T.; Cristina, V.-H.; Lacal Arantegui, R.; et al. *EU Coal Regions: Opportunities and Challenges Ahead*; Publications Office of the European Union: Luxembourg, 2018.
7. Ingram, T.; Bartuś, K.; Baron, M.; Bielecki, Ł. Sytuacja Przedsiębiorstw Okołogórniczych w Polsce, Ekspertyza Dotycząca Likwidacji Kopalń Węgla Kamiennego Dla Sektora Okołogórniczego Oraz Sytuacji Społeczno-Gospodarczej w Polsce, Etap I, Uniwersytet Ekonomiczny w Katowicach Na Zlecenie Górniczej Izby Przemysłowo Handlowej. 2020. Available online: http://www.giph.com.pl/files/Publikacje/Ekspertyza.pdf (accessed on 11 November 2021).
8. Kretschmann, J. Stakeholder Orientated Sustainable Land Management: The Ruhr Area as a Role Model for Urban Areas. *Vestnik* **2014**, *2*, 127–132. [CrossRef]
9. Smith, F.; Underwood, B. Mine Closure: The Environmental Challenge. *Min. Technol.* **2000**, *109*, 202–209. [CrossRef]
10. Müller, B.; Finka, M.; Lintz, G. (Eds.) The challenge of structural change for industrial cities and regions in the cee countries. In *Rise and Decline of Industry in Central and Eastern Europe: A Comparative Study of Cities and Regions in Eleven Countries*; Central and Eastern European Development Studies (CEEDES); Springer: Berlin/Heidelberg, Germany, 2005; ISBN 978-3-540-40478-1.
11. Wirth, P.; Mali, B.; Fischer, W. Problems and potentials of post-mining regions. In *Post-Mining Regions in Central Europe. Problems, Potentials, Possibilities*; OEKOM: Munich, Germany, 2012; ISBN 978-3-86581-294-0.
12. Kovalev, V.; Gerike, B.; Khoreshok, A.; Gerike, P. Preventive Maintenance of Mining Equipment Based on Identification of Its Actual Technical State. In Proceedings of the Taishan Academic Forum—Project on Mine Disaster Prevention and Control, Qingdao, China, 17–20 October 2014; pp. 184–189.
13. Tyulenev, M.; Zhironkin, S.; Litvin, O. The Low-Cost Technology of Quarry Water Purifying Using the Artificial Filters of Overburden Rock. *Pollut. Res.* **2015**, *34*, 825–830.
14. Tyulenev, M.; Gvozdkova, T.; Zhironkin, S.; Garina, E. Justification of Open Pit Mining Technology for Flat Coal Strata Processing in Relation to the Stratigraphic Positioning Rate. *Geotech. Geol. Eng.* **2016**, *35*, 203–212. [CrossRef]
15. Kosmaty, J. Wałbrzyskie tereny pogórnicze po 15 latach od zakończenia eksploatacji węgla (EN: Wałbrzych post-mining land 15 years after coal extraction was ended). *Górnictwo Geol.* **2011**, *6*, 131–148.

16. *Statistical Yearbooks of the Wałbrzych Voivodeship: 1977, R.1; 1979, R 3; 1981 R. 5; 1985 R.7; 1987 R. 9; 1991, R. 11; 1993, R. 12; 1995, R.9; 1997, R. 16*; Wojewódzki Urząd Statystyczny w Wałbrzychu (Śląska Biblioteka Cyfrowa): Wałbrzych, Poland. Available online: https://sbc.org.pl/access (accessed on 4 November 2021).
17. Shekov, V.; Ivanov, A.; Jalas, P.; Laaksoharju, M.; Horner, D. BSUIN—A Unique Innovation Programme for Underground Space Development to Be Tested in the Ruskeala Marble Quarry and Sortavala Geopark. In Proceedings of the 18th International Multidisciplinary Scientific GeoConference, Albena, Bulgaria, 2–8 July 2018.
18. Shekov, V.; Jalas, P.; Joutsenvaara, J.; Kisiel, J.; Laaksoharju, M.; Pytel, W.; Horner, D.; Mischo, H.; Giese, R.; Mockus, V.; et al. BSUIN Project as an Innovative Platform for Mining&Industrial Underground Heritage Investigation in Russia. In Proceedings of the International conference "InterCarto. InterGIS", Petrozavodsk, Russia, 19–22 July 2018; Volume 24, pp. 285–296.
19. Chugh, Y.P. Concurrent Mining and Reclamation for Underground Coal Mining Subsidence Impacts in China. *Int. J. Coal Sci. Technol.* **2018**, *5*, 18–35. [CrossRef]
20. Dodd, W.E. Evaluating Underground Mine Reclamation Projects in North Dakota. In Proceedings of the 35th Annual Conference of the National Association of Abandoned Mine Land Programs, Danields, WV, USA, 22–25 September 2013.
21. Pytel, W.; Fuławka, K.; Pałac-Walko, B.; Mertuszka, P.; Kisiel, J.; Jalas, P.; Joutsenvaara, J.; Shekov, V. Universal Approach for Risk Identification and Evaluation in Underground Facilities. *Min. Sci.* **2020**, *27*, 165–181. [CrossRef]
22. Shekov, V.; Shekov, K.; Fuławka, K.; Pytel, W. Safe Use of Mining-and-Industrial Heritage and Underground Space in Tourism Sector. In Proceedings of the 19th International Multidisciplinary Scientific GEO Conference, Vienna, Austria, 28 June–7 July 2019; SGEM WORLD SCIENCE, Technology Ltd.: Albena, Bulgaria, 2019; Volume 19, pp. 569–577.
23. Shekov, K.; Muller, T.; Fuławka, K.; Joutsenvaara, J. Legislative Control over Use of Old Mine Workings for Learning Purposes. *Gorn. Zhurnal* **2019**, *3*, 11–16. [CrossRef]
24. Pactwa, K.; Woźniak, J.; Dudek, M. Sustainable Social and Environmental Evaluation of Post-Industrial Facilities in a Closed Loop Perspective in Coal-Mining Areas in Poland. *Sustainability* **2021**, *13*, 167. [CrossRef]
25. Morseletto, P. Targets for a Circular Economy. *Resour. Conserv. Recycl.* **2020**, *153*, 104553. [CrossRef]
26. Xie, H.; Zhao, J.W.; Zhou, H.W.; Ren, S.H.; Zhang, R.X. Secondary Utilizations and Perspectives of Mined Underground Space. *Tunn. Undergr. Space Technol.* **2020**, *96*, 103129. [CrossRef]
27. Cui, C.-Q.; Wang, B.; Zhao, Y.-X.; Xue, L.-M. Waste Mine to Emerging Wealth: Innovative Solutions for Abandoned Underground Coal Mine Reutilization on a Waste Management Level. *J. Clean. Prod.* **2020**, *252*, 119748. [CrossRef]
28. Edwards, J.A.; Coit, J.C.L. Mines and Quarries: Industrial Heritage Tourism. *Ann. Tour. Res.* **1996**, *23*, 341–363. [CrossRef]
29. Olszewski, J.; Chruścielewski, W.; Jankowski, J. Radon on Underground Tourist Routes in Poland. *Int. Congr. Ser.* **2005**, *1276*, 360–361. [CrossRef]
30. Olszewski, J.; Zmyślony, M.; Wrzesień, M.; Walczak, K. Occurrence of radon in the Polish underground tourist routes. *Med. Pr.* **2015**, *66*, 557–563. [CrossRef]
31. Dos Santos Costa, S.S.; Santos, E. Mining Tourism and Geotourism: Alternatives Solutions to Mine Closure and Completion. In *24th World Mining Congress Proceedings: Sustainability in Mining*; IBRAM: Rio de Janeiro, Brazil, 2016; pp. 301–309, ISBN 978-85-61993-11-5.
32. World Travel & Tourism Council (WTTC). *Travel & Tourism Economic Impact 2021: Global Economic Impact&Trends 2021*; World Travel & Tourism Council (WTTC): London, UK, 2021.
33. Alexandrowicz, Z.; Urban, J.; Andreychouk, V. Crystal Caves in the 'Wieliczka' Salt Mine—Unique Cave Site. *Z. Geomorphol. Suppl. Issues* **2021**, *62*, 235–254. [CrossRef]
34. Ćopić, S.; Đorđevića, J.; Lukić, T.; Stojanović, V.; Đukičin, S.; Snežana, B.; Igor, S.; Aleksandar, T. Transformation of Industrial Heritage: An Example of Tourism Industry Development in the Ruhr Area (Germany). *Geogr. Pannonica* **2014**, *18*, 43–50. [CrossRef]
35. Armis, R.; Kanegae, H. The Attractiveness of a Post-Mining City as a Tourist Destination from the Perspective of Visitors: A Study of Sawahlunto Old Coal Mining Town in Indonesia. *Asia-Pac. J. Reg. Sci.* **2019**, *4*, 443–461. [CrossRef]
36. Hattori, K.; Kaido, K.; Matsuyuki, M. The Development of Urban Shrinkage Discourse and Policy Response in Japan. *Cities* **2017**, *69*, 124–132. [CrossRef]
37. Akhtar, N.; Khan, N.; Mahroof Khan, M.; Ashraf, S.; Hashmi, M.S.; Khan, M.M.; Hishan, S.S. Post-COVID 19 Tourism: Will Digital Tourism Replace Mass Tourism? *Sustainability* **2021**, *13*, 5352. [CrossRef]
38. Wycieczki Panoramiczne-Muzeum Górnictwa Węglowego w Zabrzu. Available online: https://www.ai360.pl/panoramy/609,50 33 (accessed on 25 July 2021).
39. Museum of Mines of Mercury Monte Amiata, Santa Fiora, Grosseto, Włochy. Available online: https://artsandculture.google. com/incognito/partner/museo-delle-miniere-di-mercurio-del-monte-amiata (accessed on 25 July 2021).
40. Sterling, R.; Admiraal, H.; Bobylev, N.; Parker, H.; Godard, J.-P.; Vähäaho, I.; Rogers, C.D.F.; Shi, X.; Hanamura, T. Sustainability Issues for Underground Space in Urban Areas. *Proc. Inst. Civ. Eng.-Urban Des. Plan.* **2012**, *165*, 241–254. [CrossRef]
41. Bartel, S.; Janssen, G. Underground Spatial Planning—Perspectives and Current Research in Germany. *Tunn. Undergr. Space Technol.* **2015**, *55*, 112–117. [CrossRef]
42. Bettini, A. New Underground Laboratories: Europe, Asia and the Americas. *Phys. Dark Universe* **2014**, *4*, 36–40. [CrossRef]
43. Best, A.; Caciolli, A.; Fülöp, Z.; Gyürky, G.; Laubenstein, M.; Napolitani, E.; Rigato, V.; Roca, V.; Szücs, T. Underground Nuclear Astrophysics: Why and How. *Eur. Phys. J. A* **2016**, *52*, 72. [CrossRef]

44. Guglielmetti, A. Nuclear Astrophysics and Underground Accelerators. *Phys. Dark Universe* **2014**, *4*, 10–13. [CrossRef]
45. Cerna, C. The Jiangmen Underground Neutrino Observatory (JUNO). *Nucl. Instrum. Methods Phys. Res. Sect. A Accel. Spectrom. Detect. Assoc. Equip.* **2020**, *958*, 162183. [CrossRef]
46. Bellerive, A.; Klein, J.R.; McDonald, A.B.; Noble, A.J.; Poon, A.W.P. The Sudbury Neutrino Observatory. *Nucl. Phys. B* **2016**, *908*, 30–51. [CrossRef]
47. Trzaska, W.H.; Bezrukov, L.; Enqvist, T.; Joutsenvaara, J.; Kuusiniemi, P.; Loo, K.; Lubsandorzhiev, B.; Sinev, V.; Slupecki, M. Possibilities for Underground Physics in the Pyhasalmi Mine. In Proceedings of the Thirteenth Conference on the Intersections of Particle and Nuclear Physics (CIPANP2018), Palm Springs, CA, USA, 28 May–3 June 2018.
48. Jalas, P.; Enqvist, T.; Isoherranen, V.; Joutsenvaara, J.; Kutuniva, J.; Kuusiniemi, P. Callio Lab, a New Deep Underground Laboratory in the Pyhäsalmi Mine. *J. Phys. Conf. Ser.* **2017**, *888*, 012156. [CrossRef]
49. Underground Experiment Sites. Available online: http://www.skbinternational.se/laboratory-services/underground-experiment-sites/ (accessed on 5 November 2021).
50. Schlueter, R.; Mischo, H. Potential and Applications of Underground in Situ Bioleaching. In Proceedings of the 3rd International Future Mining Conference, Sydney, Australia, 4–6 November 2015.
51. Eisen, N.; Schlömann, M.; Schopf, S. Bioleaching of Indium-Bearing Sphalerite under Underground Mining Temperatures. In Proceedings of the IMWA 2016—Mining Meets Water—Conflicts and Solutions, Freiberg, Germany, 1 July 2016.
52. Wiatowski, M.; Stańczyk, K.; Świądrowski, J.; Kapusta, K.; Cybulski, K.; Krause, E.; Grabowski, J.; Rogut, J.; Howaniec, N.; Smoliński, A. Semi-Technical Underground Coal Gasification (UCG) Using the Shaft Method in Experimental Mine "Barbara". *Fuel* **2012**, *99*, 170–179. [CrossRef]
53. Prostański, D. Experimental Study of Coal Dust Deposition in Mine Workings with the Use of Empirical Models. *J. Sustain. Min.* **2015**, *14*, 108–114. [CrossRef]
54. Kramarczyk, B.; Pytlik, M.; Mertuszka, P. Effect of Aluminium Additives on Selected Detonation Parameters of a Bulk Emulsion Explosive. *Mater. Wysokoenerg. High Energy Mater.* **2020**, *12*, 99–113. [CrossRef]
55. Mertuszka, P.; Fuławka, K.; Pytlik, M.; Szastok, M. The Influence of Temperature on the Detonation Velocity of Selected Emulsion Explosives. *J. Energ. Mater.* **2020**, *38*, 336–347. [CrossRef]
56. Saigustia, C.; Robak, S. Review of Potential Energy Storage in Abandoned Mines in Poland. *Energies* **2021**, *14*, 6272. [CrossRef]
57. Hahn, F.; Jagert, F.; Bussmann, G.; Nardini, I.; Bracke, R.; Seidel, T.; Konig, T. The Reuse of the Former Markgraf II Colliery as a Mine Thermal Energy Storage. In Proceedings of the European Geothermal Congress 2019, Den Haag, The Netherlands, 11–14 June 2019.
58. Menendez, J.; Ordóñez, M.A.; Alvarez, R.; Loredo, J. Energy from Closed Mines: Underground Energy Storage and Geothermal Applications. *Renew. Sustain. Energy Rev.* **2019**, *108*, 498–512. [CrossRef]
59. Li, B.; Zhang, J.; Ghoreishi-Madiseh, S.A.; Brito, M.; Xuejie, D.; Kuyuk, A. Energy Performance of Seasonal Thermal Energy Storage in Underground Backfilled Stopes of Coal Mines from China. *J. Clean. Prod.* **2020**, *275*, 122647. [CrossRef]
60. GreenForges Website. Available online: https://www.greenforges.com/team (accessed on 10 May 2021).
61. Van Hooijdonk, R. Vertical Farming Goes High-Tech and Underground. Available online: https://www.hortibiz.com/newsitem/?tx_news_pi1%5Bnews%5D=37302&cHash=4a0918ebe1ad8b0653f72a5be5c27757 (accessed on 10 May 2021).
62. Callio, P. Edible Insects. Available online: https://callio.info/natural-resources-business/edible-insects/ (accessed on 10 May 2021).
63. Osterwalder, A.; Pigneur, Y. *Business Model Generation: A Handbook for Visionaries, Game Changers, and Challengers | Wiley*; Wiley & Sons: Hoboken, NJ, USA, 2010; ISBN 978-0-470-87641-1.
64. Moritz, S. Service Design—Practical Access to an Evolving Field. *Köln Int. Sch. Des.* **2005**, *246*.
65. Stickdorn, M.; Edgar Hormess, M.; Lawrence, A.; Schneider, J. *This Is Service Design Doing. Applying Service Design Thinking in the Real World*; A Practitioner's Handbook; O'Reilly Media, Inc.: Sebastopol, CA, USA, 2018; ISBN 978-1-4919-2718-2.
66. Stickdorn, M.; Schneider, J. *This Is Service Design Thinking: Basics, Tools, Cases*; BIS Publishers: Amsterdam, The Netherlands, 2011; ISBN 978-1-118-15630-8.
67. Clardy, A. Strengths vs. Strong Position: Rethinking the Nature of SWOT Analysis. *Mod. Manag. Sci. Eng.* **2013**, *1*, 100–122.

 *energies*

Article

Modern Solution for Fast and Accurate Inventorization of Open-Pit Mines by the Active Remote Sensing Technique—Case Study of Mikoszów Granite Mine (Lower Silesia, SW Poland)

Jaroslaw Wajs *, Paweł Trybała , Justyna Górniak-Zimroz , Joanna Krupa-Kurzynowska and Damian Kasza

Faculty of Geoengineering, Mining and Geology, Wrocław University of Science and Technology, Wyb. Wyspianskiego 27 St., 50-370 Wrocław, Poland; pawel.trybala@pwr.edu.pl (P.T.); justyna.gorniak-zimroz@pwr.edu.pl (J.G.-Z.); joanna.krupa-kurzynowska@pwr.edu.pl (J.K.-K.); damian.kasza@pwr.edu.pl (D.K.)
* Correspondence: jaroslaw.wajs@pwr.edu.pl; Tel.: +48-320-6823

Abstract: Mining industry faces new technological and economic challenges which need to be overcome in order to raise it to a new technological level in accordance with the ideas of Industry 4.0. Mining companies are searching for new possibilities of optimizing and automating processes, as well as for using digital technology and modern computer software to aid technological processes. Every stage of deposit management requires mining engineers, geologists, surveyors, and environment protection specialists who are involved in acquiring, storing, processing, and sharing data related to the parameters describing the deposit, its exploitation and the environment. These data include inter alia: geometries of the deposit, of the excavations, of the overburden and of the mined mineral, borders of the support pillars and of the buffer zones, mining advancements with respect to the set borders, effects of mining activities on the ground surface, documentation of landslide hazards and of the impact of mining operations on the selected elements of the environment. Therefore, over the life cycle of a deposit, modern digital technological solutions should be implemented in order to automate the processes of acquiring, sharing, processing and analyzing data related to deposit management. In accordance with this idea, the article describes the results of a measurement experiment performed in the Mikoszów open-pit granite mine (Lower Silesia, SW Poland) with the use of mobile LiDAR systems. The technology combines active sensors with automatic and global navigation system synchronized on a mobile platform in order to generate an accurate and precise geospatial 3D cloud of points.

Keywords: mobile laser scanning; Velodyne LiDAR; Riegl scanning system; open pit mine

Citation: Wajs, J.; Trybała, P.; Górniak-Zimroz, J.; Krupa-Kurzynowska, J.; Kasza, D. Modern Solution for Fast and Accurate Inventorization of Open-Pit Mines by the Active Remote Sensing Technique—Case Study of Mikoszów Granite Mine (Lower Silesia, SW Poland). *Energies* **2021**, *14*, 6853. https://doi.org/10.3390/en14206853

Academic Editors: Sergey Zhironkin and Dawid Szurgacz

Received: 28 September 2021
Accepted: 17 October 2021
Published: 19 October 2021

1. Introduction

Modern land surveying in the mining industry is based on input data acquired from both classical methods (leveling, tacheometry) and modern solutions: digital photogrammetry, measurements with the use of Global Navigation Satellite Systems (GNSS), and laser scanning. These data allow the preparation of maps documenting the deposit, situational plans illustrating the advancement of mining operations, or 3D visualizations showing planned reclamation forms. Commonly used worldwide, these surveying techniques have both advantages and disadvantages. Their implementation typically depends on the level of detail expected in the mining map, and thus on the accuracy of a particular measurement method and on the duration of the measurement process [1–7].

Classical surveying methods allow highly accurate and precise measurement results. The vertical displacements can be recorded with the use of classical and precise leveling with accuracy levels of several millimeters [8] or under 1 mm [9], respectively. Displacements on the surface of a mining area can be observed in 3D. For this purpose, total station measurements are combined with static GNSS measurements [10].

One of the interesting solutions, which is an extension of satellite measurements, is the use of the so-called pseudolites. Pseudoliths are terrestrial transmitters that are transmitting satellite-like signals to assist satellite navigation in areas depleted in terms of signal availability from traditional satellites. The areas of application of this solution are opencast mines with steep and high slopes, where poor satellite availability due to obscured horizon limits the ability to receive the GNSS signal [11].

Recent years have also been marked by the dynamic development of measurement technology based on ground-based radar. In mining, both underground and opencast, it is mainly used to monitor the stability of slopes or to measure surface deformation [12], rarely to monitor the exploitation progress or build the 3D models of mining area.

Observations in the form of a 3D continuous surface allow the construction of a 3D model from the land or aerial level [13,14]. Such measurements typically involve UAV photogrammetry techniques [15] and LiDAR, mainly ALS, and increasingly often UAV [16]. With the use of such techniques, the model can be quickly reconstructed with an accuracy of below 7 cm [17], which, of course, depends on laser scanner accuracy itself as well as resolution of the scan. In the case of UAVs, the main factor determining the accuracy of the calculated model is so-called ground sampling distance (GSD; it is the distance between pixel centers measured on the ground). Updating 3D mine models does not require as high accuracies as in the case of monitoring subsidence movements or slope stability [18].

Limitations of the above methods also need to be stressed. Most importantly, leveling, total station (excluding situations when prisms are stabilized on the measured points) or photogrammetry survey methods cannot be used without the natural and/or an artificial source of light, which fact practically disqualifies these methods from being used at night or would entail the need to provide prohibitively expensive artificial lighting. Atmospheric conditions (e.g., clouds, wind or atmospheric precipitation) are also a considerable limitation, which may contribute to a lower quality of the measurement results or even prevent the measurements entirely. On the other hand, the GNSS measurements—and more specifically their precision—depend on the availability of satellite constellations, and the use of the so-called differential corrections also increases the cost of the entire procedure. Importantly, in the case of relatively deep open-pit excavations, which may have obstructed views of the satellites, the measurement accuracy is lowered. Recent developments in geomatics have allowed the use of a wide range of sensors to record the geometry of both objects and other features on land [19]. Currently, hybrid sensors such as Mobile LiDAR Systems (MLS) provide additional quality to the inventorying process of mining facilities, as they offer solutions which prove flexible in terms of accuracy, resolution and access to areas which are otherwise inaccessible to vehicles [20].

The aim of this article is to present an MLS-based measurement solution for open-pit mining industry. The proposed technique employs a Riegl VMZ 400i measurement platform and a Velodyne LiDAR sensor in the Simultaneous Localization and Mapping (SLAM) approach to data acquisition and localization. The main concept presented in this paper utilizes a high-resolution, precise geometrical data source (MLS) for creating the initial model of the mine and updating it periodically using low-cost sensor and open-source algorithms. Such solution could provide information about near real-time tracking of the progress of mining works and allow e.g. volumetric calculations of the excavated material.

The measurements from both systems provide data for developing a digital representation (the so-called Digital Twin) of the geometry of a mining excavation, which may be used over the mine life cycle in such applications as monitoring the exploitation state of the mining excavation, performed in the form of cyclical measurements in order to control and optimize (improve) the efficiency of the mining process, constructing a management model of an active or a closed excavation, providing information about the state of the closed excavation and constructing its revitalization model. The methods here proposed allow quick access to the excavation-related data with a minimum workload required from an operator to acquire and process the data and, as such, they reflect well the principles of Industry 4.0 [21–25].

2. Materials and Methods

The main idea behind this study is based on the use of a point cloud obtained from a RIEGL VMZ commercial hybrid laser-scanning platform equipped with a VZ400i scanner mounted on an off-road vehicle. The system allows a quick acquisition of three-dimensional data about the analyzed object, in this case about a granite quarry. The concept of integrating a high-resolution MLS approach with full georeferencing is a base for further analyses of sensors used in acquisitions of 3D geometrical data related to the mine. The experiment was performed with a low-cost Velodyne scanning sensor was, in a handheld SLAM MLS approach. Tests of such a solution consist in obtaining three-dimensional information about the mine (a point cloud) over a time t_0 and in comparing the results with the results from the Velodyne instrument. Consecutive measurements performed in time $t_1, t_2, \ldots t_n$ may be in the form of the so-called additional measurements with the use of low-cost handheld MLS sensor. Figure 1 is a schematic diagram of the measurements.

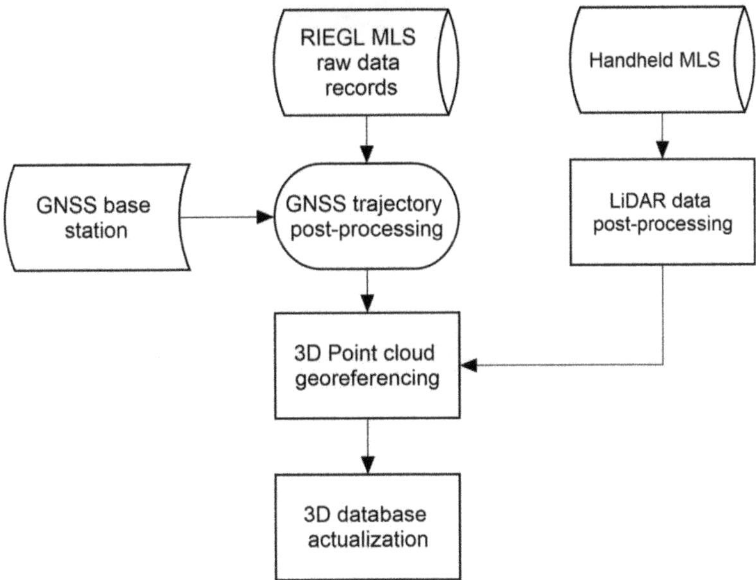

Figure 1. Schematic diagram of the 3D modeling methodology based on two Mobile LiDAR Systems (MLS) systems.

2.1. Description of the Study Area

This experiment was performed on the Mikoszów granite deposit located south-east of the town of Strzelin (Lower Silesia, SW Poland; Figure 2). Based on lease no. 10/2001, until 2016 Mineral Polska Sp. z o.o. mined the deposit for granite and gneiss. The company is planning to renew its lease to mine the deposit with the same technology. The planned output is 800,000 Mg per year. The geological resources of the Mikoszów deposit are 23,249,840 Mg (as per 31 December 2015). Until 2016, the deposit was mined with the use of a mixed wall-shortwall system with parallel advancement of the mining front. The deposit was extracted by drilling and blasting with the use of explosive materials and short and long drillholes. The mined material was loaded with loaders or excavators into mobile hoppers of crushing/sorting machines or into technological vehicles which transported it to processing devices located outside the mining plant. The processed stone was loaded with a loader from the storage site onto vehicles provided by the clients [26–29].

Figure 2. (**A**)—location of the study area; (**B**)—view of the quarry in the NW direction, as of Nov. 2020 (Photo by J. Górniak-Zimroz).

2.2. Mobile Laser Scanning

The literature mentions a number of mobile and autonomous mapping platforms which can collect data form indoor mapping [30]. The main advantage of the MLS system mounted on vehicle lies in the sensor fusion. The mobile mapping platform is equipped with GNSS, IMU (Inertial Measurement Unit) and DMI (Distance Measurement Indicator) sensors. The GNSS observations are essential in the kinematic mode of Lidar data acquisition. For a perfect trajectory, both static and dynamic alignment is required. If this condition is not met, the derived point clouds are distorted and lose spatial consistency. MLS data acquisition in urban areas may be affected by multipath effects and by signal obstruction due to buildings. This can lead to inaccurate GNSS measurements and, therefore, errors in the estimated trajectory [31]. Kukko et al. (2012) presented on-board sensors integration and MLS platform data acquisition from a vehicle, boat-mounted MLS for mapping fluvial processes and snowmobile application for studying the characteristics of and changes in snow cover. The main advantage of such mobile scanning platforms include fast and smart data collection. In MLS, the slightly elevated point of view gives the advantage of observing vertical surfaces with angle of incidence close to $0°$.

The mobile laser scanning technique allows fast and rapid 3D data acquisition in mining areas. In this technique, the measurement is performed with a scanner, and the time-dependent positions of the scanner are also recorded. In comparison to the standard laser scanning technique (Terrestrial Laser Scanning—TLS), in which the measuring instrument is located on an elevating tripod, the MLS has a similar incidence angle of the laser beam with respect to the scanned surface. However, as the system is mobile, it allows the acquisition of data for areas that were not visible from the perspective of the previous location of the scanning platform. The point cloud thus obtained has a relatively smaller number of occlusions and gaps. Examples of the integration of MLS data for the purpose of geological structure mapping were described in [32]. What is more, the recordings of LiDAR MLS datasets provide an alternative point of view. The average height above ground is greater than in the case of TLS scanning stations mounted on tripods. The density of the scan is similar to that of the TLS scan, and the density of the MLS records depends on the movement speed of the scanning platform.

2.2.1. The Riegl VMZ Mobile Scanning Platform

The MLS Riegl VMZ400i system used in the study comprised: a GNSS system based on simultaneous trajectory measurements from two antennas, an IMU, a DMI and a laser scanner (Figure 3).

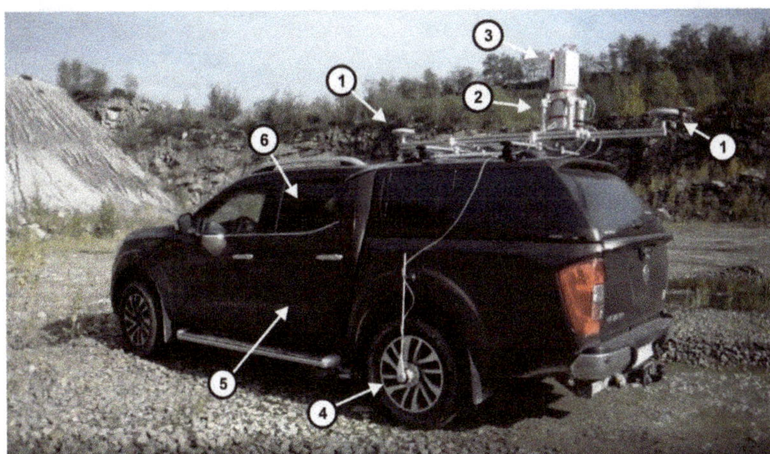

Figure 3. Mobile laser scanning system operated in the excavation: (1) Global Navigation Satellite Systems (GNSS) antenna; (2) Inertial Measurement Unit (IMU); (3) Riegl VZ400i scanner; (4) DMI; (5) power unit (inside the vehicle); (6) driver and control unit—computer with installed software and its operator (inside the vehicle).

2.2.2. Acquisition of MLS Data

The MLS data were recorded for the entire area of the Mikoszów mine and its vicinity. The recording process was performed with the use of the Riegl VMZ 400i hybrid laser scanning system set in the radar mode. The data acquisition process in this mode takes place while the scanner rotates 360° (around it's Z axis) and the whole MLS system drives along the planned trajectory as well. For the purpose of this article, an area of interest (AOI) was defined and indicated in yellow in Figure 4.

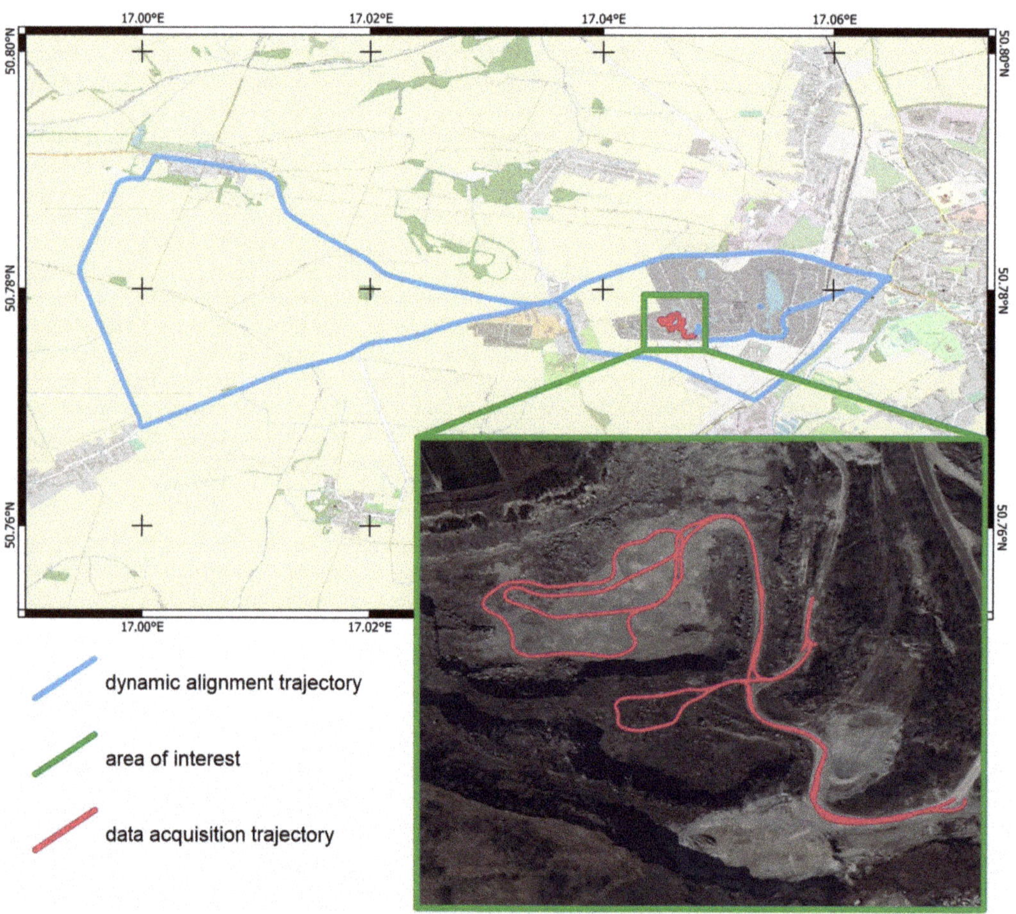

Figure 4. Sketch of the MLS data acquisition trajectory. The presented trajectory (blue line) covers the part of the route of the scanning system used for georeference of the acquired point cloud as well as (purple line) the part of the route within the excavation during the recording process of the data.

The procedure of recording MLS data, which is based on GNSS measurements, required the position of the MLS platform to be acquired around the analyzed mine in the form of a dynamic alignment trajectory loop. All of the works related to the GNSS trajectory measurements were performed in the Applanix PosPac MMS software, using the In-Fusion single base adjustment solution. The procedure allowed the GNSS observations of the platform to be linked with the BASE receiver, which was located in the central part of the mine pit, on its southern slope. The In-Fusion solution integrates the GNSS sensor with displacements recorded by the IMU and the DMI. The DMI allows precise information on the start and stop of the platform to be obtained from a source different to the GNSS. Figure 5 shows a schematic diagram of the MLS data processing procedure.

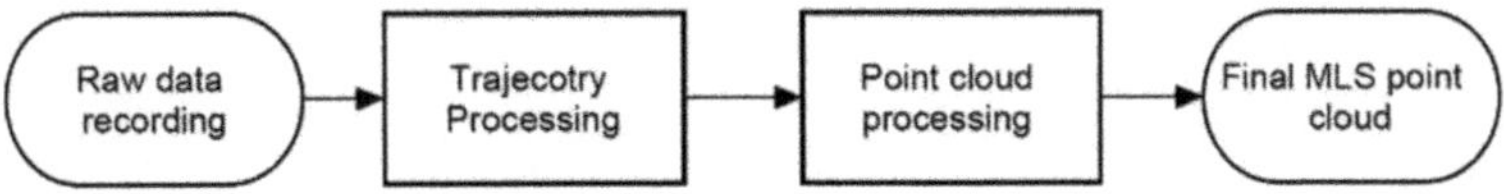

Figure 5. Diagram of MLS data processing.

The MLS measurements and their processing procedure are based on a precise measurement of the time recorded by the measurement system. The processing of a point cloud consists in overlaying individual scan lines on the 3D trajectory in time. The measurement accuracy of the MLS Riegl VMZ 400i is ensured owing to two GNSS antennas that measure the GAMS azimuth (GNSS Azimuth Measurement System). This approach helps eliminate inertial drift errors that are typically produced in a single-antenna approach). GAMS determines the movement direction of the vehicle very accurately regardless of its speed, resulting in the best possible heading accuracy and the best performance in any demanding environment, e.g., with an insufficient number of satellites. The full technical data are shown in Table 1.

Table 1. Riegl VMZ MLS technical data [33].

MLS Parameter	Value
Scanner Min. Range	0.5 [m]
Scanner Max. Range	800 [m]
Accuracy/Precission	5/3 [1] [mm]
Measurement Rate	0.5 M [measurement/s]
Scan field of view—Vertical	100 [°]
Scan field of view—Horizontal	360 [°]
Max Lines per Second (lps)	240 [lps]
GNSS Position Accuracy	20–50 [mm]
Roll/Pith/Heading Accuracy	0.015/0.05 [°]

[1] one sigma @ 100 m.

2.3. Handheld Mobile Laser Scanning

Another MLS technique is based on small mobile laser scanners (originally used in robotics) and on the SLAM technology. The technology consists in simultaneous, iterative determining the position of the observer and in constructing a map (or a 3D model) of the surrounding area. SLAM algorithms are mainly based on data obtained from laser scanners, stereoscopic cameras or monocular cameras. Solutions of this type are normally used in GNSS-denied environments, but if such limitations are not present, the position indicated by the GNSS receiver can be additionally used to improve the quality of SLAM-based positioning [34,35]. The pose of the observation unit and its movement can be additionally determined from the input data provided by other sources, such as IMU sensors [36,37] or wheel odometry [38]. SLAM algorithms are found in numerous implementations and are frequently based on the Kalman filters [35,39], graphs [40,41], or voxels [42].

SLAM functions by solving two basic problems:

- Estimation of consecutive transformations from the coordinate systems (related to the operator), containing point clouds obtained at times t_i and t_{i+1}, into a uniform global coordinate system—this process is referred to as laser odometry,
- Identification of the so-called loop closure, or return visit locations. In the case when point clouds obtained in non-consecutive time points t_i and t_j correspond to identical actual locations, another condition, different than the transformations calculated from

the consecutive scans, can be added to the bundle adjustment of the measurement trajectory. This fact significantly improves the quality and the robustness of the results obtained from the SLAM algorithm, and eliminates errors related to the position drifting in time due to the relative identification of consecutive observing positions.

MLS systems based on small portable laser scanners are now most typically offered as backpack systems, frequently integrated with cameras and GNSS receivers (Figure 6A) or as handheld scanners (Figure 6B). The measurement system can be further simplified by using only a small LiDAR sensor carried by an operator (Figure 6C). Despite a different approach to the sensor arrangement, the systems are, in fact, similar and share the measurement methodology. Therefore, later in this article they will be synonymously referred to as handheld laser scanners.

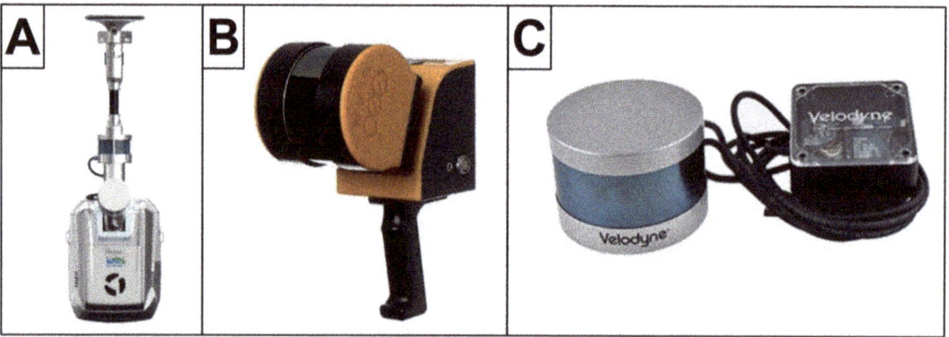

Figure 6. Examples of LiDAR-based scanning systems: (**A**) GreenValley International LiBackpack DGC50, (**B**) GeoSLAM ZEB Horizon, (**C**) Velodyne VLP-16 with an interface box.

The measurement is performed by an operator who carries a handheld or a backpack scanner and walks around the surveyed area. The latter factor most significantly distinguishes this solution from mobile or stationary laser scanners. On the one hand, it is a limitation, as the range and speed of the measurement is smaller than in the case of mobile scanning performed from a wheeled vehicle, drone or robot. Unlike in the above solutions, the measurement process cannot be automatized. Nevertheless, the operator can easily and naturally adjust the density of the point cloud in desired areas, by prolonging the data acquisition time. Modern SLAM algorithms also allow consecutive scans to be recorded and 3D models of the area to be constructed in real time. In combination with visualization techniques (e.g., on a tablet), this function allows a more effective and precise coverage of the surveyed area. Another advantage lies in the fact that the operator does not have to be highly qualified. Data processing is automatic, and therefore the survey can be performed by a person not familiar with the SLAM technology, or by an autonomous vehicle. Only the post-processing of the data (correctness verification and adding loop closures, georeferencing, improving the quality of defining the measurement trajectory), which allows an improved quality and accuracy of the resultant point cloud, requires higher competences and the ability to use a particular software.

Importantly, the measurement trajectory and thus the resultant point cloud, is biased with a drift error, i.e., a measurement uncertainty which increases with time. As already mentioned, this error is limited by revisiting the previously scanned locations and by allowing the SLAM algorithm to perform loop closure. However, for this to be possible, the measurement path needs to be planned in such a manner that the already surveyed locations are revisited at a sufficient frequency [43,44].

Many of the SLAM algorithms which process 3D data provide lidar odometry with the use of the Generalized Iterative Closest Point (GICP) algorithm. Ren et al. [45] proposed a modification of this algorithm, allowing for the extraction of the ground plane from individual scans and using them as landmarks in order to increase the robustness of the algorithm. Subsequently, the researchers compared the proposed method with other state-of-the-art algorithms which use a lidar sensor only (VLP-16). These included Lidar Odometry and Mapping (LOAM) [46], Lightweight and Ground-Optimized Lidar Odometry (LeGO-LOAM) [47] and Berkeley Localization and Mapping (BLAM) [48]. The proposed algorithm was designed for the consistent localization of autonomous vehicles in roadways and tunnels of an underground mine. Nevertheless, tests performed on two paths inside a building and on two paths in an underground mine indicate that the results in the mining environment are significantly worse. The authors stress the significance of the loop closure and of introducing plane constraints in obtaining satisfactory results from the tested algorithms.

Vasenna & Clerici [49] introduce a concept of integrating data obtained from classical point cloud construction methods (TLS and UAV photogrammetry) with data from a commercial, backpack SLAM system manufactured by Heron. The aim of the study was to verify the possibility of locating the SLAM operator in an open-pit mine environment previously modeled with the use of classical methods. Having successfully verified this possibility, the authors proposed a methodology for using SLAM in detecting changes of excavation geometry. The estimated accuracy of such detections is above 3–4 cm.

In this study, HDL-SLAM algorithm framework, proposed by Koide et al. [50], has been used. It is based on pose graph optimization and allows a significant elasticity in selecting parameters (for both the LiDAR odometry and the loop closure) and additional conditions in the SLAM algorithm, such as ground plane constraint, GNSS constraint, LiDAR odometry, estimation method and numerous robust kernels. Moreover, the authors provided an interactive graph editing program, which allows the resultant trajectory and point cloud to be improved in the post-processing stage with the use of manual edition tools, loop closure densification and edge refinement. The general concept of the data processing acquired by handheld SLAM is presented in Figure 7. The framework was selected due to its multiple options, which enable adjustments to the conditions of a particular surveyed object, and also due to the open-source implementation in the Robot Operating System (ROS) [51], which allows seamless integration with the LiDAR sensor and with the remaining software installed on a Linux-run laptop computer. The measurements were performed with the use of the Velodyne VLP-16. Its parameters are presented in Table 2.

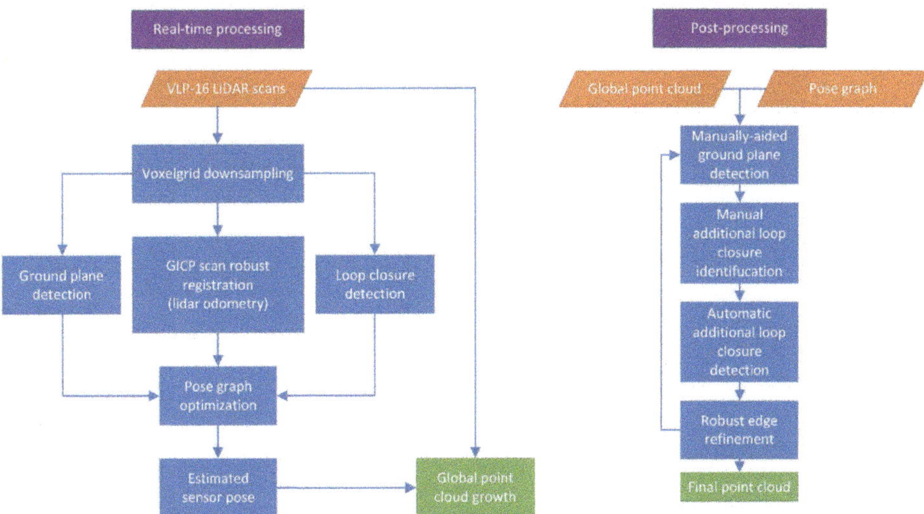

Figure 7. Handheld LiDAR SLAM real-time (**left**) and post-processing (**right**) workflows.

Table 2. Velodyne VLP-16 technical data.

Handheld LiDAR Parameter	Value
Laser Max. Range	100 [m]
Range Accuracy	3 [cm]
Measurement Rate	0.3 M [measurement/s]
Scan Angle Resolution—Vertical	2.0 [°]
Scan Angle Resolution—Horizontal	0.2 [°]
Lines in each scan (channels)	16

3. Results

The fact that the hybrid Riegl VMZ 400i scanner was installed on a Nissan Navara 4×4 off-road vehicle enabled the operator to reach the lowest level of the Mikoszów mine pit, which was the area subjected to further analysis, and to enter the otherwise least accessible locations. The system functions owing to the fact that the scanner was installed above the vehicle and the view of the satellites remained unobstructed. The unfavorable Positional Dilution Of Precision (PDOP) and the insufficient number of satellites visible in the vicinity of the vertical slopes was compensated for by data from IMU and DMI. The geometrical processing of the data was performed in parallel with preliminary filtration. The extraction parameters consisted of such LiDAR signal attributes as reflectance, distance, deviation. The next stage consisted of data filtering. The resultant point cloud represented all classes and comprised above 9.7×10^7 points (Table 3). The resultant point cloud was not filtered in order to classify the point cloud in accordance with the ISPRS .las recommendations. Figure 8 is a visualization in the form of a raster of the measurement range. The recorded MLS point cloud is a set of XYZ 3D data and a series of parameters such as amplitude and reflectance.

The handheld LiDAR measurements were performed by an operator who carried the LiDAR scanner (Velodyne VLP-16) in his hand. The scanner was connected to a laptop in the backpack. The data were recorded on the laptop, in rosbag format in ROS. The measurement covered the area of the lowest level of the open-pit and was performed by an operator walking around the mine. Attention was paid to intentionally form numerous smaller measurement loops and to stop the measurement in the vicinity of the start location in order to achieve a robust point cloud by forming a loop closure. The measurement continued for 16 min and 54 s, during which a total of 10,676 scans were acquired.

The measurement results were subsequently processed offline with the use of SLAM algorithms in order to combine them in a consistent point cloud. The preliminary estimation of the measurement trajectory was performed as the scans were replayed with actual speed. In the next step, the trajectory was manually improved in the post-processing mode by additionally indicating points which correspond to a flat terrain and by indicating clear revisit moments not identified automatically. Subsequently, automatic trajectory improvement methods were used to detect additional loop closures and robust refinement of pose graph edges. This process was iterated until satisfying results were obtained (no deviating observations (at 3 sigma), the trajectory visually corresponds to the actual measurement path). In effect, the obtained trajectory allowed the point clouds to be recorded from all scans. The resultant cloud, comprising 2,241,746 points, was filtered in Cloud Compare to remove measurement noise. The final result comprising a set of 1,968,367 points is shown in Figure 8. The entire post-processing lasted for approximately 30 min.

Table 3. Riegl and Velodyne mls measurement data volume.

Data Source	Number of Points	Mean Point Density [points/m^2]
MLS Riegl VMZ 400i	97,018,764	1500
Handheld SLAM LiDAR	2,241,746	50

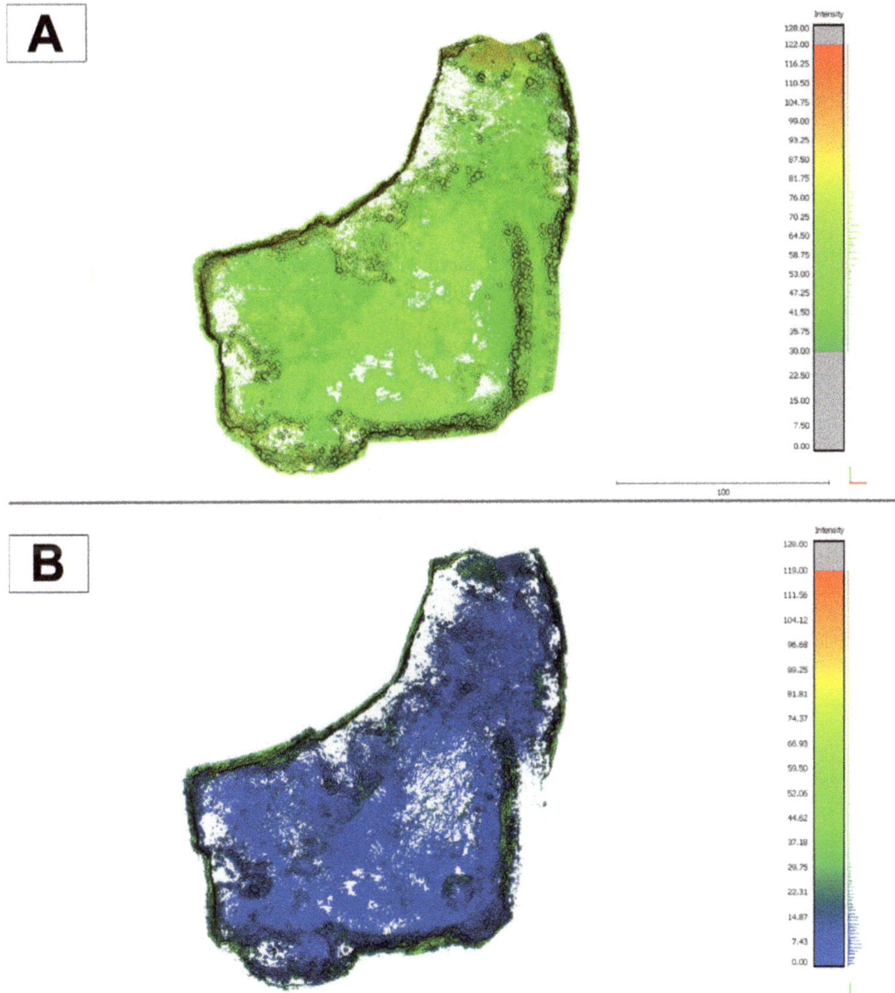

Figure 8. Top view of the point clouds: Riegl MLS VMZ 400i (**A**) and handheld Velodyne SLAM (**B**). Colors represent intensity value in a 7-bit range.

4. Discussion

The analyses demonstrate that mobile LiDAR measurement techniques provide input data that ensure that the constructed 3D mine models are georeferenced. The developed methodology for the processing of SLAM data is based on georeferencing them to the T_0 model built using the MLS Riegl VMZ 400i technique. The results of the experiment also indicate that it is possible to record a point cloud with the Iterative Closest Point (ICP) method, using a low-cost handheld scanner and with reference to the MLS point cloud.

The developed data processing method is based on overlapping the point clouds produced with the use of LiDAR MLS and SLAM techniques. The advantage of the use of MLS Riegl VMZ 400i lies in the fact that the 3D model has a full georeference in the chosen EPSG coordinate system. The SLAM point cloud recorded in the local system allows a continuous representation of the 3D surface in an open-pit mine. The MLS Riegl sensor is also advantageous due to its range, which reaches 800 m, with the SLAM point cloud recording at up to approx. 100 m. Another analyzed aspect is the scanning resolution. As presented in Table 3, average resolution of the MLS measurement was 1500 points/m^2 while the SLAM resolution was at an average of 50 points/m^2. An important part of this study was to verify whether the SLAM technique can be used as a low-cost approach to 3D modeling of mine geometry on ground. The results and the developed methodology clearly demonstrate that the implementation of the SLAM technique allows updates to the 3D model of an open-pit mine. Measurements of the identical ground surface showed the LiDAR data coherence level to be at 0.05 m. Figure 9 shows the spatial distribution of the distances between the points in the SLAM-based cloud, while Figure 10 is their histogram. A more detailed analysis of the locations with the highest individual deviations between the models leads to a conclusion that they occur in areas not covered by the MLS scanning or on areas covered with scattered greenery. In practical applications typical of an active open-pit mine, both cases should not affect the quality of the obtained geometrical data. The obtained accuracy levels (the distance between SLAM point cloud—MLS delivered mesh model is at a level of ±5 cm) is sufficient to calculate volumes in open-pit mines of similar scale and to update their 3D models, since the geometry changes associated with the periodically monitored advances of the mining operation would usually be in the order of meters.

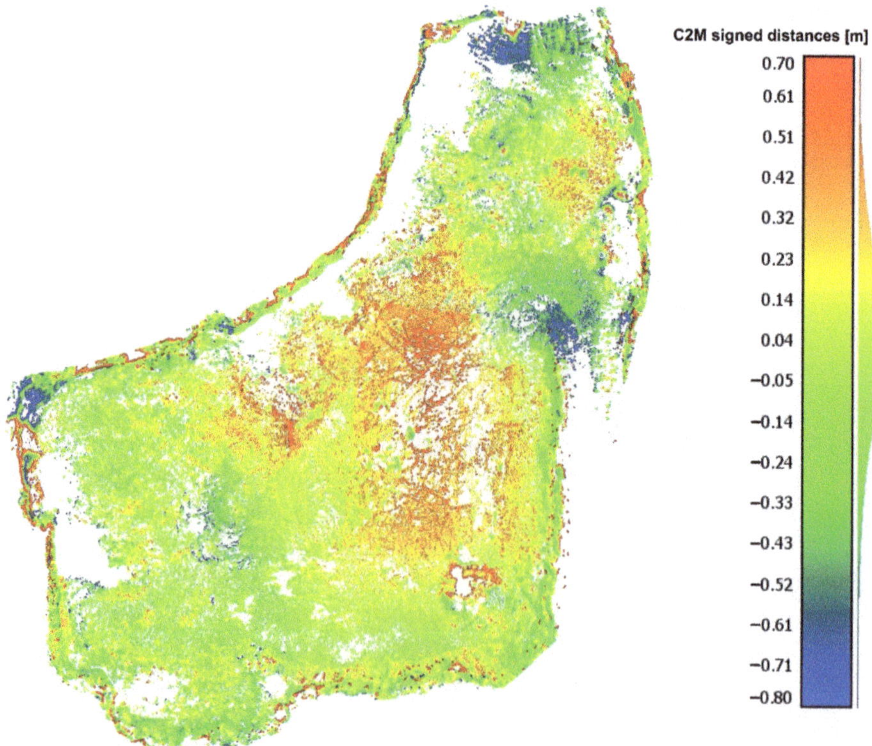

Figure 9. The spatial distribution of the distances between MLS Riegl and Velodyne SLAM points. Colors represent the distances from SLAM cloud to reference mesh.

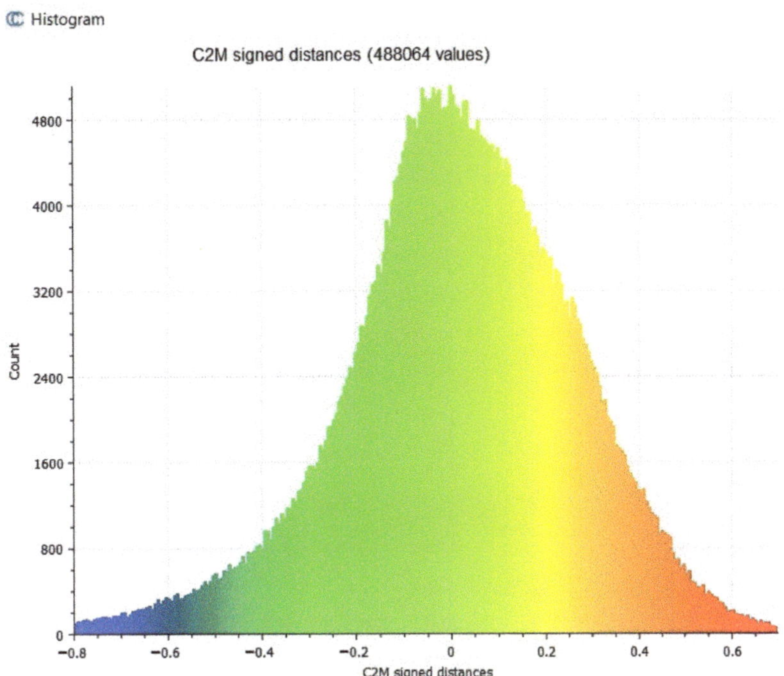

Figure 10. Distribution of the distances from SLAM cloud to reference mesh.

LiDAR observations in the implemented MLS approach allow the angles and distances to be measured from multiple measurement stations represented by the trajectory of the vehicle movement. In comparison to a classic approach to LiDAR measurements, TLS offers a possibility to record a continuous surface in a DSM and to reconstruct it closely representing the 3D surface with high measurement resolution. Importantly, in both the classic TLS approach and the MLS, the measuring instrument has an identical incidence angle at the analyzed object.

The LiDAR MLS method is seemingly limited by the impossibility to acquire data from objects which are covered with water. The observed gaps in the data are due to the absorption of the active LiDAR beam by the air-water medium, which causes the laser beam to become deflected and the LiDAR measurement point not to be recorded (white empty pixels in Figure 11). Another limitation of the LiDAR methods lies in the so-called shadows cast by objects which obstruct the laser beam. The above limitations are minimized by implementing the MLS technique, which allows a continuous measurement along an MLS trajectory.

Figure 11. Isometric visualization of intensity MLS LiDAR data with white gaps.

5. Conclusions

Mobile laser scanning technology is experiencing a dynamic growth in surveying. This growth is observed in the precision of the measurements, in the amounts of obtained data (scanning resolution and speed), and in the number of the transport platforms being employed.

The in-field tests performed in the Mikoszów granite mine employed two types of laser scanning systems to demonstrate their usefulness and the potential of using the results in the modeling and monitoring of the geometry changes in an open-pit excavation. The novelty of the presented solution lies in the integration of spatial data acquired with sensors that vary in accuracy, measurement platform and procedure and data processing to optimize the effort and cost of maintaining a time-varying 3D mine model.

The solution proposed by the authors is based on performing the first scan with the use of a precise Riegl VMZ 400i system—this is indicated as state (point cloud) T_0. The most significant advantages of the method include the measurement range, data acquisition speed and the quality of the MLS trajectory adjustment, which allows precision in the order of single centimeters. Moreover, our method does not require the use of ground control points in the measurement area or outside of it. The results of the bundle adjustment of the MLS measurements were similar to the accuracy of RTK GNSS.

Data obtained with the use of this method may be successfully used in 3D modeling the geometry of an excavation, or in planning or monitoring the progress of mining operations, with respect to both compact rock (such as granite in this case, which may be mined in the form of both blocks and aggregate), and bulk minerals.

Consecutive measurement sessions are performed at times $T_1 - \ldots - T_n$ with the use of Velodyne VLP-16, which is a tool operated in the SLAM approach. The experiment, which consisted of acquiring data and then in combining the acquired data into a local point cloud and in georeferencing the data into an MLS cloud, demonstrated that the data are fitted at a precision sufficient to use them in the modeling of the advancement of mining works. From the perspective of mining-related surveying, the data are of adequate quality. Also, the solution proves economical, as it is based on the Velodyne VLP-16 low-cost approach. Importantly, a scanner of this type does not need a highly skilled operator. Additionally, the procedure of data processing and visualization is partially automated (on the basis of low-cost and open-access software solutions), which is a factor limiting the cost related to buying the software and employing a trained operator.

A novel procedure of carrying out 3D measurements was developed that can be applied to the entire life cycle of an open-pit mine. The experiment was successfully carried out to test it in the real mining environment. The core element of the proposed procedure is the creation of a base mine model with a precise MLS platform. Subsequent stages of work, i.e., periodic measurement sessions with a SLAM-based system, are utilized to update the base model and perform calculations of the excavated volume. It is worth noting that the advantage of using Velodyne VLP-16 or similar lidar in the SLAM solution is that there is no need for a highly skilled operator, costly equipment, or software. Such low-cost solution could enable employing digital twin concepts in the small mining companies, e.g., quarries.

Another advantage of the proposed solution is the lack of need to establish the ground control points. The base model is georeferenced using simultaneous GNSS RTK measurements and the subsequent SLAM models are registered to it in cloud-to-cloud manner, since the majority of the mine model is stable over time.

The authors would especially like to emphasize that the results of the experiment performed in the Mikoszów open-pit mine clearly indicate that the integration of the MLS LiDAR technique and of the handheld SLAM LiDAR technique according to the method here described allows a 3D model of the excavation to be constructed and updated in real time. The resultant high-resolution point cloud allows the designing, inventorying, and feeding of big data databases in mines, according to the idea of Industry 4.0. LiDAR techniques used in the analyzed mine enable a fast and comfortable acquisition of 3D information about the excavation over the life cycle of a mine. This information may be used at every stage of the mining operations in open-pit mines, including in prospecting and exploration works, in access and development works, in the exploitation of the mineral, in the reclamation works, and in the management of the post-mining area.

Author Contributions: Conceptualization, J.G.-Z., J.W. and D.K.; methodology, J.W. and P.T.; software, P.T. and J.W.; validation, J.W. and P.T.; formal analysis, J.W. and P.T.; investigation, J.W. and P.T.; resources, J.G.-Z. and J.K.-K.; data curation, J.G.-Z. and J.K.-K.; writing—original draft preparation, J.W., P.T., J.G.-Z., J.K.-K. and D.K.; writing—review and editing, J.W., P.T., J.G.-Z., J.K.-K. and D.K.; visualization, J.W. and P.T.; supervision, J.G.-Z. and J.K.-K.; project administration, J.G.-Z., J.K.-K. and D.K. All authors have read and agreed to the published version of the manuscript.

Funding: This research received no external funding.

Institutional Review Board Statement: Not applicable.

Informed Consent Statement: Not applicable.

Data Availability Statement: The data presented in this study are available on request from the corresponding author.

Acknowledgments: Authors would like to thank Mineral Polska Ltd and the mining plant operations manager, Mirosław Mróz, for the opportunity to perform measurements session in the Mikoszów open-pit mine.

Conflicts of Interest: The authors declare no conflict of interest.

References

1. Battulwar, R.; Winkelmaier, G.; Valencia, J.; Naghadehi, M.Z.; Peik, B.; Abbasi, B.; Parvin, B.; Sattarvand, J. A Practical Methodology for Generating High-Resolution 3D Models of Open-Pit Slopes Using UAVs: Flight Path Planning and Optimization. *Remote Sens.* **2020**, *12*, 2283. [CrossRef]
2. Nieto, J.I.; Monteiro, S.T.; Viejo, D. 3D geological modelling using laser and hyperspectral data. In Proceedings of the 2010 IEEE International Geoscience and Remote Sensing Symposium, Honolulu, HI, USA, 25–30 July 2010. [CrossRef]
3. Werner, T.; Bebbington, A.; Gregory, G. Assessing impacts of mining: Recent contributions from GIS and remote sensing. *Extr. Ind. Soc.* **2019**, *6*, 993–1012. [CrossRef]
4. Deliormanli, A.H.; Maerz, N.H.; Otoo, J. Using terrestrial 3D laser scanning and optical methods to determine orientations of discontinuities at a granite quarry. *Int. J. Rock Mech. Min. Sci.* **2014**, *66*, 41–48. [CrossRef]
5. Buckley, S.J.; Howell, J.; Enge, H.; Kurz, T. Terrestrial laser scanning in geology: Data acquisition, processing and accuracy considerations. *J. Geol. Soc.* **2008**, *165*, 625–638. [CrossRef]

6. Shahbazi, M.; Sohn, G.; Théau, J.; Menard, P. Development and Evaluation of a UAV-Photogrammetry System for Precise 3D Environmental Modeling. *Sensors* **2015**, *15*, 27493–27524. [CrossRef] [PubMed]
7. Keawaram, B.; Dumrongchai, P. Comparisons of Surveying with Terrestrial Laser Scanner and Total Station for Volume Determination of Overburden and Coal Excavations in Large Open-Pit Mine. *World Acad. Sci. Eng. Technol. Int. J. Geol. Environ. Eng.* **2017**, *11*, 964–972.
8. Zaky, K.; Ghonem, A.; El Semary, H. Accuracy, time and cost of different leveling types. *Eng. Res. J.* **2002**, *79*, 55–67.
9. Can, E.; Mekik, C.; Kuşçu, S. Akcın, H. Computation of subsidence parameters resulting from layer movements post-operations of underground mining. *J. Struct. Geol.* **2013**, *47*, 16–24. [CrossRef]
10. Brown, N.; Kaloustian, S.; Roeckle, M. Monitoring of open pit mines using combined GNSS satellite receivers and robotic total stations. In Proceedings of the 2007 International Symposium on Rock Slope Stability in Open Pit Mining and Civil Engineering, Perth, Australia, 12–14 September 2007; Australian Centre for Geomechanics: Crawley, Australia, 2007; pp. 417–429.
11. Einicke, G.; Martin, S.; von Voithenberg, M.V.; Enderle, W. Pseudolite Options for Improved Mining Navigation. In Proceedings of the International Global Navigation Satellite Systems Society IGNSS Symposium 2006, Holiday Inn Surfers Paradise, Gold Coast, QL, Australia, 17–21 July 2006.
12. McHugh, E.L.; Dwyer, J.; Long, D.G.; Sabine, C. *Applications of Ground-Based Radar to Mine Slope Monitoring*; Report of Investigations 9666; NIOSH-Publications Dissemination: Cincinnati, OH, USA, 2006.
13. Wajs, J. Research on surveying technology applied for DTM modelling and volume computation in open pit mines. *Min. Sci.* **2015**, *22*, 75–83. [CrossRef]
14. Xu, Z.; Xu, E.; Wu, L.; Liu, S.; Mao, Y. Registration of Terrestrial Laser Scanning Surveys Using Terrain-Invariant Regions for Measuring Exploitative Volumes over Open-Pit Mines. *Remote Sens.* **2019**, *11*, 606. [CrossRef]
15. Tong, X.; Liu, X.; Chen, P.; Liu, S.; Luan, K.; Li, L.; Liu, S.; Liu, X.; Xie, H.; Jin, Y.; et al. Integration of UAV-Based Photogrammetry and Terrestrial Laser Scanning for the Three-Dimensional Mapping and Monitoring of Open-Pit Mine Areas. *Remote Sens.* **2015**, *7*, 6635–6662. [CrossRef]
16. Jozkow, G.; Totha, C.; Grejner-Brzezinska, D. UAS topographic mapping with Velodyne LiDAR sensor. *ISPRS Ann. Photogramm. Remote Sens. Spat. Inf. Sci.* **2016**, *3*, 201–208. [CrossRef]
17. Toth, C.; Jozkow, G.; Grejner-Brzezinska, D. Mapping with Small UAS: A Point Cloud Accuracy Assessment. *J. Appl. Geod.* **2015**, *9*, 213–226. [CrossRef]
18. Jaboyedoff, M.; Abellán, A.; Carrea, D.; Derron, M.H.; Matasci, B.; Michoud, C. Mapping and monitoring of landslides using LiDAR. In *Natural Hazards*; CRC Press: Boca Raton, FL, USA, 2018; pp. 397–420.
19. Kukko, A.; Kaartinen, H.; Hyyppä, J.; Chen, Y. Multiplatform Mobile Laser Scanning: Usability and Performance. *Sensors* **2012**, *12*, 11712–11733. [CrossRef]
20. Okhotin, A.L. Application of Laser scanning in mine surveying. *Proc. FIG Comm.* **2009**, *6*, 53–62.
21. Bertayeva, K.; Panaedova, G.; Natocheeva, N.; Belyanchikova, T. Industry 4.0 in the mining industry: Global trends and innovative development. In *E3S Web of Conferences*; EDP Sciences: Les Ulis, France, 2019; Volume 135, p. 04026.
22. Lööw, J.; Abrahamsson, L.; Johansson, J. Mining 4.0—The Impact of New Technology from a Work Place Perspective. *Min. Metall. Explor.* **2019**, *36*, 701–707. [CrossRef]
23. Gackowiec, P.; Podobińska-Staniec, M.; Brzychczy, E.; Kühlbach, C.; Özver, T. Review of Key Performance Indicators for Process Monitoring in the Mining Industry. *Energies* **2020**, *13*, 5169. [CrossRef]
24. Sukiennik, M. Challenges Faced by Businesses in the Mining Industry in the Context of the Industry 4.0 Philosophy. *Multidiscip. Asp. Prod. Eng.* **2018**, *1*, 621–626. [CrossRef]
25. Sishi, M.N.; Telukdarie, A. Implementation of industry 4.0 technologies in the mining industry: A case study. In Proceedings of the 2017 IEEE International Conference on Industrial Engineering and Engineering Management (IEEM), Singapore, 10–13 December 2017. [CrossRef]
26. Gruszecki, J. (Geological Company Proxima S.A., Wroclaw, Poland). Geological documentation of the granite and gneiss deposit in category C1 Mikoszow in Strzelin, province. Wroclaw. Unpublished work. 1993.
27. Kominowski, K. (A-Z Geometr s.c., Walbrzych, Poland). Geological documentation of the Mikoszow granite and gneiss deposit in cat. C1, places Mikoszow, commune and over Strzelin, voivodeship Lower Silesia. Unpublished work. 2001.
28. Majkowska, U. (Majkowska Geological Services, Wroclaw, Poland). Appendix No. 1 to the geological documentation of the Mikoszow granite and gneiss deposit in cat. C1 in Strzelin, Strzelin commune, province Wroclaw. Unpublished work. 1996.
29. Szkudlarek, L.; Bernatowicz, W.; Ryng-Duczmal, W.; Bernatowicz, M.; Koltowska, M.; Gil, R.; Filipowska, I. *Renewal of the Mining License for the Mikoszow Deposit*; Environmental Impact Report; Ekovert Szkudlarek: Wroclaw, Poland, 2017.
30. Adán, A.; Quintana, B.; Prieto, S. Autonomous Mobile Scanning Systems for the Digitization of Buildings: A Review. *Remote Sens.* **2019**, *11*, 306. [CrossRef]
31. Hussnain, Z.; Elberink, S.O.; Vosselman, G. Enhanced trajectory estimation of mobile laser scanners using aerial images. *ISPRS J. Photogramm. Remote Sens.* **2021**, *173*, 66–78. [CrossRef]
32. Singh, S.K.; Raval, S.; Banerjee, B.P. Automated structural discontinuity mapping in a rock face occluded by vegetation using mobile laser scanning. *Eng. Geol.* **2021**, *285*, 106040. [CrossRef]
33. Riegl Hybrid Mobile Laser Scanning System Brochure. Available online: http://www.riegl.com/uploads/tx_pxpriegldownloads/RIEGL_VMZ_at-a-glance_brochure_2018-11-28.pdf (accessed on 26 January 2021).

34. Chang, L.; Niu, X.; Liu, T.; Tang, J.; Qian, C. GNSS/INS/LiDAR-SLAM Integrated Navigation System Based on Graph Optimization. *Remote Sens.* **2019**, *11*, 1009. [CrossRef]
35. Hening, S.; Ippolito, C.A.; Krishnakumar, K.S.; Stepanyan, V.; Teodorescu, M. *3D LiDAR SLAM Integration with GPS/INS for UAVs in Urban GPS-Degraded Environments*; AIAA Information Systems-AIAA Infotech@Aerospace; American Institute of Aeronautics and Astronautics: Reston, VA, USA, 2017. [CrossRef]
36. Karam, S.; Lehtola, V.; Vosselman, G. Strategies to integrate imu and lidar slam for indoor mapping. ISPRS Annals of the Photogrammetry. *Remote Sens. Spat. Inf. Sci.* **2020**, *1*, 223–230. [CrossRef]
37. Sadruddin, H.; Mahmoud, A.; Atia, M.M. Enhancing Body-Mounted LiDAR SLAM using an IMU-based Pedestrian Dead Reckoning (PDR) Model. In Proceedings of the 2020 IEEE 63rd International Midwest Symposium on Circuits and Systems (MWSCAS), Springfield, MA, USA, 9–12 August 2020. [CrossRef]
38. Quan, M.; Piao, S.; Tan, M.; Huang, S.S. Tightly-Coupled Monocular Visual-Odometric SLAM Using Wheels and a MEMS Gyroscope. *IEEE Access* **2019**, *7*, 97374–97389. [CrossRef]
39. Brossard, M.; Bonnabel, S.; Barrau, A. Invariant Kalman Filtering for Visual Inertial SLAM. In Proceedings of the 2018 IEEE 21st International Conference on Information Fusion (FUSION), Cambridge, UK, 10–13 July 2018. [CrossRef]
40. Mendes, E.; Koch, P.; Lacroix, S. ICP-based pose-graph SLAM. In Proceedings of the 2016 IEEE International Symposium on Safety, Security, and Rescue Robotics (SSRR), Lausanne, Switzerland, 23–27 October 2016. [CrossRef]
41. Pierzchała, M.; Giguère, P.; Astrup, R. Mapping forests using an unmanned ground vehicle with 3D LiDAR and graph-SLAM. *Comput. Electron. Agric.* **2018**, *145*, 217–225. [CrossRef]
42. Muglikar, M.; Zhang, Z.; Scaramuzza, D. Voxel Map for Visual SLAM. In Proceedings of the 2020 IEEE International Conference on Robotics and Automation (ICRA), Paris, France, 31 May–31 August 2020. [CrossRef]
43. Guclu, O.; Can, A.B. Fast and Effective Loop Closure Detection to Improve SLAM Performance. *J. Intell. Robot. Syst.* **2017**, *93*, 495–517. [CrossRef]
44. Sammartano, G.; Spanò, A. Point clouds by SLAM-based mobile mapping systems: Accuracy and geometric content validation in multisensor survey and stand-alone acquisition. *Appl. Geomat.* **2018**, *10*, 317–339. [CrossRef]
45. Ren, Z.; Wang, L.; Bi, L. Robust GICP-Based 3D LiDAR SLAM for Underground Mining Environment. *Sensors* **2019**, *19*, 2915. [CrossRef] [PubMed]
46. Zhang, J.; Singh, S. Low-drift and real-time lidar odometry and mapping. *Auton. Robot.* **2016**, *41*, 401–416. [CrossRef]
47. Shan, T.; Englot, B. LeGO-LOAM: Lightweight and Ground-Optimized Lidar Odometry and Mapping on Variable Terrain. In Proceedings of the 2018 IEEE/RSJ International Conference on Intelligent Robots and Systems (IROS), Madrid, Spain, 1–5 October 2018. [CrossRef]
48. Nelson, E. Blam—Berkeley Localization and Mapping. Available online: https://github.com/erik-nelson/blam (accessed on 26 January 2021).
49. Vassena, G.; Clerici, A. Open pit mine 3d mapping by tls and digital photogrammetry: 3d model update thanks to a slam based approach. *Int. Arch. Photogramm. Remote Sens. Spat. Inf. Sci.* **2018**, *42*, 1145–1148. [CrossRef]
50. Koide, K.; Miura, J.; Menegatti, E. A portable three-dimensional LIDAR-based system for long-term and wide-area people behaviour measurement. *Int. J. Adv. Robot. Syst.* **2019**, *16*, 172988141984153. [CrossRef]
51. Quigley, M.; Conley, K.; Gerkey, B.; Faust, J.; Foote, T.; Leibs, J.; Bergerm, E.; Wheeler, R.; Ng, A.Y. *ROS: An open-source Robot Operating System*; ICRA: Kobe, Japan, 2009; Volume 3, p. 5.

 *energies*

Article

Influence of Driving Direction on the Stability of a Group of Headings Located in a Field of High Horizontal Stresses in the Polish Underground Copper Mines [†]

Karolina Adach-Pawelus *[iD] and Daniel Pawelus

Faculty of Geoengineering, Mining and Geology, Wroclaw University of Science and Technology, 50-370 Wroclaw, Poland; daniel.pawelus@pwr.edu.pl
* Correspondence: karolina.adach@pwr.edu.pl
† This paper is an extended version of the paper published in 2018 World Multidisciplinary Earth Sciences Symposium, Prague, Czech Republic, 3–7 September 2018, 012097.

Abstract: This paper investigates the problem of stability in a group of headings driven in high horizontal stress fields in the copper ore mines of the Legnica-Glogow Copper Belt (LGCB). The headings are protected with the roof bolting system. This problem is of high importance due to special safety regulations which apply in mining workings serving as airways and haulageways. The analysis was performed for a group of four headings driven in the geological and mining conditions of the Polkowice-Sieroszowice mine. The stability of the headings was evaluated with the use of Finite Element Method (FEM). The parameters of the rocks used in the numerical modeling have been determined on the basis of the Hoek–Brown classification, with the use of the RocLab 1.0 software. The parameters of the stress field have been identified on the basis of in situ measurements, which were performed in the Polkowice-Sieroszowice mine in 2012. The measurements were carried out with the use of the overcoring method, which is a stress relief method. A CSIRO HI probe was used as the measuring device. The tests were carried out on three measuring points, on which six successful tests were performed. The measurements confirmed the presence of high horizontal stresses in the rock mass. Numerical modeling was performed using the Phase2 v.8.0 software, in a triaxial stress state and in a plane strain state. The rock mass was described with an elastic-plastic model with softening. Numerical analyses were based on the Mohr–Coulomb failure criterion. It was assumed that the optimal measure of the stability of the group of headings is the range of the formed zone of yielded rock mass in the excavation roof. Numerical simulations have shown that the direction of driving the headings in the field of increased horizontal stresses may be of key importance for the stability of the headings in LGOM mines. The greatest extent of the yielded rock mass zone in the excavation roof occurred when the group of headings was driven in the direction perpendicular to the direction of the maximum horizontal stress component σ_H. The obtained results served to provide an example of the application of a roof bolting system to protect headings driven in unfavorable conditions in a high horizontal stress field.

Keywords: numerical modeling; high horizontal stress; excavations stability assessment

 **check for updates**

Citation: Adach-Pawelus, K.; Pawelus, D. Influence of Driving Direction on the Stability of a Group of Headings Located in a Field of High Horizontal Stresses in the Polish Underground Copper Mines. *Energies* **2021**, *14*, 5955. https://doi.org/10.3390/en14185955

Academic Editors: Sergey Zhironkin and Dawid Szurgacz

Received: 5 August 2021
Accepted: 16 September 2021
Published: 19 September 2021

Publisher's Note: MDPI stays neutral with regard to jurisdictional claims in published maps and institutional affiliations.

1. Introduction

Numerous observations and measurements performed both worldwide [1–13] and in Poland [14–16] suggest that in many cases the value of primary horizontal stresses is significantly greater than the values accepted to date, which are based solely on the Poisson's ratio ν. Another observation also indicated that the maximum component of the horizontal stresses in the rock mass was frequently up to several times higher than the vertical component. Therefore, primary stresses in the rocks forming the Earth's crust are believed to result from the accumulation of gravitational and tectonic stress fields [17], and the intensity of horizontal stresses is the function of

- interaction of individual tectonic units,
- the terrain surface,
- tectonic features of the rock mass,
- depth of the rock mass, and
- stiffness of rock material, expressed, among others through Poisson's ratio v and the modulus of linear deformation (longitudinal modulus of elasticity) E.

Research conducted by many scientists shows that high horizontal stress has a significant impact on underground excavations stability [9,18–21].

In a number of mines worldwide, some of the cases in which the stability of a heading was lost (Figure 1) or some phenomena having an impact observed in the heading were caused by an inadequate knowledge of the extents and directions of horizontal stresses in the mined rock mass [2,4–7]. The direction in which a heading is driven and in which horizontal stresses propagate was also observed to influence the stability of the heading (Figure 2). In the most advantageous scenario, the direction of the maximum increased horizontal stress component is parallel to the longer axis of symmetry of the heading. In such case, the heading does not lose its stability due to horizontal stresses (Figure 2a). In the case when the heading is driven at an angle to the direction of the maximum horizontal stress component, the roof may suffer damage, and the floor may become uplifted at the left or right wall (Figure 2b,c). When the direction of the maximum component is perpendicular to the direction in which the heading is driven, the least advantageous scenario occurs and the heading loses its stability (Figure 2d). In the central part of the heading, the roof collapses and the floor is uplifted [22].

Figure 1. Collapse of immediate roof due to increased horizontal stresses in the Beckley mine [7].

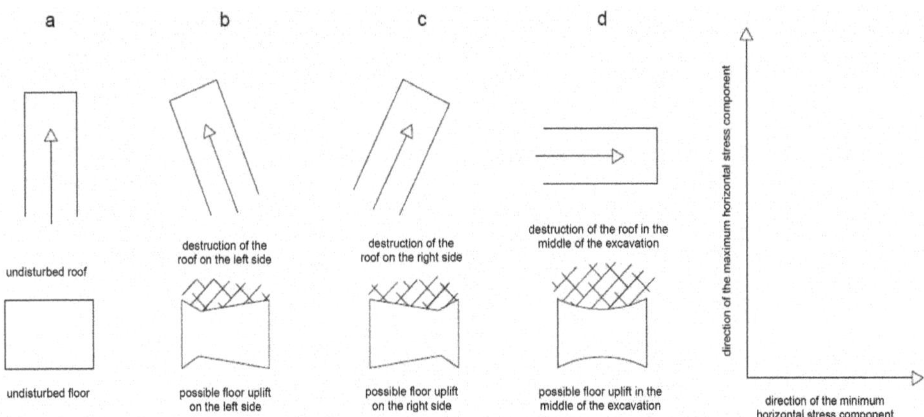

Figure 2. Influence of horizontal stress on heading stability (based on the works in [3,23]). (**a**) undisturbed floor; (**b**) possible floor uplift on the left side; (**c**) possible floor uplift on the right side; (**d**) possible floor uplift in the middle of the excavation.

In Polish underground mines, no efforts are made towards identifying the directions and values of primary horizontal stresses in the rock mass for the purpose of designing optimal heading systems and heading support structures. Roof bolting is the most commonly used roof support system in Polish copper mines of the Legnica-Glogow Copper Belt (LGCB) region. Prior to the choice of a support type for a mining excavation, roof class is determined in accordance with the "Instructions on determining the geomechanical parameters of roof rocks with respect to roof classes in copper mines", as required in the selection of a roof bolting system design [24]. The classification of roof rocks (from class 1—the worst, to class 5—the best) is based on such parameters as

- roof bedding (vertical split),
- concentration of mineralized foigs,
- fault concentration,
- average fault throw, and
- tensile strength of the roof rock beam.

The roof bolting system is selected after identifying the roof class and in accordance with the "Regulations on the selection, construction and control of excavation support in the KGHM Polska Miedz S.A. mines" [25]. The roofs in the headings are protected with bolts at least 1.6 m long. The distance between the bolts (the rock bolt pattern) is adjusted depending on the class of the roof and on the width of the heading below the roof. Supports are also installed on the heading walls. The wall bolts have a length of at least 1.6 m and are spaced in the side walls at 1.5×1.5 m. The lower row of the bolts is situated at a distance of approximately 1.8 m from the floor. The side walls are provided with a support system if

- excavation height is greater than 4.5 m (regardless of the inclination angle of the side walls) or
- excavation height is not greater than 4.5 m and moving the side walls outwards by approximately 10° is not possible.

As copper ore extraction depth in the LGCB region has exceeded the depth of 1200 m below ground level. and is performed in increasingly difficult geological and mining conditions, the copper ore mines in the region may suffer from a growing number of problems related to the stability of excavations located in the field of high horizontal stresses. Therefore, further extensive research needed to be performed into the directions and values of main stresses in the mines operated by KGHM Polska Miedz S.A. The next step should involve a research program which would allow identifying the influence of the directions and values of high horizontal stresses on the stability of mining excavations supported with roof bolting systems and located at a depth below 1200 m below ground level.

Numerical methods significantly expand research possibilities related to the analysis and evaluation of the stability of excavations and their support systems in underground mines. Numerical modeling gives the possibility to predict the stress concentration zones and the potential locations in which the rock mass may become unstable in the vicinity of a mining excavation. The numerical, FEM-based computer analyses performed to date on the stability of mining excavations have demonstrated the influence of horizontal stress direction on the stability of excavations in the LGCB mines [22,26].

This research, which focuses on the influence of the direction in which headings are driven in the "Polkowice-Sieroszowice" mine in the presence of high horizontal stress fields, is a continuation and development of research performed as part of a program titled "Implementation of numerical methods in the analysis of selected natural hazards in underground mines". This research was performed between 2017 and 2018 at the Mining Division, Wroclaw University of Science and Technology. Some results of this research were presented in scientific conferences and published [26,27]. Further research is needed however, as the results of numerical simulations may be used to plan and design inter alia access, preparatory and production excavations of various shapes and dimensions, as well as to aid the choice of adequate primary and secondary support systems. It is of special importance for the copper ore extraction operations performed in the area of the LGCB at depths exceeding 1200 m below ground level.

2. In-Situ Tests of Stresses in the Polkowice–Sieroszowice Mine

In 2012, the LGCB copper ore mines (the Rudna mine and the Polkowice-Sieroszowice mine) were the sites of a test program—the most extensive program in the history of Polish mining industry—aimed at in situ examinations of stresses in the rock mass. The measurements were conducted as part of a research project entitled "Determination of the impact of the primary stress directions and magnitudes on the optimal geometry of mining fields" [16]. The measurements were performed by a team consisting of the employees of KGHM Cuprum Ltd. R&D Center, the Rudna mine and Golder Associates Ltd. The aim of the measurements was to identify the distribution of primary stresses in the regions planned for operation within the Glogow Gleboki-Przemyslowy mining area. The analysis of the progress of the access and preparatory works indicated that the measurement points should be located in the northern and north-western parts of the Polkowice-Sieroszowice mine. Eventually, three locations for the measurement points were identified:

- measurement point 1: located in the "Sieroszowice I" mine area, between crosscuts 13 and 12 in main haulage roadway T-360, at a depth of 966.0 m below ground level;
- measurement point 2: located in division G-62, in main haulage roadway T-357, in the vicinity of crosscut P-84; and
- measurement point 3: located in the "Sieroszowice I" mine area, in the recess of main incline E-1, in the vicinity of crosscut 63, at a depth of 906.0 m below ground level.

The in situ measurements in the Polkowice-Sieroszowice mine were performed with the use of the overcoring method, which is a stress relief method. The method consists in cutting a rock sample from the rock mass in which a high horizontal stress field is observed and in simultaneously measuring the strain or displacement due to stress relief in the sample [28]. The stages of the overcoring method are shown in Figure 3. The most important of them are

- boring a hole in the rock which has a large-diameter (60–220 mm) and a length sufficient to treat the influence of the mining excavation as negligible (Figure 3a),
- driving a pilot hole typically 38 mm in diameter (Figure 3b),
- placing a measurement device (probe) in the pilot hole, and
- effecting stress relief in the cut-out cylinder, whose deformations are recorded with a measurement device (Figure 3c).

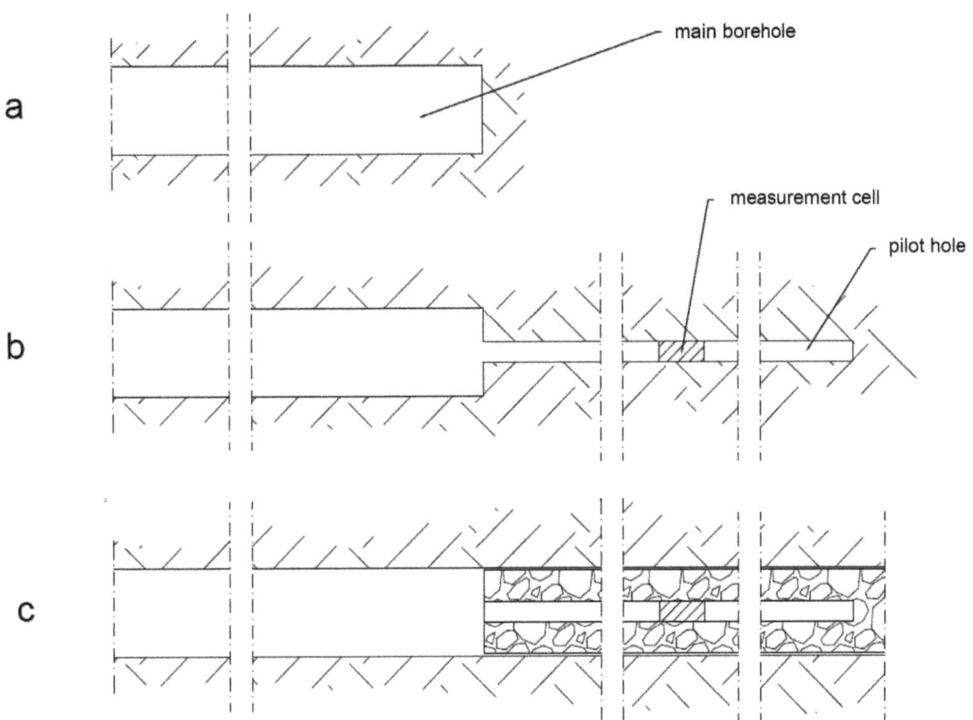

Figure 3. Stages of the overcoring method. (**a**) boring a hole in the rock which has a large-diameter (60–220 mm) and a length sufficient to treat the influence of the mining excavation as negligible; (**b**) driving a pilot hole typically 38 mm in diameter and placing a measurement device (probe) in the pilot hole; (**c**) effecting stress relief in the cut-out cylinder, whose deformations are recorded with a measurement device.

The stress tensor components in the rock mass are calculated from the values of strain measured during the process of relieving the sample in the vicinity of the measurement device. After the measurement, the rock sample is recovered for laboratory testing in order to determine Poisson's coefficient v and longitudinal modulus of elasticity E. Examinations also include visual inspection of the measuring cell and the surrounding rock, as well as the quality of the adhesive bond and the degree of rock disturbance [28].

The measurement device used in the Polkowice-Sieroszowice mine was a CSIRO HI test probe (Figure 4). It has 12 independent strain gauges glued in an arrangement of three "rosettes", three gauges in each "rosette", and three gauges glued circumferentially. Owing to the above, a single measurement provides a sufficient amount of data and allows the components of the stress tensor in the three-axis system to be fully determined. The probe is secured in the borehole with a special adhesive of known deformation parameters. The disadvantages of the overcoring measurements with the use of the CSIRO HI probe include the sensitivity to loose rock fragments in the borehole which hinder correct insertion of the probe, as well as thermal effects generated during the drilling process and the presence of water in the borehole [16,28]. The obtained results allowed determining horizontal stress distribution in the rock mass for the Polkowice-Sieroszowice mine (Table 1).

Figure 4. CSIRO HI test probe.

Table 1. Parameters of horizontal stress fields at the Polkowice-Sieroszowice mine [16,26].

Measurement Test No.	Stress Field Parameters				
	σ_H [MPa]	α_H [°]	σ_h [MPa]	α_h [°]	σ_v [MPa]
SP1T2	29.9	160.0	22.3	70.0	27.7
SP1T3	29.9	157.0	24.4	67.0	27.9
SP1T4	20.6	158.0	16.9	68.0	22.7
SP2T2	32.2	6.0	26.1	96.0	27.7
SP3T2	27.7	156.0	14.5	66.0	27.6
SP3T3	19.2	139.0	12.8	49.0	18.2

The symbols used in the above table are as follows: σ_H—maximum component of horizontal stress, α_H—azimuth of the maximum horizontal stress component, σ_h—minimum component of horizontal stress, α_h—azimuth of the minimum horizontal stress component, σ_v—vertical stress.

3. Stability Prediction for a Group of Headings at the Polkowice-Sieroszowice Mine

The influence of high horizontal stresses on the stability of headings in the Polkowice-Sieroszowice mine was modeled with the use of the Phase2 v. 8.0 software (Rocscience, Toronto, ON, Canada). This computer application is based on Finite Element Method (FEM). In FEM, a solution to a typical problem is searched for in the following stages:

- dividing an area into subareas,
- determining FEM equations for the elements,
- gluing (aggregating) the elements,
- allowing for boundary conditions,
- solving the equations, and
- calculating additional values in other (than nodes) points of the area [22].

The parameters of the rocks used in the numerical modeling are shown in Table 2. The parameters were determined from the geomechanical tests of rock samples. Rock samples for laboratory tests were obtained from the Jm-06 To-1 borehole, which is located in the Sieroszowice I mining area, division G-62, at the crossing of main incline E-1 and crosscut 33. The analysis of the data obtained from the Jm-06 To-1 borehole indicated that the rock mass represents a geological structure typical for the Fore Sudetic Monocline in which the access and preparatory excavations of the Polkowice-Sieroszowice mine are driven. The immediate roof is built of carbonate formations (Calcareous dolomite II) having high strength and strain parameters. The rocks forming the mined deposit height and the formations in the floor have various values of strength and strain parameters.

Table 2. Mean strength and strain rock parameters determined in laboratory tests for the Jm-06 To-1 borehole [26].

Location	Rock Type	h [m]	ρ [kg/dm^3]	R_c [MPa]	R_r [MPa]	E_i [GPa]	v [–]
Roof	Anhydrite	16.00	2.94	92.16	6.69	53.22	0.26
	Calcareous dolomite II	9.00	2.82	236.10	14.59	113.17	0.25
	Calcareous dolomite I	0.50	2.47	98.43	6.04	38.77	0.26
	Streaky dolomite	0.90	2.77	140.57	9.33	40.73	0.24
Excavation	Clay dolomite	0.55	2.63	79.50	5.70	28.75	0.23
	Quartz sandstone IV	0.25	2.40	39.97	2.78	16.53	0.19
	Quartz sandstone III	0.30	2.25	16.57	0.80	7.23	0.13
Floor	Quartz sandstone II	3.90	2.07	20.67	1.22	8.63	0.14
	Quartz sandstone I	1.10	2.02	16.95	0.75	6.65	0.12

The Hoek–Brown failure criterion, which is broadly used in geomechanical analyses of rock mass deformations and effort, was assumed for the rock mass. The generalized Hoek–Brown failure criterion for a fractured rock mass may be described with the following equation [29]:

$$\sigma_1 = \sigma_3 + \sigma_{ci} \cdot \left(m_b \cdot \frac{\sigma_3}{\sigma_{ci}} + s \right)^a,\qquad(1)$$

where

σ_1 and σ_3—values of the maximum and minimum principal effective stress at failure,

m_b—the Hoek–Brown constant for the rock mass,

s and a—constants depending on the rock mass properties,

σ_{ci}—the uniaxial compressive strength of the rock sample.

When rock mass tensile strength σ_{tm} is exceeded, the equation for $a = 0.5$ can be formulated as follows:

$$\sigma_{tm} = \frac{\sigma_{ci}}{2} \cdot \left(m_b - \sqrt{m_b^2 + 4s} \right).\qquad(2)$$

After the failure criterion was assumed, the following rock mass parameters have been determined for each of the rock layers obtained from the Jm-06 To-1 borehole (Table 3):

– uniaxial tensile strength of the rock mass σ_t,
– cohesion c,
– internal friction angle φ, and
– rock mass modulus of elasticity E_{rm}.

Table 3. Rock mass parameters determined with the RocLab 1.0 application the Jm-06 To-1 measurement borehole [26].

Location	Rock Type	σ_t [MPa]	c [MPa]	ϕ [°]	E_{rm} [MPa]
Roof	Anhydrite	0.738	6.896	38.66	39,000.37
	Calcareous dolomite II	5.226	21.535	39.00	99,628.96
	Calcareous dolomite I	1.495	7.853	37.69	31,649.89
	Streaky dolomite	3.112	12.821	39.00	35,856.57
Excavation	Clay dolomite	0.828	5.653	36.31	21,068.41
	Quartz sandstone IV	0.103	2.784	39.06	8595.60
	Quartz sandstone III	0.043	1.154	39.06	3759.60
Floor	Quartz sandstone II	0.053	1.439	39.06	4487.60
	Quartz sandstone I	0.044	1.180	39.06	3458.00

The calculations were performed with the use of the RocLab 1.0 software, which employs the Hoek–Brown classification [29–32].

Numerical modeling was performed using the Phase2 v. 8.0, in a triaxial stress state and in plane strain state. Numerical simulations were performed for an isotropic and for a uniform medium. The rock medium was described with an elastic-plastic model with softening. The strength–strain parameters of the rocks in the model are shown in Table 4. The numerical modeling was performed on the basis of the Mohr–Coulomb failure criterion, which states that rock may reach threshold effort if the following condition is met:

$$\sigma_1 = \sigma_3 \cdot \frac{1 + sin\varphi}{1 - sin\varphi} + \frac{2ccos\varphi}{1 - sin\varphi},\tag{3}$$

or

$$\sigma_3 = -\sigma_t\tag{4}$$

where

σ_1 and σ_3—effective maximum and minimum stress at failure,
ϕ—internal friction angle,
c—cohesion, and
σ_t—uniaxial tensile strength of the rock mass.

Table 4. Rock mass parameters adopted for the elastic-plastic numerical model with softening.

Location	Rock Type	h [m]	E_s [MPa]	ν [–]	σ_t [MPa]	ϕ_{peak} [°]	c_{peak} [MPa]	ϕ_{dyl} [°]	ϕ_{resid} [°]	c_{resid} [MPa]
Roof	Anhydrite	16.00	39,000.37	0.26	0.738	38.66	6.896	2.00	36.73	1.379
	Calcareous dolomite II	9.00	99,628.96	0.25	5.226	39.00	21.535	2.00	37.05	4.307
Excavation (h = 3.5 m)	Deposit mined in dolomite-sandstone	2.50	25,184.11	0.22	1.617	38.16	7.847	2.00	36.25	1.569
Floor	Quartz sandstone II	3.90	4487.60	0.14	0.053	39.06	1.439	2.00	37.11	0.288
	Quartz sandstone I	1.10	3458.00	0.12	0.044	39.06	1.180	2.00	37.11	0.236

The numerical analyses were performed for a group of four headings. The excavations have a trapezoidal shape. The inclination angle of the side walls was assumed at 10°. The roof was identified to be class four [24] and was protected with bolts 1.6 m in length with the bolting pattern 1.5 × 1.5 m [25]. Table 5 contains the dimensions of headings in the assumed cross-sections.

Table 5. Dimensions of the analyzed headings.

Excavation Height h [m]	Excavation Width Below the Roof d_{wst} [m]	Excavation Width at the Floor d_{wsp} [m]	Mean Excavation Width $d_{wśr}$ [m]	Excavation Surface Area S_r [m²]	Side Wall Inclination Angle α [°]
3.5	7.0	5.8	6.4	22.4	10.0

The numerical model was based on the values of primary stresses recorded in the Polkowice-Sieroszowice mine, at measurement point 2 (Table 1, measurement number SP2T2). Two variants of loads acting on the group of headings were assumed for the numerical calculations. The flat, rectangular plate comprising four openings (located inside and shaped to correspond to the shapes of the analyzed excavations) was loaded on its edges:

- load variant 1 (maximum horizontal stress component σ_H is in the direction parallel to the longer axis of the headings):
  - side edges: p_x = 26.10 MPa,
  - upper edge and bottom edge: p_z = 27.70 MPa,
  - direction perpendicular to plate surface: p_y = 32.20 MPa,
- load variant 2 (maximum horizontal stress component σ_H is in the direction perpendicular to the longer axis of the headings):

- side edges: $p_x = 32.20$ MPa,
- upper edge and bottom edge: $p_z = 27.70$ MPa,
- direction perpendicular to plate surface: $p_y = 26.10$ MPa.

The edges of the analyzed plate were equipped with supports which do not slide either in the vertical or in the horizontal direction. The numerical analysis employed finite elements having three nodes and triangular shape. The plate edges were assumed to be at a 100.0 m distance from the extreme points on each side of the analyzed headings (the roof, the floor and the side walls). In the middle of the plate, in the location the headings, smaller size finite elements were used (finite element grid density region) in order to increase the accuracy of numerical calculations. Based on the numerical calculations of heading stability for each model (calculation variant), the following parameters were determined:

- distribution of principal stresses σ_1,
- distribution of principal stresses σ_3,
- horizontal stress distribution σ_{xx},
- vertical stress distribution σ_{yy},
- total displacements,
- yielded element area (yielded rock mass zone).

The analysis of the results indicated that the optimal measure of the stability of the group of headings is the range of the yielded rock mass zone in the roof of the heading.

4. Results of Numerical Simulations and Selection of the Bolting System

The numerical models of the stability of headings located in the field of high horizontal stresses in the Polkowice-Sieroszowice mine confirmed the results obtained in previous research [22,26]. The numerical simulations demonstrated among other things that

- the direction in which headings are driven in a field of high horizontal stresses may be of key importance to the stability of headings in the LGCB mines, and
- problems with stability may occur when the yielded rock zone in the roof is larger than the bolted zone.

Figures 5–8 show the maximum range of yielded rock mass (between 50% and 100%) in the roofs of the headings. In the case of a group of headings driven parallel to the direction of the maximum horizontal stress component σ_H (load variant 1), the maximum range was 1.36 m (Figure 7). At the same time, the maximum range of yielded rock mass in the roofs of the headings driven in perpendicular to the direction of the maximum horizontal stress component σ_H (load variant 2) was greater by 0.58 m and reached 1.94 m (Figure 8). The numerical calculations also demonstrated that:

- The surface of the relaxed area around a heading increases together with the increase of the horizontal stress in the rock mass (high stress field in the rock mass). Meanwhile, stress concentration in the analyzed headings occurs only in the corners of the roofs and of the side walls.
- The greatest total displacements occur in the floors of the analyzed headings (formed of sandstones having low strength and strain parameters). The range of the zone affected by displacements increases together with the increase of horizontal stresses in the rock mass.
- The surface of the yielded area in the rock mass around headings driven in the heading group increases together with the increase of horizontal stresses in the rock mass (high stress field in the rock mass). This phenomenon is strictly related to the stress and strain parameters of the rock layers surrounding the excavations and negatively influences their stability.
- The verification of the results of the numerical simulations obtained for the plastic-elastic model with rock softening confirmed that they correspond most accurately to the observed cases of stability losses in the mining excavations of Polish copper mines.

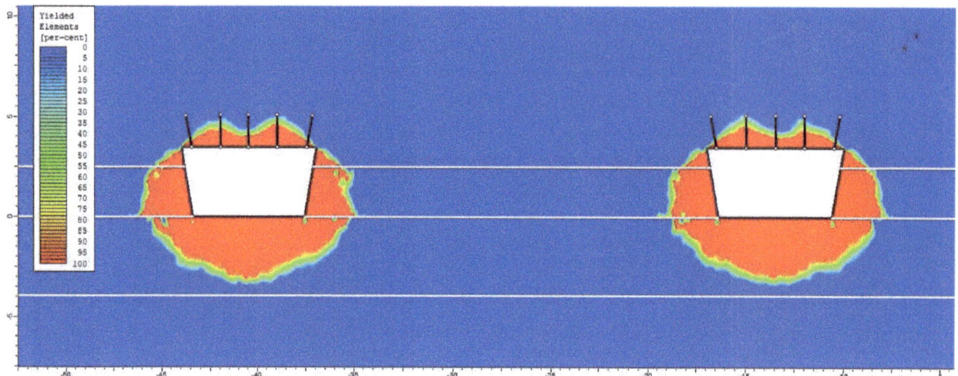

Figure 5. Yielded element area around heading 1 and heading 2, load variant 1.

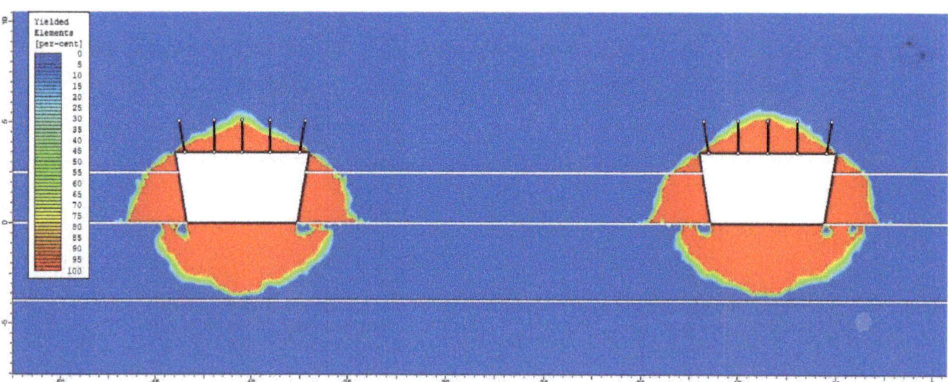

Figure 6. Yielded element area around heading 1 and heading 2, load variant 2.

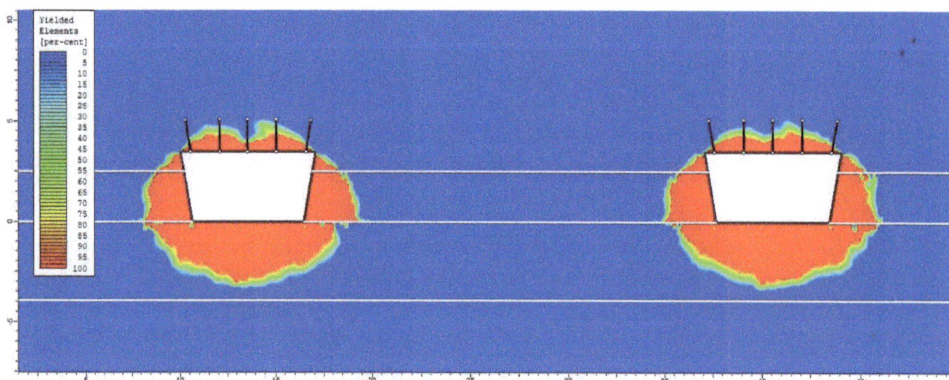

Figure 7. Yielded element area around heading 3 and heading 4, load variant 1.

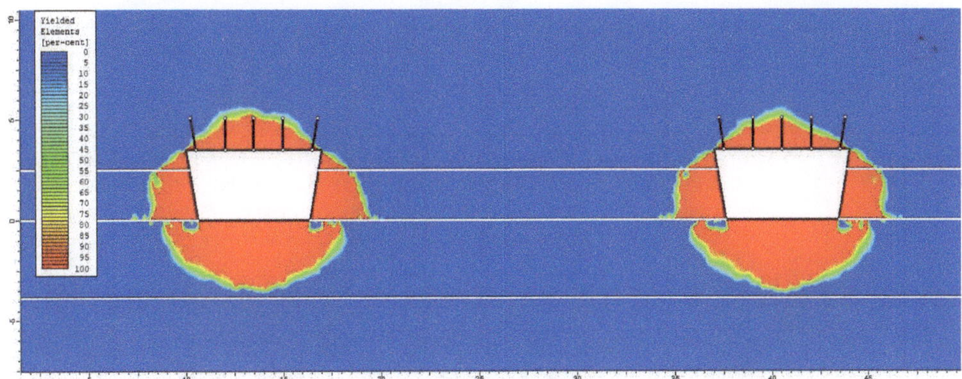

Figure 8. Yielded element area around heading 3 and heading 4, load variant 2.

Table 6 lists the range of yielded rock mass (yield between 50% and 100%) in the roofs of the analyzed headings. For the assumed geological and mining conditions of the Polkowice-Sieroszowice mine, the simulated change of direction in which the heading group is driven resulted in an increased yielded zone in the roofs within the range of $0.45 \div 0.58$ m, which translates into an increase by 34.09% to 42.65%, respectively. The maximum range of yielded rock mass in the roofs of the headings driven in perpendicular to the direction of the maximum horizontal stress component σ_H (load variant 2) was greater than the bolted zone of 1.6 m. The excavations may suffer from the loss of roof stability (collapse of the immediate roof in the excavation).

Table 6. List of the yielded rock mass range in the roofs of the analyzed headings (yield between 50% and 100%).

Excavation	Yield Range in the Roof [m]		Increase of Yield Range in the Roof	
	Load Variant 1	Load Variant 2	[m]	[%]
1	1.32	1.77	0.45	34.09
2	1.34	1.91	0.57	42.54
3	1.36	1.94	0.58	42.65
4	1.30	1.76	0.46	35.38

The numerical modeling allowed an optimal selection of the roof bolting system design for a group of headings located in the field of high horizontal stresses. The support system was based on resin-grouted bolts, which are intended to protect the excavation over the entire period if its operation. Depending on the direction in which the heading group is driven, the following bolt lengths and positions were selected:

— in the case of a group of headings driven parallel to the direction of the maximum horizontal stress component σ_H, the bolts were 1.6 m in length and the bolting pattern (distance between the bolts) was 1.5×1.5 m,

— and in the case of a group of headings driven perpendicular to the direction of the maximum horizontal stress component σ_H, the bolts were 2.2 m in length and the bolting pattern (distance between the bolts) was 1.5×1.5 m,

The simulations were based on an assumption that the bolted zone in the roof must be larger by at least 0.5 m than the maximum range of the 100% yielded zone.

5. Conclusions

Some observed cases of stability loss in mining excavations are caused by a lack of knowledge about the size and direction in which horizontal stresses act in the mined rock mass. The in situ measurements and the numerical simulations confirmed that the driving direction of a heading and the directions of horizontal stresses affect heading stability in the geological and mining conditions of the Polkowice-Sieroszowice mine.

The stability of headings depends the direction in which they are driven in a field of high horizontal stresses, in relation to the direction of the maximum horizontal stress component σ_H in the rock mass. Therefore, this parameter should be given appropriate consideration in the process of selecting an optimal roof support system for a heading. The additional parameters which influence heading stability include: the shape of the heading, its cross section surface area (heading width under the roof), heading depth (the value of stresses in the rock mass), and the stress and strain parameters of the rocks surrounding the heading.

An accurate recognition of the stress field in the mining area allows the development of optimal preventive methods. Numerical methods prove very useful in such tasks, as they allow broad and extensive analyses of heading stability. The in situ observations of actual stability loss cases in the LGCB copper ore mines and the numerical simulations of heading stability fully confirm the need to use numerical modelling in preparing designs of mining excavations and of their support systems.

Author Contributions: Conceptualization, D.P.; methodology, K.A.-P. and D.P.; software, D.P.; validation, D.P.; formal analysis, K.A.-P.; investigation, D.P.; resources, D.P.; data curation, D.P.; writing—original draft preparation, K.A.-P. and D.P.; writing—review and editing, K.A.-P.; visualization, D.P.; supervision, D.P.; project administration, K.A.-P. and D.P.; funding acquisition, K.A.-P. All authors have read and agreed to the published version of the manuscript.

Funding: The research work was co-founded with the research subsidy of the Polish Ministry of Science and Higher Education granted for 2021.

Institutional Review Board Statement: Not applicable.

Informed Consent Statement: Not applicable.

Data Availability Statement: The data presented in this study are available on request from the corresponding author.

Conflicts of Interest: The author declares no conflict of interest.

References

1. Hast, N. The state of stresses in the upper part of the earth's crust. *Eng. Geol.* **1967**, *2*, 5–17. [CrossRef]
2. Mark, C. Horizontal stress and its effects on longwall ground control. *Min. Eng.* **1991**, *11*, 1356–1360.
3. Daws, G.; Hons, B. Roof bolting in coal mining—Design and implementation. *Min. News* **1992**, *1*, 27–32. (In Polish)
4. Mark, C.; Mucho, T.P. *Longwall Mine Design for Control of Horizontal Stress, New Technology for Longwall Ground Control*; US Government Printing Office: Washington, DC, USA, 1994; pp. 53–73.
5. Mark, C.; Mucho, T.P.; Dolinar, D. Horizontal stress and longwall headgate ground control. *Min. Eng.* **1998**, 61–68.
6. Mark, C. Focus on Ground Control: Horizontal Stress. *Coal Age* **2001**, *106*, 47–50.
7. Agapito, J.; Gilbride, L. Horizontal stresses as indicators of roof stability. In Proceedings of the SME Annual Meeting, Phoenix, AZ, USA, 25–27 February 2002.
8. Heidbach, O.; Tingay, M.; Barth, A.; Reinecker, J.; Kurfeß, D.; Müller, B. Global stress pattern based on the World Stress Map database release. *Tectonophysics* **2008**, *482*, 3–15. [CrossRef]
9. Trinh, N.; Jonsson, K. Design considerations for an underground room in a hard rock subjected to a high horizontal stress field at Rana Gruber, Norway. *Tunn. Undergr. Space Technol.* **2013**, *38*, 205–212. [CrossRef]
10. Zhaoa, X.G.; Wanga, J.; Caib, M.; Maa, L.K.; Zonga, Z.H.; Wanga, X.Y.; Sua, R.W.; Chena, M.; Zhaoa, H.G.; Chenc, Q.C.; et al. In-situ stress measurements and regional stress field assessment of the Beishan area, China. *Eng. Geol.* **2013**, *163*, 26–40. [CrossRef]
11. Haghi, A.H.; Kharrat, R.; Asef, M.R. A case study for HCL-based fracturing and stress determination: A Deformation/Diffusion/Thermal approach. *J. Petrol. Sci. Eng.* **2013**, *112*, 105–116. [CrossRef]
12. Haghi, A.H.; Chalaturnyk, R.; Ghobadi, H. The state of stress in SW Iran and implications for hydraulic fracturing of a naturally fractured carbonate reservoir. *Int. J. Rock. Mech. Min. Sci.* **2018**, *105*, 28–43. [CrossRef]
13. Salmachi, A.; Rajabi, M.; Wainman, C.; Mackie, S.; McCabe, P.; Camac, B.; Clarkson, C. History, Geology, In Situ Stress Pattern, Gas Content and Permeability of Coal Seam Gas Basins in Australia: A Review. *Energies* **2021**, *14*, 2651. [CrossRef]
14. Fabjanczyk, M.; Bryja, Z.; Bugajski, W.; Katulski, A. Measurement of pre-exploitation stress field in KGHM Polska Miedz, O/ZG Rudna. *Proc. Undergr. Min. School* **1997**, *1*, 67–75. (In Polish)

15. Fabich, S.; Lis, J.; Pytel, W.; Szadkowski, T.; Szlązak, M. *Calculation of Stress in Rock Mass in Various Geological and Mining Conditions on the Basis of In-Situ Measurements, Stage I: Development of Technologies Allowing the Measurement of Primary and Production Stress Tensor in the Rock Mass, and Completing the First Stage of Measurements*; Cuprum R&D Centre: Wroclaw, Poland, 2003. (In Polish)
16. Butra, J. (Ed.) *Magnitude and Directions of In-Situ Stress in the Deep Part of a Copper ore Deposit with an Appendix "Determination of the Impact of the Primary Stress Directions and Magnitudes on the Optimal Geometry of Mining Fields*; KGHM Cuprum sp. z o.o. R&D Centre: Wroclaw, Poland, 2012. (In Polish)
17. Kidybiński, A. *Basics of Mining Geotechnics*; Slask: Katowice, Poland, 1982.
18. Ming, J.; Hongjun, G. Influence of in-situ rock stress on the stability of roadway surrounding rock: A case study. *J. Geophys. Eng.* **2020**, *17*, 138–147. [CrossRef]
19. Esterhuizen, G.S.; Dolinar, D.R.; Iannacchione, A. Field observations and numerical studies of horizontal stress effects on roof stability in US limestone mines. *J. S. Afr. Inst. Min. Metall.* **2008**, *108*, 345–352.
20. Abdel-Meguid, M.; Rowe, R.K.; Lo, K.Y. Three-dimensional analysis of unlined tunnels in rock subjected to high horizontal stress. *Can. Geotech. J.* **2003**, *40*, 6. [CrossRef]
21. Li, H.; Lin, B.; Hong, Y.; Gao, Y.; Yang, W.; Liu, T.; Wang, R.; Huang, Z. Effects of in-situ stress on the stability of a roadway excavated through a coal seam. *Intern. J. Min. Sci. Technol.* **2017**, *27*, 917–927. [CrossRef]
22. Pawelus, D. Influence of Horizontal Stress on the Stability of Underground Excavations in Copper Mines. Ph.D. Thesis, Wroclaw University of Science and Technology, Wroclaw, Poland, 2010. (In Polish).
23. Cała, M.; Flisiak, J.; Tajduś, A. *Mechanism of Engagement between Bolts and Rock Mass of Various Formation*; Library of Undergr. Min. School: Krakow, Poland, 2001. (In Polish)
24. Collaborative Publication. *Instructions on Determining the Geomechanical Parameters of Roof Rocks with Respect to Roof Classes in Copper Mines, as Required in the Selection of a Roof Bolt System Design*; KGHM Polska Miedz S.A.: Lubin, Poland, 2017. (In Polish)
25. Collaborative Publication. *Regulations on the Selection, Construction and Control of Excavation Support in the KGHM Polska Miedz S.A. Mines*; KGHM Polska Miedz S.A.: Lubin, Poland, 2017. (In Polish)
26. Pawelus, D. Stability assessment of headings situated in a field of high horizontal stress in Polish copper mines by means of numerical methods. In *IOP Conference Series—Earth & Environmental Science, Proceedings of the World Multidisciplinary Earth Sciences Symposium, Prague, Czech Republic, 3–7 September 2018*; IOP Publishing Ltd.: Bristol, UK, 2019; Volume 221, p. 221.
27. Pawelus, D. The azimuths difference method as an effective method of determining the value of horizontal stress acting on mining excavations in underground mines. In *Science and Technologies in Geology, Exploration and Mining 1.3, Exploration and Mining, Proceedings of the 18th International Multidisciplinary Scientific GeoConference, Albena, Bulgaria, 2–8 July 2018*; Curran Associates, Inc.: New York, NY, USA, 2018; Volume 18, p. 18.
28. Amadei, B.; Stephansson, O. *Rock Stress and Its Measurement*; Chapman & Hall: London, UK, 2009.
29. Hoek, E.; Carranza-Torres, C.T.; Corkum, B. Hoek-Brown failure criterion—2002 edition. In Proceedings of the North American Rock Mechanics Society Meeting, Toronto, ON, Canada, 7–10 July 2002.
30. Hoek, E. Strength of rock and rock masses. *ISRM News J.* **1994**, *2*, 4–16.
31. Hoek, E.; Brown, E.T. Practical estimates of rock mass strength. *Int. J. Rock Mech. Min. Sci.* **1997**, *34*, 1165–1186. [CrossRef]
32. Hoek, E.; Marinos, P. GSI: A geologically friendly tool for rock mass strength estimation. In Proceedings of the ISRM International Symposium, Melbourne, Australia, 19–24 November 2000.

 *energies*

Article

A Method for Structure Breaking Point Detection in Engine Oil Pressure Data

Aleksandra Grzesiek [1,*], Radosław Zimroz [1], Paweł Śliwiński [2], Norbert Gomolla [3] and Agnieszka Wyłomańska [4]

1 Faculty of Geoengineering, Mining and Geology, Wroclaw University of Science and Technology, Na Grobli 15, 50-421 Wroclaw, Poland; radoslaw.zimroz@pwr.edu.pl
2 KGHM Polska Miedź S.A., M. Skłodowskiej-Curie 48, 59-301 Lubin, Poland; pawel.sliwinski@kghm.com
3 DMT GmbH & Co. KG, Am Technologiepark 1, 45307 Essen, Germany; norbert.gomolla@dmt-group.com
4 Faculty of Pure and Applied Mathematics, Hugo Steinhaus Center, Wrocław University of Science and Technology, Wyspiańskiego 27, 50-370 Wrocław, Poland; agnieszka.wylomanska@pwr.edu.pl
* Correspondence: aleksandra.grzesiek@pwr.edu.pl

Abstract: In this paper, a heavy-duty loader operated in an underground mine is discussed. Due to extremely harsh operational conditions, an important maintenance problem is related to engine oil pressure. We have found that when the degradation process appears, the nature of variation of pressure engine oil changes. Following this observation, we have proposed a data analysis procedure for the structure break point detection. It is based on specific data pre-processing and further statistical analysis. The idea of the paper is to transform the data into a nearly monotonic function that describes the variation of machine condition or in the statistical language—change of the regime inside the process. To achieve that goal we proposed an original data processing procedure. The dataset analyzed in the paper covers one month of observation. We have received confirmation that during that period, maintenance service has been done. The purpose of our research was to remove ambiguity related to direct oil pressure analysis and visualize oil pressure variation in the diagnostic context. As a fleet of machines in the considered company covers more than 1000 loaders/trucks/drilling machines, the importance of this approach is serious from a practical point of view. We believe that it could be also an inspiration for other researchers working with industrial data.

Keywords: machine diagnostics; LHD; engine oil pressure data; oil pump wear; statistical analysis; convergence functions

 check for updates

Citation: Grzesiek, A.; Zimroz, R.; Śliwiński, P.; Gomolla, N.; Wyłomańska A. A Method for Structure Break Point Detection in Engine Oil Pressure Data. *Energies* **2021**, *14*, 5496. https://doi.org/10.3390/en14175496

Academic Editors: Sergey Zhironkin and Dawid Szurgacz

Received: 27 July 2021
Accepted: 1 September 2021
Published: 3 September 2021

Publisher's Note: MDPI stays neutral with regard to jurisdictional claims in published maps and institutional affiliations.

1. Introduction

Maintenance procedures for a fleet of the load-haul dump (LHD) mobile machines operated in underground mine are critically important and challenging. To reach the required level of efficiency, modern maintenance is supported by on-board monitoring systems and advanced analytics. The monitoring system is acquiring data in every 1 s (or more, depending on the type of variable) and measured values are stored in local memory. After the shift, when the machine is going back to the so-called machine chamber, the data are automatically transferred via WIFI to the server on the surface (data storage and processing center). Specific analysis is automatically performed and some reports are available via the website. Unfortunately, the variables are very specific and exhibit different characteristics, thus it is required to apply different algorithms to their analysis. Till now, simple comparison with threshold is used, for example, for low frequency temperature data. In this paper, we present an example related to the engine oil pressure measurement. A base for all analysis is the need to detect the structure break point (called also the anomaly or the structure/regime change point), interpret data automatically, and provide simple information to appropriate staff. Engine oil pressure is very different than mentioned temperature data. It is relatively high-frequency variable. It varies from zero to ca. 700 kPa, but we have found that distribution of the process is more informative than just simple particular sample values. It is hard to make any conclusion from the raw signal. If engine oil pressure is too high, there is an automatic safety valve that releases excess oil from the system. A much worse case is when oil pressure is too small—such a situation may lead to the damage of the engine. However, as we mentioned, the simple diagnosis via comparison with the threshold level is impossible. Thus, in this paper, we propose a statistical procedure for the detection of the so-called structure break point in the highly variable random data.

Our strategy is based on the statistical approach that minimizes the impact of local changes in the examined data. We consider the machine as a time-varying system, so we decide not to analyze the data sample by sample but to estimate its characteristics in the work shifts perspective. A daily operation in the considered mine is divided in four shifts, each 6 h. It is due to mining law regulation, specific processes required for mining production and extremely harsh condition underground. Even if some experts may notice changes in the raw signal, it is very hard to provide objective rules to detect anomalies in practice. Due to the mentioned specific operation of the machine in the mine, firstly we segment the data according to the work shifts. We received advice, that operation of LHD during the first and second work shifts is very different than during the third and fourth work shifts (during the night). The 1st work shift in the day is between 6 a.m.–12 p.m.; the 2nd work shift—12 p.m.–6 p.m.; the third work shift—6 p.m.–0 a.m. and the four work shift—0 a.m.–6 a.m. It appears that shifts are different, indeed, however, from our perspective, it is not critical. From the statistical point of view changes in the data related to change of condition are much stronger than changes related to different work shifts. Our approach allow us to process data to the form that contains information about the condition of the system, not about its daily operation parameters variation as speed, load, etc. From a highly time-varying process related to oil pressure variation, we receive a nearly monotonic function describing the change of condition in the system. Such data transformation is readable for maintenance staff and may be used as a basis for maintenance decisions. It has great practical importance as the population of the LHD fleet exceeds 1000 machines.

A novelty of the approach is related to the appropriate processing of raw data. In the data analysis, we use the fact that the empirical distribution of the engine oil pressure time series may change in the subsequent work shifts, which is strongly related to the degradation process of the machine. Thus, we examine the distance measures between distributions of the data corresponding to the work shifts. Here we utilize the divergence functions that measure the distance between two probability distributions expressed in the means of their probability density functions (pdfs). Based on this characteristic we apply the classical test statistics (i.e., Student's *t* and Wilcoxon) used in the problem of testing if two samples have equal means. The final result is the detection of the structure break point which corresponds to the moment of the repair registered by the maintenance staff in the considered mine. From the signal processing perspective, the novelty of the approach is related to statistical processing of long term, historical industrial data to achieve unequivocal information about the change of condition. We propose analyzing data in segments to cancel out local disturbances and extract patterns seen as the probability density function. Then, the difference between segments is estimated by the distance between distributions and such a new feature is used to establish the structural break point in the process. We propose the original data processing procedure that covers pre-processing step (data validation, missing values handling, re-sampling, reshaping data to associate them with specific nature of the process), data analysis (probability density function of data instead of the sample by sample analysis, measures of distance in multidimensional space and finally structural break point detection—i.e., the identification of moment when the process change its nature) and the presentation of the results (transforming obtained information back to real samples domain to visualize change moment). The general concept of the procedure is demonstrated in Figure 1. Namely, the raw data is subjected to a three-stage procedure (successive blocks of the diagram), as a result of which we detect the structural break point. The steps highlighted in each of three stages are described in detail in the following sections, namely, the data pre-processing is presented in Section 2.4, the analysis is described in Section 3, and the visualization is presented in Section 4.

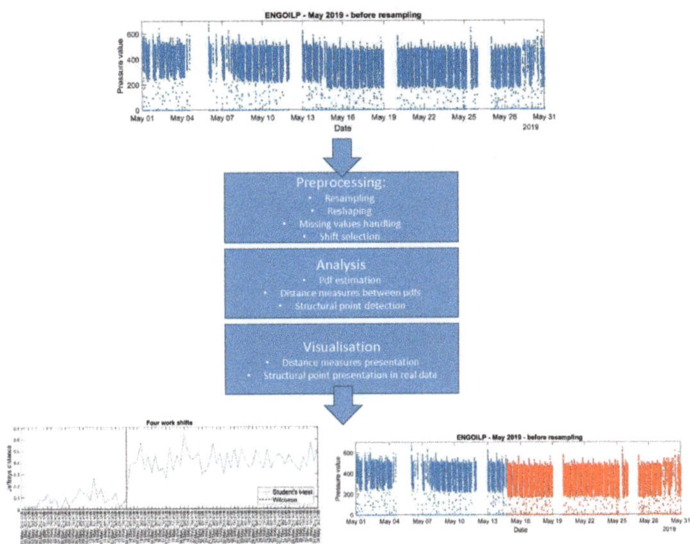

Figure 1. The general concept of the proposed diagnostic procedure. The steps highlighted in the schematic blocks are described in detail in the following sections of the paper.

1.1. Brief State of the Art

LHD machines are commonly used in the mining industry and are critically important for the production process. There are many papers related to various aspects of LHDs. Most of them are focused on reliability analysis [1–6], prediction and assessment of machines breakdowns [7], machine performance measures [8–12], utilisation analysis, optimisation, production analysis [11,13–15], risk evaluation [16], residual life estimation [17], etc. Just a few works have mentioned condition monitoring [18–22], technical and operational aspects related to LHD machines [23–26] and real measurement of output torque, identification of operation regimes, etc. [27–32]. LHD machines are equipped with on-board monitoring systems. They are used for control as well as for maintenance purposes. As one receives real data from the machine, it opens many opportunities for their analysis.

As it was mentioned, from the mathematical/statistical perspective, the considered problem is related to the segmentation of the data which, is strongly connected with the structure break point detection. In recent years, substantial works on segmentation methods for different applications appeared in the literature. A few interesting applications include condition monitoring [33,34] (where structural break detection method based on the adaptive regression splines technique has been proposed to recognize a change of operational regime in copper ore crusher vibration and local maxima method has been proposed in the time–frequency domain for spike detection in bearings vibration, respectively), biomedical signals (e.g., electrocardiogram) [35–39] (where hidden Markov models, moving average and Savitzky-Golay filter, cepstral analysis, wavelet transform, envelope-based segmentation, etc., have been used), speech analysis [40–42] (where nonlinear speech analysis based on the microcanonical multiscale formalism, adaptation of Appel and Brandt algorithm, innovation (Shur) adaptive filter have been discussed), econometrics [43,44] (where the regime switching model is applied), and seismic signals [45–49] (where, among others, cumulative sum of Gaussian probability density functions and Markov regime-switching models as well as the empirical second moment of given raw signal have been proposed for seismic signal segmentation in order to extract seismic events).

Many segmentation methods are based on simple statistics in time domain, the cumulative squared data or empirical second moment [50–53]. However, one can also find methods based on the representation of the data in different domains, such as time-frequency [34,54]. See also the effective segmentation methods used in the physical sciences, like the methods based on the so-called recurrence statistic [55–59]. The method proposed in this paper is based on the relatively simple statistics, however, they are applied to the characteristics of the data describing the distribution in the work shift.

The problem discussed in the paper could be seen as a process diagnosis (fault detection in the process). Various approaches to fault detection and diagnosis in process data have been developed over decades. They can be divided into three main categories: data-driven approaches, deep-knowledge-based approaches, and analytical-model-based approaches [60]. An analytical-model- and deep-knowledge-based methods rely heavily on fundamental knowledge of the process [60]. Unfortunately, in our data, it is very difficult to provide a model of the process as a pressure variation is depending on the behavior of the operator, the environment, performed task, etc. It leads to—using data science language—highly non-Gaussian, nonlinear, non-stationary data. A data-driven methods specifically refer to methods that rely purely on operational data without using process knowledge. Data-driven approaches learn from history and place no requirements on models or expert knowledge [60].

Data-driven process monitoring or statistical process monitoring applies multivariate statistics and machine learning methods to fault detection and diagnosis for industrial process [61]. Various variants of PCA as recursive PCA (RPCA), dynamic PCA (DPCA), and kernel PCA (KPCA), have been used for process diagnosis. The main idea is that structure of PCA will change when an abnormal situation will appear in the data. For data-driven approaches, methods based on the hidden Markov model, Ddynamic neural network, kernel independent component analysis, Gaussian mixture model, hidden semi-Markov model have been applied. A deep review has been provided in [60].

In the context of real-time analysis one can consider two approaches: decision "sample by sample" or decision "segment by segment". In the first case, we can compare the amplitude of the incoming sample with the threshold or we can test if the sample belongs to the same distribution as previous samples. One may search for pre-defined events (zero value, flat line, min/max, outliers detection, etc.) [62]. It is very difficult to apply this approach in our case. In the second-mentioned technique, one may play with more reliable statistical approaches, instead of samples the statistics or the distributions may be compared. We are doing this in our research. We test if the mean value estimated from at least two new samples is different than the historical one.

It should be also mentioned that there is a class of data-mining based solutions, which could be adopted here, for example-time-series clustering, see review [63], anomaly/novelty detection in time series, [64], see review [65,66], process mining [67]. We consider checking this in near future, however, we believe that the proposed techniques are simple and quick and do not require advanced training algorithms as most of the data mining techniques.

As an interesting approach one may find in [68] where historical SCADA data for the normal condition from wind turbines were used to train this multi-layer network model layer-wise to extract the relationships between SCADA variable trained for the healthy case, the model will not be able to reconstruct real data if input will be abnormal. Thus, residual data will be higher, and easy to notice an anomaly. As mentioned it used an advanced neural network-based model rather than simple mean testing as we proposed.

1.2. Structure of the Paper

The paper is organized as follows: in Section 2 we describe the machine, experiment, and the data as well as we give the step by step pre-processing procedure used for the engine oil pressure. In Section 3 we present the statistical methodology applied for the data processing, it consists of the calculation of the distance measures for empirical probability density functions corresponding to work shifts and finally the application of the Student's *t* and Wilcoxon statistics for the structure break point detection. In Section 4 we demonstrate the results for the real engine oil pressure data. The last section contains the summary and concludes the paper.

2. Machine, Obds, Experiment and Data Description

2.1. Machine Description

The machine used in the experiment is a Loader LKP-0903 produced by KGHM ZANAM. Basic parameters of the truck are: length 10,600 mm, width 3150 mm, height 1750 mm total weight 28,700 kg, standard bucket 4.6 m^3, tramming capacity 96 kN, power rating 181 kW, driving speed 4.5 km/h (1st gear) up to 19.0 km/h (4th gear). The photo of the machine is shown in Figure 2. An LHD loader is quite a complex machine that consists of several subsystems as the drive unit, transmission system, hydraulic system for lifting the loader bucket, etc. The heart of the machine is the combustion engine. One of the most critical parameters of engine operation is Engine Oil Pressure. For more details see [69]. The loader is used to transport copper ore from the so-called mining face (mining front, where extraction of ore is performed) to the screen (reloading point, from loader to continuous belt conveyor system for transporting bulk material for long distances up to mining shaft and then to the surface). The mining process is complicated and consists of several steps. It implicates the way of the usage of LHD machines. Daily operation is divided into four shifts, but due to several factors (i.e., blasting procedures), the LHD cannot operate continuously. If LHD is not in operation, the monitoring system will produce NaNs (Not a Number values). It may happen that most of the data for a given shift are NaNs. During the weekend LHD is not used, either. All these specific cases lead to dedicated data processing and analysis methodology related to particular data stream from SCADA.

Figure 2. Machine LK3 used in underground mine.

2.2. On Board Diagnostic System Description

To maintain such types of machines a list of critical parameters has been defined to be measured on the machine. The list of parameters is case-depended. In Table A1, see Appendix A, we have present the parameters that are most important for our case study. Parameters are associated with the various components. Their dynamics, variability, etc, are different. Some of the data are sampled with 1 Hz, some others considered as low-frequency variables may be sampled with every 5 s or more. In this paper, we will use oil pressure signal only sampled every 5 s.

The data considered in this paper described one month of LHD operation. We received them from the data server as historical data.

2.3. Experiment Description

The machine considered in the paper is a regular example taken from the mine (LK3 419R machine). The data covers one month of operation. We have selected such a period because we received information that during that period some repair action has been done. Our experiment is a passive one. We just observe the operation of the machine using OBD monitoring data. According to information received from maintenance staff, turbocharger replacement has been done on 14 May 2019. The diagnostic task it to identify the replacement moment based on proposed feature obtained thorough raw pressure data processing. Once again, we would like to highlight that based on observation of raw data we cannot notice this point.

2.4. Data Pre-Processing

Raw engine oil pressure data has been presented in Figure 3 (top). As one can observe the pressure values vary from 0 up to 700 kPa. The engine oil pressure was acquired every 5 s, however, for each day the data acquisition process could start with slightly different time points (exactly at midnight or 1, 2, 3 .. seconds after midnight). The empty spaces indicate the weekends. Thus, only the working days were taken to the analysis. To be able to compare the variability of data during a single shift we re-sampled data to the same time-basis using the linear interpolation. As one may see, it is a cosmetic change—the shape of the signal has not been changed see Figure 3 (bottom).

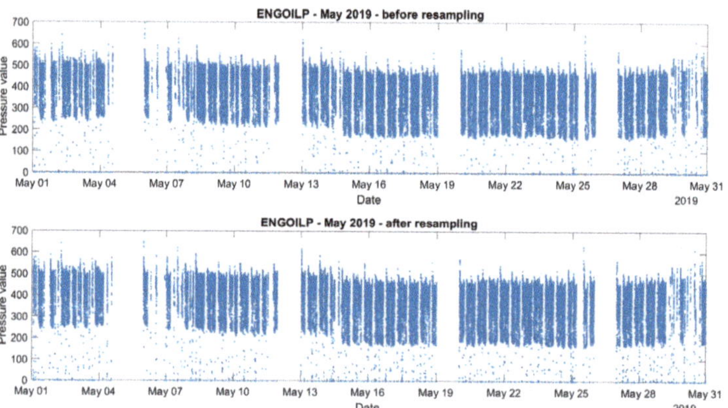

Figure 3. Engine oil pressure data from the database (**top panel**) and re-sampled data (**bottom panel**).

As mentioned, we know that turbocharger has been replaced on 14 of May but it is still difficult to identify a significant change in data on that date. Similar data were analyzed in [21], where the specific representation of the oil pressure was proposed, namely, the authors reshaped the data into the two-dimensional array where the x-axis describes the number of shift or date for a given day, and on the y-axis we have the so-called local shift time (0–6 h) or time corresponding to a single day (0–24 h). In this paper, we propose to apply the local shift time representation of the one-month measurements, see Figure A1 given in Appendix B. As one can see, it is difficult to clearly indicate the structure break point (14th of May) on the map presented in Figure A1. Thus, we propose to seperately analyze also the first, second, third, and fourth shifts. Moreover, to have the same time for each shift we re-sampled the data according to the most frequent time-basis using linear interpolation. The representation of the oil pressure measurements broken down into four sub-sets (corresponding to four shifts) after re-sampling is demonstrated in Figure A2 given in Appendix B.

However, as was mentioned, the monitoring system produces NaNs. If in the analyzed sample a significant amount of data are NaNs, then the analysis and corresponding interpretation may not be reliable. Thus, we examined how much data (in percent) are NaNs and before the further analysis, we removed the work shifts with more than 40% of NaNs. The oil pressure after removing the days with more than 40% NaNs represented as shift by shift is demonstrated in Figure 4. The oil pressure broken down into four shifts after removing the shifts with more than 40% NaNs is presented in Figure 5. The data demonstrated in Figures 4 and 5 are analyzed using the procedures described in the next section. The consecutive steps of the pre-processing scheme described above are demonstrated in Figure 6.

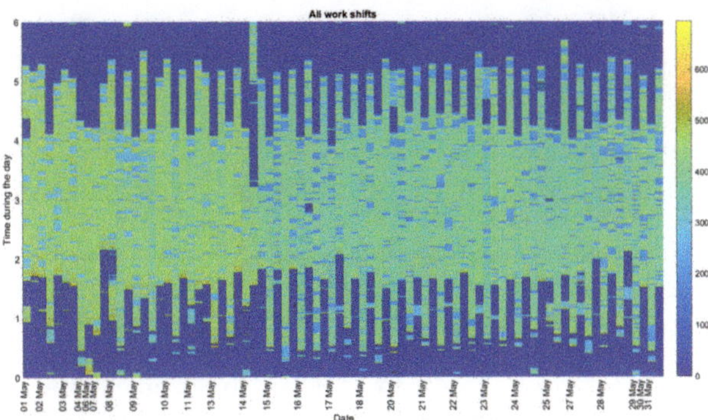

Figure 4. Map of re-sampled data (presented shift by shift) after removing the days with more than 40% of NaNs.

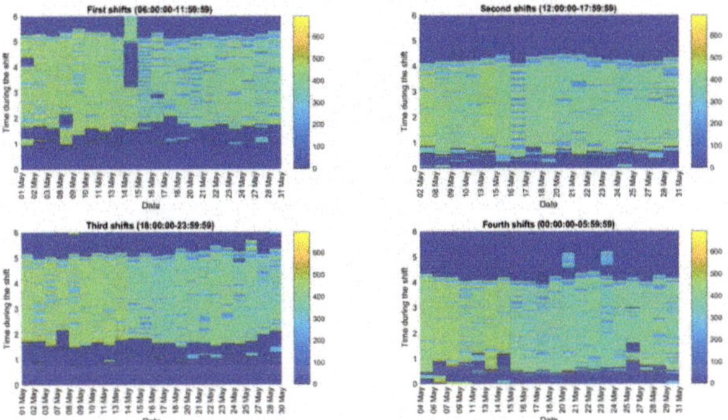

Figure 5. Maps of re-sampled data corresponding to four work shifts after removing the days with more than 40% of NaNs.

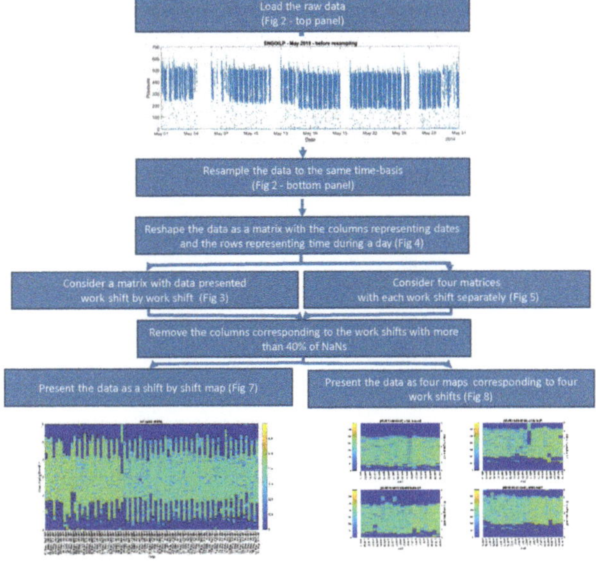

Figure 6. The concept of the pre-processing scheme.

3. Methodology

In this section, we present the methodology applied for the real data presented in Section 2. At first, we set the notation used further in the paper. Let m_{ij} denote the i-th measurement during the j-th work shift where $i = 1, \ldots, N$ and $j = 1, \ldots, M$, and consequently let $\mathbf{m}_j = (m_{1j}, m_{2j}, \ldots, m_{Nj})$ denote a vector of measurement corresponding to the j-th work shift. Moreover, let $p_j(x)$ indicate the theoretical probability density function of $\mathbf{m}_j$. To detect the moment when the character of the data changes, we consider the pdfs corresponding to subsequent work shifts, namely $p_1(x), p_2(x), \ldots, p_M(x)$.

In probability theory and statistics, to quantify the similarity of two distributions one can use the so-called divergence (or contrast) functions that measure the distance of one probability distribution to another. In general, the divergence is not a concept as strong as distance, because it does not have to be symmetric in arguments or satisfy the triangle inequality. Among various contrast functions, we distinguish one very important class of divergence coefficients, namely the class of the so-called f-divergences of the following form [70–72]

$$I_{f,g}(p_{k^*}(x), p_{k^{**}}(x)) = g\left(\int p_{j^{**}}(x) f\left(\frac{p_{k^*}(x)}{p_{k^{**}}(x)} \right) dx \right), \tag{1}$$

where $p_{k^*}(x)$, $p_{k^{**}}(x)$ are the probability density functions corresponding to two variables, $f(t)$ is a continuous convex real function on $\mathbb{R}_+$ and $g(t)$ is an increasing function on $\mathbb{R}$. It is important to mention that the f-divergences are always non-negative and they are equal to zero if and only if the densities $p_{k^*}(x)$ and $p_{k^{**}}(x)$ coincide. For more properties of the divergence functions defined in Equation (1) we refer the readers to [72,73]. Depending on the choice of $f(t)$ and $g(t)$ one can obtain different forms of the contrast functions. In this paper, we consider three specific measures belonging to the class defined above.

The first measure, called the Hellinger distance, corresponds to the case of $g(t) = \sqrt{0.5\,t}$ and $f(t) = \left(\sqrt{t} - 1 \right)^2$ in Equation (1) and it is given by the following formula [72]

$$H(p_{k^*}(x), p_{k^{**}}(x)) = \sqrt{0.5 \int \left(\sqrt{p_{k^*}(x)} - \sqrt{p_{k^{**}}(x)} \right)^2 dx}. \tag{2}$$

Let us notice that the Hellinger distance is symmetric with respect to the arguments and obeys the triangle inequality. It satisfies the property that $0 \leq H(p_{k^*}(x), p_{k^{**}}(x)) \leq 1$ with the minimum value corresponding to the case when $p_{k^*}(x) = p_{k^{**}}(x)$ for every $x \in \mathbb{R}$, and the maximum value achieved when $p_{k^*}(x)$ is equal to zero for every x for which $p_{k^{**}}(x)$ is nonzero and vice versa.

As the second divergence measure, we consider a modification of the Hellinger distance defined above. Namely, for $g(t)$ being an identity function and $f(t)$ defined as in Hellinger case, we obtain the so-called Jeffreys distance of the following form [74]

$$J(p_{k^*}(x), p_{k^{**}}(x)) = 2H^2(p_{k^*}(x), p_{k^{**}}(x)) = \int \left(\sqrt{p_{k^*}(x)} - \sqrt{p_{k^{**}}(x)} \right)^2 dx. \tag{3}$$

Let us notice that similarly to the Hellinger distance, the Jeffreys measure is symmetric in the arguments. Moreover, it takes values between 0 and 2.

The Chernoff distance, the third example of the divergences conisidered in this paper, corresponds to the case when $g(t) = -\log(-t)$ and $f(t) = -t^{1-\alpha}$ with $0 < \alpha < 1$ in Equation (1) and takes the following form [73,74]

$$CH(p_{k^*}(x), p_{k^{**}}(x)) = -\log\left(ch(p_{k^*}(x), p_{k^{**}}(x)) \right), \tag{4}$$

where

$$ch(p_{k^*}(x), p_{k^{**}}(x)) = \int p_{k^*}(x)^\alpha p_{k^{**}}(x)^{1-\alpha} dx, \tag{5}$$

is called the Chernoff coefficient. In general, expect the case of $\alpha = 0.5$, the Chernoff distance given in Equation (4) is not symmetric in the arguments nor does it satisfy the triangle inequality. Moreover, let us notice that since $0 \leq ch(p_{k*}(x), p_{k**}(x)) \leq 1$ we have that $0 \leq CH(p_{k*}(x), p_{k**}(x)) \leq +\infty$ with the minimum value corresponding to the instance when $p_{k*}(x) = p_{k**}(x)$ for every $x \in \mathbb{R}$. The special case of the Chernoff coefficient given in Equation (5), i.e., the case of $\alpha = 0.5$, is called the Bhattacharyya coefficient related directly to the Hellinger distance given in Equation (2), namely

$$H(p_{k*}(x), p_{k**}(x)) = \sqrt{1 - ch(p_{k*}(x), p_{k**}(x))}.$$

Let us note that from the practical point of view to calculate the empirical counterparts of the divergences defined in Equations (2)–(4) there is a need to estimate the probability density functions and use them instead of the theoretical pdfs in the above definitions. In our case we will apply the empirical divergences to the measurements corresponding to the work shifts, namely to the vectors $\mathbf{m}_j = (m_{1j}, m_{2j}, \ldots, m_{Nj})$ where $j = 1, \ldots, M$. For this purpose, one can use the kernel density estimator of $p_j(x)$ defined as follows [75,76]

$$\hat{p}_j(x) = \frac{1}{Nh} \sum_{i=1}^{N} K\left(\frac{x - m_{ij}}{h}\right) \tag{6}$$

for any $j = 1, \ldots, M$ and $x \in \mathbb{R}$, where $K(\cdot)$ is the non-negative kernel smoothing function, and h is the bandwidth. The choice of kernel smoothing function determines the shape of the curve used to estimate the probability density function. In our case, we take the normal kernel that is simply the standard normal pdf of the form

$$K(x) = \frac{1}{\sqrt{2\pi}} \exp\left\{-\frac{x^2}{2}\right\} \text{ for } x \in \mathbb{R}, \tag{7}$$

and the bandwidth is chosen using the Silverman's rule of thumb to be optimal for estimating normal densities [77]. The procedure is implemented in many programming languages and it is available in numerous mathematical packages, e.g., the function "ksdensity" in Matlab.

Since the distribution of the subsequent samples $\mathbf{m}_1, \mathbf{m}_2, \ldots, \mathbf{m}_M$ changes for certain $j^* \in \{1, 2, \ldots, M\}$, we can identify the moment of change by examining the empirical contrast functions describing the similarity of the pdf corresponding to the first work shift and the pdfs corresponding to all the other work shifts in the sample, namely we analyze the following vector

$$\hat{I}_{f,g}(p_1(x), p_1(x)), \hat{I}_{f,g}(p_1(x), p_2(x)), \ldots, \hat{I}_{f,g}(p_1(x), p_M(x)),$$

where the f-divergence are specified in Equations (2)–(4). We expect that the values taken by the similarity measures in the sub-samples

$$\hat{I}_{f,g}(p_1(x), p_1(x)), \ldots, \hat{I}_{f,g}(p_1(x), p_{j^*}(x))$$

and

$$\hat{I}_{f,g}(p_1(x), p_{j^*+1}(x)), \ldots, \hat{I}_{f,g}(p_1(x), p_M(x))$$

differ significantly. The preliminary analysis of the vector of measurements corresponding to the work shifts indicates the corresponding distributions are different, more precisely, we observe that some characteristics responsible for the location change with respect to time. We refer the reader to the next section for more details. Thus, to determine the structure break point in the data, we decided to apply the test statistics used commonly in the problem of testing if two samples have equal means. However, we apply this methodology not to the raw vector of observations, but to the empirical convergence distance described above. The procedure used to detect the structure break point j^* in the convergence distances corresponding to the work shifts is described below.

Let us consider M independent random variables $I_{f,g}(p_1(x), p_1(x))$, $I_{f,g}(p_1(x), p_2(x))$, $\ldots$, $I_{f,g}(p_1(x), p_M(x))$ with the cumulative distribution functions denoted as $F_{I,1}$, $F_{I,2}$, $\ldots$, $F_{I,M}$ and the following means $\mu_{I,1}, \mu_{I,2}, \ldots, \mu_{I,M}$. Now, let us examine the location testing problem based on the expected value with the null and alternative hypotheses defined as follows

$$H_0 : \mu_{I,1} = \mu_{I,2} = \ldots = \mu_{I,M}$$
$$H_1 : \exists\, M^* \in \{2, \ldots, M-1\} \text{ such that } \mu_{I,1} = \ldots = \mu_{I,M^*} \neq \mu_{I,M^*+1} = \ldots = \mu_{I,M}.$$

Let us notice that the alternative hypothesis can be also presented as

$$H_1 = \bigcup_{M^{**}=2}^{M-1} H_{1,M^{**}},$$

where M^{**} is a fixed value and $H_{1,M^{**}}$ is the alternative hypothesis corresponding to the two-sample test comparing means in two populations, i.e.,

$$H_{1,M^{**}} : \mu_1 = \mu_{I,1} = \ldots = \mu_{I,M^{**}} \text{ and } \mu_2 = \mu_{I,M^{**}+1} = \ldots = \mu_{I,M}.$$

Now, to verify H_0 against H_1 we can use the maximum-based statistic of the following form

$$S_M(\epsilon) = \max_{\lfloor \epsilon M \rfloor \leq M^{**} \leq \lfloor (1-\epsilon)M \rfloor} |T(M^{**})|, \tag{8}$$

where $T(M^{**})$ can be any statistic for testing H_1 against H_{1,M^*} and $\epsilon \in (0, 0.5)$ is used to guarantee that there are at least $\lfloor n\epsilon \rfloor$ elements in both sub-samples. Let us notice that the choice of ϵ is crucial since a small value allows us to detect the very early or very late change in the distribution, but at the same time, it involves considering a small size sample which hinders the proper statistical inference. Moreover, it is important to mention that the distribution of the max-type statistic given in Equation (8) under H_0 does not depend on the analyzed random sample distribution. In our method, we want to locate the moment of the most significant change in data, i.e., we want to identify the value of M^{**} corresponding to the max-type statistic given in Equation (8).

For our purpose, we use two exemplary statistics from two-sample tests for equal means in independent groups, namely the one corresponding to the parametric Student's t-test and the one corresponding to the non-parametric Wilcoxon test. For our data, the Student's t statistic $T_S(M^{**})$ takes the following form

$$T_S(M^{**}) = \frac{(\overline{I_1} - \overline{I_2})}{I_{std}\sqrt{1/M^{**} + 1/(M - M^{**})}}, \tag{9}$$

where

$$I_{std}^2 = \frac{(M^{**} - 1)I_{1,std}^2 + (M - M^{**} - 1)I_{2,std}^2}{M - 2},$$

and $\overline{I_1}$, $\overline{I_2}$ and $I_{1,std}^2$, $I_{2,std}^2$ are the empirical means and empirical variances corresponding to the sub-samples

$$I_1 = (\hat{I}_{f,g}(p_1(x), p_1(x)), \ldots, \hat{I}_{f,g}(p_1(x), p_{M^{**}}(x)))$$

and

$$I_1 = (\hat{I}_{f,g}(p_1(x), p_{M^{**}+1}(x)), \ldots, \hat{I}_{f,g}(p_1(x), p_M(x))),$$

respectively. In turn, the Wilcoxon statistic denoted by $T_W(M^{**})$ is given by the following formula

$$T_W(M^{**}) = \frac{W(M^{**}) - EW(M^{**})}{\sqrt{Var(W(M^{**}))}},\qquad(10)$$

where $W(M^{**}) = \sum_{i=1}^{M^{**}} R_i$, $E[W(M^{**})] = M^{**}(M+1)/2$, $Var[W(M^{**})] = M^{**}(M - M^{**})(M+1)/12$ and R_i denotes the number of elements in a combined vector $I = (\hat{I}_{f,g}(p_1(x), p_1(x)), \ldots, \hat{I}_{f,g}(p_1(x), p_M(x)))$ which are smaller or equal to $i - th$ element of I_1. For more information regarding the Student's t-test and the Wilcoxon test we refer the readers to [78–83]. The successive stages used in the data analysis are presented as the diagram in Figure 7. Similar methodology, also based on the above-mentioned tests, is applied by the authors in [84].

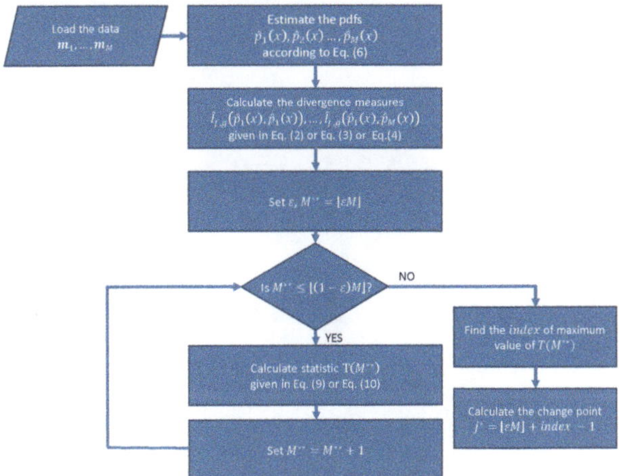

Figure 7. The diagram showing the successive stages of the methodology described.

4. Real Data Analysis

In this section, we analyze the real data introduced in Section 2 using the methodology described above. At the same time, we consider the data presented shift by shift in Figure 4, and the data divided into the separate classes corresponding to four work shifts presented in Figure 5. To indicate the difference in the distributions, we calculate the empirical probability density functions for the subsequent columns of the data matrices given in Figures 4 and 5. The corresponding plots are presented in Figures 8 and 9 as the two-dimensional graphs where the probability density functions are evaluated using the kernel density estimation as described in the previous section. As one can notice, the shape of the empirical pdfs changes at a certain point. As a consequence, the empirical pdfs take smaller values with greater probability.

Since in Figures 8 and 9 one can see the difference in the shape of the probability density function, to identify the structure break point we use the characteristics based on the distance between the empirical pdfs, namely the Hellinger distance, the Jeffreys distance and the Chernoff distance with $\alpha = 0.5$ introduced in Section 3. At first, we plot the maps presenting the matrices of the values taken by the above characteristics where each two empirical pdfs are compared to each other. For the Hellinger distance, the matrices are given in Figures 10 and 11. As one can see they are symmetric to the diagonal. Besides, the calculated Helligner distance divides the work shifts into two groups, between which the values taken by the distance measure differ significantly. The analogous plots corresponding to the Jeffreys and Chernoff distances are given in Appendix B, see Figures A3, A4, A7 and A8.

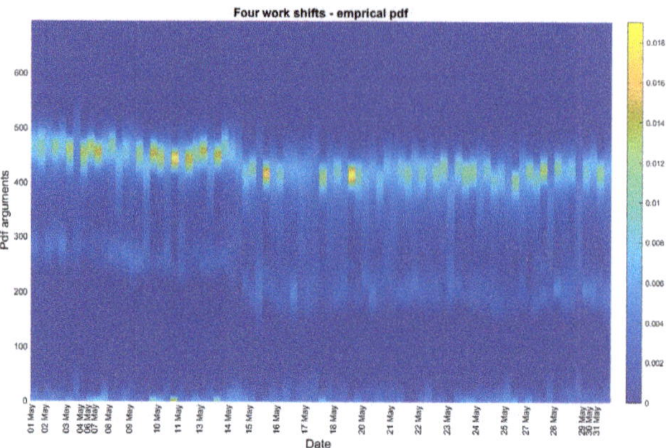

Figure 8. The empirical probability density functions (two-dimensional plot) for the pre-processed data presented shift by shift.

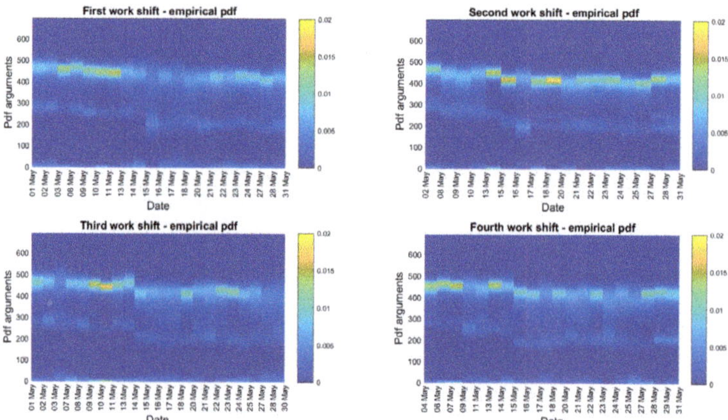

Figure 9. The empirical probability density functions (two-dimensional plot) for the pre-processed data corresponding to four work shifts.

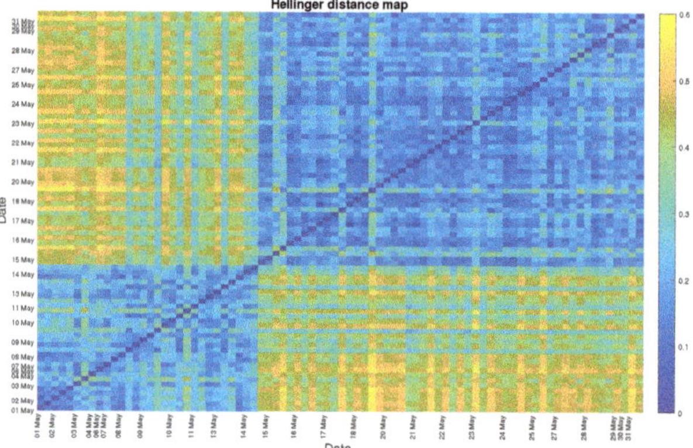

Figure 10. The matrix presenting the values of the Hellinger distance calculated based on the empirical probability density functions corresponding to the data presented shift by shift.

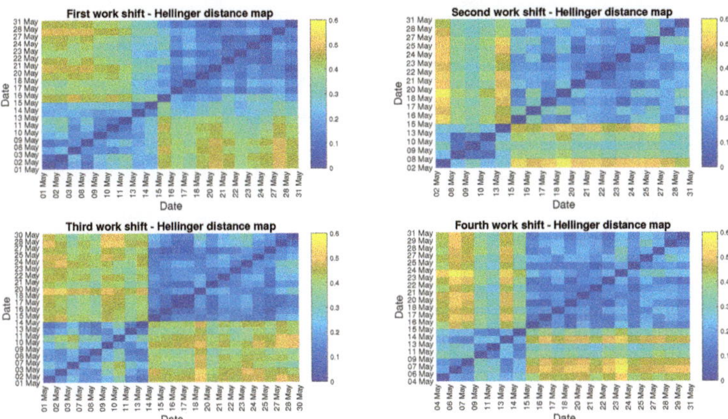

Figure 11. The matrices presenting the values of the Hellinger distance calculated based on the empirical probability density functions corresponding to the data representing four work shifts.

Now, we apply the Student's *t*-based and Wilcoxon-based procedures leading to the identification of the structure break point in the data to the values representing one row of the Hellinger distance matrices presented in Figures 10 and 11. The selected row corresponds to the Hellinger distance between the subsequent pdfs and the base pdf chosen here as the one corresponding to the first shift in a sample. In other words, we identify the structure break point based on the comparison of each empirical pdf with the empirical pdf of the first shift. The values taken by the distance measure and the identified structure break point are presented in Figures 12 and 13 which correspond to the case of the data presented shift by shift and each work shift treated separately, respectively. As one can see, in most of the cases (except the fourth work shift in Figure 13) the procedures based on the Student's *t*-test and the Wilcoxon test are consistent and indicate the same structure break point. We mention here that the shift marked on the plots is the last shift before the change in distribution occurs. This means that the nature of the data changes with subsequent work shifts taking place after the one marked on the plot. The similar graphs corresponding to the Jeffreys and Chernoff distances are presented in Appendix B, see Figures A5, A6, A9 and A10. In Tables A2–A5 we present the summary of the results together with the values taken by the test statistics corresponding to both procedures (Student's *t*-based and Wilcoxon-based) and the data presented shift by shift and each work shift treated separately. As one can see, for the data presented shift by shift, both procedures applied to all three considered measures of distance between pdfs indicate that the structure break point occurs after the 1st work shift on 14 May. On this day, from the 3rd work shift the character of the data changes. It is important to mention that the structure break point occurs between the 1st and 3rd work shift on 14 May because during the 2nd work shift the LHD machine was not operating. The same conclusions can be drawn when we analyze the results corresponding to four work shifts treated separately, besides the case of the Wilcoxon-based procedure applied to the data from the fourth work shifts where the structure break point is indicated only after 15 May. It is important to notice that the choice of divergence measure does not cause any change in the final result for the analyzed data. Although the values taken by the calculated distances differ, the overall behaviour of the functions is similar. As a consequence, we obtain exactly the same output for both procedures (based on Students *t*-test and on Wilcoxon test) applied to the values taken by Hellinger, Jeffreys and Chernoff distances.

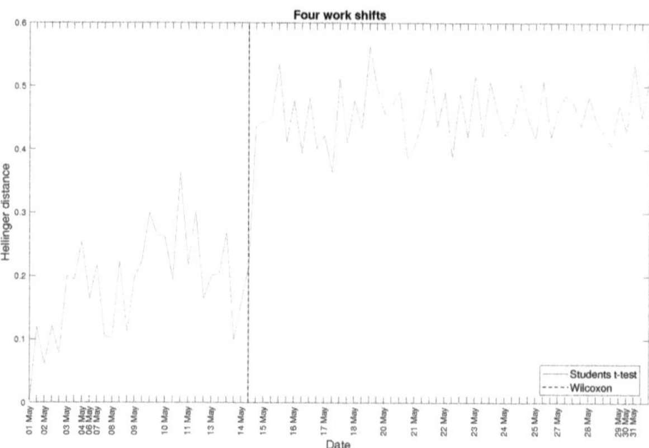

Figure 12. The Hellinger distance between the subsequent pdfs and the base pdf chosen as the first one in a sample and the structure break point identified by the methods based on the Student's *t*-test and Wilcoxon test for the data presented shift by shift.

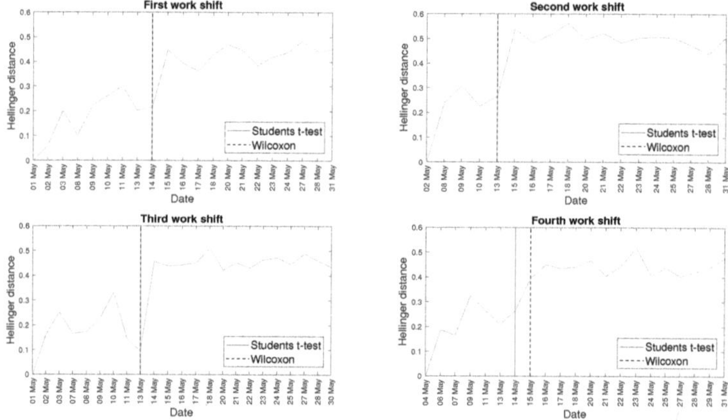

Figure 13. The Hellinger distance between the subsequent pdfs and the base pdf chosen as the first one in a sample and the structure break points identified by the methods based on the Student's *t*-test and Wilcoxon test corresponding to the data representing four work shifts.

5. Discussion

The developed procedure allows us to transform the one-month data from highly variable to a nearly monotonic feature. It should be said that the behaviour/shape of our feature is different than a bath-curve known from reliability theory. Here, there is a "switch" from good to bad condition. Therefore, using this data we are able just to detect the moment of change, without the possibility to detect the early stage of damage or tracking its development. It is not related to the proposed method, but to specific data.

The a priori knowledge that a repair action has been performed is necessary to validate the results. The detected date of structure break point is the same as the date of maintenance action done in the company. The extracted feature—i.e., the distance between the distributions estimated for each shift is convincing and easy to interpret. From an industrial perspective, it is essential. We believe that it has also appropriate quality from a scientific point of view as we proposed a new method for structure break point detection, i.e., the method for identification of the moment when our feature is changing (that corresponds to the change in the process). This method is based on the Student's *t* and Wilcoxon statistics that are commonly applied in testing if two samples have the same means. The base of the procedure is very intuitive, we expect that the distribution of the time series is changing along with the degradation process. Thus, we proposed analyzing the characteristics of the empirical distributions of the data corresponding to the work shifts, i.e., the measures of the distance between empirical probability distribution functions. Using the characteristics, we could obtain the monotonic feature of the data that was further segmented.

The problem considered here is essential from the practical point of view. Even if various SCADA data are increasingly available, practical information for an end-user is still doubtful. It is especially the case for highly variable data acquired from time-varying systems such as mining machines, wind turbines, etc. [85–87]. To develop novel diagnostic procedures, historical data are used to assure the "training" process. Often the data are normalized as they depend on operating conditions [85,87]. Unfortunately, we cannot use normalization—engine oil pressure cannot be higher if the machine is more loaded. However, we have noticed that the statistical properties of the process (engine oil pressure variation) are changed when the machine changes the condition. It was the base for our approach.

6. Conclusions

As the conclusion we need to highlight few important issues:

- We have proposed a novel multistep procedure that covers pre-processing, statistical analysis, and visualization.
- The proposed procedure is the novel one. The innovation is related to the combination of the crucial steps of the proposed methodology, namely, the initial segmentation and the representation as a matrix in the work shift perspective as well as the analysis of the characteristics of the data (probability density functions) and the distance measures based on them. Finally, the last step (segmentation) is performed not for the real data, but for the distance measures of the time series' pdfs. According to our knowledge, this approach is rarely used in real applications.
- The utilization of the distance measures of time series' pdfs causes that we do not consider here the problem when one parameter of the data changes (like mean or variance). The examined issue is much more general. The analysis of the pdf's changes causes the algorithm is sensitive to the dynamics of various characteristics of the data, not only the single one.
- The proposed approach is a universal one. It can be used for any cycle that corresponds to the considered phenomena (in our case it is a work shift), to any characteristics of the data (in our case it is the probability density function), and to any distance measure applied to the characteristics (in our case there are distance measures based on the probability density functions).
- The whole procedure is automatic, thus we believe it could be implemented in the monitoring systems used in the company. When the new data corresponding to the next day (four work shifts) come, then using the introduced procedure we can test if they belong to the current regime. In our case, the new sample means the data corresponding to the next day. This approach is often used in monitoring systems. Thus, in some sense, the methodology can be used in a continuous manner.
- The historical data with precise knowledge about replacement has been used for training and validation. Implementation of the proposed method as an automatic data processing procedure should not be a problem for any new machine. Small dataset from a couple of shifts from a new machine (new data set) will be enough to establish the averaged picture of the signature of good condition. If the damage will appear (change of regime), the method will be able to detect it after a few work shifts (min. 2), that is much better than the current situation. Note that a machine with such damage was able to operate for two weeks as there was not a tool to detect the problem.
- It should be highlighted, the proposed methodology has also some limitations. One of this is related to the special requirements of the data. More precisely, there is a

need to consider data that could be arranged as work shifts (or any other cycles). This influences the identified structure break point corresponds to the work shift, not the real time point (like hour).

Author Contributions: conceptualization, A.G., A.W., R.Z., N.G.; investigation, A.G., A.W., R.Z.; methodology, A.G., A.W., R.Z.; resources, P.Ś., R.Z., N.G.; software, A.G.; supervision, A.W., R.Z., P.Ś., N.G.; validation, A.G., A.W., R.Z.; visualization, A.G.; writing—original draft, A.G., A.W., R.Z.; writing—Review and editing, A.G., A.W., R.Z., N.G. All authors have read and agreed to the published version of the manuscript.

Funding: This research received no external funding.

Institutional Review Board Statement: Not applicable.

Informed Consent Statement: Not applicable.

Data Availability Statement: Data are not available due to non-disclosure agreements.

Conflicts of Interest: The authors declare no conflict of interest.

Appendix A. Tables

Table A1. A list of monitored parameters for LHD machines.

Variable Name	Description
	Temperatures
'ENGCOOLT'	temperature of the cooling liquid of the internal combustion engine
'GROILT'	oil temperature of transmission and torque
'HYDOILT'	hydraulic oil temperature
	Pressures
'ENGOILP'	oil pressure of the internal combustion engine
'BREAKP'	breaking pressure
'GROILP'	transmission oil pressure
'HYDOILP'	pressure in the hydraulic system
	Others
'ENGRPM'	engine speed
'FUELUS'	instant fuel consumption
'SELGEAR'	direction and current gear
'SPEED'	average speed every 1s

Table A2. The identified date corresponding to the last shift before the change in data and the values of test statistics for the data presented shift by shift.

Hellinger Distance	
Student's t	Wilcoxon
1st shift on 14th of May $\mathcal{T}(m^*) = -19.76$	1st shift on 14th of May $\mathcal{T}(m^*) = -7.54$
Jeffreys distance	
Student's t	Wilcoxon
1st shift on 14th of May $\mathcal{T}(m^*) = -19.72$	1st shift on 14th of May $\mathcal{T}(m^*) = -7.54$
Chernoff distance	
Student's t	Wilcoxon
1st shift on 14th of May $\mathcal{T}(m^*) = -19.76$	1st shift on 14th of May $\mathcal{T}(m^*) = -7.53$

Table A3. The identified dates corresponding to the last shifts before the change in data and the values of test statistics corresponding to the data representing four work shifts—Hellinger distance.

Hellinger Distance	
First Shift	
Student's t	Wilcoxon
14th of May $\mathcal{T}(m^*) = -8.39$	14th of May $\mathcal{T}(m^*) = -3.83$
Second shift	
Student's t	Wilcoxon
13th of May $\mathcal{T}(m^*) = -8.53$	13th of May $\mathcal{T}(m^*) = -3.20$
Third shift	
Student's t	Wilcoxon
13th of May $\mathcal{T}(m^*) = -10.73$	13th of May $\mathcal{T}(m^*) = -3.97$
Fourth shift	
Student's t	Wilcoxon
14th of May $\mathcal{T}(m^*) = -7.83$	15th of May $\mathcal{T}(m^*) = -3.77$

Table A4. The identified dates corresponding to the last shifts before the change in data and the values of test statistics corresponding to the data representing four work shifts—Jeffreys distance.

Jeffreys Distance	
First Shift	
Student's t	**Wilcoxon**
14th of May $\mathcal{T}(m^*) = -11.21$	14th of May $\mathcal{T}(m^*) = -3.84$
Second shift	
Student's t	Wilcoxon
13th of May $\mathcal{T}(m^*) = -11.93$	13th of May $\mathcal{T}(m^*) = -3.20$
Third shift	
Student's t	Wilcoxon
13th of May $\mathcal{T}(m^*) = -14.76$	13th of May $\mathcal{T}(m^*) = -3.97$
Fourth shift	
Student's t	Wilcoxon
14th of May $\mathcal{T}(m^*) = -9.68$	15th of May $\mathcal{T}(m^*) = -3.77$

Table A5. The identified dates corresponding to the last shifts before the change in data and the values of test statistics corresponding to the data representing four work shifts - Chernoff distance.

Chernoff Distance	
First Shift	
Student's *t*	Wilcoxon
14th of May $\mathcal{T}(m^*) = -13.94$	14th of May $\mathcal{T}(m^*) = -3.84$
Second shift	
Student's *t*	Wilcoxon
13th of May $\mathcal{T}(m^*) = -11.55$	13th of May $\mathcal{T}(m^*) = -3.20$
Third shift	
Student's *t*	Wilcoxon
13th of May $\mathcal{T}(m^*) = -10.52$	13th of May $\mathcal{T}(m^*) = -3.97$
Fourth shift	
Student's *t*	Wilcoxon
14th of May $\mathcal{T}(m^*) = -9.68$	15th of May $\mathcal{T}(m^*) = 3.77$

Appendix B. Additional Graphs

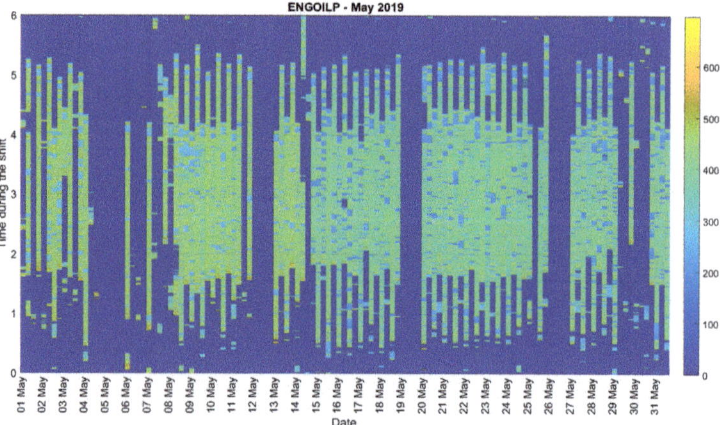

Figure A1. Re-sampled data presented as a shift-by-shift map of engine oil pressure values for subsequent days.

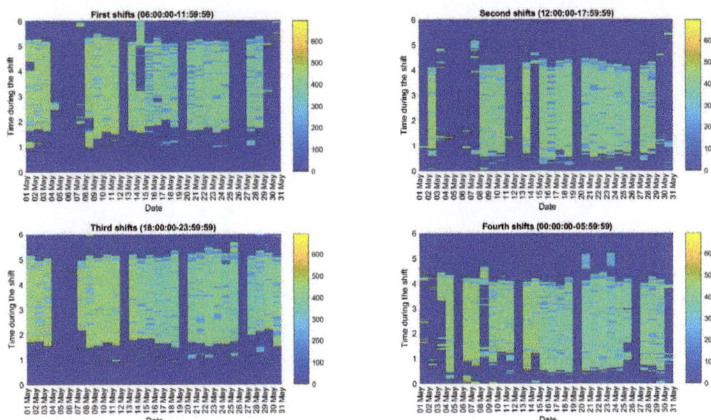

Figure A2. Re-sampled data presented as the maps of engine oil pressure values corresponding to four work shifts during a day.

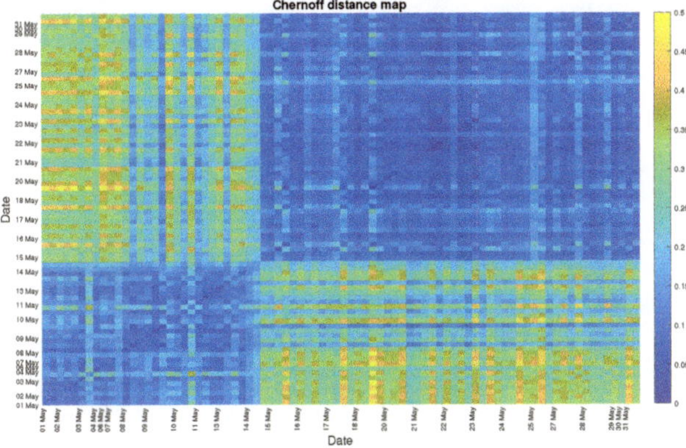

Figure A3. The matrix presenting the values of the Chernoff distance calculated based on the empirical probability density functions corresponding to the data presented shift by shift.

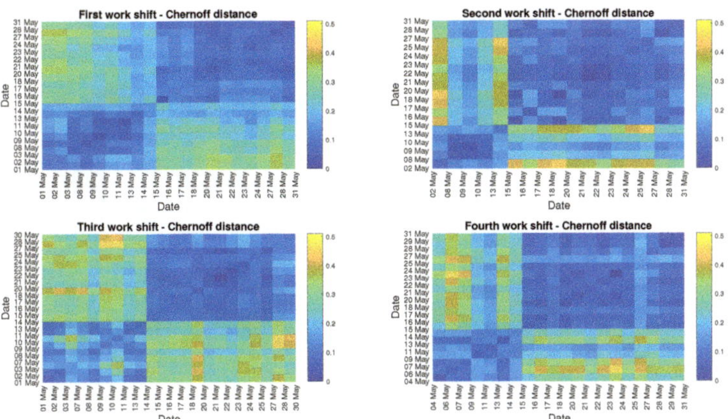

Figure A4. The matrices presenting the values of the Chernoff distance calculated based on the empirical probability density functions corresponding to the data representing four work shifts.

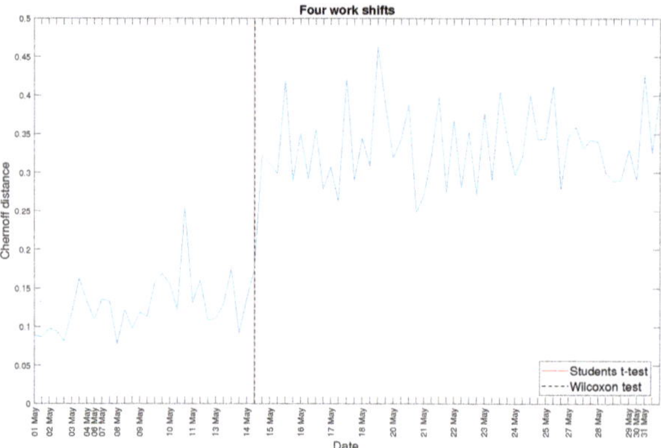

Figure A5. The Chernoff distance between the subsequent pdfs and the base pdf chosen as the first one in a sample and the structure break point identified by the methods based on the Student's *t*-test and Wilcoxon test for the data presented shift by shift.

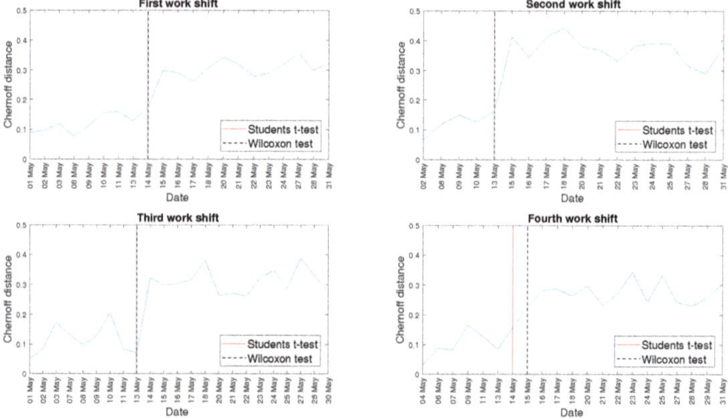

Figure A6. The Chernoff distance between the subsequent pdfs and the base pdf chosen as the first one in a sample and the structure break points identified by the methods based on the Student's *t*-test and Wilcoxon test corresponding to the data representing four work shifts.

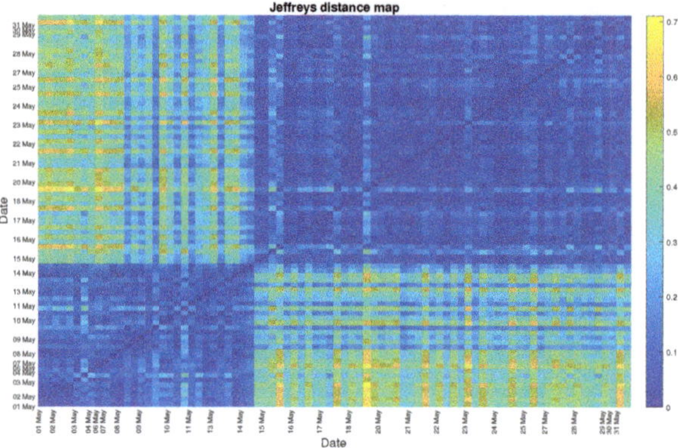

Figure A7. The matrix presenting the values of the Jeffreys distance calculated based on the empirical probability density functions corresponding to the data presented shift by shift.

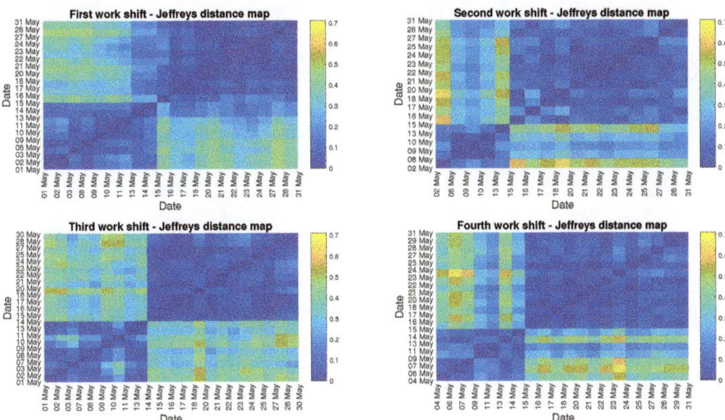

Figure A8. The matrices presenting the values of the Jeffreys distance calculated based on the empirical probability density functions corresponding to the data representing four work shifts.

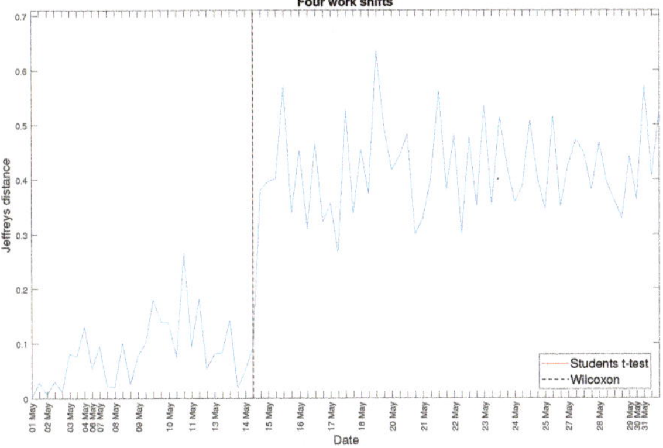

Figure A9. The Jeffreys distance between the subsequent pdfs and the base pdf chosen as the first one in a sample and the structure break point identified by the methods based on the Student's *t*-test and Wilcoxon test for the data presented shift by shift.

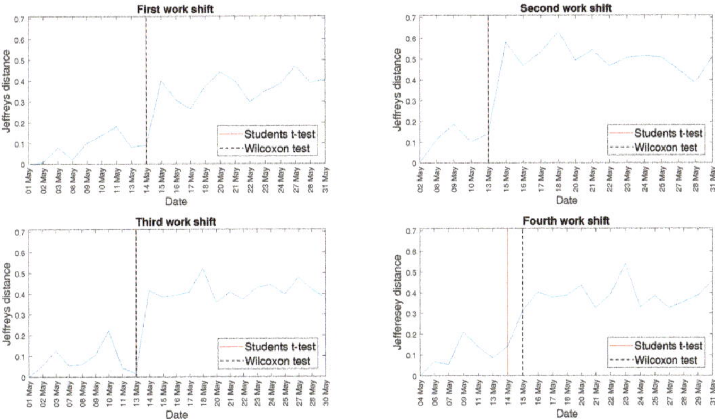

Figure A10. The Jeffreys distance between the subsequent pdfs and the base pdf chosen as the first one in a sample and the structure break points identified by the methods based on the Student's *t*-test and Wilcoxon test corresponding to the data representing four work shifts.

References

1. Vashistha, S.; Kumar Agrawal, A.; Siddiqui, M.; Chattopadhyaya, S. Reliability and Maintainability Analysis of LHD Loader at Saoner Mines, Nagpur, India. *IOP Conf. Ser. Mater. Sci. Eng.* **2019**, *691*, 012013. [CrossRef]
2. Jakkula, B.; Govinda Raj, M.; Murthy, C. Maintenance management of load haul dumper using reliability analysis. *J. Qual. Maint. Eng.* **2019**, *26*, 290–310. [CrossRef]
3. Chatterjee, S.; Bandopadhyay, S. Reliability estimation using a genetic algorithm-based artificial neural network: An application to a load-haul-dump machine. *Expert Syst. Appl.* **2012**, *39*, 10943–10951. [CrossRef]
4. Dindarloo, S. Reliability forecasting of a load-haul-dump machine: A comparative study of ARIMA and neural networks. *Qual. Reliab. Eng. Int.* **2016**, *32*, 1545–1552. [CrossRef]
5. Bala, R.; Govinda, R.; Murthy, C. Reliability analysis and failure rate evaluation of load haul dump machines using Weibull distribution analysis. *Math. Model. Eng. Probl.* **2018**, *5*, 116–122. [CrossRef]
6. Paithankar, A.; Chatterjee, S. Forecasting time-to-failure of machine using hybrid Neuro-genetic algorithm–a case study in mining machinery. *Int. J. Mining Reclam. Environ.* **2018**, *32*, 182–195. [CrossRef]
7. Balaraju, J.; Govinda Raj, M.; Murthy, C.S.N. Prediction and Assessment of LHD Machine Breakdowns Using Failure Mode Effect Analysis (FMEA). In *Reliability, Safety and Hazard Assessment for Risk-Based Technologies*; Varde, P.V., Prakash, R.V., Vinod, G., Eds.; Springer: Singapore, 2020; pp. 833–850.
8. Jakkula, B.; Mandela, G.; Chivukula, S. Application ANN Tool for Validation of LHD Machine Performance Characteristics. *J. Inst. Eng. (India) Ser. D* **2020**, *101*, 27–38. [CrossRef]
9. Jakkula, B.; Mandela, G.; Chivukula, M. Improvement of overall equipment performance of underground mining machines- a case study. *Adv. Model. Anal. A* **2018**, *79*, 6–11. [CrossRef]
10. Jakobs, A. The Sandvik LH621, from hardrock loader to high-performance machine in German salt and potash mining. *World Min. Surf. Undergr.* **2018**, *70*, 276–279.
11. Mkhwanazi, D. Optimizing LHD utilization. *J. South Afr. Inst. Min. Metall.* **2011**, *111*, 273–280.
12. Krot, P.; Śliwiński, P.; Zimroz, R.; Gomolla, N. The identification of operational cycles in the monitoring systems of underground vehicles. *Measurement* **2020**, *151*, 107111. [CrossRef]
13. Mbhalati, W. LHD optimization at an underground chromite mine. *J. South Afr. Inst. Min. Metall.* **2015**, *115*, 313–320. [CrossRef]
14. Fukui, R.; Kusaka, K.; Nakao, M.; Kodama, Y.; Uetake, M.; Kawai, K. Production analysis of functionally distributed machines for underground mining. *Int. J. Min. Sci. Technol.* **2016**, *26*, 477–485. [CrossRef]
15. Stefaniak, P.; Zimroz, R.; Obuchowski, J.; Śliwiński, P.; Andrzejewski, M. An Effectiveness Indicator for a Mining Loader Based on the Pressure Signal Measured at a Bucket's Hydraulic Cylinder. *Procedia Earth Planet. Sci.* **2015**, *15*, 797–805. [CrossRef]
16. Balaraju, J.; Govinda Raj, M.; Murthy, C. Fuzzy-FMEA risk evaluation approach for LHD machine—A case study. *J. Sustain. Min.* **2019**, *18*, 257–268. [CrossRef]
17. Ghodrati, B.; Hoseinie, S.; Kumar, U. Context-driven mean residual life estimation of mining machinery. *Int. J. Mining Reclam. Environ.* **2018**, *32*, 486–494. [CrossRef]
18. Laukka, A.; Saari, J.; Ruuska, J.; Juuso, E.; Lahdelma, S. Condition-based monitoring for underground mobile machines. *Int. J. Ind. Syst. Eng.* **2016**, *23*, 74–89. [CrossRef]
19. Zimroz, R.; Wodecki, J.; Król, R.; Andrzejewski, M.; Śliwiński, P.; Stefaniak, P. Self-propelled Mining Machine Monitoring System—Data Validation, Processing and Analysis. In *Mine Planning and Equipment Selection*; Drebenstedt, C., Singhal, R., Eds.; Springer International Publishing: Cham, Switzerland, 2014; pp. 1285–1294.
20. Wodecki, J.; Stefaniak, P.; Michalak, A.; Wyłomańska, A.; Zimroz, R. Technical condition change detection using Anderson-Darling statistic approach for LHD machines—engine overheating problem. *Int. J. Mining Reclam. Environ.* **2018**, *32*, 392–400. [CrossRef]
21. Michalak, A.; Śliwiński, P.; Kaniewski, T.; Wodecki, J.; Stefaniak, P.; Wyłomańska, A.; Zimroz, R. Condition Monitoring for LHD Machines Operating in Underground Mine—Analysis of Long-Term Diagnostic Data. In *Proceedings of the 27th International Symposium on Mine Planning and Equipment Selection—MPES 2018*; Widzyk-Capehart, E., Hekmat, A., Singhal, R., Eds.; Springer International Publishing: Cham, Switzerland, 2019; pp. 471–480.
22. Stefaniak, P.; Śliwiński, P.; Poczynek, P.; Wyłomańska, A.; Zimroz, R. The Automatic Method of Technical Condition Change Detection for LHD Machines—Engine Coolant Temperature Analysis. In *Advances in Condition Monitoring of Machinery in Non-Stationary Operations*; Fernandez Del Rincon, A., Viadero Rueda, F., Chaari, F., Zimroz, R., Haddar, M., Eds.; Springer International Publishing: Cham, Switzerland, 2019; pp. 54–63.
23. Paraszczak, J.; Gustafson, A.; Schunnesson, H. Technical and operational aspects of autonomous LHD application in metal mines. *Int. J. Mining Reclam. Environ.* **2015**, *29*, 391–403.
24. Gustafson, A.; Paraszczak, J.; Tuleau, J.; Schunnesson, H. Impact of technical and operational factors on effectiveness of automatic load-haul-dump machines. *Trans. Institutions Min. Metall. Sect. A Min. Technol.* **2017**, *126*, 185–190.
25. Kaniewski, T.; Śliwiński, P.; Hebda-Sobkowicz, J.; Zimroz, R. Comprehensive, experimental verification of the effects of the lock-up function implementation in LHD haul trucks in the deep underground mine. In Proceedings of the Mining Goes Digital: Proceedings of the 39th International Symposium on Application of Computers and Operations Research in the Mineral Industry (APCOM 2019), Wrocław, Poland, 4–6 June 2019; pp. 506–514.

26. Śliwiński, P.; Kaniewski, T.; Hebda-Sobkowicz, J.; Zimroz, R.; Wyłomańska, A. Analysis of dynamic external loads to haul truck machine subsystems during operation in a deep underground mine. In Proceedings of the Mining Goes Digital: Proceedings of the 39th International Symposium on Application of Computers and Operations Research in the Mineral Industry (APCOM 2019), Wrocław, Poland, 4–6 June 2019; pp. 515–524.

27. Wang, Y.; Jin, T.; Liu, L. Output torque prediction of hybrid underground LHD motor based on least square support vector machine. *Meitan Xuebao J. China Coal Soc.* **2017**, *42*, 619–625.

28. Saari, J.; Odelius, J. Detecting operation regimes using unsupervised clustering with infected group labelling to improve machine diagnostics and prognostics. *Oper. Res. Perspect.* **2018**, *5*, 232–244. [CrossRef]

29. Wyłomańska, A.; Zimroz, R. Signal segmentation for operational regimes detection of heavy duty mining mobile machines-a statistical approach. *Diagnostyka* **2014**, *15*, 33–42.

30. Wodecki, J.; Michalak, A.; Stefaniak, P. Review of smoothing methods for enhancement of noisy data from heavy-duty LHD mining machines. *E3S Web Conf.* **2018**, *29*, 00011. [CrossRef]

31. Stefaniak, P.K.; Zimroz, R.; Śliwiński, P.; Andrzejewski, M.; Wyłomańska, A. Multidimensional Signal Analysis for Technical Condition, Operation and Performance Understanding of Heavy Duty Mining Machines. In *Advances in Condition Monitoring of Machinery in Non-Stationary Operations*; Chaari, F., Zimroz, R., Bartelmus, W., Haddar, M., Eds.; Springer International Publishing: Cham, Switzerland, 2016; pp. 197–210.

32. Śliwiński, P.; Andrzejewski, M.; Kaniewski, T.; Hebda-Sobkowicz, J.; Zimroz, R. Selection of variables acquired by the on-board monitoring system to determine operational cycles for haul truck vehicle. In Proceedings of the Mining Goes Digital: Proceedings of the 39th International Symposium on Application of Computers and Operations Research in the Mineral Industry (APCOM 2019), Wrocław, Poland, 4–6 June 2019; pp. 525–533.

33. Kucharczyk, D.; Wyłomańska, A.; Zimroz, R. Structural break detection method based on the Adaptive Regression Splines technique. *Physica A* **2017**, *471*, 499–511. [CrossRef]

34. Obuchowski, J.; Wyłomańska, A.; Zimroz, R. The local maxima method for enhancement of time-frequency map and its application to local damage detection in rotating machines. *Mech. Syst. Signal Process.* **2014**, *46*, 389–405. [CrossRef]

35. Andreao, R.V.; Dorizzi, B.; Boudy, J. ECG signal analysis through hidden Markov models. *IEEE Trans. Biomed. Eng.* **2006**, *53*, 1541–1549. [CrossRef] [PubMed]

36. Azami, H.; Mohammadi, M.; Bozorgtabar, B. An Improved Signal Segmentation Using Moving Average and Savitzky-Golay Filter. *J. Signal Inf. Process.* **2012**, *3*, 39–44.

37. Bhagavatula, C.; Jaech, A.; Savvides, M.; Bhagavatula, V.; Friedman, R.; Blue, R.; O Griofa, M. Automatic segmentation of cardiosynchronous waveforms using cepstral analysis and continuous wavelet transforms. In Proceedings of the 19th IEEE International Conference on Image Processing, Orlando, FL, USA, 30 September–3 October 2012; pp. 2045–2048.

38. Choi, S.; Jiang, Z. Comparison of envelope extraction algorithms for cardiac sound signal segmentation. *Expert Syst. Appl.* **2008**, *34*, 1056–1069. [CrossRef]

39. Micó, P.; Mora, M.; Cuesta-Frau, D.; Aboy, M. Automatic segmentation of long-term ECG signals corrupted with broadband noise based on sample entropy. *Comput. Methods Programs Biomed.* **2010**, *98*, 118–129. [CrossRef]

40. Khanagha, V.; Daoudi, K.; Pont, O.; Yahia, H. Phonetic segmentation of speech signal using local singularity analysis. *Digit. Signal Process.* **2014**, *35*, 86–94. [CrossRef]

41. Lovell, B.; Boashash, B. Segmentation of non-stationary signals with applications. In Proceedings of the International Conference on Acoustics, Speech, and Signal Processing, New York, NY, USA, 11–14 April 1988; Volume 5, pp. 2685–2688.

42. Makowski, R.; Hossa, R. Automatic speech signal segmentation based on the innovation adaptive filter. *Int. J. Appl. Math. Comput. Sci.* **2014**, *24*, 259–270. [CrossRef]

43. Janczura, J.; Weron, R. Goodness-of-fit testing for the marginal distribution of regime-switching models with an application to electricity spot prices. *AStA Adv. Stat. Anal.* **2013**, *97*, 239–270. [CrossRef]

44. Janczura, J. Pricing electricity derivatives within a Markov regime-switching model: A risk premium approach. *Math. Methods Oper. Res.* **2014**, *79*, 1–30. [CrossRef]

45. Chen, C. On a segmentation algorithm for seismic signal analysis. *Geoexploration* **1984**, *23*, 35–40. [CrossRef]

46. Gaby, J.E.; Anderson, K.R. Hierarchical segmentation of seismic waveforms using affinity. *Geoexploration* **1984**, *23*, 1–16. [CrossRef]

47. Kucharczyk, D.; Wyłomańska, A.; Obuchowski, J.; Zimroz, R.; Madziarz, M. Stochastic Modelling as a Tool for Seismic Signals Segmentation. *Shock Vib.* **2016**, *2016*, 8453426. [CrossRef]

48. Popescu, T.D. Signal segmentation using changing regression models with application in seismic engineering. *Digit. Signal Process.* **2014**, *24*, 14–26. [CrossRef]

49. Sokołowski, J.; Obuchowski, J.; Zimroz, R.; Wyłomańska, A.; Koziarz, E. Algorithm Indicating Moment of P-Wave Arrival Based on Second-Moment Characteristic. *Shock Vib.* **2016**, *2016*, 4051701. [CrossRef]

50. Gajda, J.; Sikora, G.; Wyłomańska, A. Regime Variance Testing—A Quantile Approach. *Acta Phys. Pol. B Proc. Suppl.* **2013**, *44*, 1015–1035. [CrossRef]

51. Makowski, R.; Zimroz, R. New techniques of local damage detection in machinery based on stochastic modelling using adaptive Schur filter. *Appl. Acoust.* **2014**, *77*, 130–137. [CrossRef]

52. Makowski, R.; Zimroz, R. A procedure for weighted summation of the derivatives of reflection coefficients in adaptive Schur filter with application to fault detection in rolling element bearings. *Mech. Syst. Signal Process.* **2013**, *38*, 65–77. [CrossRef]

53. Tsay, R.S. Outliers, level shifts, and variance changes in time series. *J. Forecast.* **1988**, *7*, 1–20. [CrossRef]
54. Urbanek, J.; Barszcz, T.; Zimroz, R.; Antoni, J. Application of averaged instantaneous power spectrum for diagnostics of machinery operating under non-stationary operational conditions. *Measurement* **2012**, *45*, 1782–1791. [CrossRef]
55. Lanoiselée, Y.; Grebenkov, D. Unraveling intermittent features in single-particle trajectories by a local convex hull method. *Phys. Rev. E* **2017**, *96*, 022144. [CrossRef]
56. Wagner, T.; Kroll, A.; Haramagatti, C.R.; Lipinski, H.G.; Wiemann, M. Classification and Segmentation of Nanoparticle Diffusion Trajectories in Cellular Micro Environments. *PLoS ONE* **2017**, *12*, 1–20.
57. Akimoto, T.; Yamamoto, E. Detection of transition times from single-particle-tracking trajectories. *Phys. Rev. E* **2017**, *96*, 052138. [CrossRef]
58. Sikora, G.; Wyłomańska, A.; Krapf, D. Recurrence statistics for anomalous diffusion regime change detection. *Comput. Stat. Data Anal.* **2018**, *128*, 380–394. [CrossRef]
59. Sikora, G.; Wyłomańska, A.; Gajda, J.; Solé, L.; Akin, E.; Tamkun, M.; Krapf, D. Elucidating distinct ion channel populations on the surface of hippocampal neurons via single-particle tracking recurrence analysis. *Phys. Rev. E* **2017**, *96*, 062404. [CrossRef]
60. Li, W.; Li, H.; Gu, S.; Chen, T. Process fault diagnosis with model- and knowledge-based approaches: Advances and opportunities. *Control. Eng. Pract.* **2020**, *105*, 104637. [CrossRef]
61. Qin, S. Survey on data-driven industrial process monitoring and diagnosis. *Annu. Rev. Control.* **2012**, *36*, 220–234. [CrossRef]
62. Yan, Y.; Li, J.; Gao, D. Condition parameter modeling for anomaly detection in wind turbines. *Energies* **2014**, *7*, 3104–3120. [CrossRef]
63. Aghabozorgi, S.; Seyed Shirkhorshidi, A.; Ying Wah, T. Time-series clustering—A decade review. *Inf. Syst.* **2015**, *53*, 16–38. [CrossRef]
64. Branisavljević, N.; Kapelan, Z.; Prodanović, D. Improved real-time data anomaly detection using context classification. *J. Hydroinform.* **2011**, *13*, 307–323. [CrossRef]
65. Chandola, V.; Banerjee, A.; Kumar, V. Anomaly detection: A survey. *ACM Comput. Surv.* **2009**, *41*, 1–58. [CrossRef]
66. Markou, M.; Singh, S. Novelty detection: A review—Part 1: Statistical approaches. *Signal Process.* **2003**, *83*, 2481–2497. [CrossRef]
67. Myers, D.; Suriadi, S.; Radke, K.; Foo, E. Anomaly detection for industrial control systems using process mining. *Comput. Secur.* **2018**, *78*, 103–125. [CrossRef]
68. Zhao, H.; Liu, H.; Hu, W.; Yan, X. Anomaly detection and fault analysis of wind turbine components based on deep learning network. *Renew. Energy* **2018**, *127*, 825–834. [CrossRef]
69. Available online: https://www.kghmzanam.com/en/kategoria/mining-machinery/loaders/ (accessed on 27 July 2021).
70. Csiszár, I. Information-Type Measures of Difference of Probability Distributions and Indirect Observations. *Stud. Sci. Math. Hung.* **1967**, *2*, 299–318.
71. Csiszár, I. *I*-Divergence Geometry of Probability Distributions and Minimization Problem. *Ann. Probab.* **1975**, *3*, 146–158. [CrossRef]
72. Basseville, M. Distance measures for signal processing and pattern recognition. *Signal Process.* **1989**, *18*, 349–369. [CrossRef]
73. Basseville, M. Divergence measures for statistical data processing - An annotated bibliography. *Signal Process.* **2013**, *93*, 621–633. [CrossRef]
74. Chung, J.; Kannappan, P.; Ng, C.; Sahoo, P. Measures of distance between probability distributions. *J. Math. Anal. Appl.* **1989**, *138*, 280–292. [CrossRef]
75. Hill, P.D. Kernel estimation of a distribution function. *Commun. Stat. Theory Methods* **1985**, *14*, 605–620.
76. Bowman, A.W.; Azzalini, A. *Applied Smoothing Techniques for Data Analysis*; Oxford University Press Inc.: New York, NY, USA, 1997.
77. Silverman, B. *Density Estimation: For Statistics and Data Analysis*; Chapman & Hall: London, UK, 1986.
78. Blair, R.C.; Higgins, J.J. A Comparison of the Power of Wilcoxon's Rank-Sum Statistic to that of Student'st Statistic Under Various Nonnormal Distributions. *J. Educ. Stat.* **1980**, *5*, 309–335. [CrossRef]
79. Rice, J.A. *Mathematical Statistics and Data Analysis*, 3rd ed.; Duxbury Press: Belmont, CA, USA, 2006.
80. Hogg, R.; Craig, A. *Introduction to Mathematical Statistics*, 4th ed.; Macmillan: New York, NY, USA, 1978.
81. Venables, W.N.; Ripley, B.D. *Modern Applied Statistics with S*, 4th ed.; Springer: New York, NY, USA, 2010.
82. Fay, M.P.; Proschan, M.A. Wilcoxon-Mann-Whitney or *t*-test? On assumptions for hypothesis tests and multiple interpretations of decision rules. *Stat. Surv.* **2010**, *4*, 1–39. [CrossRef] [PubMed]
83. Conover, W.J. *Practical Nonparametric Statistics*, 2nd ed.; Wiley: New York, NY, USA, 1980; pp. 225–226.
84. Grzesiek, A.; Zimroz, R.; Śliwiński, P.; Gomolla, N.; Wyłomańska, A. Long term belt conveyor gearbox temperature data analysis—Statistical tests for anomaly detection. *Measurement* **2020**, *165*, 108124. [CrossRef]
85. Urbanek, J.; Barszcz, T.; Straczkiewicz, M.; Jablonski, A. Normalization of vibration signals generated under highly varying speed and load with application to signal separation. *Mech. Syst. Signal Process.* **2017**, *82*, 13–31. [CrossRef]
86. Schmidt, S.; Heyns, P.; Gryllias, K. A methodology using the spectral coherence and healthy historical data to perform gearbox fault diagnosis under varying operating conditions. *Appl. Acoust.* **2020**, *158*, 107038. [CrossRef]
87. Schmidt, S.; Heyns, P. Normalisation of the amplitude modulation caused by time-varying operating conditions for condition monitoring. *Measurement* **2020**, *149*, 106964. [CrossRef]

Article

Measurement and Simulation of Flow in a Section of a Mine Gallery

Jakub Janus * and Jerzy Krawczyk

Strata Mechanics Research Institute, The Polish Sciences Academy, Reymonta 27, 30-059 Kraków, Poland; krawczyk@imgpan.pl
* Correspondence: janus@imgpan.pl

Abstract: Research work on the air flow in mine workings frequently utilises computer techniques in the form of numeric simulations. However, it is very often necessary to apply simplifications when building a geometrical model. The assumption of constant model geometry on its entire length is one of the most frequent simplifications. This results in a substantial shortening of the geometrical model building process, and a concomitant shortening of the time of numerical computations; however, it is not known to what extent such simplifications worsen the accuracy of simulation results. The paper presents a new methodology that enables precise reproduction of the studied mine gallery and the obtaining of a satisfactory match between simulation results and in-situ measurements. It utilises the processing of data from laser scanning of a mine gallery, simultaneous multi-point measurements of the velocity field at selected gallery cross-sections, unique for mine conditions, and the SAS turbulence model, recently introduced to engineering analyses of flow issues.

Keywords: computational fluid dynamics; numerical model geometry; laser scanning

check for
updates

Citation: Janus, J.; Krawczyk, J. Measurement and Simulation of Flow in a Section of a Mine Gallery. *Energies* **2021**, *14*, 4894. https://doi.org/10.3390/en14164894

Academic Editor: Sergey Zhironkin

Received: 11 June 2021
Accepted: 4 August 2021
Published: 10 August 2021

Publisher's Note: MDPI stays neutral with regard to jurisdictional claims in published maps and institutional affiliations.

1. Introduction

Proper operation of mine ventilation is one of the necessary conditions for the safe and effective functioning of underground mines. Ventilation systems ensure such a condition of atmosphere that enables normal mine operations, and also prevent or possibly minimise the effects of failures and disasters, if any. For many years, mine ventilation used methods of computational fluid dynamics, which were developed for other fields, such as aviation, construction of fluid-flow machines, the automotive industry and meteorology. In particular, it was related to applications of the finite volume method to resolve equations of fluid mechanics [1]. In each of the aforementioned fields, and even individual issues, to achieve a satisfactory quality of numerical modelling, it was necessary to find or develop a specific methodology and to verify it experimentally. An example of the methodology for solving particular problems such as the effect of machine-mounted dust scrubbers on the performance of face ventilation systems can be found in [2]. Other applications of CFD have been described by Ren and Balusu in [3]. An updated review of CFD applications in mining can be found in [4]. More recent work refers to similar supports, but in roadways of rectangular shape [5] Despite the above-mentioned solutions, for many cases, the issues of numerical modelling of flow effects in mines encounter numerous problems, which so far have not been resolved to a satisfactory degree. They result inter alia from the vastness of such ventilation systems, frequently comprising hundreds of kilometers of workings with complicated and irregular shapes. Such irregularities result from, for example, difficult to predict local deformations caused by the pressure of surrounding strata. Previous measuring techniques did not provide a practical possibility to take such irregularities into consideration.

The non-stationary nature of a flow is another issue. An irregular shape of the workings and the presence of numerous objects that are significant for the operation of a mine, which must fit into the limited space of underground workings, causes substantial local flow disturbances. This results in difficulties in achieving satisfactory mapping of the flow by means of previous generation turbulence models [6]. Furthermore, so far, there has been a very small number of papers addressing specific flow conditions in galleries with the arched roof support.

Mines frequently operate under conditions of coexisting hazards and feature difficult environmental conditions [7]. This results in particular difficulties when acquiring experimental data, necessary for verification and validation of modelling. This paper presents the application of novel measuring and numerical modelling techniques, which allow one to achieve a satisfactory match between simulation results and measurement data. They were applied to a section of a gallery with a shape and environmental conditions representative of Polish coal mines. The obtained results are a point of reference for further analyses of the influence of geometry representation simplifications on the accuracy of numerical simulations. This work aims at the optimisation of numerical modelling in order to achieve sufficient reliability whilst minimising the computation time and the level of power involved.

Accurate reproduction of the geometry of the gallery was obtained by means of terrestrial laser scanning [8–10]. The computation area geometry was determined based on this. This method has so far been applied for the CFD modeling of problems that are remote from mine ventilation, such as the flows around Formula I cars or around the human body. The results of recent work [11] show application to much closer problems. This paper both confirms the methodology used and presents solutions that can reduce the workload involved in transferring the scaning data into meshable geometry.

The data on the velocity field were obtained by means of a system of simultaneous multi-point measurements of flow velocity using intrinsically safe vane anemometers, developed with the authors' contribution.

A sequence of flow simulations, using several turbulence models from the classical k-e, through k-wSST, to the recently introduced Scale Adaptive Simulation method, has been performed in search of the best method. The latter, described in [12] has turned out to provide remarkably better fit to the experimental data then the alternative.

2. Description of the Object of Studies and of the Method for Computation Area Geometry Determination

A fragment of the Grodzisko crosscut located at the Sobieski coal mine was selected. It consisted of two apparently straight sections connected with a bend (see Figure 1). The total length of the analysed section was 132 m. The section was almost horizontal. The floor was covered with mildly waved mud and some water pools. The gallery had an arched support with an initial cross-section of 13 m^2. It was formed several years ago, and the pressure of the surrounding rocks introduced irregular deformations of the initial shape [13], resulting in approx. 12% variability of the cross-sectional area (see Figure 2). The gallery was free from obstacles except for two pipelines and rails on the floor. This object can be considered as being representative of Polish collieries, and its location allowed for several hours of measurement sessions without interfering with production activities.

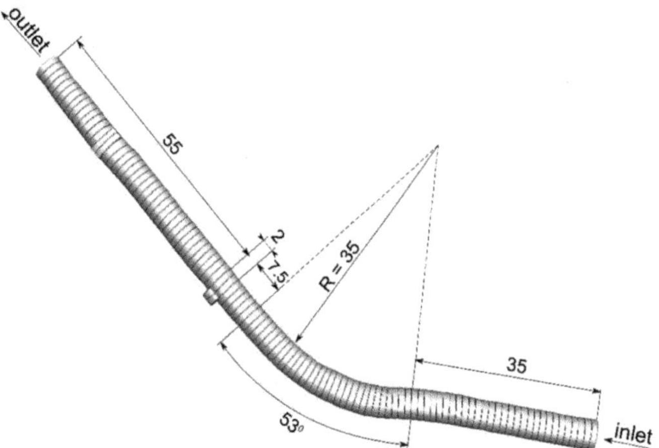

Figure 1. Numerical model geometry dimensioning—an 'outside' view.

Figure 2. 'Inside' view of a mine gallery section before a bend–point cloud.

2.1. Determination of the Object Geometry by Means of the Laser Scanning Technique

To obtain a complete spatial model of the workings, the entire section of gallery was divided into 11 scanning cross-sections, which provided 22 measuring positions. The first scanning cross-section was set approx. 23.0 m before the inlet to the bend, while the last scanning section was approx. 56.0 m behind the outlet from the bend (Figure 3).

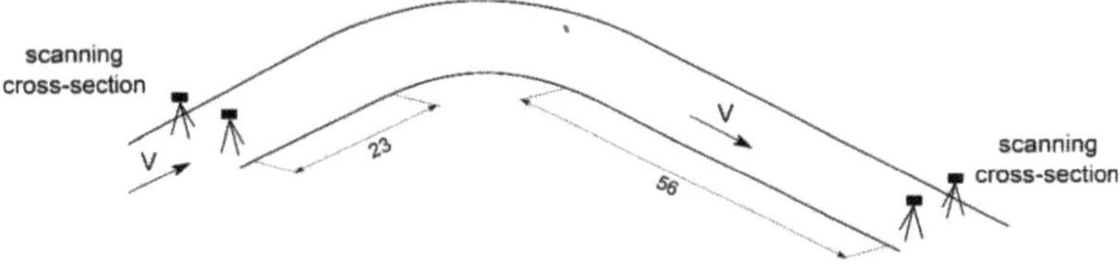

Figure 3. Diagram of arrangement of scanning cross-sections.

The processing of sets acquired by the laser scanner can be referred to as pre-processing. The measurement data processing required preparation of the data for further processing. At this stage, the most important processes should include the orientation of a point cloud through the combination of a few point clouds (obtained from individual measuring positions) into one data set [9]. The point cloud was then filtered, which consisted in cleaning and removing measurement noise and discontinuities [14]. The laser scanning of the Grodzisko crosscut geometry, the obtained point cloud pre-processing and the reduction in the number of points resulted in the obtaining of a very good digital reproduction of the entire space of the scanned gallery fragment, consisting of 152 million points (Figure 2) [8].

This cloud, however, was far too complex to be directly used by available comoutational mesh generation software. For computational mesh generation, it is recommended to simplify the geometry, neglecting details unnecessary for the flow simulation. Further simplification of boundary geometry could not be conducted automatically without losing important details such as ribs. Therefore, the shape of working had to be manually redrawn using the point cloud as a template. As a result, the geometry of the geometrical model was obtained with accurate reproduction of the actual object shape (Figure 4).

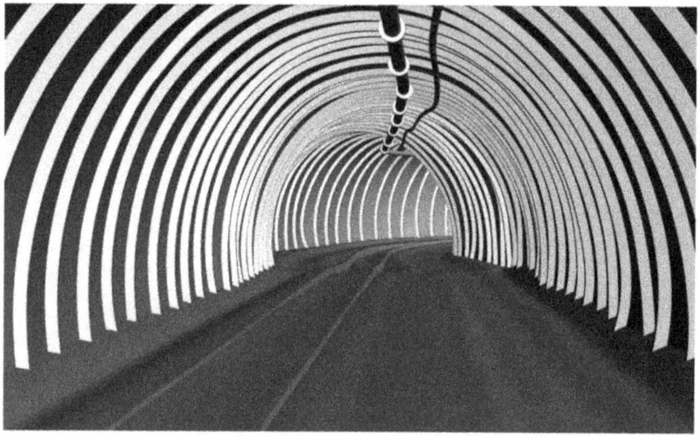

Figure 4. Geometry of the numerical mode—an 'inside' view of a straight section of a mine gallery before a bend.

The computation area, comprising 110 arches of ŁP support, a hydraulic pipe, a pipeline and rails of a mine railway, consisted of (Figure 1):

- Straight section before the bend, 35 m long;
- Section of the bend, with a turning radius of 35 m, and an angle of rotation of 53°;
- Straight section behind the bend, 64.5 m long, taking into account the existence of a stable, 2.0 m long at a distance of 7.5 m behind the outlet from the bend.

2.2. Generation of the Numerical Grid

Application of the finite volume method required division of the computational domain into a set of interconnected volumes, which is referred to as generation of the numerical grid. The geometries of the area and the arrangement of flow directions, which are difficult to predict, in the entire numerical model justified application of tetrahedral or polyhedral volumes of a uniform size in every direction. The grid density increased towards the boundaries due to application of the size function, which allowed for control of the computational grid size close to a selected point, edge or surface. Numerical models were initially meshed with tetrahedral non-structural grids, which were then converted to a polyhedral grid (Figure 5). Thanks to the conversion, the number of cells decreased, which significantly shortened the time of the numerical calculations.

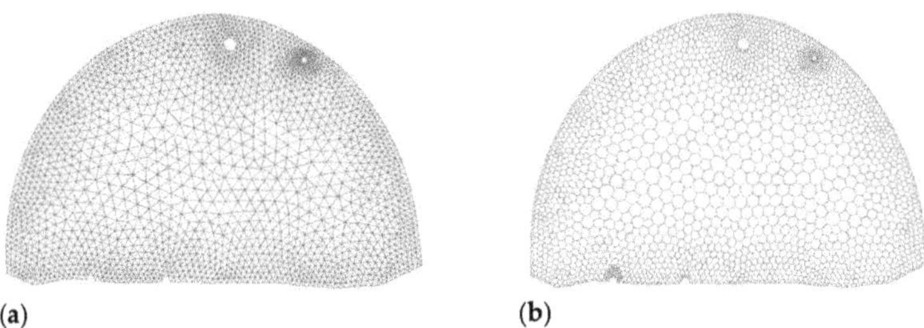

(a)   **(b)**

Figure 5. Computational grid at the inlet cross-section: (**a**) tetrahedral grid, (**b**) after conversion to a polyhedral grid.

A dimensionless parameter of length y^+ for the computational area falls within a range of $10 \leq y^+ \leq 440$. The number of cells within a range of $10 \leq y^+ \leq 30$ is small, which allows for acceptance of the grid quality in terms of a dimensionless parameter of length y^+.

The skewness factor for the numerical grid, with the parameters also presented, is shown in Table 1 in the form of a percentage share of grid cells in a specific range of skewness factor.

Table 1. Percentage share of numerical grid cells for skewness factor ranges.

Skewness Factor	Percentage Share (%)	Cell Quality
1	0.00	degenerate
0.9 < 1	0.00	bad
0.75–0.9	0.00	poor
0.5–0.75	10.17	fair
0.2–0.5	52.37	good
0 > 0.25	29.24	excellent
0	8.22	equilateral

3. Measurements of the Air Velocity Profile in the Mine Workings

In situ measurements provided data for setting the flow boundary conditions and verification of simulation results. For this purpose, three cross-sections of the gallery were selected. Specific mine conditions allowed for the application of an in-house designed and manufactured the Multipoint System of Flow Velocity Measurements (known as the SWPPP system) [15].

Multipoint measurements have been carried out by several researchers; for example, simultaneous methane concentration sampling by Wala et al. [2] or sequential velocity measurements by Martikainen et al. [16]. In most cases, however, this measurement was sequential, conducted with a single instrument repositioned from one point to another. This method provides proper information on the velocity field if the flow is sufficiently steady and the point averaging periods are sufficiently long. Simultaneous measurement eliminates this source of uncertainty and enables the obtaining of the data in a shorter time.

Measurements with the use of the SWPPP system consist of placing the appropriate number of vane anemometers in the mine drift cross-section and then simultaneously measuring the air flow velocity. To calculate the volumetric flow rate, the cross section of a mine drift, where the measurement takes places, needs to be integrated. In reality, it is impossible to measure the whole velocity field in cross-section. That is why the flow velocity is measured in a finite number of points on this surface and then, based on the results, the velocity distribution is estimated using the linear triangulation. This method consists of dividing the cross-section area into triangular areas so that the vertices of the triangles are points at which the flow velocity is known from the measurements.

To place the anemometric sensors in the selected cross-section points of the gallery, it was necessary to use an appropriate bearing structure consisting of four vertical beams, to which the anemometers were fixed, and one or two horizontal stiffening beams. The vane anemometers were fixed on vertical beams. The number of sensors could be adapted to the working's cross-section, with this being a compromise between ensuring the required precision of distribution determination and achieving the smallest possible interference into the flow itself.

The properties of the velocity sensors are listed below:

- Measuring range of flow velocities: $\pm$(0.2–20 m/s);
- Measuring error of flow velocities: $\pm$(0.5% rdg + 0.02 m/s);
- Measurement resolution: 0.01 m/s;
- Frequency of measurements: 1 Hz.

Before measurements, each of the 18 methane anemometers used in the SWPPP system [17] were adjusted in the aerodynamic tunnel of the Calibration Laboratory for Ventilation Measuring Instruments of the IMG-PAN, which has received accreditation from the Polish Centre for Accreditation.

Measurements by means of the SWPPP system were performed at three measuring cross-sections, presented in Figure 6. After recording flow conditions at Position I, the set of 16 anemometers was moved to Position II and then to Position III. Simultaneously, laser scanning was performed in such a manner as to exclude interference with the flow measurements. Cross-section II was to provide the initial boundary conditions, and Positions III and I were used for verification. Vane anemometers measure the velocity component parallel to their axis; therefore, all positions were in places where a dominant velocity component parallel to the gallery axis may be expected.

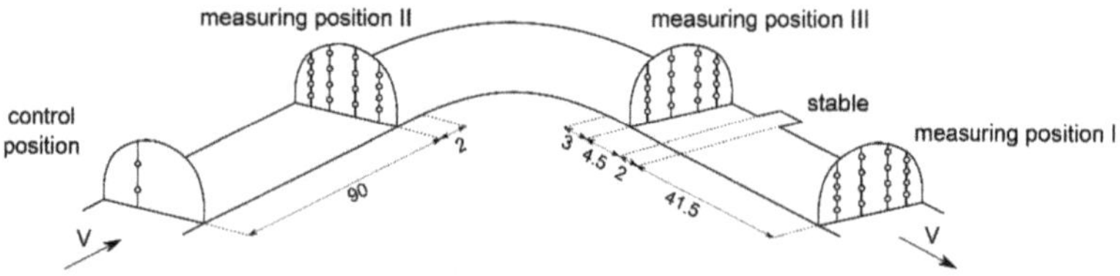

Figure 6. Diagram of arrangement of measuring cross-sections in the mine workings.

The arrangement of measuring points in the cross-sections of measuring positions, together with their coordinates, is presented in Figure 7. In addition, each measuring point was marked with a number corresponding to the anemometer number.

Two anemometers on a vertical line, 90 m before Position II, continuously recorded velocity to detect variations of the flow rate, if any. Their readings were used to trim the results from Positions I to III to a uniform flow rate.

For each position, 16 values of flow velocity were recorded for periods of approx. 15 min. After excluding periods of flow disturbances, the values of the velocities were averaged for periods that were approx. five minutes long. The averaged flow rate for Position I was 18.05 m^3/s, and for Position II, it was 18.12 m^3/s. For Position III, the flow rate rose to 20.55 m^3/s, which was confirmed by readings of the control anemometers at the inlet. Therefore, the velocities for Position III were scaled down by a factor of 0.88.

measuring position II

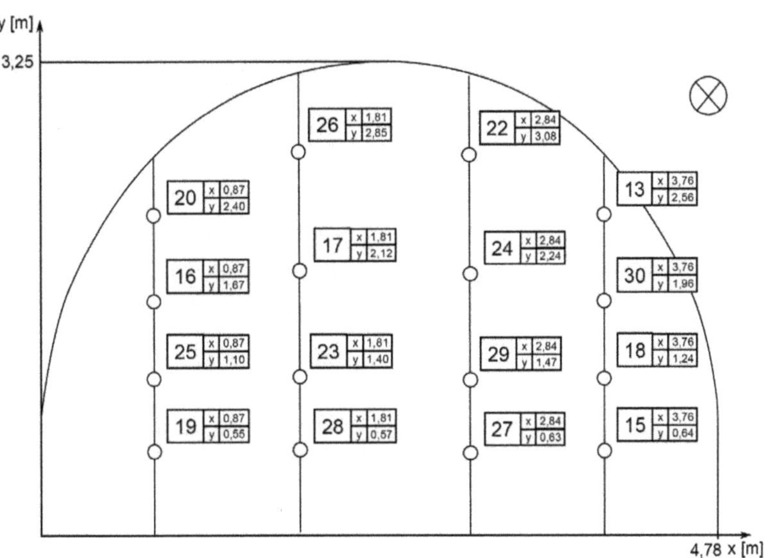

measuring position III

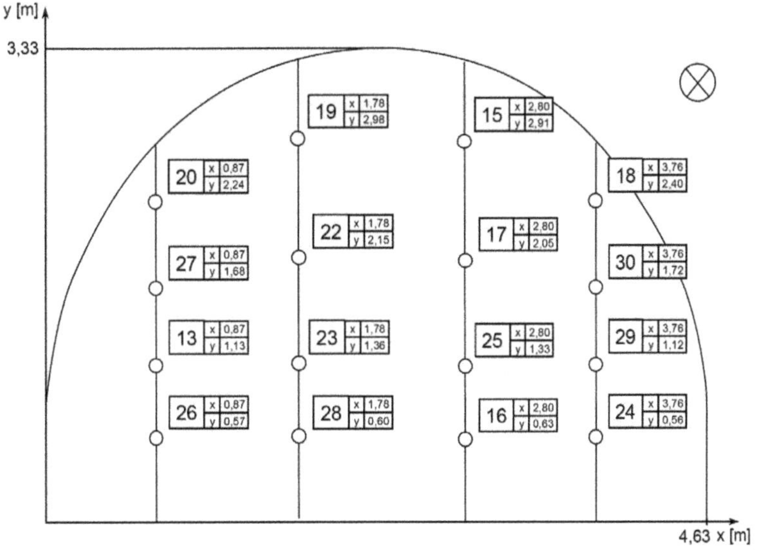

Figure 7. *Cont.*

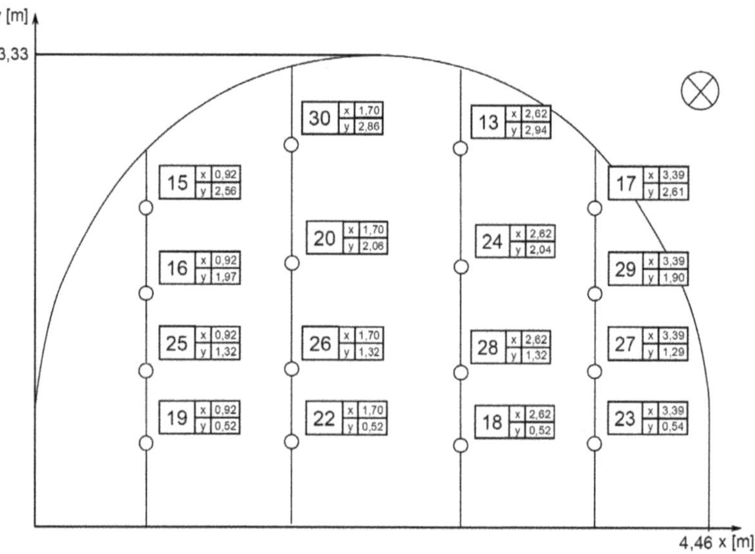

Figure 7. Diagram of arrangement of anemometers on measuring positions.

4. Description of the Numerical Modelling Method

The flow in the gallery was turbulent (Re $\approx 4 \times 10^5$). Additional flow irregularities could be generated by random geometric distortions. In near wall regions, quasi periodic ribs of the arched roof support generated specific non-stationary vortex structures [18]. Such a level of complication justified the consideration of unsteady flow models.

4.1. Turbulence Model Selection

For modelling the finite volume method, ANSYS Fluent software [19] was applied. In particular, the SAS turbulent model is a semi-empirical model based on equations of transport of turbulence kinetic energy (k) and kinetic energy dissipation (ω). The model assumes the addition of an additional source, Q_{SAS}, in the kinetic energy dissipation rate equation:

$$\frac{\partial}{\partial t}(\rho k) + \frac{\partial}{\partial x_j}(\rho k u_i) = P_k - \beta_k \rho \omega k \frac{\partial}{\partial x_j}\left[(\mu + \sigma_k \mu_T)\frac{\partial k}{x_j}\right] \tag{1}$$

$$\frac{\partial}{\partial t}(\rho \omega) + \frac{\partial}{\partial x_j}(\rho \omega u_j) = \frac{\partial}{\partial x_j}\left[(\mu + \sigma_\omega \mu_T)\frac{\partial \omega}{x_j}\right] + \frac{\gamma \omega}{k} P_k - \beta_\omega \rho \omega^2 \\ +(1 - F_1)\sigma_{\omega 2}\frac{2\rho}{\omega}\frac{\partial k}{\partial x_i}\frac{\partial \omega}{\partial x_i} + Q_{SAS} \tag{2}$$

An additional source, according to Menter and Egorov [12], can be presented in the form:

$$Q_{SAS} = \left[\frac{1}{2}\rho \eta 2\kappa \left(\frac{l_s}{l_m}\right)^2 D_{ik}D_{ki} - 2C_s\frac{\rho k}{\sigma_s}\max\left\{\frac{1}{\omega^2}\frac{\partial \omega}{\partial x_i}\frac{\partial \omega}{\partial x_i}, \frac{1}{K^2}\frac{\partial k}{\partial x_i}\frac{\partial k}{\partial x_i}\right\}\right] \tag{3}$$

where:

η_2, σ_s, C_s—model constant parameters;
κ—von Karman constant.

The constant parameters are:

$$\eta_2 = 3.51 \quad \sigma_s = \tfrac{2}{3} \quad C_s = 2 \quad \kappa = 0.41$$

The characteristic length scale of the turbulence model, l_s, is related to parameters $k - \omega$:

$$l_s = \frac{\sqrt{k}}{\sqrt[4]{C_\mu}\,\omega} \tag{4}$$

The von Karman length scale in the one-dimensional model is defined as:

$$l_m = \kappa \frac{\partial u}{\partial y}\left(\frac{\partial^2 u}{\partial y^2}\right)^{-1} \tag{5}$$

While, in the three-dimensional models, it is defined by the formula:

$$l_m = \kappa \frac{\sqrt{\tfrac{1}{2}D_{jk}D_{kj}}}{\sqrt{\sum_{i=1}^{3}\left(\frac{\partial^2 u_i}{\partial x_j \partial x_j}\right)^2}} = \kappa \sqrt{\frac{D_{jk}D_{kj}}{2\sum_{i=1}^{3}\left(\frac{\partial^2 u_i}{\partial x_j \partial x_j}\right)^2}} \tag{6}$$

The turbulent viscosity is determined from the dependence containing the first mixing function:

$$\mu_T = \frac{a_1 k}{\mathrm{Max}\left\{a_1\omega,\ F_1(k,\omega)\sqrt{\tfrac{1}{2}D_{ik}D_{ki}}\right\}} \tag{7}$$

The information provided by the von Karman length scale allows the SAS model to adapt dynamically to resolved structures in a URANS simulation, which results in the fact that in unstable flow areas, it behaves similarly to the LES computational model. At the same time, in stable flow areas, the model ensures the RANS simulation using the *k-ω* SST turbulence model. This model provides a chance to capture interesting flow phenomena with reasonably limited computational effort. Final selection of the turbulence model was preceded by a sequence of case studies with the application of several turbulence models. For initial calculations, the *k-e* model was applied. Simulations were then continued with the *k-ω* SST model, providing a better agreement with field measurements. Finally, the SAS model was applied, providing the best fit to the experimental results.

4.2. Determination of Boundary Conditions

At the computational domain inlet, a velocity profile based on in-situ measurements was defined. Actual measurements provided the velocity in 16 points of the cross-section, a much smaller number than the computational nodes of the inlet. Due to their metrological properties, vane anemometers did not provide data on turbulence. This is why the inlet profile of velocity and turbulence was generated by auxiliary calculations. The inlet rib-wall shape was extruded into a 30 m long gallery. At the inlet of this gallery, an actual measured profile slightly smoothed by a linear interpolation was applied. A uniform 10% turbulence and actual hydraulic diameter were assumed. The intensity value was taken from hot wire measurements taken in similar conditions [20,21]. The flow calculation for this auxiliary case resulted in a smooth profile of velocity and turbulence. The measured and calculated velocity values in the measurement points were sufficiently close (see Figure 8 and Table 2).

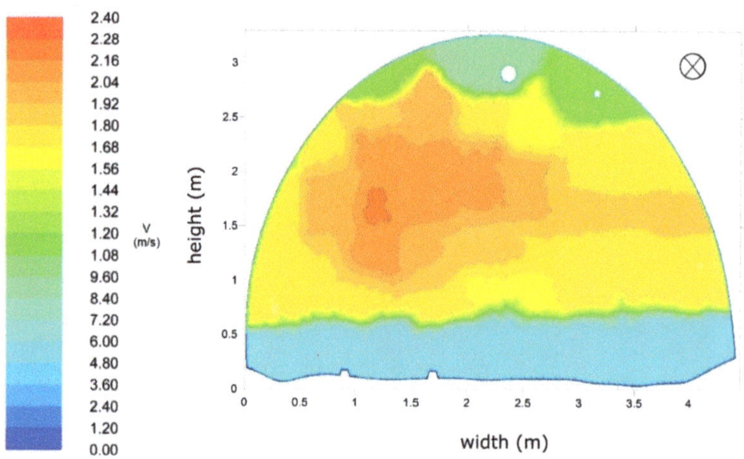

Figure 8. Velocity profile obtained by linear interpolation of the experimental data from Position II.

Table 2. Comparison of numerical computation results with the results of measurements for measuring positions together with the calculated relative error.

Sensor	Measuring Position II			Measuring Position III			Measuring Position I		
	Vav (m/s)		Relative Error (%)	Vav (m/s)		Relative Error (%)	Vav (m/s)		Relative Error (%)
	SWPPP	CFD		SWPPP	CFD		SWPPP	CFD	
13	1.18	0.86	26.77	2.12	1.95	7.74	1.35	1.23	9.38
15	1.7	1.38	19.33	1.19	1.34	12.22	1.41	1.35	4.37
16	1.76	1.77	0.79	1.62	1.41	12.97	1.83	1.61	11.87
17	1.91	1.81	5.57	1.71	1.33	22.22	1.29	1.33	3.41
18	1.93	1.61	16.51	1.14	1.13	0.61	1.68	1.67	0.85
19	1.79	1.59	11.46	1.44	1.59	10.04	1.85	1.7	8.27
20	0.91	1.07	17.81	1.8	1.81	0.34	2	1.64	17.95
22	0.99	1.07	7.75	1.85	1.51	18.61	1.97	1.75	10.97
23	2.26	2.04	9.78	1.87	1.51	19.52	1.63	1.62	0.71
24	1.8	1.78	1.33	1.21	0.98	18.53	2.04	1.56	23.68
25	1.79	1.72	4.36	1.58	1.43	9.29	2.03	1.73	14.46
26	1.15	1.43	24.51	2.09	2.05	2.03	2.21	1.86	15.73
27	1.83	1.35	26.41	2.14	1.96	8.28	1.86	1.69	9.41
28	1.85	1.36	26.19	1.96	1.85	5.87	2.06	1.81	11.73
29	2.14	1.88	12.05	1.31	1.11	14.9	1.79	1.57	12.3
30	1.83	1.34	26.93	1.36	1.2	11.71	1.39	1.38	0.66
mean relative error (%)			14.85			10.93			9.73

The boundary condition at the inlet to the model was defined as the velocity inlet, in which the velocity profiles were set (velocity profile from measurements, Figure 8). Using the velocity inlet boundary condition, the total pressure was not constant, but was matched to ensure a specific velocity distribution.

The outlet has been defined as outflow, corresponding to the outflow model in which it does not define the velocity nor the pressure conditions. The floor, arches, rails and pipelines were defined as wall surfaces. Inequalities of the floor in the model were treated as roughness with a height of 0.05 m, and in the case of arches, rails and pipelines, a roughness height of 0.001 m was chosen.

For the SAS turbulence model, a vortex generator provided more realistic flow conditions at the inlet. In this cases, unsteady flow calculations with a time step of 0.01 s were applied. Solutions converge after less than 20 iterations per time step. Residuals were below 1×10^{-5} for continuity and much smaller for other variables. Recommended methods for the SAS model solutions, including second order pressure and bounded central differencing for transient formulation, have been applied. For each model, the calculations were performed until stabilization of the time averaged values. Averaging was then restarted, and the new average was used for comparisons. The time of the introductory calculations was equal to a few multiples of the flow passage time.

4.3. Grid Sensitivity Study

For the grid sensitivity study, a shorter section of the gallery, containing the first 32 m plus a 17 m long part-length of the bend, was considered. This section had properties that were representative of the whole domain, but was considerably shorter, which facilitated the denser grid calculations. The coarser grid model consisted of 9.7 million tetrahedral cells, and after conversion to a polyhedral grid, the number of cells went down to 2.1 million. The second model consisted of 13 million tetrahedral cells; after conversion, there were 2.8 million polyhedral cells. The third grid model consisted of 22.2 million tetrahedral cells, and after conversion, there were 4.1 million polyhedral cells. The velocity components V_x, V_y and V_z at the measuring line at height $z = 1.5$ m were used for the comparison of solutions. The measuring line was situated 32 m behind the inlet to the numerical model, at the place of installation of measuring Position II (Figure 9).

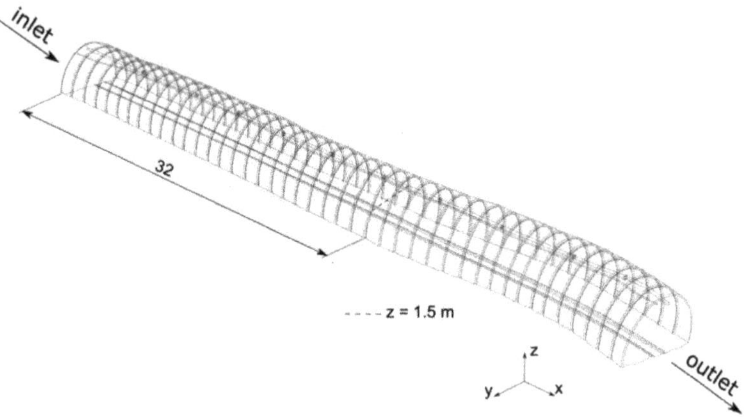

Figure 9. Location of measuring line.

An example of the velocity comparison is shown in Figure 10 The low differences of values for the 4.1 million and 2.8 million cell models justify the selection of the second model's grid density.

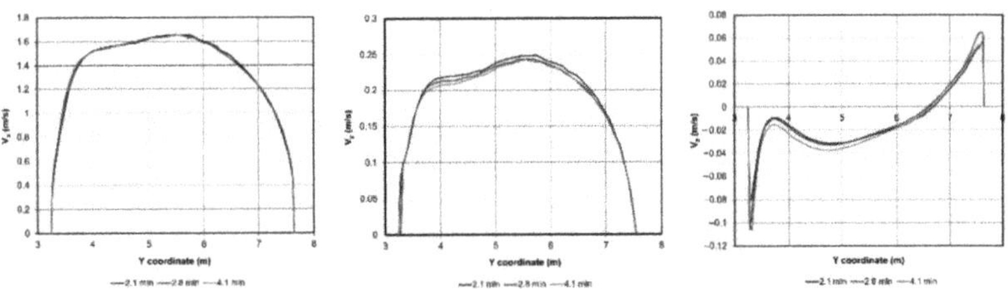

Figure 10. Comparison of velocity vector components for three grid sizes.

5. Comparison of Numerical Simulation Results with Results of Velocity Profile Shape Measurements

The obtained results of numerical computations and the results of measurements in the mine gallery are presented in the form of velocity profile contours for individual measuring positions. In addition, the values of average velocities (obtained from measurements and CFD computations) are presented in a tabular form for each measuring point (corresponding to the anemometer position in the measuring cross-section) for individual measuring positions.

To present a quantitative assessment of the precision measure of the numerical simulations, a decision was made to use a relative error measure [22]:

$$\delta = \frac{\Delta x}{x} \cdot 100\% = \frac{|x - x_0|}{x} \cdot 100\% \tag{8}$$

where:

x—measured quantity;
x_0—quantity computed numerically.

An average relative error for the entire measuring cross-section is calculated as a sum of relative errors at individual measuring points divided by the number of measuring points:

$$\delta_{sr} = \frac{1}{n}\sum_{i=1}^{n} \frac{|x - x_0|}{x} \cdot 100\% \tag{9}$$

where:

n—number of measuring points in the cross-section.

The results of numerical computations show a high qualitative and quantitative convergence as compared with the measurement results for all three measuring cross-sections.

The recorded values of average air flow velocities at measuring Position II present a developed velocity profile, which is confirmed by the velocity profile obtained by means of numerical computations (Figure 11). The mean relative error, being a comparison of measurement results with numerical computations, for measuring Position II is 14.85% (Table 2).

For measuring Position III, situated behind the bend, the air flow due to the action of inertial force sticks to the outer wall of the bend, resulting in the origination of vortices and flow recirculation zones at the opposite wall (Figure 11). Because of this, in both the measurement results and the numerical computation results, it is possible to observe the location of the highest velocities, situated approx. 1 m from the left wall. The obtained mean value of relative error for measuring Position III is 10.93% (Table 2).

In the case of Position I, the velocity contours obtained from numerical computations are made in the cross-section plane situated 56.0 m behind the bend outlet. The analysis of the measurement results shows a reshaping of the velocity profile, but the effect of the bend in the form of a slight asymmetry is still visible. The largest velocity range was recorded in the central part of the gallery cross-section (Figure 11). The results of numerical simulation reflect the shape of the velocity profile, which is also confirmed by a mean relative error equal to 9.73% (Table 2).

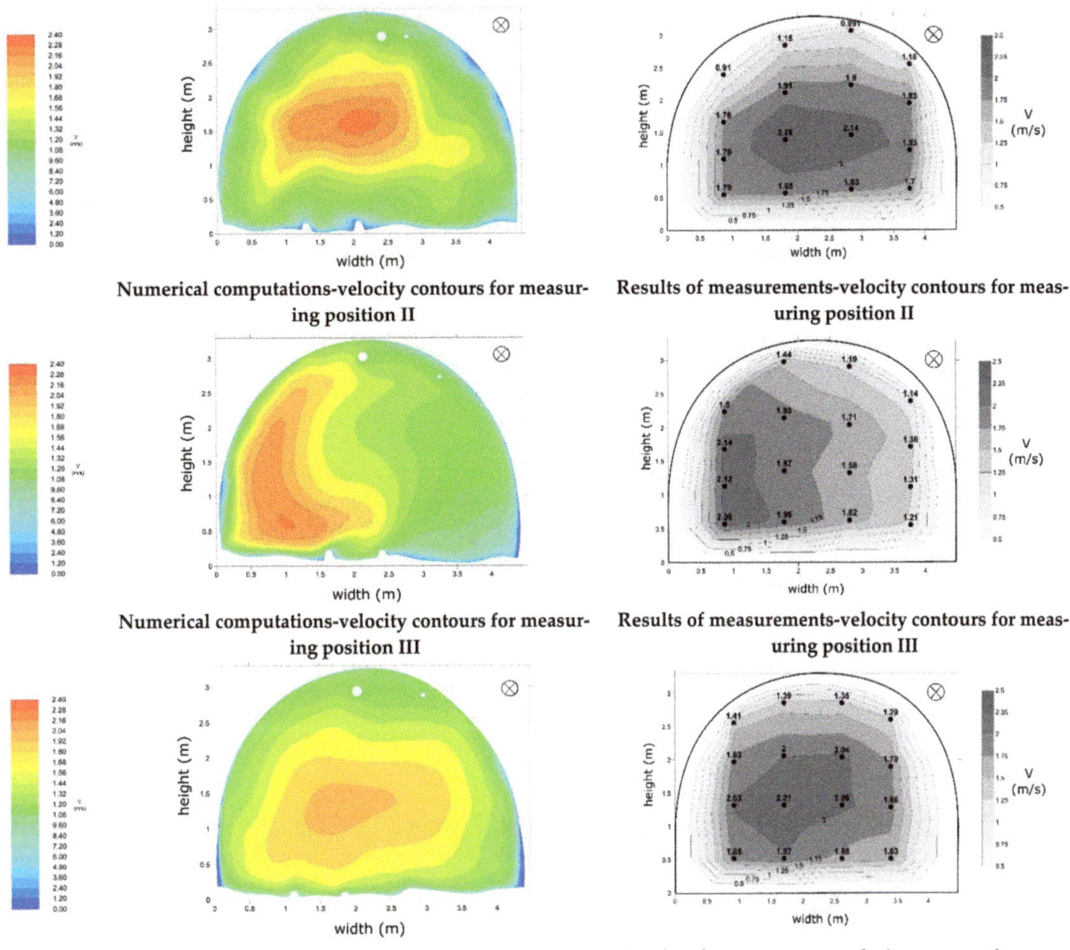

Numerical computations-velocity contours for measuring position II

Results of measurements-velocity contours for measuring position II

Numerical computations-velocity contours for measuring position III

Results of measurements-velocity contours for measuring position III

Numerical computations-velocity contours for measuring position I

Results of measurements-velocity contours for measuring position I

Figure 11. Comparison of numerical computations with the results of measurements for three measuring positions.

6. Conclusions and Discussion

The paper presents a broad range of performed work aimed at the numerical modelling of turbulent air flow in a section of a mine gallery. This comprised air flow velocity measurements in the mine gallery and the use of modern measuring techniques to accurately represent the shape of the gallery. All of the undertaken actions led to the most precise possible acquisition of the information that was necessary to carry out numerical simulations in a mine environment.

Measuring data acquired from the Multi-Point Velocity Field Measurement System (SWPPP), allowing simultaneous measurement of flow velocity at sixteen measuring points of a mine gallery cross-section, were used to determine boundary conditions and to verify simulations. The database was acquired in this way, containing results of instantaneous measurements and of average air flow velocities at individual measuring points for three measuring positions, as well as velocity profiles obtained through estimation.

In addition, to design an accurate representation of the shape of the mine gallery in which the flow velocity measurements were carried out, a decision was made to use the laser scanning technique. Based on the performed measurements of the shape of the mine gallery, a geometrical model was designed, with this being a copy of the actual model, taking into consideration the existence of a hydraulic pipe and a pipeline in the vicinity of the roof, floor unevenness, and deformations of individual arches of the support.

A numerical model for the simulation of non-stationary turbulent air flow was developed based on the obtained data. The finite volume method was used in simulations. Target simulations were preceded with a grid sensitivity study. A sequence of case studies used several turbulence models, starting from the classical k-e, through k-w SST, and ending with the SAS model, which provided the best fit to the field measurements. The obtained results of numerical computations show a high qualitative and quantitative convergence as compared with measurement results. The obtained velocity profiles for all three measuring positions represent the shape of profiles obtained by means of vane anemometers. A quantitative analysis—the analysis of the results of the average air flow velocity measurements for measuring points—shows a small mean error. The highest mean error between the measurement and numerical computation results is 14.85%, while the smallest is 9.73%. Discrepancies are comparable with the velocity measurement uncertainty, estimated at approx. 10% for flows featuring such significant fluctuations.

The results presented in this paper provided a point of reference for studies carried out by J. Janus [23] on the influence of simplifications of geometrical representation of the computational area on the precision of simulation results. In particular, it turned out that the impact of neglecting the presence of pipelines, floor unevenness and cross-section fluctuations causes a dozen or so percent deterioration of the accuracy of the obtained results. Determination of the velocity profile at the inlet is more important for simulation accuracy. This issue will be the subject of further analyses.

A 3D scanner was used to generate a cloud of points determining the shape of gallery walls. Hand-held scanners may be used for this, which substantially accelerate and simplify the scanning. However, the problem of point cloud conversion into a shape that is useful for computational grid generators remains. A computational grid with centimeter sizes corresponds to the flow conditions in galleries. The data from the point cloud should be filtered, removing details that are insignificant for the flow modelling. So far, the presence of support ribs has practically made automation of this process impossible. This issue requires further work, perhaps with the use of artificial intelligence to distinguish the ribs.

Data on a single velocity component at a few dozen points were obtained in field measurements. These provide a rough outline of velocity profiles. They do not, however, provide sufficient information about fluctuation components and flow phenomena in the vicinity of walls. These issues are the subject of a separate study using hot wire probes for the velocity vectors and fluctuation component measurements [24]. Implementation of their results may further improve the quality of numerical modelling.

The point cloud data may also be used for building scale models with additive manufacturing technologies (3D printing). These models may be used for laboratory experiments. In laboratory conditions, far more informative PIV velocity field measurement methods may be applied [25].

Author Contributions: Conceptualization, J.J.; Methodology, J.J. and J.K.; Software, J.J.; Validation, J.J. and J.K.; Formal Analysis, J.J. and J.K.; Investigation, J.J.; Resources, J.J.; Data Curation, J.J.; Writing—Original Draft Preparation, J.J.; Writing—Review and Editing, J.J. and J.K.; Visualization, J.J.; Supervision, J.J. and J.K.; Project Administration, J.J.; Funding Acquisition, J.J. Both authors have read and agreed to the published version of the manuscript.

Funding: This paper is financed from the statutory funds of the Strata Mechanics Research Institute of the Polish Sciences Academy.

Institutional Review Board Statement: Not applicable.

Informed Consent Statement: Not applicable.

Data Availability Statement: Not applicable.

Acknowledgments: This paper presents results of the statutory research of the Strata Mechanics Research Institute of the Polish Sciences Academy in years 2017–2018.

Conflicts of Interest: The authors declare no conflict of interest.

References

1. De Souza, A. *How to—Understand Computational Fluid Dynamics Jargon*; NAFEMS Ltd.: Hamilton, UK, 2005.
2. Wala, M.A.; Vytla, S.; Taylor, C.D.; Huang, G. Mine face ventilation: A comparison of CFD results against benchmark experiments for the CFD code validation. *Min. Eng.* **2007**, *59*, 49–55.
3. Ren, T.; Balusu, R. The Use of CFD modelling as a tool for solving mining health and safety problems. In Proceedings of the 10th Underground Coal Operators' Conference, Sydney, Australia, 23–24 November 2010.
4. Wang, Z.; Ren, T.; Ma, L.; Zhang, J. Investigations of Ventilation Airflow Characteristics on a Longwall Face—A Computational Approach. *Energies* **2018**, *11*, 1564. [CrossRef]
5. Luo, Y.; Zhao, Y. Field and experimental research on airflow velocity boundary layer in coal mine roadway. *Arch. Min. Sci.* **2020**, *65*, 255–270. [CrossRef]
6. Branny, M.; Karch, M.; Wodziak, W.; Jaszczur, M.; Nowak, R.; Szmyd, J. An experimental validation of a turbulence model for air flow in a mining chamber. *J. Phys. Conf. Ser.* **2014**, *530*, 012029. [CrossRef]
7. Wierzbiński, K. Wpływ geometrii chodnika wentylacyjnego i sposobu jego likwidacji na rozkład stężenia metanu w rejonie wylotu ze ściany przewietrzane sposobem U w świetle obliczeń numerycznych CFD. *Zesz. Nauk. Inst. Gospod. Surowcami Miner. Energią Pol. Akad. Nauk* **2016**, *94*, 217–228.
8. Janus, J. Construction the numerical models geometry by using terrestrial laser scanning. In *Selected Issues Related to Mining and Clean Coal Technology*; AGH: Kraków, Poland, 2016; pp. 235–242.
9. Janus, J. Assessment of the possibilities of using laser scanning for numerical models constructions. *Trans. Strat. Mech. Res. Inst.* **2016**, *18*, 27–34.
10. Warneke, J.; Dwyer, J.G.; Orr, T. Use of a 3-D scanning laser to quantify drift geometry and overbreak due to blast damage in underground manned entries. Presented at the 1st Canada—U.S. Rock Mechanics Symposium, Vancouver, BC, Canada, 27–31 May 2007.
11. Bouchiba, H.; Santoso, S.; Deschaud, J.-E.; Rocha-Da-Silva, L.; Goulette, F.; Coupez, T. Computational fluid dynamics on 3D point set surfaces. *J. Comput. Phys. X* **2020**, *7*, 100069. [CrossRef]
12. Menter, F.R.; Egorov, Y. A scale adaptative simulation model using two equations models. In Proceedings of the 43 AIAA Aerospace Sciences Meeting and Exhibit, Reno, NV, USA, 10–13 January 2005.
13. Walentek, A. Analysis of the applicability of the convergence control method for gateroad design based on conducted underground investigations. *Arch. Min. Sci.* **2019**, *64*, 765–783. [CrossRef]
14. Sokoła-Szewioła, V.; Wiatr, J. Application of laser scanning method for the elaboration of digital spatial representation of the shape of underground mining excavation. *Przegląd Górniczy* **2013**, *8*, 206–211.
15. Krach, A.; Krawczyk, J.; Kruczkowski, J.; Pałka, T. Variable of the Velocity Field and Volumetric Flow Rate in Air Ways of Underground Mines. *Arch. Min. Sci.* **2006**, *1*.
16. Martikainen, A.L.; Dougherty, H.N.; Taylor, C.D.; Mazzella, A.L. Sonic anemometer airflow monitoring technique for use in underground mines. In Proceedings of the 13th U.S./North American Mine Ventilation Symposium, Sudbury, ON, Canada, 13–16 June 2010; pp. 217–224.
17. Kruczkowski, J.; Ostrogórski, P. *Metanoanemometr SOM 2303: Nowoczesne Metody Zwalczania Zagrożeń Aerologicznych w Podziemnych Wyrobiskach Górniczych*; GIG: Katowice, Poland, 2015; pp. 117–127.
18. Cardwell, N.D.; Vlachos, P.P.; Thole, K. Developing and fully developed turbulent flow in ribbed channels. *Exp. Fluids* **2011**, *50*, 1357–1371. [CrossRef]
19. Ansys Inc. *Ansys Fluent Theory Guide*; Ansys Inc.: Canonsburg, PA, USA, 2019.
20. Ligęza, P.; Poleszczyk, E.; Skotniczny, P. Method and the system of spatial measurement of velocity field of air flow in a mining heading. *Arch. Min. Sci.* **2009**, *54*, 419–440.
21. Skotniczny, P.; Ostrogórski, P. Three-dimensional air velocity distributions in the vicinity of a mine heading's sidewall. *Arch. Min. Sci.* **2018**, *63*, 335–352. [CrossRef]
22. Roszkowski, W.; Trutwin, W.; Wacławik, J. *Mine Ventilation Measurements*; Wydawnictwo "Śląsk": Katowice, Poland, 1992.
23. Janus, J. Numerical Modeling of Flow Phenomena in Mine Drifts Using the Results of Terrestrial Laser Scanning. Ph.D. Thesis, Strata Mechanics Research Institute of the Polish Sciences Academy, Kraków, Poland, 2018.
24. Skotniczny, P.; Ostrogórski, P.; Krawczyk, J.; Janus, J. Determination and acquisition of parameters which characterize the turbulent flow in vicinity of sidewall in deep coal mine. *Trans. Strat. Mech. Res. Inst.* **2018**, *21*, 61–73.
25. Kumar, A.; Arya, S.; Wedding, W.C.; Novak, T. Examination of Capture Efficacies of a Shearer Mounted Flooded Bed Dust Scrubber Using Experiments and Computational Fluid Dynamics (CFD) Modelling on a Reduced Scaled Model. In Proceedings of the 16th North American Mine Ventilation Symposium, Golden, CO, USA, 17–22 June 2017; pp. 20-1–20-8.

Article

Influence of Heavy Weight Drill Pipe Material and Drill Bit Manufacturing Errors on Stress State of Steel Blades

Oleg Bazaluk [1], Andrii Velychkovych [2], Liubomyr Ropyak [3], Mykhailo Pashechko [4], Tetiana Pryhorovska [5] and Vasyl Lozynskyi [6,*]

1   Belt and Road Initiative Institute for Chinese-European Studies (BRIICES), Guangdong University of Petrochemical Technology, Maoming 525000, China; bazaluk@ukr.net

2   Department of Construction and Civil Engineering, Ivano-Frankivsk National Technical University of Oil and Gas, 076019 Ivano-Frankivsk, Ukraine; a_velychkovych@ukr.net

3   Department of Computerized Engineering, Ivano-Frankivsk National Technical University of Oil and Gas, 076019 Ivano-Frankivsk, Ukraine; l_ropjak@ukr.net

4   Department of Fundamentals of Technology, Lublin University of Technology, 20618 Lublin, Poland; m.pashechko@pollub.pl

5   Department of Engineering and Computer Graphics, Ivano-Frankivsk National Technical University of Oil and Gas, 076019 Ivano-Frankivsk, Ukraine; tetiana.pryhorovska@nung.edu.ua

6   Department of Mining Engineering and Education, Dnipro University of Technology, 49005 Dnipro, Ukraine

*   Correspondence: Lozynskyi.v.h@nmu.one

Abstract: Drilling volumes should be increased in order to increase hydrocarbon production, but this is impossible without the usage of high-quality drilling tools made of modern structural materials. The study has to analyze the design, technological and operational methods to increase the performance of drilling tools made of various materials and has highlighted prospects of technological method applications. The scientific novelty of the study consists in the development of a new analytical model of PDC drill bit–well interaction. The developed model takes into account the drill bit manufacturing errors in the form of bit body–nipple axes misalignment on the drill bit strength. This result makes it possible to determine the permissible manufacturing errors to provide safe operation of the drill bit. It is established that there is an additional transverse force that presses the drill bit to the well wall in the rock due to manufacturing errors. It is determined that the magnitude of this clamping force can be significant. The material effect has been analyzed on additional clamping force. It is established that geometric imperfection of the drill bit causes the minimal effect for the elastic system of the pipe string, which includes a calibrator and is composed of drill pipes based on composite carbon fiber material, and the maximal effect—for steel drill pipes. Polycrystalline diamond compact (PDC) drill bit and well wall contact interaction during operation in non-standard mode is considered. Non-standard stresses are determined, and the strength of the blades is estimated for different values of drilling bit manufacturing error.

Keywords: drilling; pipe; drill bit; manufacturing error; material; model; elastic modulus

Citation: Bazaluk, O.; Velychkovych, A.; Ropyak, L.; Pashechko, M.; Pryhorovska, T.; Lozynskyi, V. Influence of Heavy Weight Drill Pipe Material and Drill Bit Manufacturing Errors on Stress State of Steel Blades. *Energies* **2021**, *14*, 4198. https:// doi.org/10.3390/en14144198

Academic Editors: Sergey Zhironkin and Dawid Szurgacz

Received: 17 June 2021
Accepted: 10 July 2021
Published: 12 July 2021

Publisher's Note: MDPI stays neutral with regard to jurisdictional claims in published maps and institutional affiliations.

1. Introduction

Nowadays, the global challenge is associated with the need to ensure sustainable development [1–4]. Despite the rapid development of green energy, oil and gas remain the main energy carriers [5,6]. In addition, coal processed by modern technologies is widely used as an energy source [7,8]. Demand for oil and gas will remain stable in the near future, and it means high oil and gas prices on the world market [9]. At the same time, the problem of providing energy to industries and the population is complicated by the fact that oil and gas deposits are deep and difficult to extract; this stipulates high requirements for materials and good quality drilling tools [10,11].

Drill bits operate under high loads, in corrosive and abrasive environments, and carbon-containing materials (steels and alloys) are the main type of material used to manufacture drilling equipment and tools. It should be noted that in addition to having an abrasive environment, rocks are heterogeneous with different physical and mechanical properties [12–15]. Crack coalescence is a significant phenomenon produced in rocks during the cutting process [16]; thus, it is necessary to provide a simulation of the crack's propagation and the effect of the in-situ stresses on the mechanical specific energy PDC cutter as shown in [17–19].

Modern smart composite materials, obtained by reinforcing the carbon matrix with fibrous carbon material (carbon-carbonic compositional materials), are used today to manufacture individual parts. These materials can potentially be used instead of drill pipes which are made out of traditional materials–aluminum alloys (duralumin), titanium alloys, and steel [20]. When drilling for oil, gas, and water and during the extraction of other minerals, rock is ruined and brought to the surface by the washing liquid because of the metal drill bit–rock interaction. The rigidity of the drill string, and hence the interaction between the drill bit and the rock, is dependent on the material of the drill pipe (the most commonly used are: steel of different strength groups, such as steel 40 KhN, aluminum alloys AD31 and duralumin D16T, titanium alloy VT1-0 as well as magnesium alloys and composite carbon materials) [21–23].

The smart drilling of wells is becoming more widespread [24,25], and increasing the depth of drilling and the construction of wells of various configurations requires new materials, structures, and advanced technologies for the manufacture of drilling tools to be used. Design, technological and operational methods are used to increase drilling tool performance [26,27]. Design methods include improving threaded joint tightness [28–30]; preventing self-unscrewing [31]; rational choice of materials by studying structural changes; conducting corrosion studies of metals [32–35], and studying the processes of destruction [36,37], including different types of coatings (if present) [38–41]; substantiation of the stress state of pipes [42]; optimization of armaments, supports and units for drill bit washing [43,44], etc. The works of [45–48] present new approaches to strength and optimization estimates according to the operational criteria of rotating bearing elements for mining machines.

Technological methods to improve drilling tool quality include a number of measures [48,49]. The most common are: increasing the accuracy of helical surfaces and surfaces of plain bearings [50–52], as well as laser processing of operational surfaces and grinding of threads [53,54]. Studies of metal structure [55–57] and diffusion processes [58–60] are necessary for the development of modern materials and hardening technologies. Coatings [60–63] occupy an important place among the methods of hardening and surfacing [64] and are used to provide high wear resistance of the operational surfaces of parts.

Various solid compounds and their mixtures are used as a coating material: oxides [65,66], nitrides, and carbides [67,68] of the composition [69–71]. The use of electrochemical chromium plating on parts with low-toxic electrolytes is promising. Operational methods for improving drilling tool performance include drill string-bit oscillation mathematical modeling [72]; rational choice of drilling modes considering forces [73–76], temperature [77–79], life cycle control of threaded connections [80], and the rational choice of drill tube material for drill string assembling [81], as well as drill tube stress control [82].

The design features and specifics of drilling tools and their contact interaction with the well wall and face cause dynamic loads. Some authors [83–86] consider the vibration protection of drilling tools as a guarantee of ensuring high technical and economic performance of well drilling. Problems with designing vibration protection devices for long structures (drill columns, pump-compressor rods) were considered in [87–89]. The modeling of contact interaction in shell-rod systems under nonmonotonic loading to estimate the damping capacity of these systems was carried out in [90–93]. The issues on the mechanical-mathematical modeling of rod relation, referring to drill string problems [94–96], are still relevant. For the clarification of rod surface—an elastic or inelastic medium interaction model is required for the safe operation of pipelines [97–99], the increase reliability and durability of drill strings, and to ensure the quality centering of casings [100,101]. Digital technologies are used to account for the key indicators of mining operations [102].

The data presented in the literature review are not enough to assess the impact of drill bit manufacturing errors and drill string rigidity composed of pipes of different materials on bit–rock force interaction. Researchers pay considerable attention to the choice of materials and hardening technologies, but do not pay enough attention to substantiate the accuracy of PDC drill bits and study the bit-rock force interaction. The results of the analysis on drill bit manufacturing errors and drill string rigidity composed of pipes of different materials are necessary for the rational design of drill bits, the development of manufacturing technology, and substantiation of technological modes of drilling.

This work aims to study the influence of drill bit manufacturing errors and drill string rigidity composed of pipes of different materials on bit–rock force interaction and the prospects of latest carbon materials for drilling tool application compared with traditional materials (aluminum, titanium alloys and carbon steel).

To achieve this goal, the following tasks were set:

— to develop a mathematical model of drill bit rock interaction taking into account manufacturing errors.
— to investigate the influence of the mechanical properties of different classes of drill pipe materials on the additional force that clamps the drill bit to the well wall and the level of stresses in the metal bit.

2. Materials and Methods

We used the bibliosemantic method and content analysis to specify the main types of errors that occur during the manufacture of PDC drill bits for the first stage of this study. The analysis of previous scientific literature research, electronic resources, as well as practical and production experience of the authors, made it possible to identify the necessary scientific data and establish relationships between them and state an unsolved problem in this field. As a result, the purpose and objectives of the study were formulated.

We formulated the main ideas and concepts of the developed analytical model of bit blade–well bore interaction, which takes into account the drill bit body–nipple axes misalignment influence on the operational strength of the blades using the methods of structural-logical and system analysis.

In order to provide the practical implementation of the intended analytical model, we used methods from the mechanics of a deformable solid to obtain an analytical result for assessing the bit blade strength.

The analytical study was conducted in two stages. In the first stage, we studied the global change in the behavior of the elastic system "drill pipes-drill bit-well wall" depending on the magnitude of the bit manufacturing errors. In the second stage of the study, we studied the drill bit–rock interaction. As a result, an analytical expression was obtained to find the maximum equivalent stresses in the drill bit blade made with manufacturing errors.

The reliability of the obtained analytical results was confirmed by the validity of the geometric-linear formulation of the problem, the strict implementation of mathematical methods tested in the literature for analytical research, and the convergence of the results of individual partial cases with known results.

In order to provide the numerical analysis of our research results, we selected the following structural materials of drill pipes: carbon steel, aluminum alloys, titanium alloy, and composite carbon fiber material. Fine-crystalline limestone was used as a rock-forming material for the well wall. It was assumed that that the elastic properties of the considered materials are determined by their Young's moduli and Poisson's ratios, and that strength is regulated by the yield strength.

The physical and mechanical properties of the materials, the geometric dimensions of the drill bit, and drill pipes are given below when considering numerical examples.

3. Results

There are a number of PDC drill bit manufacturing errors. In particular, as a result of the assembly operations of the welding blades with the body, screwing, and subsequent welding of the nipple with the bit body, there may be a radial runout of the blades and misalignment of the nipple–bit body which would worsen the operation conditions of the rock, destroying elements of the bit.

Let us determine the influence of the errors mentioned above on magnitude of the cutting force, the level of stresses, and the strength of the bit blades during rock drilling (Figure 1). Let the drill bit, which has a manufacturing error (nipple-body misalignment) of Δ (Figure 1b), operate in combination with a calibrator. The bit and the calibrator are connected by drill pipes made of different materials with a length of l.

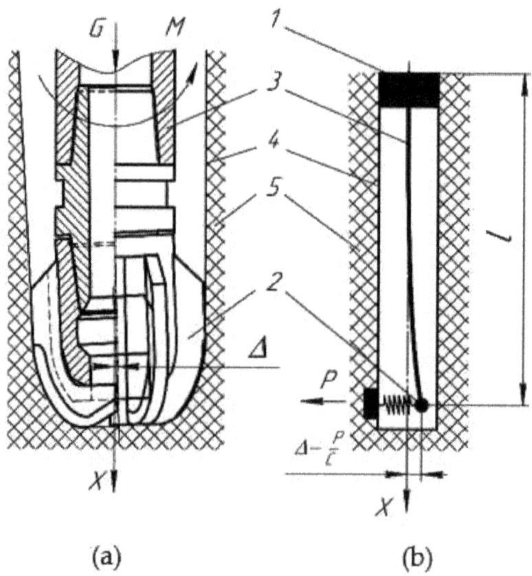

Figure 1. Interaction of the drill bit with the bottom hole of the well (**a**) and the calculation scheme for finding the force clamping the drill bit to the well wall (**b**): 1—calibrator; 2—PDC drill bit; 3—weighted drill pipe; 4—well wall; 5—rock.

To simplify the model, we assume that there is a clamp in position of the calibrator in the well. The manufacturing error will lead to a freelance kinematic load of the elastic system, as a result of which an additional transverse clamping force will appear on the bit (Figure 1a). The pliability, C, of the rock (see Equation (1)) to be contacted by the PDC bit is simulated by an elastic support with rigidity, so in comparison to it we will consider the bit absolutely rigid.

Using the principle of independence of forces for linear systems, we present the canonical equation for the method of forces:

$$\delta_{11}P + \Delta_{1p} = \Delta - \frac{1}{C}P, \tag{1}$$

where $\delta_{11} = l^3/3EJ$ and $\Delta_{1p} = 0$—transverse movements from the right end of the elastic system (released from the ligaments) caused by a unit force and an external force load, respectively; E is the modulus of elasticity of the drill pipe material; $J = 0.05D^4[1 - (d/D)^4]$ is the axial moment of inertia of the drill pipe cross-section; d, D are the inner and outer diameters of the drill pipes.

Solving Equation (1), we obtain the expression for the additional clamping force on the bit:

$$P = \frac{\Delta}{\frac{1}{3}\frac{l^3}{EJ} + \frac{1}{C}},\tag{2}$$

Let us estimate a possible range for the clamping force P depending on the nipple-body misalignment Δ. To make the estimation more concrete, we need to choose a specific size for the weighted drill pipe with an outer diameter of 203 mm and an inner diameter of 80 mm and assume that the bit is in contact with an absolutely rigid rock (we obtain the upper range of possible clamping force). The distance between the bit and the first calibrator is 1 m. We consider that the drill pipes are made of different metal alloys: steel alloy 40 KhN, aluminum alloy AD31, duralumin alloy D16, and titanium alloy VT1-0 with the following modules of elasticity: steel alloy—2.1×10^5 MPa, aluminum alloy—6.7×10^4 MPa, duralumin alloy—7.3×10^4 MPa, and titanium alloy—1.12×10^5 MPa. It should be noted that today, modern technologies used for the manufacture of structural materials allow the use of smart composite materials for the production of drilling tools obtained by reinforcing the carbon matrix with carbon fiber material (carbon-carbonic compositional materials). In particular, drill pipes made of such materials demonstrate high economic efficiency in the oil and gas sector [102–106]. Therefore, for analysis, we chose another option—drill pipe based on composite carbon fiber material (modulus of elasticity in the axial direction of such material—5.6×10^4 MPa).

Analysis of the dependences presented in Figure 2 shows that the geometric imperfection of the drill bit causes minimal effect on the elastic system of the pipe string, which includes a calibrator and is composed of drill pipes based on composite carbon fiber material, and the maximal effect—for steel drill pipes.

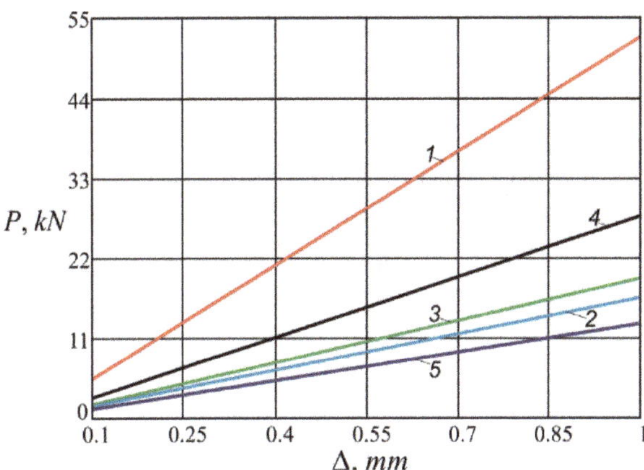

Figure 2. Abnormal transverse clamping force on the bit (upper range): 1—carbon steel; 2—aluminum alloy; 3—duralumin alloy; 4—titanium alloy; 5—composite carbon fiber material.

In the case of pipes made of light alloys of aluminum and titanium, the magnitude of the clamping force is intermediate. Now we will consider in more detail the features of the drill bit (blade)-rock interaction.

The normal mode of operation requires the bit to be perfectly made (i.e., has no nipple-body misalignment, $\Delta = 0$) and the well hole to be ideal, in that case each blade deeps into the rock for the same value of radial recess a_0 (Figure 3).

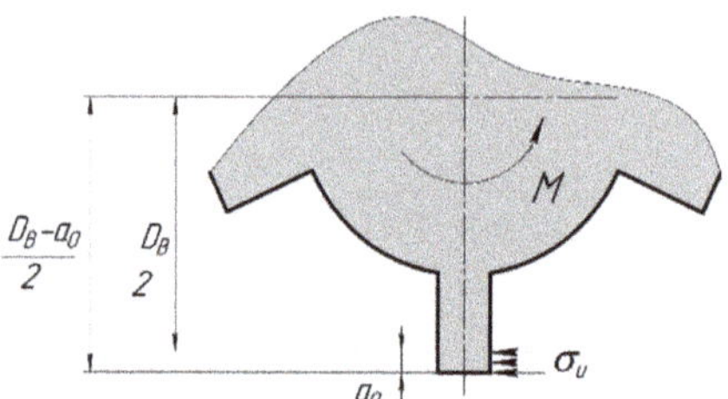

Figure 3. Scheme of blade–well wall contact interaction (normal mode of operation).

So long as a_0 is a small value, we assume that the contact stresses are evenly distributed over the blade-rock contact area. The cutting forces are distributed equally between all the blades of the bit. This way, the global equilibrium equation of the system, which refers to the longitudinal axis Ox, can be represented as follows (for a six-blade bit $n = 6$):

$$M = \sum_{i=1}^{n} \sigma_u a_0 H \frac{D_b - a_0}{2} = 3\sigma_u A_0 (D_b - a_0), \tag{3}$$

where σ_u is the rock strength, H is the blade length, D_b is the diameter of the bit, A_0 is the blade-rock contact area, M is the torque on the bit.

An abnormal mode of operation is when the bit is made with errors (i.e., has nipple-body misalignment $\Delta \neq 0$), and there is an inhomogeneous distribution of cutting forces between the blades of the bit. If $\Delta > a_0$, some blades lose contact with the rock, and heterogeneity of force distribution becomes significant and can cause a loss of drill bit strength. Figure 4 shows the two most unfavorable cases of blade-rock interaction.

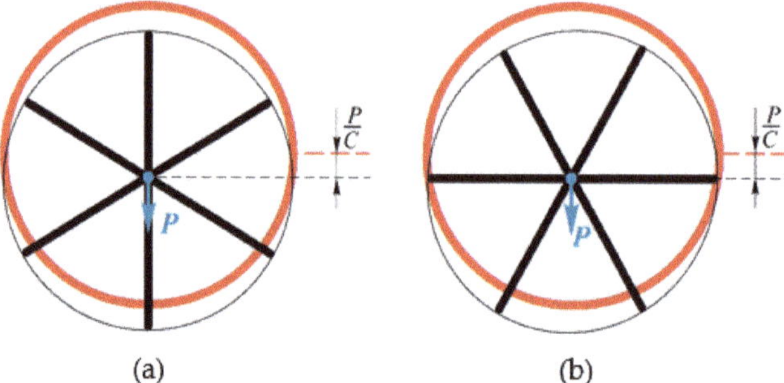

(a)   (b)

Figure 4. Contact interaction of the metal blades of the drill bit with the well wall (abnormal mode of operation): (**a**)—three blades operate; (**b**)—two blades operate.

Let us determine the additional radial recess Δ_r^i of the blade-rock interaction and radial forces P_i associated with the abnormal case.

Assuming that the clamping force P enforces the drill bit to move for $P/C = \delta$ directionally to this force action (Figure 5). Assuming the drill bit as a solid body, we neglect its deformations:

$$b_0 k_0 = b_1 k_1 = b_2 k_2 = \frac{P}{C} = \delta.$$

for an arbitrarily located blade, one can always consider a triangle similar to the triangle $b_2 s k_2$ (Figure 4) and express the radial displacements of the blades Δ_r^i using δ. In this case:

$$\Delta_r^0 = \delta; \; \Delta_r^1 = \Delta_r^2 = \delta \cos \vartheta.$$

radial P_i and tangential P_{ti} forces in the respective directions are found assuming that the reaction of the rock is proportional to the magnitude of the displacement. For the case shown in (Figure 5) we will have:

$$P_0 = \frac{1}{3}P; \ P_1 = P_2 = \frac{1}{3}P\cos\vartheta, \ P_{t0} = 0, \ P_{t1} = P_{t2} = \frac{1}{3}P\sin\vartheta.$$

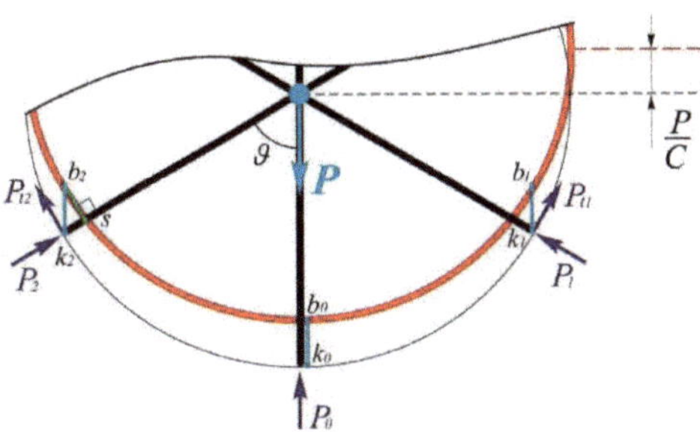

Figure 5. Calculation scheme for determining the additional radial recess of the rock Δ_r^i and the radial forces P_i.

This way, the global equilibrium equation of the system, written relative to the longitudinal axis of the bit, will take the following form (Figure 6):

$$M = \sum_{i=1}^{n*}\left[\sigma_u(a_0 + \Delta_r^i)H\frac{D_b - (a_0 + \Delta_r^i)}{2} + fP_i\frac{D_b}{2}\right], \tag{4}$$

where $n*$—number of blades contacting with the rock; f—the blade-rock friction coefficient.

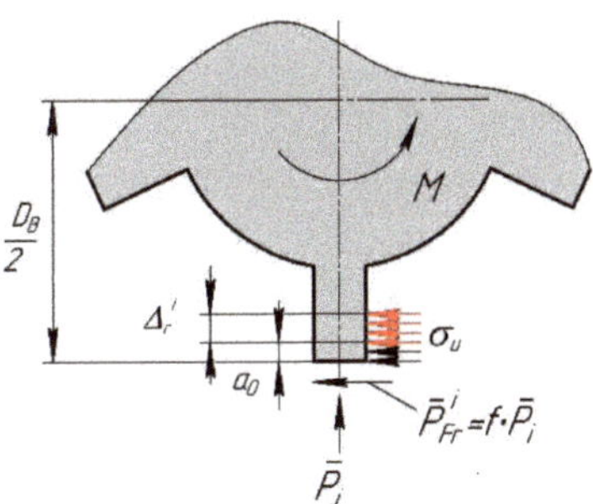

Figure 6. Scheme of the drill bit blade loading for abnormal mode of operation (black shows the standard loads, and red—additional, which arise due to an abnormal operation).

When performing numerical calculations, the stiffness of the model elastic support (see Figure 2) will be determined using the known values of the coefficient of subgrade resistance k. For this case:

$$C = kn \cdot hH, \tag{5}$$

where h is the width of the blade.

Let us determine the maximum stresses that can occur in the material of the blade. We neglect the shear stresses due to the predominant influence of bending strength and axial load. Then for the normal case (Figure 7a):

$$\begin{cases} \sigma_{max}^+ \\ \sigma_{max}^- \end{cases} = \pm\frac{M_z}{W_z} = \pm\frac{(\sigma_u a_0 H)\left(L - \frac{a_0}{2}\right)}{\frac{Hh^2}{6}} = \pm\frac{6}{h^2}\sigma_u a_0\left(L - \frac{a_0}{2}\right), \tag{6}$$

where σ_{max}^{+} and σ_{max}^{-} are the max tensile and compressive stresses of the blade; M_z is the max bending moment; W_z is the axial moment of resistance; and L and h are the height and width of the blade. $F - \sigma \, \Delta \cdot$.

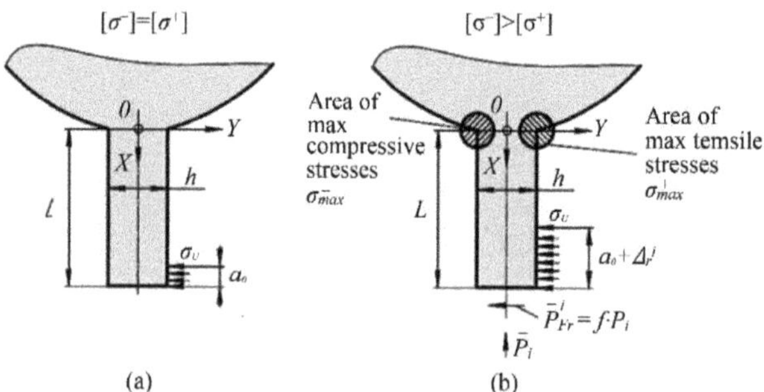

Figure 7. Calculation scheme for assessing the strength of the blade: (**a**) is normal mode of operation, (**b**) is abnormal mode of operation.

The vertical blade was the most loaded one for the abnormal mode of operation (Figure 4a). Figure 7b presents the calculation scheme for PDC drill bit operation in the non-standard mode. The max tensile and compressive stresses in the blade material are:

$$
\begin{cases} \sigma_{max}^{+} \\ \sigma_{max}^{-} \end{cases} = -\frac{N_x}{A_{bl}} \pm \frac{M_z}{W_z} = -\frac{P_i}{Hh} \pm \frac{\left(\sigma_u(a_0 + \Delta_r^i)H\right)\left(L - \frac{a_0 + \Delta_r^i}{2}\right) + fP_iL}{\frac{Hh^2}{6}}. \tag{7}
$$

where A_{bl} is the cross-sectional area of the blade; N_x the axial force.

The max stresses (modulus):

$$
\left|\sigma_{max}^{-}\right| = \frac{1}{A_{bl}}\left[P_i + \frac{6}{h}\left(\left(\sigma_u(a_0 + \Delta_r^i)H\right)\left(L - \frac{a_0 + \Delta_r^i}{2}\right) + fP_iL\right)\right]. \tag{8}
$$

If the blade is made of plastic material, it is necessary for strength ensuring to provide the stresses calculated by Formula (8) less than the allowable stresses.

If the blade is placed at an angle in the direction of the clamping force (see Figure 5), it is necessary to take into account the effect on the magnitude of the tangential force P_{ti}:

$$
\begin{cases} \sigma_{max}^{+} \\ \sigma_{max}^{-} \end{cases} = -\frac{P_i}{Hh} \pm 6\frac{\left(\sigma_u(a_0 + \Delta_r^i)H\right)\left(L - \frac{a_0 + \Delta_r^i}{2}\right) + (fP_i \pm P_{ti})L}{Hh^2}.
$$

Let us consider a specific numerical example. Consider a specific numerical example. Let the distance between the bit and the first calibrator be $l = 1$ m. The inner and outer diameters of the weighted drill pipes are $d = 147$ mm and $D = 190.5$ mm, the modulus of elasticity of the pipe material is $E = 2.1 \times 10^{11}$ Pa. Drilling is carried out with a six-bladed bit with the following parameters: $D_b = 292$ mm is the diameter of the bit; $H = 226$ mm is the blade length; $h = 29.8$ mm is the blade width; $L = 49$ mm is the blade height. The main rock that forms the well wall is fine-crystalline limestone with subgrade resistance $k = 490$ MN/m^3 and tensile strength $\sigma_u = 90$ MPa. The rock-blade friction coefficient is $f = 0.12$.

The axial moment of inertia of the cross-section of the weighted drill pipes is:

$$
J = 0.05D^4\left[1 - \left(\frac{d}{D}\right)^4\right] = 0.05 \cdot 190.5^4\left[1 - \left(\frac{147}{190.5}\right)^4\right] = 4.25 \times 10^{-5} \text{ m}^4.
$$

According to (5), the stiffness of the model elastic support for the case in (Figure 4a) is:

$$C = kn \times hH = 490 \cdot 10^6 \cdot 3 \cdot 29.8 \cdot 10^{-3} \cdot 226 \cdot 10^{-3} = 9.89 \cdot 10^6 \text{ N/m}.$$

assuming that the nipple-body misalignment Δ ranges from 0.2 mm to 1 mm, by Equation (2), we determine the possible range of the additional transverse clamping force. For example, for $\Delta = 0.4$ mm.

$$P = \frac{\Delta}{\frac{1}{3}\frac{l^3}{EJ} + \frac{1}{C}} = \frac{0.4 \cdot 10^{-3}}{\frac{1}{3} \cdot \frac{l^3}{2.1 \cdot 10^{11} \cdot 4.25 \cdot 10^{-5}} + \frac{1}{9.89 \cdot 10^6}} = 2.89 \times 10^3 \text{ N}$$

next, using the scheme presented in Figure 5, we calculate the additional radial recess of the blade in the rock:

$$\Delta_r = \frac{P}{C} \cos \vartheta = \frac{2.89 \cdot 10^3}{9.89 \cdot 10^6} \cos 0 = 0.292 \text{ mm}.$$

and the radial force $P_0 = P/3 = 0.96$ kN caused by misalignment (if the blade is at an angle to the direction of the clamping force, it is also necessary to calculate the tangential force P_{ti}).

Next, we set the external torque (rotary drilling) $M = 3$ kN $\times$ m, and from Equation (3), we determined the radial recess of the bit blades in the rock of the well wall during the ideal drilling process. Equation (8) calculated the maximum stresses that occur in the blades for the abnormal case (Table 1).

Table 1. Maximum stresses at different values of the bit manufacturing error.

Parameters	Misalignment Δ, mm				
	0.2	0.4	0.6	0.8	1.0
Additional transverse clamping force P, kN	1.45	2.89	4.33	5.78	7.22
Additional radial recess Δr, mm	0.146	0.292	0.438	0.584	0.730
Radial force caused by misalignment, P_0, kN	0.48	0.96	1.44	1.92	2.41
Radial force caused by the misalignment $\left| \bar{\sigma}_{max} \right|$	202.2	233.5	264.8	296.2	327.3

Note: the maximum stresses in the blades of the perfectly made bit are 167.5 MPa for the selected operating mode $M = 3$ kNm.

Analysis of the calculations given in Table 1 shows that drill bit manufacturing error increasing causes an increase in clamping force, additional radial recess, additional radial force, and maximum stresses in the blade. The growth of these indicators is significant, so it is necessary to provide technological methods to increase the accuracy of PDC drill bit manufacturing.

4. Discussion

The high standards of well construction envisage improved performance of modern drilling tools. Now, PDC drill bits are widely used, and their popularity in the world continues to grow. The main advantage of these drill bits over competitors is the lack of moving elements in the structure; in addition, they are technological in terms of manufacture and efficient in operation. To get all the possible benefits of PDC drill bits in practice, one needs to ensure the proper accuracy of the manufacture. Analysis of several series of PDC bits revealed that the main manufacturing errors are the magnitude of body-nipple misalignment.

The technology of PDC drill bit manufacturing envisages drill bit body and nipple assembling and further their welding as the final stage. Actually, the aforementioned manufacturing error occurs at this stage due to temperature deformations caused by welding. Such errors within an operation can affect the controllability of the bit. When using the bit with errors, additional stresses occur in drill string bottom hole assembling.

Moreover, additional stresses occur in the elements of the bit. First, when the axes of the nipple and the body on the bit do not match, there is an additional transverse clamping force. It is determined that the aforementioned clamping force depends on the error magnitude, bottom hole assembly, pipe materials, and subgrade resistance of the rock.

If the PDC drill bit is manufactured with errors, cutting forces are distributed inhomogeneously between the blades of the bit. In some cases, some blades may lose contact with the rock, and in this case, the heterogeneity of force distribution becomes very significant and can cause a loss of structural strength. In particular, the provided numerical analysis showed that a drill bit body-nipple misalignment error is associated with an additional clamping force which can exceed 7 kN (for using steel pipes of wide range). This clamping force can significantly affect the bit—wellbore contact interaction due to the additional radial recess of the blades in the rock. The maximum equivalent stress in the bit blades can be more than the standard one by almost two times.

When performing numerical testing of the results, we considered the bottom hole assemblies made of different metal alloys (steel alloy, aluminum alloy, duralumin alloy, titanium alloy) and also considered drill Pipe Based on Composite Carbon Fiber Material.

Among a number of currently known materials used for weighted drill pipes, composite carbon drill pipes [102] have shown minimal sensitivity to the geometric imperfection of the bit. Researchers [103–107] study the mechanical properties of polymers and the stress state, but there is no data on the operation of the drill string of pipes made of composite carbon materials and the bit. Analysis of the bit–well contact interaction during operation in non-standard mode allowed assessing the non-standard stresses that occur in the blades of the bit. Such results allow estimating the durability of blades and defining admissible errors of manufacturing at which operation of the bit will remain safe.

5. Conclusions

Analysis of previous studies has identified the main types of errors associated with PDC drill bit technology. In particular, body–nipple screwing, and subsequent welding can cause a body–nipple misalignment error—a mismatch between the axes of the nipple and the body.

We have developed an analytical model of the bit blade–rock contact interaction that takes into account the aforementioned nipple–body axes mismatch influence on the maximum equivalent stresses in the material of the blades. We considered that the error in the manufacture of the bit leads to a free kinematic load of the elastic system "drill pipes–bit–wellbore", resulting in an additional transverse clamping force acting on the bit. We simulated the behavior of the rock in contact with the bit as elastic support, the rigidity of which is expressed through the coefficient of subgrade resistance. After that, we considered the features of the rock–drill bit contact interaction for the standard and non-standard modes. It was found that the transverse clamping force on the bit causes additional radial recess of the blades into the rock if the blade operates in the non-standard mode. As a result, there is a significantly heterogeneous distribution of cutting forces between the blades of the bit, which in certain circumstances can lead to a loss of structural strength and an emergency during the construction of the well. As a result, we have obtained an analytical relation to specify the maximum equivalent stresses in the bit blade material for drill bits made with errors.

We specified the PDC drill bit nipple-body misalignment effect on the load on the steel blades for concrete numerical samples. Simulations have shown that the nipple-body misalignment increasing causes an increase in the clamping force. It is determined that the magnitude of the clamping force can become significant.

Peculiarities of drill bit-well wall contact interaction during operation in non-standard mode are considered. Extraordinary stresses were determined, and the strength of the blades with bit manufacturing errors was estimated.

The influence of the mechanical properties of different materials of weighted drill pipes on the value of free-force clamping force is analyzed. It is established that the least sensitivity to the geometric imperfection of the drill bit is shown by the elastic system of the pipe string, which includes a calibrator and is composed of composite carbon drill pipes. (Drill Pipe Based on Composite Carbon Fiber Material), and the largest-from steel drill pipes.

Further research is planned to take into account the stiffness of the threaded joints when calculating the transverse force that clamps the metal drill bit to the well wall.

Author Contributions: Conceptualization, A.V. and O.B.; methodology, A.V.; software, T.P.; validation, L.R., A.V. and M.P.; formal analysis, A.V.; investigation, V.L.; resources, M.P.; data curation, A.V.; writing—original draft preparation, V.L.; writing—review and editing, A.V.; visualization, T.P.; supervision, L.R.; project administration, O.B. and L.R.; funding acquisition, O.B. All authors have read and agreed to the published version of the manuscript.

Funding: This study was carried out as part of the project "Belt and Road Initiative Institute for Chinese-European studies (BRIICES)" and was funded by the Guangdong University of Petrochemical Technology.

Institutional Review Board Statement: Not applicable.

Informed Consent Statement: Not applicable.

Data Availability Statement: Data are contained within the article.

Acknowledgments: The team of authors express their gratitude to the reviewers for valuable recommendations that have been taken into account to improve significantly the quality of this paper. The authors are grateful to the Ministry of Science and Education of Ukraine for the grant to implement the project D-8-21-P.

Conflicts of Interest: The authors declare no conflict of interest.

References

1. Zhironkin, S.; Demchenko, S.; Kayachev, G.; Taran, E.; Zhironkina, O. Convergent and nature-like technologies as the basis for sustainable development in the 21st Century. *E3S Web Conf.* **2019**, *105*, 03008. [CrossRef]
2. Ursul, A.; Ursul, T.; Dugarova, M. Towards a global sustainable future. *Philos. Cosmol.* **2017**, *15*, 37–50.
3. Zhironkin, S.; Khloptsov, D.; Skrylnikova, N.; Petinenko, I.; Zhironkina, O. Economic principles of mining region sustainable development. *E3S Web Conf.* **2018**, *41*, 04010. [CrossRef]
4. Rozin, V. From engineering and technological process to post-cultural technology. *Future Hum. Image* **2020**, *15*, 99–109. [CrossRef]
5. IEA. *World Energy Outlook. Flagship Report—2019*; International Energy Agency: Paris, France, 2019. Available online: https://www.iea.org/reports/world-energy-outlook-2019 (accessed on 20 May 2021).
6. Krichevskiy, S. Evolution of technologies, "green" development and grounds of the general theory of technologies. *Philos. Cosmol.* **2015**, *14*, 120–139.
7. Falshtynskyi, V.; Saik, P.; Lozynskyi, V.; Dychkovskyi, R.; Petlovanyi, M. Innovative aspects of underground coal gasification technology in mine conditions. *Min. Miner. Depos.* **2018**, *12*, 68–75. [CrossRef]
8. Lozynskyi, V.; Medianyk, V.; Saik, P.; Rysbekov, K.; Demydov, M. Multivariance solutions for designing new levels of coal mines. *Rud. Geol. Naft. Zb.* **2020**, *35*, 23–31. [CrossRef]
9. Le Billon, P.; Kristoffersen, B. Just cuts for fossil fuels? Supply-side carbon constraints and energy transition. *Environ. Plan. A Econ. Space* **2020**, *52*, 1072–1092. [CrossRef]
10. Bazaluk, O.; Slabyi, O.; Vekeryk, V.; Velychkovych, A.; Ropyak, L.; Lozynskyi, V. A technology of hydrocarbon fluid production intensification by productive stratum drainage zone reaming. *Energies* **2021**, *14*, 3514. [CrossRef]
11. Svyrydenko, D.; Mozgin, W. The soft power of the state as a dialectic of contemporary dependencies in the international arena. *Ukrainian Policymaker* **2019**, *5*, 89–97. [CrossRef]
12. Olovyannyy, A.; Chantsev, V. Numerical experiments concerning long-term deformation of rock samples. *Min. Miner. Depos.* **2019**, *13*, 18–27. [CrossRef]
13. Babets, D.V.; Sdvyzhkova, O.O.; Larionov, M.H.; Tereshchuk, R.M. Estimation of rock mass stability based on probability approach and rating systems. *Nauk. Visnyk Natsionalnoho Hirnychoho Universytetu* **2017**, *2*, 58–64.
14. Gorova, A.; Pavlychenko, A.; Borysovs'ka, O. The study of ecological state of waste disposal areas of energy and mining companies. In *Annual Scientific-Technical Colletion—Mining of Mineral Deposits*; CRC Press: Boca Raton, FL, USA, 2013; pp. 169–172. [CrossRef]
15. Khomenko, O.; Barna, T. Zonal-and-wave structure of open systems on micro, mega- and macrolevels of the universe. *Philos. Cosmol.* **2019**, *22*, 24–32. [CrossRef] [PubMed]

16. Haeri, H.; Marji, M.F. Simulating the crack propagation and cracks coalescence underneath TBM disc cutters. *Arab. J. Geosci.* **2016**, *9*, 124. [CrossRef]

17. Marji, M.F. Modeling of Cracks in Rock Fragmentation with a Higher Order Displacement Discontinuity Method. Ph.D. Thesis, Middle East Technical University, Ankara, Turkey, 1996.

18. Marji, M.F. Simulation of crack coalescence mechanism underneath single and double disc cutters by higher order displacement discontinuity method. *J. Cent. South Univ.* **2015**, *22*, 1045–1054. [CrossRef]

19. Behnia, M.; Goshtasbi, K.; Marji, M.F.; Golshani, A. Numerical simulation of crack propagation in layered formations. *Arab. J. Geosci.* **2013**, *7*, 2729–2737. [CrossRef]

20. Bondarenko, V.; Kovalevs'ka, I.; Ganushevych, K. *Progressive Technologies of Coal, Coalbed Methane, and Ores Mining*; CRC Press: London, UK, 2014; 523p. [CrossRef]

21. Dreus, A.Y.; Sudakov, A.K.; Kozhevnikov, A.A.; Vakhalin, Y.N. Study on thermal strength reduction of rock formation in the diamond core drilling process using pulse flushing mode. *Nauk. Visnyk Natsionalnoho Hirnychoho Universytetu* **2016**, *3*, 5–10.

22. Sobko, B.; Lozhnikov, O.; Levytskyi, V.; Skyba, G. Conceptual development of the transition from drill and blast excavation to non-blasting methods for the preparation of mined rock in surface mining. *Rud. Geol. Naft. Zb.* **2019**, *34*, 21–28. [CrossRef]

23. Moisyshyn, V.; Voyevidko, I.; Tokaruk, V. Design of bottom hole assemblies with two rock cutting tools for drilling wells of large diameter. *Min. Miner. Depos.* **2020**, *14*, 128–133. [CrossRef]

24. Araujo, A.C.; Landon, Y.; Lagarrigue, P. Smart drilling for Aerospace Industry: State of art in research and education. *Procedia CIRP* **2021**, *99*, 387–391. [CrossRef]

25. Dychkovskyi, R.O.; Lozynskyi, V.H.; Saik, P.B.; Petlovanyi, M.V.; Malanchuk, Y.Z.; Malanchuk, Z.R. Modeling of the disjunctive geological fault influence on the exploitation wells stability during underground coal gasification. *Arch. Civ. Mech. Eng.* **2018**, *18*, 1183–1197. [CrossRef]

26. Falshtynskyi, V.S.; Dychkovskyi, R.O.; Lozynskyi, V.G.; Saik, P.B. Determination of the technological parameters of borehole underground coal gasification for thin coal seams. *J. Sustain. Min.* **2013**, *12*, 8–16. [CrossRef]

27. Bazaluk, O.; Sai, K.; Lozynskyi, V.; Petlovanyi, M.; Saik, P. Research into dissociation zones of gas hydrate deposits with a heterogeneous structure in the Black Sea. *Energies* **2021**, *14*, 1345. [CrossRef]

28. Onysko, O.R.; Kopey, V.B.; Panchuk, V.G. Theoretical investigation of the tapered thread joint surface contact pressure in the dependence on the profile and the geometric parameters of the threading turning tool. *IOP Conf. Ser. Mater. Sci. Eng.* **2020**, *749*, 012007. [CrossRef]

29. Onysko, O.; Borushchak, L.; Kopei, V.; Lukan, T.; Medvid, I.; Vryukalo, V. Computer studies of the tightness of the drill string connector depending on the profile of its tapered thread. *Lect. Notes Netw. Syst.* **2020**, *128*, 720–729. [CrossRef]

30. Onysko, O.; Kopei, V.; Medvid, I.; Pituley, L.; Lukan, T. Influence of the Thread Profile Accuracy on Contact Pressure in Oil and Gas Pipes Connectors. *Lect. Notes Mech. Eng.* **2020**, 432–441. [CrossRef]

31. Shatskyi, I.; Ropyak, L.; Velychkovych, A. Model of contact interaction in threaded joint equipped with spring-loaded collet. *Eng. Solid Mech.* **2020**, *8*, 301–312. [CrossRef]

32. Striletskyi, Y.Y.; Melnychuk, S.I.; Gryga, V.M.; Pashkevych, O.P. Using broadband signals for structural change detection in metal details. *Nauk. Visnyk Natsionalnoho Hirnychoho Universytetu* **2020**, *3*, 19–26. [CrossRef]

33. Saakiyan, L.S.; Efremov, A.P.; Ropyak, L.Y.; Gorbatskii, A.V. A method of microelectrochemical investigations. *Sov. Mater. Sci.* **1987**, *23*, 267–269. [CrossRef]

34. Saakiyan, L.S.; Efremov, A.P.; Ropyak, L.Y. Effect of stress on the microelectrochemical heterogeneity of steel. *Prot. Met.* **1989**, *25*, 185–189.

35. Chudyk, I.; Poberezhny, L.; Hrysanchuk, A.; Poberezhna, L. Corrosion of drill pipes in high mineralized produced waters. *Procedia Struct. Integr.* **2019**, *16*, 260–264. [CrossRef]

36. Shats'kyi, I.P.; Makoviichuk, M.V. Contact interaction of crack lips in shallow shells in bending with tension. *Mater. Sci.* **2005**, *41*, 486–494. [CrossRef]

37. Guillal, A.; Abdelbaki, N.; Bensghier, M.E.A.; Betayeb, M.; Kopei, B. Effect of shape factor on structural reliability analysis of a surface cracked pipeline-parametric study. *Frat. Integrita Strutt.* **2019**, *13*, 341–349. [CrossRef]

38. Shatskyi, I.P.; Makoviichuk, M.V.; Shcherbii, A.B. Equilibrium of cracked shell with flexible coating. In Proceedings of the 11th International Conference on Shell Structures: Theory and Applications, Gdansk, Poland, 11–13 October 2017; pp. 165–168. [CrossRef]

39. Shats'kyi, I.P.; Makoviichuk, M.V.; Shcherbii, A.B. Influence of a flexible coating on the strength of a shallow cylindrical shell with longitudinal crack. *J. Math. Sci.* **2019**, *238*, 165–173. [CrossRef]

40. Shatskyi, I.P.; Perepichka, V.V.; Ropyak, L.Y. On the influence of facing on strength of solids with surface defects. *Metallofiz. Noveishie Tekhnologii* **2020**, *42*, 69–76. [CrossRef]

41. Ropyak, L.Y.; Makoviichuk, M.V.; Shatskyi, I.P.; Pritula, I.M.; Gryn, L.O.; Belyakovskyi, V.O. Stressed state of laminated interference-absorption filter under local loading. *Funct. Mater.* **2020**, *27*, 638–642. [CrossRef]

42. Mandryk, O.; Artym, V.; Shtohry, M.; Zaytsev, V. Scientific Rationale for the Movable Pipeline Technology for Transporting CNG by Sea. *Manag. Syst. Prod. Eng.* **2020**, *28*, 168–177. [CrossRef]

43. Yakym, R.S.; Petryna, D.Y. Analysis of causes and preventing ways of early workability loss of three-cone rock bit cutters. *Metallofiz. Noveishie Tekhnologii* **2020**, *42*, 731–751. [CrossRef]

44. Vytyaz, O.; Chudyk, I.; Mykhailiuk, V. Study of the effects of drilling string eccentricity in the borehole on the quality of its cleaning. *New Dev. Min. Eng.* **2015**, 591–595.

45. Krol, R.; Kawalec, W.; Gladysiewicz, L. An effective belt conveyor for underground ore transportation systems. *IOP Conf.Ser. Earth Environ.Sci.* **2017**, *95*, 042047. [CrossRef]

46. Król, R. Studies of the durability of belt conveyor idlers with working loads taken into account. *IOP Conf. Ser. Earth Environ. Sci.* **2017**, *95*. [CrossRef]

47. Król, R.; Kisielewski, W. Research of loading carrying idlers used in belt conveyor-practical applications. *Diagnostyka* **2014**, *15*, 67–74.

48. Ropyak, L.Y.; Pryhorovska, T.O.; Levchuk, K.H. Analysis of Materials and Modern Technologies for PDC Drill Bit Manufacturing. *Prog. Phys. Met.* **2020**, *21*, 274–301. [CrossRef]

49. Nurpeissova, M.; Bekbassarov, S.; Bek, A.; Kyrgizbaeva, G.; Turisbekov, S.; Ormanbekova, A. The geodetic monitoring of the engineering structures stability conditions. *J. Eng. Appl. Sci.* **2020**, *12*, 9151–9163. [CrossRef]

50. Onysko, O.; Panchuk, V.; Kopei, V.; Havryliv, Y.; Schuliar, I. Investigation of the influence of the cutter-tool rake angle on the accuracy of the conical helix in the tapered thread machining. *J. Phys. Conf. Ser.* **2021**, *1781*, 012028. [CrossRef]

51. Medvid, I.; Onysko, O.; Panchuk, V.; Pituley, L.; Schuliar, I. Kinematics of the Tapered Thread Machining by Lathe: Analytical Study. *Lect. Notes Mech. Eng.* **2021**, 555–565. [CrossRef]

52. Danil'chenko, Y.M. Increasing the accuracy of rotation of high-speed spindle units on anti-friction bearings. *Sov. Eng. Res.* **1987**, *7*, 61–63.

53. Muthukumaran, G.; Dinesh Babu, P. Laser transformation hardening of various steel grades using different laser types. *J. Braz. Soc. Mech. Sci. Eng.* **2021**, *43*, 1–529. [CrossRef]

54. Ropyak, L.Y.; Vytvytskyi, V.S.; Velychkovych, A.S.; Pryhorovska, T.O.; Shovkoplias, M.V. Study on grinding mode effect on external conical thread quality. *IOP Conf. Ser. Mater. Sci. Eng.* **2021**, *1018*, 012014. [CrossRef]

55. Ostapovets, A.; Molnár, P.; Jäger, A. Visco-plastic self-consistent modelling of a grain boundary misorientation distribution after equal-channel angular pressing in an AZ31 magnesium alloy. *J. Mater. Sci.* **2013**, *48*, 2123–2134. [CrossRef]

56. Kuanyshbekovna, M.M.; Krupnik, L.; Koptileuovich, Y.K.; Mukhtar, E.; Roza, A. The system is "roof bolting-mountain". *Int. J. Appl. Eng. Res.* **2016**, *11*, 10454–10457.

57. Zeynullin, A.A.; Abeuov, E.A.; Demin, V.F.; Aliev, S.B.; Kaynazarova, A.S.; Kaynazarov, A.S. Estimation of ways to maintain mining works based on the application of anchor anchoring in the mines of the Karaganda coal basin. *Ugol* **2021**, *2*, 4–9. [CrossRef]

58. Tatarenko, V.A.; Radchenko, T.M.; Nadutov, V.M. Parameters of Interatomic Interaction in a Substitutional Alloy F.C.C. Ni-Fe According to Experimental Data about the Magnetic Characteristics and Equilibrium Values of Intensity of a Diffuse Scattering of Radiations. *Metallofiz. Noveishie Tekhnologii* **2003**, *25*, 1303–1319.

59. Radchenko, T.M.; Tatarenko, V.A.; Bokoch, S.M. Diffusivities and kinetics of short-range and long-range orderings in Ni-Fe permalloys. *Metallofiz. Noveishie Tekhnologii* **2006**, *28*, 1699–1720.

60. Levchuk, K.H.; Radchenko, T.M.; Tatarenko, V.A. High-Temperature Entropy Effects in Tetragonality of the Ordering Interstitial–Substitutional Solution Based on Body-Centred Tetragonal Metal. *Metallofiz. Noveishie Tekhnologii* **2021**, *43*, 1–26. [CrossRef]

61. Tarel'nik, V.B.; Konoplyanchenko, E.V.; Kosenko, P.V.; Martsinkovskii, V.S. Problems and Solutions in Renovation of the Rotors of Screw Compressors by Combined Technologies. *Chem. Pet. Eng.* **2017**, *53*, 540–546. [CrossRef]

62. Tarelnik, V.B.; Gaponova, O.P.; Konoplyantschenko, E.V.; Yevtushenko, N.S.; Gerasimenko, V.A. Analysis of the structural state of the surface layer after electro-erosive alloying. II. Peculiarities of the formation of electroerosive coatings on special steels and stops by hard wear-resistant and soft antifriction materials. *Metallofiz. Noveishie Tekhnologii* **2018**, *11*, 795–815. [CrossRef]

63. Tarelnyk, V.B.; Gaponova, O.P.; Konoplianchenko, Y.V.; Martsynkovskyy, V.S.; Tarelnyk, N.V.; Vasylenko, O.O. Improvement of quality of the surface electroerosive alloyed layers by the combined coatings and the surface plastic deformation. I. Features of formation of the combined electroerosive coatings on special steels and alloys. *Metallofiz. Noveishie Tekhnologii* **2019**, *41*, 47–69. [CrossRef]

64. Ivanov, O.; Prysyazhnyuk, P.; Lutsak, D.; Matviienkiv, O.; Aulin, V. Improvement of Abrasion Resistance of Production Equipment Wear Parts by Hardfacing with Flux-Cored Wires Containing Boron Carbide/Metal Powder Reaction Mixtures. *Manag. Syst. Prod. Eng.* **2020**, *28*, 178–183. [CrossRef]

65. Cizek, J.; Dlouhy, I.; Siska, F.; Khor, K.A. Modification of Plasma-sprayed TiO2 Coatings Characteristics via Controlling the In-flight Temperature and Velocity of the Powder Particles. *J. Therm. Spray Technol.* **2014**, *23*, 1339–1349. [CrossRef]

66. Duryahina, Z.A.; Kovbasyuk, T.M.; Bespalov, S.A.; Pidkova, V.Y. Micromechanical and Electrophysical Properties of Al2O3 Nanostructured Dielectric Coatings on Plane Heating Elements. *Mater. Sci.* **2016**, *52*, 50–55. [CrossRef]

67. Matei, A.A.; Pencea, I.; Stanciu, S.G.; Hristu, R.; Antoniac, I.; Ciovica, E.; Sfat, C.E.; Stanciu, G.A. Structural characterization and adhesion appraisal of TiN and TiCN coatings deposited by CAE-PVD technique on a new carbide composite cutting tool. *J. Adhes. Sci. Technol.* **2015**, *29*, 2576–2589. [CrossRef]

68. Prysyazhnyuk, P.; Lutsak, D.; Shlapak, L.; Aulin, V.; Lutsak, L.; Borushchak, L.; Shihab, T.A. Development of the composite material and coatings based on niobium carbide. *East. Eur. J. Enterp. Technol.* **2018**, *6*, 43–49. [CrossRef]

69. Duryagina, Z.A.; Bespalov, S.A.; Pidkova, V.Y.; Polockyj, D.Y. Examination of the dielectric layers on the structural materials formed by hybrid ion-plasma discharge system. *Metallofiz. Noveishie Tekhnologii* **2011**, *33*, 393–400.

70. Pashechko, M.; Montusiewicz, J.; Dziedzic, K.; Jozwik, J. Multicriterion Assessment of Wear Resistance of Fe–Mn–C–B Eutectic Coatings Alloyed with Si, Ni, and Cr. *Powder Metall. Met. Ceram.* **2017**, *56*, 316–322. [CrossRef]
71. Pashechko, M.I.; Dziedzic, K.; Mendyk, E.; Jozwik, J. Chemical and Phase Composition of the Friction Surfaces Fe-Mn-C-B-Si-Ni-Cr Hardfacing Coatings. *J. Tribol.* **2018**, *140*, 021302. [CrossRef]
72. Chudyk, I.; Raiter, P.; Grydzhuk, Y.; Yurych, L. Mathematical model of oscillations of a drill tool with a drill bit of cutting-scraping type. *Nauk. Visnyk Natsionalnoho Hirnychoho Universytetu* **2020**, *1*, 52–57. [CrossRef]
73. Pryhorovska, T. Rock heterogeneity numerical simulation as a factor of drill bit instability. *Eng. Solid Mech.* **2018**, *6*, 315–330. [CrossRef]
74. Pryhorovska, T.O. Study on rock reaction force depending on PDC cutter placement. *Mach. Sci. Technol.* **2017**, *21*, 37–66. [CrossRef]
75. Pryhorovska, T.A.; Chaplinskiy, S.S. Finite element modeling of rock mass cutting by cutters for PDC drill bits. *Neftyanoe Khozyaystvo-Oil Ind.* **2018**, *1*, 38–41. [CrossRef]
76. Pryhorovska, T.O.; Chaplinskyy, S.S. Probabilistic estimate of pdc drill bit wear rate. *Naukovyi Visnyk Natsionalnoho Hirnychoho Universytetu* **2014**, *5*, 39–45.
77. Tatsii, R.M.; Pazen, O.Y. Direct (Classical) Method of Calculation of the Temperature Field in a Hollow Multilayer Cylinder. *J. Eng. Phys. Thermophys.* **2018**, *91*, 1373–1384. [CrossRef]
78. Tatsiy, R.M.; Pazen, O.Y.; Vovk, S.Y.; Ropyak, L.Y.; Pryhorovska, T.O. Numerical study on heat transfer in multilayered structures of main geometric forms made of different materials. *J. Serb. Soc. Comput. Mech.* **2019**, *13*, 36–55. [CrossRef]
79. Tatsiy, R.M.; Pazen, O.Y.; Vovk, S.Y.; Kharyshyn, D.V. Direct method of studying heat exchange in multilayered bodies of basic geometric forms with imperfect heat contact. *Naukovyi Visnyk Natsionalnoho Hirnychoho Universytetu* **2021**, 60–67. [CrossRef]
80. Kopei, V.B.; Onysko, O.R.; Panchuk, V.G. Principles of development of product lifecycle management system for threaded connections based on the Python programming language. *J. Phys. Conf. Ser.* **2020**, *1426*, 012033. [CrossRef]
81. Levchuk, K.G.; Moisyshyn, V.M.; Tsidylo, I.V. Influence of mechanical properties of a material on dynamics of the stuck drilling pipes. *Metallofiz. Noveishie Tekhnologii* **2016**, *38*, 1655–1668. [CrossRef]
82. Petlovanyi, M.; Lozynskyi, V.; Saik, P.; Sai, K. Predicting the producing well stability in the place of its curving at the underground coal seams gasification. *E3S Web Conf.* **2019**, *123*, 01019. [CrossRef]
83. Velichkovich, A.; Dalyak, T.; Petryk, I. Slotted shell resilient elements for drilling shock absorbers. *Oil Gas Sci. Technol. Rev. IFP Energ. Nouv.* **2018**, *73*, 34. [CrossRef]
84. Moisyshyn, V.M.; Lyskanych, M.V.; Borysevych, L.V.; Kolych, N.B.; Zhovniruk, R.A. Integral Indicators of Change of Drilling Column Vibration-Criterion for Assessing of Roller Cone Bit Wear. *Metallofiz. Noveishie Tekhnologii* **2019**, *41*, 1087–1102. [CrossRef]
85. Mikhlin, Y.V.; Zhupiev, A.L. An application of the ince algebraization to the stability of non-linear normal vibration modes. *Int. J. Non-Linear Mech.* **1997**, *32*, 393–409. [CrossRef]
86. Grydzhuk, J.; Chudyk, I.; Velychkovych, A.; Andrusyak, A. Analytical estimation of inertial properties of the curved rotating section in a drill string. *East. Eur. J. Enterp. Technol.* **2019**, *1*, 6–14. [CrossRef]
87. Dutkiewicz, M.; Gołębiowska, I.; Shatskyi, I.; Shopa, V.; Velychkovych, A. Some aspects of design and application of inertial dampers. *MATEC Web of Conf.* **2018**, *178*, 06010. [CrossRef]
88. Shatskyi, I.; Velychkovych, A. Increase of compliance of shock absorbers with cut shells. *IOP Conf.Ser. Mater. Sci. Eng.* **2019**, *564*, 012072. [CrossRef]
89. Velychkovych, A.; Petryk, I.; Ropyak, L. Analytical study of operational properties of a plate shock absorber of a sucker-rod string. *Shock Vib.* **2020**, 3292713. [CrossRef]
90. Shatskyi, I.; Popadyuk, I.; Velychkovych, A. Hysteretic Properties of Shell Dampers. *Springer Proc. Math. Stat.* **2018**, *249*, 343–350. [CrossRef]
91. Bulbuk, O.; Velychkovych, A.; Mazurenko, V.; Ropyak, L.; Pryhorovska, T. Analytical estimation of tooth strength, restored by direct or indirect restorations. *Eng. Solid Mech.* **2019**, *7*, 193–204. [CrossRef]
92. Velichkovich, A.S.; Dalyak, T.M. Assessment of stressed state and performance characteristics of jacketed spring with a cut for drill shock absorber. *Chem. Pet. Eng.* **2015**, *51*, 188–193. [CrossRef]
93. Levchuk, K.G. Investigation of the vibration transfer process to a stuck drill string. *SOCAR Proc.* **2017**, *2*, 23–33. [CrossRef]
94. Moisyshyn, V.; Levchuk, K. Investigation on Releasing of a Stuck Drill String by Means of a Mechanical Jar. *Oil Gas Sci. Technol.* **2017**, *72*, 27–35. [CrossRef]
95. Levchuk, K.G. Diagnosis of catches of metal drill pipes by their stress-strain state in a sloping well. *Metallofiz. Noveishie Tekhnologii* **2018**, *40*, 701–712. [CrossRef]
96. Shatskyi, I.; Perepichka, V. Problem of Dynamics of an Elastic Rod with Decreasing Function of Elastic-Plastic External Resistance. *Springer Proc. Math. Stat.* **2018**, *249*, 335–342. [CrossRef]
97. Shatskyi, I.; Vytvytskyi, I.; Senyushkovych, M.; Velychkovych, A. Modelling and improvement of the design of hinged centralizer for casing. *IOP Conf. Ser. Mater. Sci. Eng.* **2019**, *564*, 12073. [CrossRef]
98. Velychkovych, A.S.; Andrusyak, A.V.; Pryhorovska, T.O.; Ropyak, L.Y. Analytical model of oil pipeline overground transitions, laid in mountain areas. *Oil Gas Sci. Technol.* **2019**, *74*, 65. [CrossRef]
99. Tyrlych, V.; Moisyshyn, V. Predicting remaining lifetime of drill pipes basing upon the fatigue crack kinetics within a pre-critical period. *Min. Miner. Depos.* **2019**, *13*, 127–133. [CrossRef]

100. Vytvytskyi, I.I.; Seniushkovych, M.V.; Shatskyi, I.P. Calculation of distance between elastic-rigid centralizers of casing. *Nauk. Visnyk Natsionalnoho Hirnychoho Universytetu* **2017**, *5*, 28–35.
101. Shatskyi, I.; Velychkovych, A.; Vytvytskyi, I.; Seniushkovych, M. Analytical models of contact interaction of casing centralizers with well wall. *Eng. Solid Mech.* **2019**, *7*, 355–366. [CrossRef]
102. Jiang, K.; Xie, R.; Yun, H. Lightweight drill pipe based on composite carbon fiber material. *J. Phys. Conf. Ser.* **2020**, *1549*, 032113. [CrossRef]
103. Benyahia, H.; Tarfaoui, M.; El Moumen, A.; Ouinas, D. Prediction of notched strength for cylindrical composites pipes under tensile loading conditions. *Compos. Part B Eng.* **2018**, *150*, 104–114. [CrossRef]
104. Fyk, M.; Biletskyi, V.; Abbood, M.; Al-Sultan, M.; Abbood, M.; Abdullatif, H.; Shapchenko, Y. Modeling of the lifting of a heat transfer agent in a geothermal well of a gas condensate deposit. *Min. Miner. Depos.* **2020**, *14*, 66–74. [CrossRef]
105. Kolosov, D.; Snihur, V.; Pysmenkova, T. Stress-strain state of rubber-cable tractive element of tubular shape. *Rozrob. Rodov.* **2020**, *14*, 43–52.
106. Galiyev, D.A.; Uteshov, E.T.; Tekenova, A.T. Digitalization of technological and organizational processes of mining operations due to the implementation of the installation system and accounting the key indicators. *News Natl. Acad. Sci. Repub. Kazakhstan Ser. Geol. Tech. Sci.* **2020**, *5*, 47–53. [CrossRef]
107. Yu, B.; Zhang, J.J.; Zhao, J.P.; Ma, T.J. Mechanical Study of Carbon Fiber Reinforced Plastic and Thin-walled Metal Liner in Bi-Material COPV Based on Grid Theory Optimization. *Mater. Sci. Forum* **2021**, *1027*, 15–21. [CrossRef]

Article

Application of UAV in Search and Rescue Actions in Underground Mine—A Specific Sound Detection in Noisy Acoustic Signal

Paweł Zimroz [1,*][iD], Paweł Trybała [1][iD], Adam Wróblewski [1][iD], Mateusz Góralczyk [1][iD], Jarosław Szrek [2][iD], Agnieszka Wójcik [1][iD] and Radosław Zimroz [1][iD]

1   Department of Mining, Faculty of Geoengineering, Mining and Geology, Wrocław University of Science and Technology, 50-421 Wroclaw, Poland; pawel.trybala@pwr.edu.pl (P.T.); adam.wroblewski@pwr.edu.pl (A.W.); mateusz.goralczyk@pwr.edu.pl (M.G.); a.wojcik@pwr.edu.pl (A.W.); radoslaw.zimroz@pwr.edu.pl (R.Z.)
2   Department of Fundamentals of Machine Design and Mechatronic Systems, Faculty of Mechanical Engineering, Wroclaw University of Science and Technology, 50-371 Wroclaw, Poland; jaroslaw.szrek@pwr.edu.pl
*   Correspondence: pawel.zimroz@pwr.edu.pl

Abstract: The possibility of the application of an unmanned aerial vehicle (UAV) in search and rescue activities in a deep underground mine has been investigated. In the presented case study, a UAV is searching for a lost or injured human who is able to call for help but is not able to move or use any communication device. A UAV capturing acoustic data while flying through underground corridors is used. The acoustic signal is very noisy since during the flight the UAV contributes high-energetic emission. The main goal of the paper is to present an automatic signal processing procedure for detection of a specific sound (supposed to contain voice activity) in presence of heavy, time-varying noise from UAV. The proposed acoustic signal processing technique is based on time-frequency representation and Euclidean distance measurement between reference spectrum (UAV noise only) and captured data. As both the UAV and "injured" person were equipped with synchronized microphones during the experiment, validation has been performed. Two experiments carried out in lab conditions, as well as one in an underground mine, provided very satisfactory results.

Keywords: UAV; acoustic signal processing; Euclidean distance; underground mine; search and rescue

check for
updates

Citation: Zimroz, P.; Trybała, P.; Wróblewski, A.; Góralczyk, M.; Szrek, J.; Wójcik, A.; Zimroz, R. Application of UAV in Search and Rescue Actions in Underground Mine—A Specific Sound Detection in Noisy Acoustic Signal. *Energies* **2021**, *14*, 3725. https://doi.org/10.3390/en14133725

Academic Editors: Francisco Manzano Agugliaro and Oh-Soon Shin

Received: 24 March 2021
Accepted: 15 June 2021
Published: 22 June 2021

Publisher's Note: MDPI stays neutral with regard to jurisdictional claims in published maps and institutional affiliations.

1. Introduction

The biggest underground mines look like underground cities. They consist of many "streets" and "street crossings". Unfortunately, each mining void looks very similar to the other. Due to the complexity of the underground corridors network, localization of an injured person or any other incident, if not reported on an emergency line, is complicated and time-consuming. It imposes a serious danger in the cases, in which rapid contact with works supervisors or rescue units is crucial. The complex structure of an underground mine makes getting lost in it much more complicated to solve and serious than in the case of an accident on the ground level. Another important issue is related to extremely harsh environmental conditions. An example of an emergency situation caused by them can be given, like one of the co-workers in a pair or small group fainting. In such a case, the presence of a drone sent to the area in which worsened environmental parameters have been reported can significantly shorten the time needed to reach the victim and others at risk.

There are many natural hazards or unfavorable environmental factors, that can impose a significant risk to any humans performing work underground, such as spontaneous combustion of coal, rock bursts, methane ejections, water hazard, high dustiness [1–4]. According to the State Mining Authority, only in 2019 did Polish mine rescue units take part in 32 rescue operations in mining plants [5]. Most of them were related to roof/wall collapse or fire occurrence. In case of such an extreme situation in the conditions of an underground mine, it is not difficult to imagine a situation in which it is necessary to search for workers. As a result of the risks present in the mining environment, such as lack of oxygen, presence of fire fumes or other harmful gases, leading to loss of consciousness, as well as other injuries, the miner may not be able to evacuate from the hazardous area. Furthermore, a person may get trapped due to a rock collapse or a fire occurring in the evacuation route. The scale of the problem can be seen in the example of the activity of Polish mining rescue units. Considering that mining rescue activities are not limited to Polish underground mines, it can be stated that on a global scale the issue of human location in emergency situations is a considerable challenge. As the area to search is large, consists of many corridors and time is a critical issue, there is a need for a quick method of finding victims of accidents.

In the paper, it is proposed to use an unmanned air vehicle (UAV) for audio signals acquisition for human voice detection. In the article, we propose a drone equipped with a microphone, capable of registering human voice and distinguishing it from the propellers' noise, to be used for detecting emergency situations in underground mines. A method using time-frequency decomposition of the signal is presented, together with the full procedure of separating the human voice and background noise. It is assumed that the person is not able to move, but can call for help by shouting. Practically, the mission may be performed using several drones searching corridor by corridor and using acoustic signals to find the location of the human.

From a signal processing perspective, the problem is related to voice detection in presence of time-varying noise with complex structure. Due to the spinning of the propellers, some deterministic and random components are present in the acoustic signal. Unfortunately, the energy of the voice signal captured by a microphone mounted below the drone is rather small and the high energy of the acoustic signal emitted by the drone makes the voice detection problem complicated. In the paper, a procedure for voice detection in presence of UAV noise is proposed. Our informative signal is in fact a human voice, but limited to specific words only like "help". So, we called it specific acoustic signal. We limit our research for "help" only, however we believe also other words might be detectable-but for sure not all. "Help" is the most intuitive in crisis situation. The experiments have been carried out in the lab as well as in the real mining corridor to test and validate the proposed procedure.

2. State of the Art

Voice activity detection (VAD), being a natural context for our sound detection method, is an inherent stage of any real-world application of speech processing. There have been numerous publications concerning various methods on this subject; however, formerly, emphasis was placed primarily on detecting voice messages between long periods of silence or low magnitude noise. If the background noise has lower energy than any material speech occurrence, then a simple thresholding criterion enables a recording or communication system to save energy, computational memory, and network bandwidth. Traditional VAD and noise estimation methods are used in VOIP (Voice over Internet Protocol), as well as various speech-based applications. Along with de-noising and feature extraction, it would be a stage of any speech recognition or voice control system [6].

More sophisticated methods were developed to deal with lower SNR (signal-to-noise ratio). Among more recent work on VAD methods, some were based on combining spectral analysis and statistical methods [7,8]. Closely related to them are the noise reduction techniques, which allow subtracting the noise to increase the intelligibility of the speech signal [9]. Extracting preliminary information from speech signals may include voiced/unvoiced speech classification [10]. Speech detection and de-noising were also considered in the context of the so-called cocktail party problem where audio sources (i.e., speech and noise) are separated through a binaural system consisting of two microphones [11].

The background noise has considerably high energy in the case of recording nearby UAV equipped with a multi-rotor propulsion system. It contains the sound of its engine and mechanical parts, such as the propellers, which cannot be damped in its natural operating mode. The methods of dealing with high-level background noise conditions are widely discussed in literature concerning acoustic measurement on machines with rotating elements such as turbochargers [12–14] among many other examples. The ego-noise of a small-size multi-rotor drone itself has been investigated by many researchers in recent years. It was thoroughly examined at static thrust [15]. Some recent research considers tracking change in noise characteristics under realistic flight conditions based on position and noise measurement of a moving drone using both camera and a multiple microphone system (which is called microphone array) [16].

There are some publications about de-noising methods concerning mainly a hovering drone. The azimuthal direction of sound arrival was analyzed (using both supervised and unsupervised methods) in [17,18] based on a microphone array mounted on a hovering drone. Though at static thrust drone was enabled to track moving sound sources. Analyzing the sound emitted by a UAV has recently acquired interest also in the detection of illegal drone activity [19], where the dynamics of a quadrotor drone was taken into account.

In our experiment, as the drone is moving from one point to another, the background noise is both high-energy and non-stationary. It originates from the variable mode of operating of a UAV and air response but also from air turbulences. We only use one microphone at a sound source, thus physical direction or distance cannot be tracked, but the distance is changing (in an approximately monotonous way) during our records.

When analyzing a non-stationary process, it is essential to perform some sort of time segmentation of the data [20]. Therefore, transformations preserving (to a certain extent) original time structure are very useful. One of the primary methods is representing a signal in the time-frequency domain via Short-Time Fourier Transform (STFT) [21].

To characterize the time intervals containing voice we adopt a method based on STFT which was discussed before in a context of cyclic impulse anomaly detection (local damage in driving units of a belt conveyor) based on calculating distance (metric) between spectral densities [22]. It was stated that informative frequency band selection may strongly affect the analysis based on the time-frequency map. To deal with it we perform both band-pass filtering and spectrogram normalization.

The application of robotics in mining industry and especially in missions related to rescue and search of the victims of an accident victims is well established and constantly developed since the beginning of the third-millennium [23–31]. The use of UAVs as a support for the search and rescue (SAR) operations has been presented among others in [32], where some significant considerations are included, such as energy limitations, the quality of sensors, and the exchange of information between UAVs and rescue teams. An example of a fixed-wing UAV-system capable of detecting accident victims employing real-time imaging, on-board processing, and assigning coordinates of the detected target can be found in [33]. Both-UAV's hardware and identification algorithm with a self-adjusting threshold were exhaustively described providing a valuable overview of the current technical tools valuable for SAR purposes in superficial conditions. A remote-controlled drone, presented by [34], is an example of how the technology of this type can not only prevent mining production obstructions but also increase safety on-site. The UAV used by the authors carried an analog thermal infrared video camera, digital video recorder, an RGB spectrum digital video camera, and a small battery power source. This solution based on commercially available devices allowed evaluation of the sub-surface temperatures in a flexible and low-cost way, nevertheless, the system was able to perform only short missions-lasting up to fifteen minutes due to the limitation of batteries weight. The character of sub-surface mining workplaces and accidents that take place there created a big demand for the introduction of urban SAR robotics in underground applications and their adaptation to solving similar problems in new-more demanding conditions. Together with maintenance-purpose robots, this technology will allow the limitation of human involvement in the most hazardous areas, contributing to the transformation of the mining sector [27,35]. The vast majority of underground mining-related applications of aerial vehicles' proposed by researchers nowadays are aimed at mapping, localization of discontinuities, and general reconnaissance of subsurface areas [36–40]. The aforementioned are undoubtedly valuable examples of how robotics can facilitate arduous and time-consuming processes, providing information about some available tools. However, recent development and increased accessibility of robotics in the raw materials sector is also aimed at increasing the safety of underground workers and providing quick and effective help in case of an accident. The detection of a human in an underground environment can be based on a video recording transmitted in real-time to an operator [41], automatic detection of humans with the use of image recognition algorithms [28], or, what we propose in this article—human voice detection. Three recent reviews of the usage of unmanned aerial vehicles [42–44] give an exhaustive insight into the spectrum of possible applications of them in both underground and open-cast mines, including rescue operations. Nonetheless, in most of the cases in which a UAV is considered to be applied in this area, the sensors in which it is equipped are the image-capturing ones. To the best authors' knowledge, the problem of voice detection and its extraction from the drone background noise have not been approached in the context described in this work (underground mining rescue action).

3. Theoretical Background of the Novel Procedure

We propose a procedure involving time-frequency decomposition of a signal. A time-frequency map (spectrogram) of a signal is generated using Short-Time Fourier Transform (STFT). The STFT of a vector of observations $x_0, ..., x_{N-1}$ is a function of discrete valued time and frequency. It is a DFT of a signal multiplied by a shifted window selecting the observations close to a given time point (equivalently in terms of filtering it is a discrete convolution of a frequency shifted signal with the window function) [45] which is given by the Equation

$$STFT(t,f) = \sum_{k=0}^{N-1} w_{t-k} x_k e^{-j2\pi fk/N}.$$

(1)

Both for further analysis and for plotting a spectrogram we will consider squared absolute value of the STFT and refer to it as a power spectral densities (PSD) matrix of size $T \times F$ containing values

$$PSD(t,f) = |STFT(t,f)|^2.$$

(2)

Its columns (time slices) we name power spectral densities and its rows (slices for given frequency) we name subsignals. There arises a problem of informative frequency band selection (see also [46–48]).

Hidden determinism associated with the motion of the drone propellers is present in the background noise. It was stated that such a character of data appears with high-energy low-frequency bands. To avoid the necessity of further investigating (see advanced filter design discussed among others by Wodecki [49]) and selecting the band of interest, we decided to use a data-driven method which proved very efficient in the context of cyclic anomaly detection. Reducing the impact of the hidden determinism involves sub-signals normalization. As a norm, we use the sum of the values of a sub-signal:

$$PSD_{\text{normalized}}(t,f) = PSD(t,f) / \sum_{u=0}^{T-1} PSD(u,f).$$

(3)

The main part of the presented method is calculating the measure of dissimilarity for every spectral density (time slice of the matrix) with a referential spectrum. Euclidean metric is used defined for power spectral densities $p(f)$ and $q(f)$ as follows:

$$D(p,q) = \sqrt{\sum_{f=0}^{F-1} \big(p(f) - q(f)\big)^2}. \tag{4}$$

The sound emitted by the drone is assumed to contain a notable noisy component. To increase SNR we do band-pass filtering in the frequency domain. According to the fact that the informative part of the human voice does not exceed 4 kHz, this value is used as a cutoff frequency. We also add a lower boundary at 0.2 kHz. Therefore, we modify the Equation (4) to obtain

$$D(p,q)_{\text{band-pass}} = \sqrt{\sum_{f=F_{\text{lower}}}^{F_{\text{upper}}} \big(p(f) - q(f)\big)^2}. \tag{5}$$

where $F_{\text{lower}} = 0.2$ kHz, $F_{\text{upper}} = 4$ kHz. For the sake of the visualization of results, we will also filter signals in the time domain.

As the referential, the mean power spectrum calculated from a reference interval (usually a few starting seconds depending on a particular sound signal) is assigned. It is assumed that in this time interval only the background noise containing the sound emitted by the drone is present. Thus for every time slice t on a spectrogram we obtain

$$D_t = D_{\text{band-pass}}\big(PSD_{\text{normalized}}(t,\cdot), PSD_{ref}(\cdot)\big). \tag{6}$$

where for a given reference interval I we have

$$PSD_{\text{ref}}(f) = \frac{1}{|I|} \sum_{t \in I} PSD_{\text{normalized}}(t,f). \tag{7}$$

To discriminate between the human voice and the drone noise we set a threshold on Euclidean distance. Beforehand, we apply moving mean smoothing (with parameter M) on the calculated metric:

$$\widetilde{D_t} = \frac{1}{M} \sum_{k=u-\lfloor \frac{M}{2} \rfloor}^{u+\lfloor \frac{M-1}{2} \rfloor} D_u. \tag{8}$$

As the threshold, the value of mode + standard error was chosen where the mode is read from the kernel smoothed empirical PDF (probability density function) as the argument of the maximum value. Values greater than the threshold are associated with the human voice therefore corresponding time interval is assigned to the voice-present cluster and marked with 1. Otherwise, a time interval is assigned to a voice-absent cluster and marked with 0. Final processing of the received output includes filling gaps (changing too short series of 0 to 1) and subsequent deleting immaterial voice (changing too short series of 1 to 0). It is accordingly necessary to beforehand define tolerance values Tol_0 and Tol_1 for both steps of this operation. Voice intervals merged by that means are marked with red color on the plots. At once, the detected intervals are exported to the output audio file. They are delimited with 1-second pause to distinguish between one another, that is, data that would be then sent by the drone to report emergency. The outline of our algorithm is as follows:

1. Calculating STFT of a signal and then its squared absolute value to obtain PSD(t,f) matrix;
2. Normalization of the subsignals (rows of the PSD matrix) in time domain and restriction of the frequency domain;
3. Defining of the referential spectrum;
4. Calculating Euclidean metric at every time interval of PSD matrix and using moving mean smoothing filter;
5. Setting a threshold on Euclidean metric;

6.   Assigning voice-present (1) and voice-absent clusters (0) to the original signal,
7.   Changing clusters in too short time intervals;
8.   Visualisation of the voice detection results on a band-pass filtered signal.

As a standard-setting, we use STFT with Kaiser window of order 5 and length 1024 which is also the number of FFT (Fast Fourier Transform) algorithm points. Other parameters which are needed to be set are the number of overlapping points in STFT, the start and end of the reference interval, the number of points in the moving mean filter, and tolerances (in seconds) for binary merging in final processing.

4. The Experiment

In the experiment, a manually controlled drone, DJI Mavic Mini was used. It is a small, commercially available quadrotor, weighing 249 g, using 4 pairs of light, 119 mm diameter propellers. This model was chosen for our case study as it represents well the wider class of micro aerial vehicles (MAV), which are, due to their small size and fairly good battery life, the most suitable type of unmanned vehicle to be used in an emergency situation in an underground mine. During a normal performance, its brushless motors are operating in the frequency range of 9000–15,000 RPM (150–250 Hz). Since the experiment was realized in indoor conditions, the propeller guard (provided by the UAV manufacturer) was mounted on the drone to avoid damage in case of the drone colliding with the walls.

To evaluate the possibility of the application of a microphone-equipped unmanned aerial vehicle in the detection of humans, two experiments have been performed. The first was a preliminary test performed in the corridors of a building (Figure 1), roughly approximating realistic conditions. Another took place in a slant of an inactive underground mine (Figure 2). In both experiments, a UAV with a microphone located underneath has been approaching a person laying in the side pit of the main corridor and calling for help. In the preliminary experiment, an additional microphone has been used, which was placed near the help-caller. The person was located behind the corner, loudly repeated the words 'HELP, HELP'. In the experiment performed in realistic conditions, a deep niche in the corridor has been selected as the 'victim's' location. The volume of help-calling, as well as the speed of the drone, were kept as constant as possible throughout the whole experiment given the technical and organizational capabilities. Some minor variations in both speed and volume of acoustic emission could be caused by a human factor, as well as variations in the speed of airflow in underground excavations.

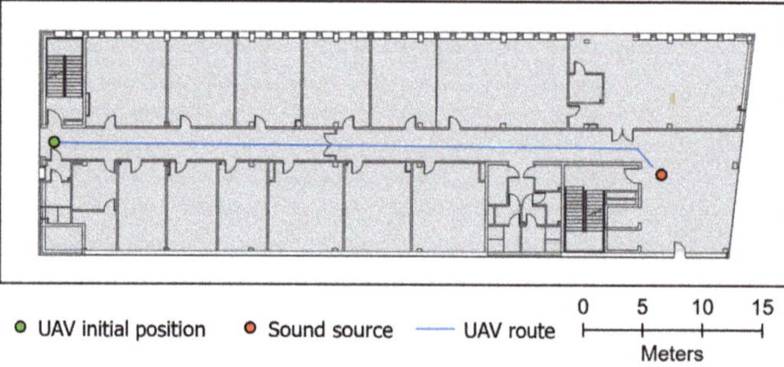

Figure 1. Drone flight path during the first experiment.

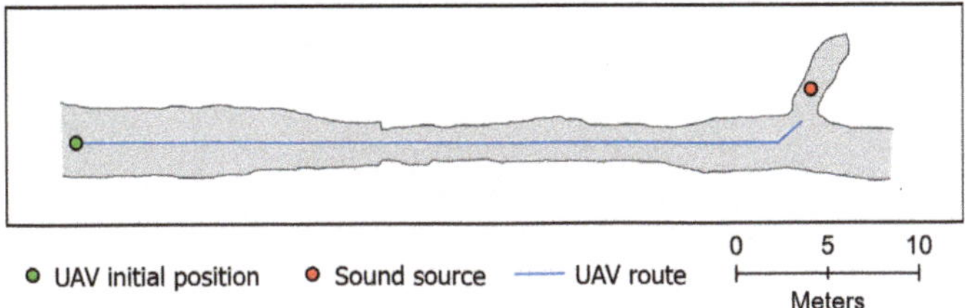

Figure 2. Drone flight path during the second experiment.

5. Results

5.1. Overground Sounds

Data acquired during the first (overground) experiment consist of sounds recorded at the same time with two microphones located at different places: first, located nearby the UAV ('next-to-drone' record); second, nearby the person who was making voice calling for help ('next-to-victim' record). The length of the signals differ (they are 62 s and 66 s long respectively) and their real-time starting moments are shifted relative to each other. Both were recorded with a sampling frequency equal to 48 kHz. The procedure presented above was applied separately to each of the records. Time waveforms of both raw sounds are presented in Figure 3. As we can see, the first of them seems a noise-like signal while in the case of the second signal the time intervals containing human voice are clearly visible.

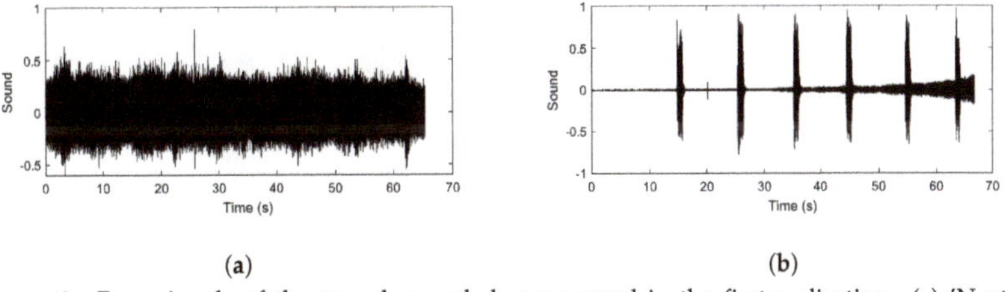

(a)   (b)

Figure 3. Raw signals of the sound recorded overground in the first realisation: (a) 'Next-to-drone' record, (b) 'Next-to-victim' record.

STFT was calculated with 80% overlapping. The resulting time-bin length is 4.3 milliseconds. Spectrograms of the raw signals which are presented in Figure 4 are records from both microphones. They illustrate the time-frequency decomposition of the sounds in terms of the log-scale PSD.

As it was stated the analysis of the data is based on the normalized spectrogram. This type of spectrogram is showed in Figure 5. It should be noted that frequencies in this case are restricted to the band 200–4000 Hz because, according to band-pass filtering of the data, we cut PSD to 0 outside of this band. In the case of the 'next-to-drone' record, we can distinguish particular moments in the evolution of the power spectrum only after normalization, the changes are visible mostly in a frequency bin located around 550–800 Hz. In the spectrograms based on the 'next-to-victim' recordings, on the other hand, the changes are by far more evident in numerous bands and valuable as the noise arises only in the ending part.

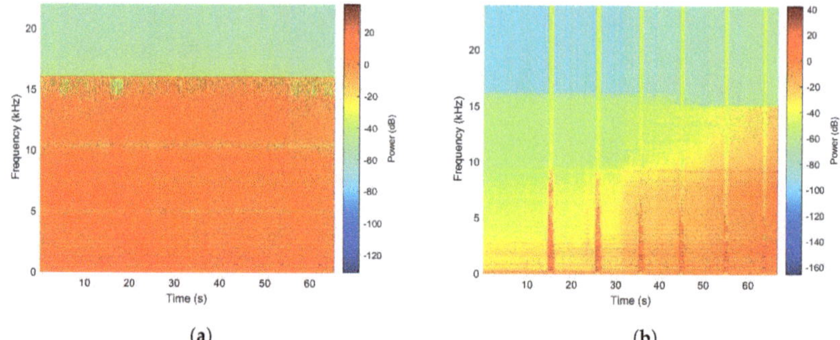

(a)   **(b)**

Figure 4. Spectrograms of the raw signal from Figure 3: (**a**) 'Next-to-drone' record, (**b**) 'Next-to-victim' record.

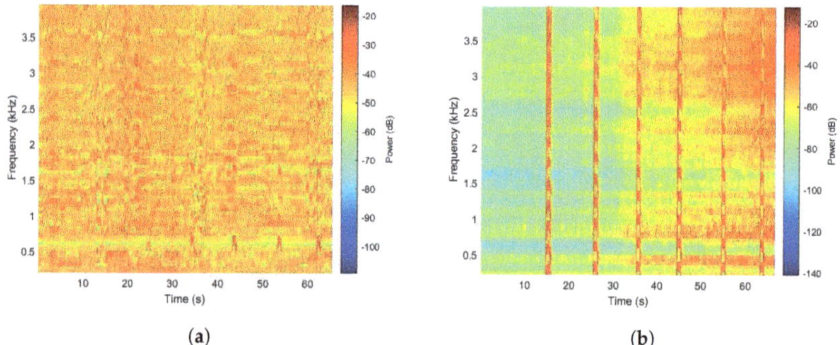

(a)   **(b)**

Figure 5. Normalized and band-pass filtered spectrograms of the signals from Figure 3: (**a**) 'Next-to-drone' record, (**b**) 'Next-to-victim' record.

We visualize band-pass filtering in the time domain in Figure 6. In the case of the 'next-to-drone' record, this processing step makes the shape of the waveform more informative compared with the waveform of the raw signal presented before.

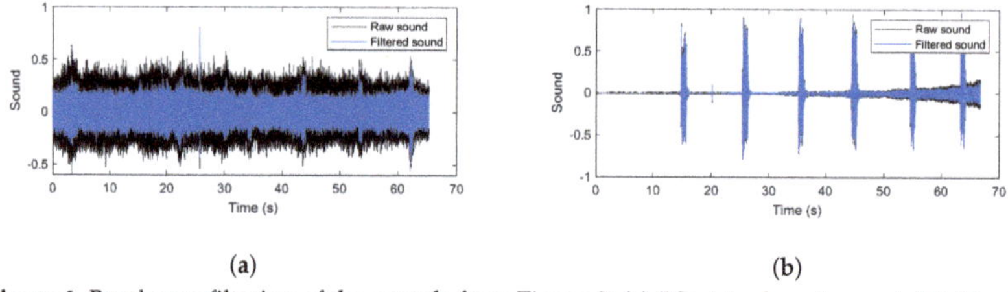

(a)   **(b)**

Figure 6. Band-pass filtering of the sounds from Figure 3: (**a**) 'Next-to-drone' record, (**b**) 'Next-to-victim' record.

The referential power spectral density was defined as a mean from 5 first seconds. Euclidean metric as a function of time is showed in Figure 7. The Euclidean metric smoothed with 20 samples moving mean is highlighted in green. The red line stands for the threshold which we obtain as mode + standard error value from the empirical distribution of our metric on the whole time domain. The empirical density is presented in Figure 8.

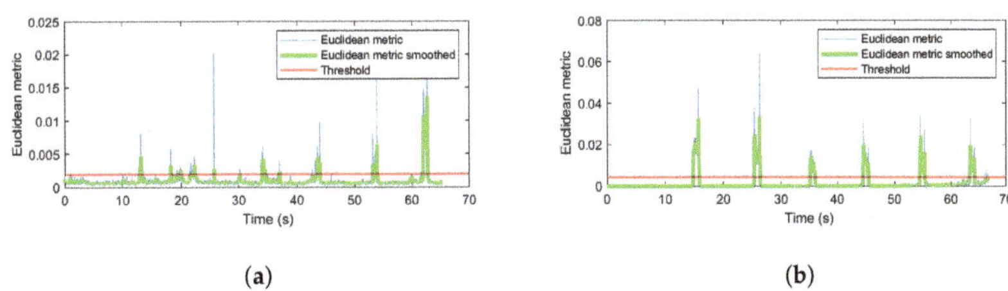

(a) **(b)**

Figure 7. Smoothed Euclidean metric calculated from normalized spectrograms from Figure 5 as a function of time: (**a**) 'Next-to-drone' record, (**b**) 'Next-to-victim' record.

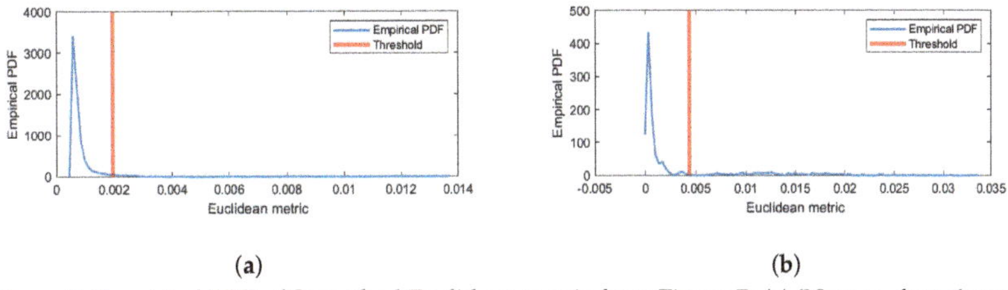

(a) **(b)**

Figure 8. Empirical PDF of Smoothed Euclidean metric from Figure 7: (**a**) 'Next-to-drone' record, (**b**) 'Next-to-victim' record.

In Figure 9 the detected intervals of the found voice are highlighted as a part of the plot of the band-pass filtered signal. Tolerances Tol_0 and Tol_1 in final processing were both set to 0.25 s. The output comprises six voice-present regions selected from the 'next-to-drone' record which correspond to the analogous segments present in the 'next-to-victim' record. The last four of them can be heard in the original raw 'next-to-drone' record.

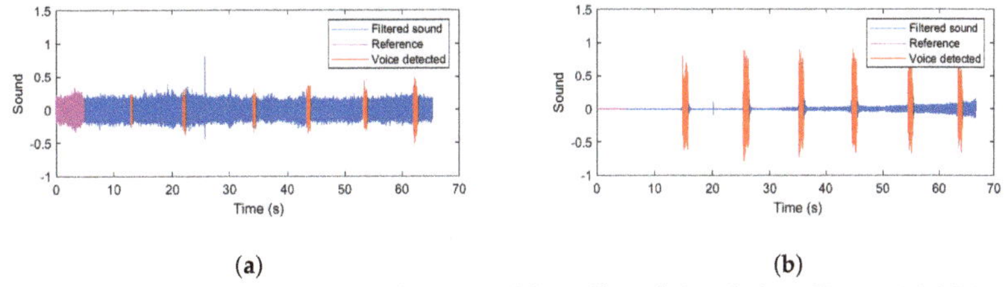

(a) **(b)**

Figure 9. Detected voice time intervals extracted from filtered signals from Figure 6: (**a**) 'Next-to-drone' record, (**b**) 'Next-to-victim' record.

5.2. Underground Sounds

Data acquired during the second (underground) experiment include only sounds recorded nearby the drone with the sampling frequency equal to 25.6 kHz. From several sounds recorded, two were chosen for the analysis.

First, we analyze a short signal with a frequent and clearly audible human voice. The signal is 34 s long and we present it in Figure 10. Its spectrogram is presented in Figure 11. Spectral structure is dominated by the noise from the drone; showever, some voice related signature is detectable so it could be a basis for automatic signal processing.

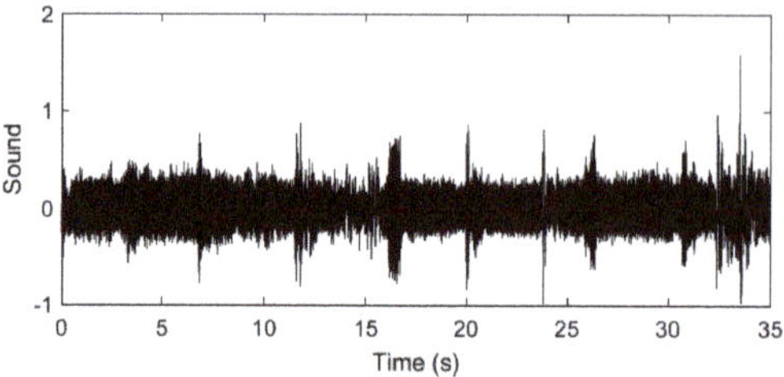

Figure 10. Raw signal of the first sound recorded underground.

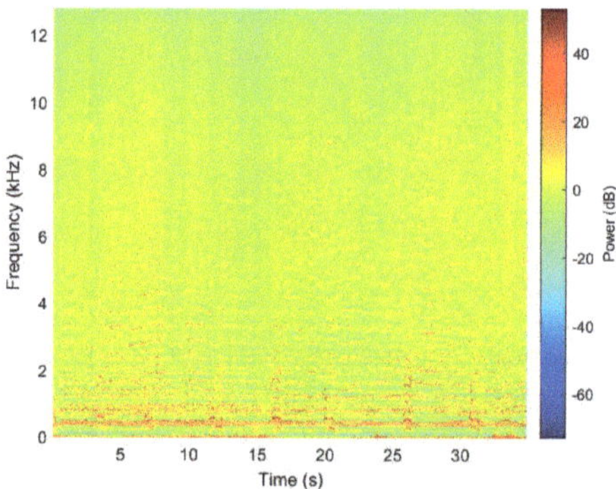

Figure 11. Spectrogram of the raw signal from Figure 10.

Since the voice regions are frequent in this realization of the experiment, this time we used lower overlapping equal to 50% so as not to blur the results. The time moments at which the voice was heard can be seen at the spectrogram, even more clearly after filtering and normalization (Figure 12).

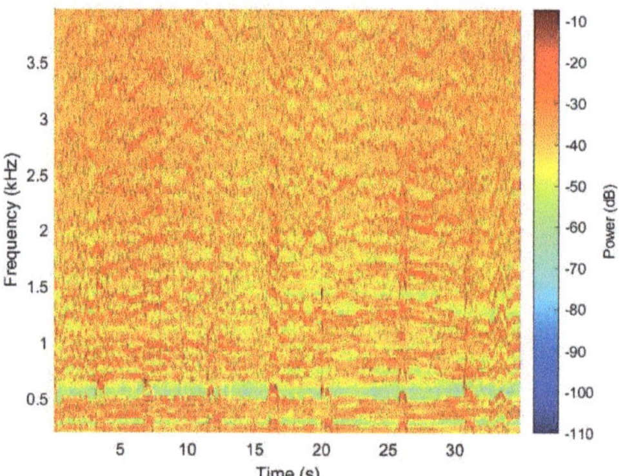

Figure 12. Normalized and band-pass filtered spectrogram of the signal from Figure 10.

A particularly nonstationary behavior of the drone alone in the ending time interval (which is also audible from the record) was successfully removed due to band-pass filtering as it can be noted at the plot visualizing band-pass filtering of the raw sound (Figure 13).

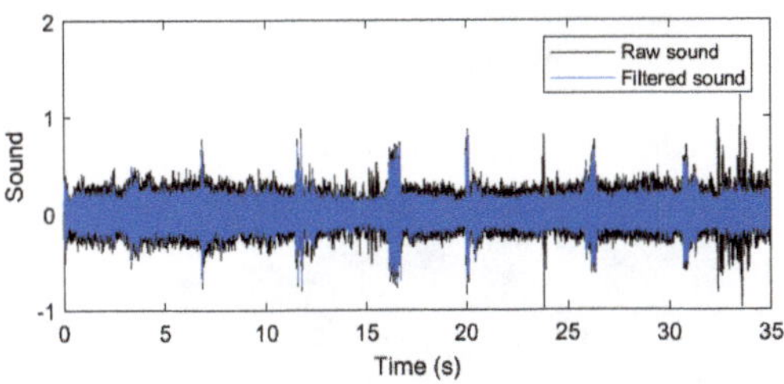

Figure 13. A band-pass filtering of the sound from Figure 10.

To determine the time intervals which correspond to the rescued person calling we calculate Euclidean metric as before. This length of the time bin is 20 ms resulting from the parameters of the STFT computation. The referential spectrum is calculated as a mean PSD from only first 2 seconds (so as to catch an interval without voice). Calculated metric and the determined threshold is presented in the time domain and on a kernel-smoothed density (Figures 14 and 15). Moving mean smoothing parameter was again set to 20. In both figures the chosen threshold equal to mode + standard error is shown. The detected voice time intervals (Figure 16) are fully consistent with all seven calls for help, which are audible from the raw sound (using Tol_0 and Tol_1 equal to 0.5 s).

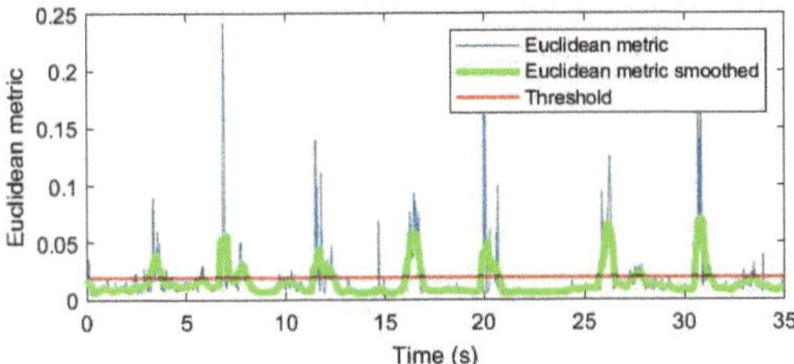

Figure 14. Smoothed Euclidean metric calculated from normalized spectrogram from Figure 12 as a function of time.

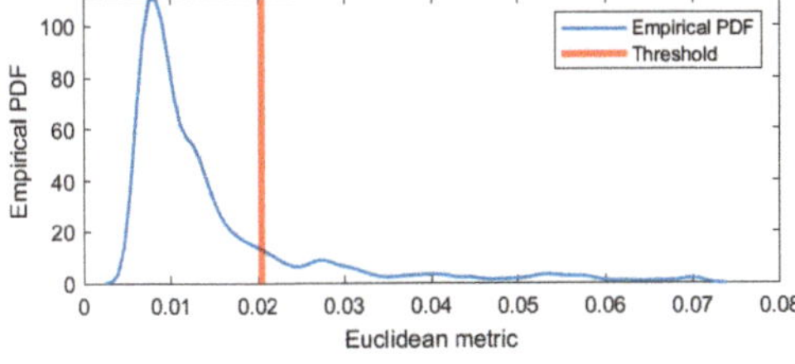

Figure 15. Empirical PDF of Smoothed Euclidean metric from Figure 14.

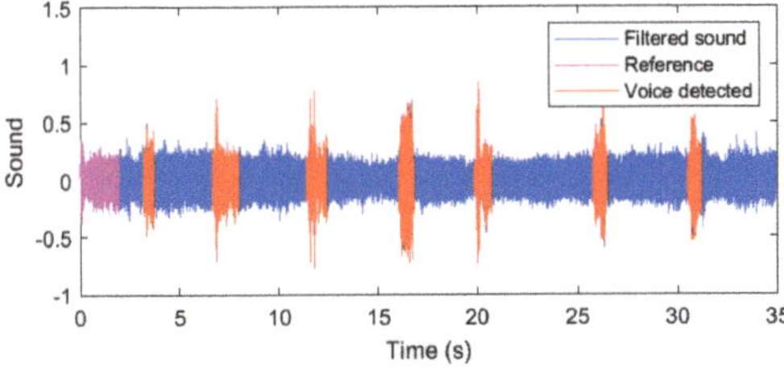

Figure 16. Detected voice time intervals extracted from filtered signals from Figure 13.

The second sound signal is 90 s long and has a low signal-to-noise ratio (Figure 17). The way of approaching the rescued person was similar to that which was performed in the overground experiment (the background noise though differ due to the underground environment and other microphone used).

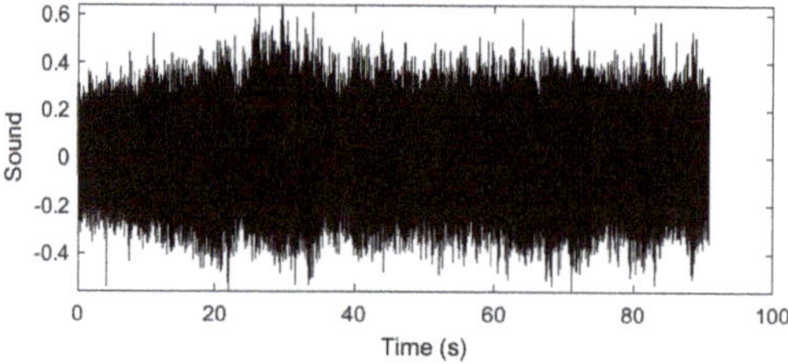

Figure 17. Raw signal of the second sound recorded underground.

STFT is computed with 80% overlapping. Its time-frequency characteristic is very volatile though harmonics, i.e., horizontal lines can be distinguished (Figures 18 and 19).

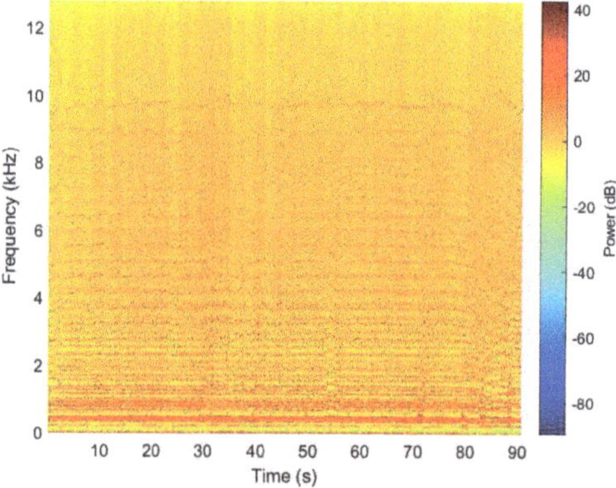

Figure 18. Spectrogram of the raw signal from Figure 17.

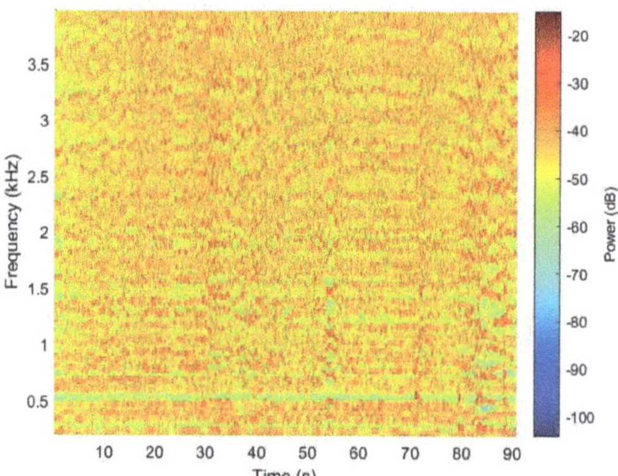

Figure 19. Normalized and band-pass filtered spectrogram of the signal from Figure 17.

The signal even after filtering has still a very noise-like shape (Figure 20). Only some slow changes in mean magnitude (corresponding also to the power) are visible.

Time bins for calculation of spectral metric are 8 ms long. Smoothed (with 20 samples moving mean as before) Euclidean metric referring to a mean power spectrum from first

5 s is particularly high in the ending seconds (Figure 21). The threshold was in this case set using mean value instead of mode value. Kernel-smoothed PDF is presented in Figure 22.

In the set of the marked intervals no clear pattern can be seen (Figure 23). Nevertheless, audition of the detection output and comparing it with the input raw sound allows us to state that all audible voice is detected (included in 4 last intervals). Parameters of Tol_0 and Tol_1 were again fixed to 0.5 s.

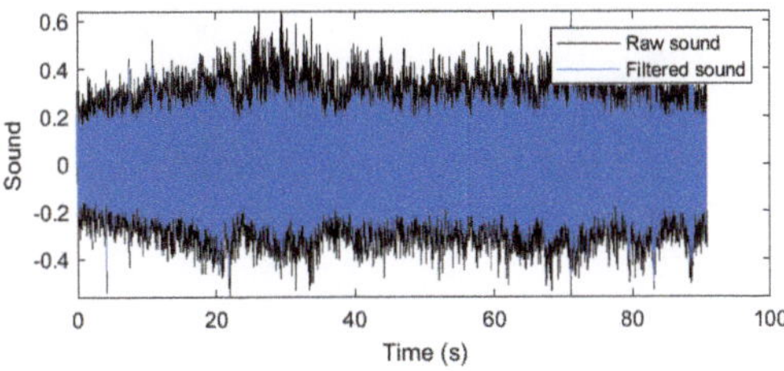

Figure 20. A band-pass filtering of the sound from Figure 17.

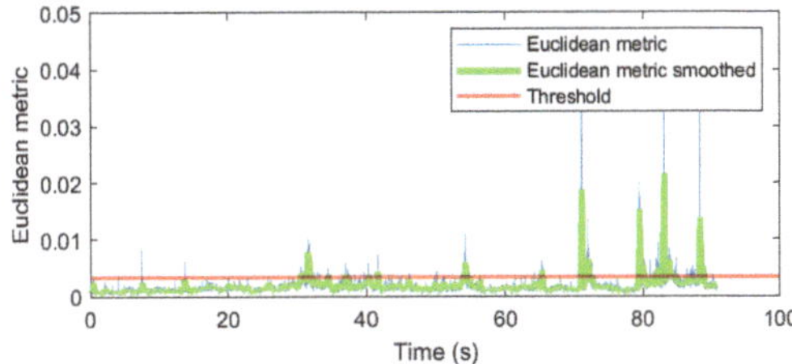

Figure 21. Smoothed Euclidean metric calculated from normalized spectrogram from Figure 19 as a function of time.

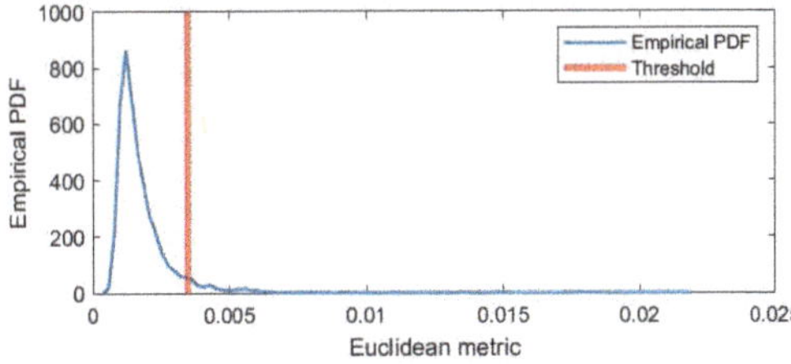

Figure 22. Empirical PDF of Smoothed Euclidean metric from Figure 21.

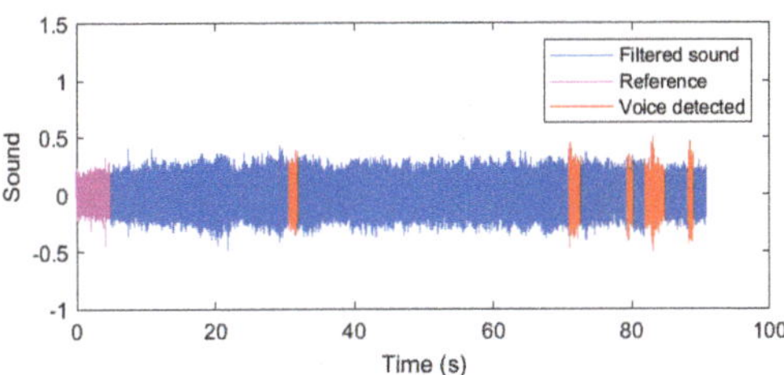

Figure 23. Detected voice time intervals extracted from filtered signals from Figure 20.

6. Validation Experiments

To evaluate the efficiency of the presented method, several lab condition experiments have been performed. We investigated the impact of the drone's flight conditions on the performance of our algorithm. During the validation experiments, we have additionally tested two other drones of different sizes (see Figure 24) to evaluate how the noise generated by the rotors of different power they are equipped with, will mask informative components related to a human voice.

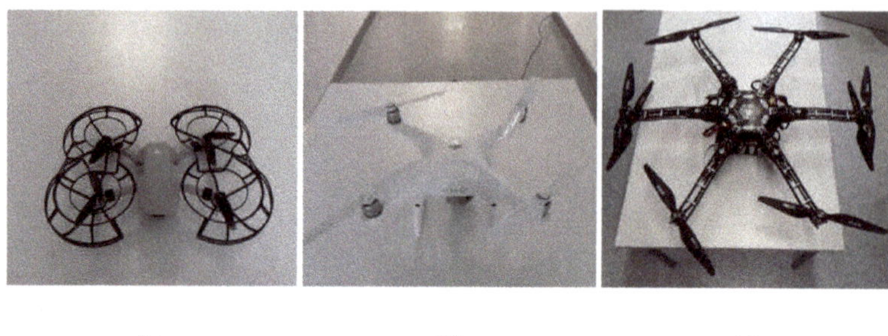

(a)   (b)   (c)

Figure 24. Drones used in the experiments: (**a**) small drone used both in mine and lab experiments, (**b**) medium size drone used in validation experiment, (**c**) large size drone used in validation experiment.

The second drone used for validation purposes is, similarly to the one used in the original experiment, a commercial quadcopter-DJI Phantom 4 Advanced, dimensions of which are 350×350 mm (the dimensions of Mavick Mini are 160×202 mm). The third one was a custom UAV, designed according to predefined specifications. The rotors' spacing is 900 mm. All sounds presented in this section were recorded using the same microphone with sampling frequency equal to 25.6 kHz.

First, almost stable flight conditions have been kept (see spectrogram in Figure 25). It means that the drone's ego-noise spectrum is more or less constant.

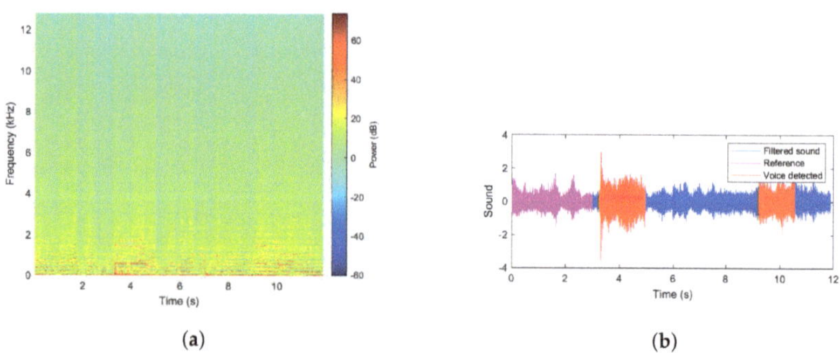

(a)   (b)

Figure 25. Example of a segment of the acoustic signal with almost constant flight conditions and neglectable spectral content fluctuation: (**a**) a raw spectrogram, (**b**) correct voice detection.

We intentionally introduced maneuvers during the flight to assess the variability of acoustic signal parameters. As we predicted, change of flight parameters will modify the spectrum significantly, including AM/FM modulations in the signal (see spectrogram in Figure 26).

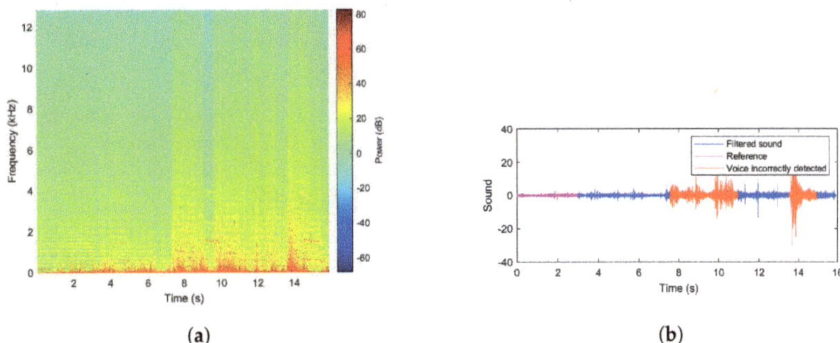

Figure 26. Example of a segment of the acoustic signal with highly variable flight conditions and significant spectral content fluctuation: (**a**) a raw spectrogram, (**b**) incorrect voice detection.

The introduced changes in flight parameters cause that human voice activity detection is way more difficult to be achieved. When the drone's flight parameters are approximately constant, the effectiveness of the algorithm used is very high (see Figure 25). As can be seen in Figure 26 the proposed method faces segmentation problems while the drone is performing the additional maneuvers. First of all, not every voice activity has been detected. Secondly, the noise generated by the drone is incorrectly detected as a human voice.

To simplify the problem, for two new drones (medium size and large size) we decided to analyze segments with stable flight conditions ignoring those with rapid maneuvers, i.e., strongly time-varying spectrum. The justification of this approach is the fact that in underground mines' conditions the dimensions of underground corridors are limited, hence the possibility of performing complex drone maneuvers during a flight is significantly limited. To reduce the risk of a situation in which the drone could be damaged as a result of a collision with excavation walls or other elements in the excavation area, the drone's flight speed must not be too high or variable, and the maneuvers should not be sudden. This translates into the type of recorded noise, the energy of which will not be high and the spectrum will not be highly variable in time. The above factors, although they have an impact on the effectiveness of the proposed method, will not occur often in real conditions in which the method is to be applied.

Results of additional experiments (focused on validation of the concept) are presented below in Figures 27 and 28. In Figure 27a, raw signal is presented. Only part of the signal plotted in black is analyzed using the proposed algorithm. Figure 27b presents the spectrogram of the analyzed signal, where a significant change in spectral content is clearly visible at ca. T = 5, 10, 15, 22, and 27 s. Figure 27c presents the Euclidean distance estimated for each spectrum. The threshold resulting from the PDF of such a measure is also plotted as a red line. Based on the proposed decision-making scheme, the final picture with detection results is presented in Figure 27d. All voice activities have been successfully detected.

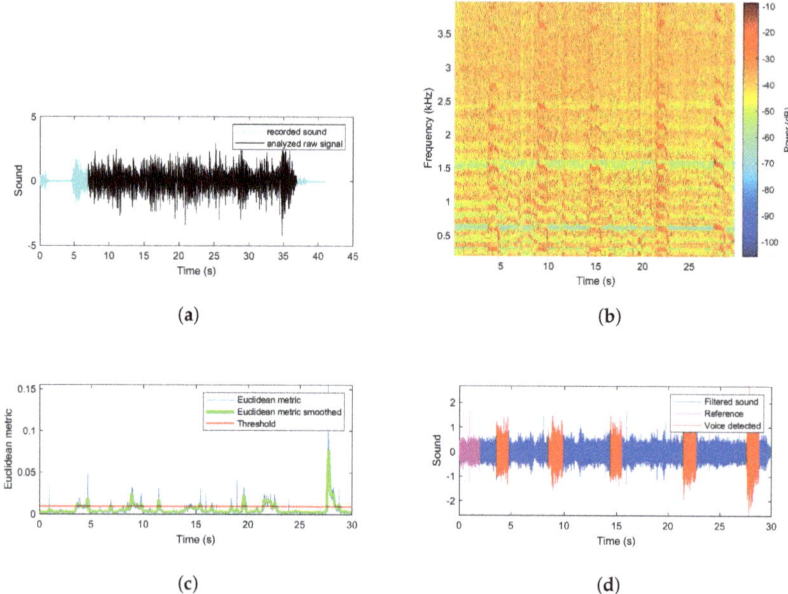

Figure 27. Results obtained with the medium size drone: (**a**) Original signal with segment selected for analysis, (**b**) normalized and band-pass filtered spectrogram selected for analysis, (**c**) estimated distance from reference spectrum with estimated threshold, (**d**) detection results.

By analogy, the same procedure has been applied in the third case (Figure 28). In Figure 28a a raw signal within the selected segment is presented. One may notice that for signals above $T = 40$ s its amplitudes are much higher; we did not select this part for further analysis. The max power of the drone and noise level were significantly higher resulting in lower SNR. We ignored this segment since in an underground mine, a drone ought to be steered very carefully and one should not risk flying with such parameters. The spectrogram (Figure 28b) for the 3rd case (selected part) is clear enough and allows detection of voice activities (Figure 28c). Final detection result visualization is presented in Figure 28d.

The results achieved in this case can raise concerns regarding the utility of the presented method when using bigger UAVs, nonetheless, taking into consideration the spatial restraints in underground mines, such drones are not recommended in this specific environment (bigger UAVs are much more cumbersome to be steered and can not reach as many locations as smaller ones).

One may conclude that based on the obtained results, the signal having its origin in a human voice is detectable in each case. Three different noise characteristics have been considered. During maneuvers (turning left/right, acceleration, etc.) the structure of drone ego-noise is time-varying and indeed it makes voice detection much more complicated. Given the aforementioned, we decided to skip the highly non-stationary segments during the detection procedure.

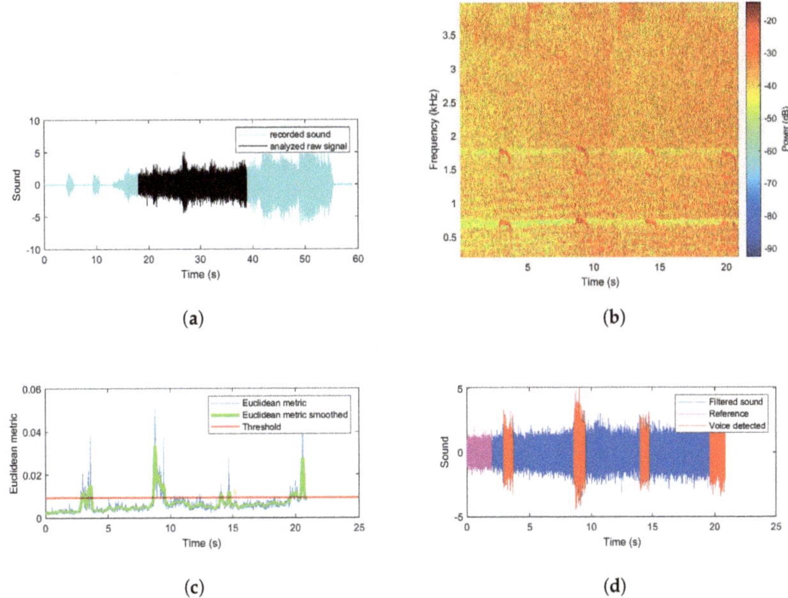

Figure 28. Results obtained with the large size drone: (**a**) raw sound recorded in the presence of a medium-size drone, (**b**) normalized and band-pass filtered spectrogram selected for analysis, (**c**) estimated distance from reference spectrum with estimated threshold, (**d**) detection results.

7. Discussion and Conclusions

A conception of drone usage for rescue operations in underground mines has been presented with the assumption of a situation in which an accident victim (being unable to rescue oneself from the accident's location) is capable of calling for help. Since the main problem to be solved in order to ensure human detection with the use of a microphone-equipped UAV was the distinction of a characteristic sound in the presence of heavy, time-varying noise, obtaining the automatic signal processing procedure is a valuable achievement, not only from the point of view of a rescue operation.

Our method used for voice detection tends to capture long phrases wholly, while a simple word /help/ is perceived mostly as /he/ or /e/. Hence high-energy voiced speech is better detected than low-energy noise-like unvoiced speech which is covered by the background noise. An improvement of the presented method can be sought in detecting voiced speech components having a harmonic structure. It should be noted that any source of the acoustic signal that is characteristic for a given situation and easily separable from drone-induced noise can be used to identify the event of interest. Taking this into consideration, future work will be conducted with the focus put on identifying other event-informative and distinguishable-from-noise sound sources, together with generalizing the signal processing technique presented. Such signals can be emitted by, for instance, electronic devices carried by underground workers. Characteristic spectral content obtained this way would be easier to detect than the human voice. The detection of human voice would then state a backup scenario in case of a lost or faulty sound emitting device.

The core of this research work—the procedure proposed in the article—is relatively simple, as it hinges on the usage of typical time-frequency representation of the registered signal and comparison of its instantaneous spectrum with a reference spectrum that is known to not contain the signal sought, but only drone's sound. It alludes directly to the novelty-detection approach, namely, if a spectrum emerges in the analyzed signal of an unknown or not expected structure, there is a high probability that it is related to an event of interest. Taking into account that during a rescue action, the presence of various noise sources is limited (due to the stopped machinery activity for instance), the abovementioned assumption can be found as legitimate. It should be underlined that what is of the highest importance is not a detected and distinguished specific human voice or a phrase, but rather, a quick and reliable segmentation of useful data (containing more than just a drone's noise) from a big dataset created with the use of a team of microphone-equipped UAVs.

Several laboratory experiments and the one conducted in realistic underground mine conditions proved that the developed procedure gives satisfactory results, and is a promising basis for the development of a useful aid tool for rescue operations. Nonetheless, its usage can be also extended to other cases in which drone-aided inspection would be safer or more efficient, which is a common case in underground mining.

The presented study is a preliminary test of a concept to be further developed, and subjected to evaluation in various use-case scenarios in different conditions. The method was tested with the use of various drones having different engine powers emitting noise of various levels. One of the key issues is the identification of drone noise. In general, it is expected that noise from the flying drone will be a non-stationary signal with a time-varying structure, especially when flight parameters are changing. At this stage, to simplify the signal processing procedure, the flight condition was as stable as possible. No maneuvers, except take-off and landing, have been considered. During the flight, the flight speed was relatively slow (noise is smaller too) and stable as much as possible. Its ultimate autonomous flight could be stabilized by the algorithm so the variability of speed will be neglectable. It simplifies the structure of the instantaneous spectrum and makes detection easier. The speed has not been measured directly, but from the spectrogram, it can be seen that it is nearly stable (frequencies related to rotor speed are not changing too much).

Even though the method can be described as 'self-calibrating' (due to the automated calculation of the referential spectrum), repeating the experiment numerous times could be beneficial since it would allow examining the method's response and accuracy giving very significant information regarding the so-called 'false alarm ratio'. In this research, we were not able to estimate the relation between signal to noise ratio and probability of detection; hence, this is assumed to be the basis of future work. Implementation of the proposed method in a real-case scenario of rescue action conducted in an underground mine must be aware of possible detection of some signal something that is not expected voice-related source. According to the specific procedures concerning rescue actions, during such a situation, all activities in the mine are supposed to be stopped until the end of rescue action. There is nearly no chance to record other noise than from the drone's operation and potential human calling for help. In addition, any false detections in such a situation are of poor probability, since the proposed method is intended to support rescue teams, so every potential detection will be verified.

Another more complicated issue is the lack of detection due to poor SNR of the acquired signal. This could be minimized by decreasing the distance between the sound source and the drone. It has been shown that if the drone is closer to the source, the signal is more visible in the noise. The task here is not to detect all signatures, so classical measures in detection efficiency analysis are not relevant. The main target is to provide data by the drone before the rescue team will be able to realize the rescue action.

Employing an autonomous UAV should be considered to obtain the benefit of reduction of the human presence in the most dangerous places as much as it is possible. It should be also stated that detection ability is related to several factors: drone noise level, human voice source level, the distance between the voice source and the drone with a microphone, etc. One could expect precise information on how close a drone should be to detect a human voice based on acoustic signals. It can be determined experimentally on the basis of many experiments in the mine. It is important to say that voice propagation in the tunnel may be slightly different due to tunnel geometry and surface properties.

Author Contributions: Conceptualization: P.Z. and R.Z.; Methodology: P.Z., P.T., J.S. and R.Z.; Software: P.Z.; Validation: P.T., J.S. and M.G.; Formal analysis: J.S., R.Z.; Investigation: J.S.; Resources: J.S.; Data analysis and visualization: J.S., A.W. (Adam Wroblewski); Writing—original draft: R.Z., M.G., A.W. (Adam Wroblewski); Writing—review & editing: J.S., R.Z., M.G., A.W. (Adam Wroblewski), P.T.; Supervision: R.Z., A.W. (Agnieszka Wójcik); Funding acquisition: R.Z., A.W. (Agnieszka Wójcik). All authors have read and agreed to the published version of the manuscript.

Funding: This activity has received funding from the European Institute of Innovation and Technology (EIT), a body of the European Union, under the Horizon 2020, the EU Framework Programme for Research and Innovation. This work is supported by EIT RawMaterials GmbH under Framework Partnership Agreement No. 19018 (AMICOS. Autonomous Monitoring and Control System for Mining Plants).

Institutional Review Board Statement: Not applicable.

Informed Consent Statement: Not applicable.

Data Availability Statement: Archived data sets cannot be accessed publicly according to the NDA agreement signed by the authors.

Acknowledgments: We would like to acknowledge colleagues from Digital Mining Center, GEO3EM Research Center and "Zloty Stok" Mine.

Conflicts of Interest: The authors declare no conflict of interest.

References

1. Tutak, M.; Brodny, J.; Szurgacz, D.; Sobik, L.; Zhironkin, S. The Impact of the Ventilation System on the Methane Release Hazard and Spontaneous Combustion of Coal in the Area of Exploitation—A Case Study. *Energies* **2020**, *13*, 4891, doi:10.3390/en13184891.
2. Szurgacz, D.; Sobik, L.; Brodny, J. Integrated method of reducing the threat of endogenous fires in hard coal mines. In Proceedings of the 14th International Innovative Mining Symposium, Kemerovo, Russian, 14–16 October 2019; Volume 105, doi:10.1051/e3sconf/201910501013.
3. Zhou, Q.; Herbert, J.H.; Hidalgo, A. Development of a stress-based approach for achieving the risk assessment of fault-related coal and gas outburst. *Int. J. Oil, Gas Coal Technol.* **2019**, *21*, 509, doi:10.1504/ijogct.2019.101473.
4. Ziętek, B.; Banasiewicz, A.; Zimroz, R.; Szrek, J.; Gola, S. A Portable Environmental Data-Monitoring System for Air Hazard Evaluation in Deep Underground Mines. *Energies* **2020**, *13*, 6331, doi:10.3390/en13236331.
5. State Mining Authority (Wyższy Urząd Górniczy), Assessment of Occupational Safety, Mine Rescue and General Safety in Relation to Mining and Geological Activities in 2019, 2020 (In Polish). Available online: https://wug.intracom.com.pl/download/WUG_Stan_bhp_19,7247.pdf (accessed on 11 June 2021).
6. Park, J.S.; Na, H.J. Front-End of Vehicle-Embedded Speech Recognition for Voice-Driven Multi-UAVs Control. *Appl. Sci.* **2020**, *10*, 6876, doi:10.3390/app10196876.
7. Cohen, I. Noise spectrum estimation in adverse environments: Improved minima controlled recursive averaging. *IEEE Trans. Speech Audio Process.* **2003**, *11*, 466–475, doi:10.1109/tsa.2003.811544.
8. Sohn, J.; Kim, N.S.; Sung, W. A statistical model-based voice activity detection. *IEEE Signal Process. Lett.* **1999**, *6*, 1–3, doi:10.1109/97.736233.
9. Zheng, B.; Hu, J.; Zhang, G.; Wu, Y.; Deng, J. Analysis of Noise Reduction Techniques in Speech Recognition. In Proceedings of the 2020 IEEE 4th Information Technology, Networking, Electronic and Automation Control Conference (ITNEC), Chongqing, China, 12–14 June 2020, doi:10.1109/itnec48623.2020.9084906.
10. Bachu, R.; Kopparthi, S.; Adapa, B.; Barkana, B. Voiced/Unvoiced Decision for Speech Signals Based on Zero-Crossing Rate and Energy. In *Advanced Techniques in Computing Sciences and Software Engineering*; Springer:Dordrecht, The Netherlands, 2009; pp. 279–282, doi:10.1007/978-90-481-3660-5_47.
11. Roman, N.; Wang, D.; Brown, G.J. Speech segregation based on sound localization. *J. Acoust. Soc. Am.* **2003**, *114*, 2236–2252, doi:10.1121/1.1610463.
12. Gagliardi, G.; Tedesco, F.; Casavola, A. An Adaptive Frequency-Locked-Loop Approach for the Turbocharger Rotational Speed Estimation via Acoustic Measurements. *IEEE Trans. Control. Syst. Technol.* **2020**, 1–13, doi:10.1109/tcst.2020.3007075.
13. Cavina, N.; Cesare, M.D.; Ravaglioli, V.; Ponti, F.; Covassin, F. Full Load Performance Optimization Based on Turbocharger Speed Evaluation via Acoustic Sensing. In *Volume 2: Instrumentation, Controls, and Hybrids*; Numerical Simulation; Engine Design and Mechanical Development; Keynote Papers; American Society of Mechanical Engineers: New York, NY, USA, 2014; doi:10.1115/icef2014-5677.
14. Ravaglioli, V.; Cavina, N.; Cerofolini, A.; Corti, E.; Moro, D.; Ponti, F. Automotive Turbochargers Power Estimation Based on Speed Fluctuation Analysis. *Energy Procedia* **2015**, *82*, 103–110, doi:10.1016/j.egypro.2015.11.889.
15. Tinney, C.E.; Sirohi, J. Multirotor Drone Noise at Static Thrust. *AIAA J.* **2018**, *56*, 2816–2826, doi:10.2514/1.j056827.

16. Zhou, T.; Jiang, H.; Sun, Y.; Fattah, R.J.; Zhang, X.; Huang, B.; Cheng, L. Acoustic characteristics of a quad-copter under realistic flight conditions. In Proceedings of the 25th AIAA/CEAS Aeroacoustics Conference. American Institute of Aeronautics and Astronautics, Delft, The Netherlands, 20–23 May 2019; doi:10.2514/6.2019-2587.

17. Wang, L.; Sanchez-Matilla, R.; Cavallaro, A. Tracking a moving sound source from a multi-rotor drone. In Proceedings of the 2018 IEEE/RSJ International Conference on Intelligent Robots and Systems (IROS), Madrid, Spain, 1–5 October 2018; doi:10.1109/iros.2018.8594483.

18. Wang, L.; Cavallaro, A. Acoustic Sensing From a Multi-Rotor Drone. *IEEE Sens. J.* **2018**, *18*, 4570–4582, doi:10.1109/jsen.2018.2825879.

19. Djurek, I.; Petosic, A.; Grubesa, S.; Suhanek, M. Analysis of a Quadcopter's Acoustic Signature in Different Flight Regimes. *IEEE Access* **2020**, *8*, 10662–10670, doi:10.1109/ACCESS.2020.2965177.

20. Wyłomańska, A.; Zimroz, R. Signal segmentation for operational regimes detection of heavy duty mining mobile machines-A statistical approach. *Diagnostyka* **2014**, *15*, 33–42.

21. Wodecki, J.; Stefaniak, P.; Obuchowski, J.; Wylomanska, A.; Zimroz, R. Combination of principal component analysis and time-frequency representations of multichannel vibration data for gearbox fault detection. *J. Vibroeng.* **2016**, *18*, 2167–2175, doi:10.21595/jve.2016.17114.

22. Zak, G.; Wylomanska, A.; Zimroz, R. Local Damage Detection Method Based on Distribution Distances Applied to Time-Frequency Map of Vibration Signal. *IEEE Trans. Ind. Appl.* **2018**, *54*, 4091–4103, doi:10.1109/tia.2018.2828787.

23. Liu, J. Current research, key performances and future development of search and rescue robot. *Chin. J. Mech. Eng.* **2006**, *42*, 1, doi:10.3901/jme.2006.12.001.

24. Murphy, R.R.; Tadokoro, S.; Nardi, D.; Jacoff, A.; Fiorini, P.; Choset, H.; Erkmen, A.M. Search and Rescue Robotics. In *Springer Handbook of Robotics*; Springer: Berlin/Heidelberg, Germany, 2008; pp. 1151–1173, doi:10.1007/978-3-540-30301-5_51.

25. Liu, Y.; Nejat, G. Robotic Urban Search and Rescue: A Survey from the Control Perspective. *J. Intell. Robot. Syst.* **2013**, *72*, 147–165, doi:10.1007/s10846-013-9822-x.

26. Said, K.O.; Onifade, M.; Githiria, J.M.; Abdulsalam, J.; Bodunrin, M.O.; Genc, B.; Johnson, O.; Akande, J.M. On the application of drones: A progress report in mining operations. *Int. J. Mining, Reclam. Environ.* **2020**, 1–33, doi:10.1080/17480930.2020.1804653.

27. Szrek, J.; Wodecki, J.; Błazej, R.; Zimroz, R. An inspection robot for belt conveyor maintenance in underground mine-infrared thermography for overheated idlers detection. *Appl. Sci.* **2020**, *10*, 4984, doi:10.3390/app10144984.

28. Szrek, J.; Zimroz, R.; Wodecki, J.; Michalak, A.; Góralczyk, M.; Worsa-Kozak, M. Application of the infrared thermography and unmanned ground vehicle for rescue action support in underground mine—the amicos project. *Remote Sens.* **2021**, *13*, 69, doi:10.3390/rs13010069.

29. Zimroz, R.; Hutter, M.; Mistry, M.; Stefaniak, P.; Walas, K.; Wodecki, J. Why Should Inspection Robots be used in Deep Underground Mines? In *Proceedings of the 27th International Symposium on Mine Planning and Equipment Selection-MPES 2018*; Widzyk-Capehart, E., Hekmat, A., Singhal, R., Eds.; Springer International Publishing: Cham, Sitzerland, 2019; pp. 497–507.

30. Szrek, J.; Trybala, P.; Goralczyk, M.; Michalak, A.; Zietek, B.; Zimroz, R. Accuracy evaluation of selected mobile inspection robot localization techniques in a gnss-denied environment. *Sensors* **2021**, *21*, 141, doi:10.3390/s21010141.

31. Skoczylas, A.; Stefaniak, P.; Anufriiev, S.; Jachnik, B. Belt Conveyors Rollers Diagnostics Based on Acoustic Signal Collected Using Autonomous Legged Inspection Robot. *Appl. Sci.* **2021**, *11*, 2299, doi:10.3390/app11052299.

32. Waharte, S.; Trigoni, N. Supporting Search and Rescue Operations with UAVs. In *Proceedings of the* 2010 International Conference on Emerging Security Technologies, Canterbury, UK, 6–7 September 2010, doi:10.1109/est.2010.31.

33. Sun, J.; Li, B.; Jiang, Y.; yung Wen, C. A Camera-Based Target Detection and Positioning UAV System for Search and Rescue (SAR) Purposes. *Sensors* **2016**, *16*, 1778, doi:10.3390/s16111778.

34. Malos, J.; Beamish, B.; Munday, L.; Reid, P.; James, C. Remote monitoring of subsurface heatings in opencut coal mines. In Proceedings of the 2013 Coal Operators' Conference. University of Wollongong, Wollongong, Australia, 14–15 February 2013; pp. 227–231.

35. Nanda, S.K.; Dash, A.K.; Acharya, S.; Moharana, A. Application of robotics in mining industry: A critical review. *Indian Min. Eng. J.* **2010**, *8*, 108–112.

36. Siebert, S.; Teizer, J. Mobile 3D mapping for surveying earthwork projects using an Unmanned Aerial Vehicle (UAV) system. *Autom. Constr.* **2014**, *41*, 1–14, doi:10.1016/j.autcon.2014.01.004.

37. Azhari, F.; Kiely, S.; Sennersten, C.; Lindley, C.; Matuszak, M.; Hogwood, S. A comparison of sensors for underground void mapping by unmanned aerial vehicles. In Proceedings of the First International Conference on Underground Mining Technology. Australian Centre for Geomechanics, Perth, Austria, 11–13 October 2017, doi:10.36487/acg_rep/1710_33_sennersten.

38. Ren, H.; Zhao, Y.; Xiao, W.; Hu, Z. A review of UAV monitoring in mining areas: Current status and future perspectives. *Int. J. Coal Sci. Technol.* **2019**, *6*, 320–333, doi:10.1007/s40789-019-00264-5.

39. Li, H.; Savkin, A.V.; Vucetic, B. Autonomous Area Exploration and Mapping in Underground Mine Environments by Unmanned Aerial Vehicles. *Robotica* **2019**, *38*, 442–456, doi:10.1017/s0263574719000754.

40. Turner, R.M.; MacLaughlin, M.M.; Iverson, S.R. Identifying and mapping potentially adverse discontinuities in underground excavations using thermal and multispectral UAV imagery. *Eng. Geol.* **2020**, *266*, 105470, doi:10.1016/j.enggeo.2019.105470.

41. Reddy, A.H.; Kalyan, B.; Murthy, C.S. Mine Rescue Robot System–A Review. *Procedia Earth Planet. Sci.* **2015**, *11*, 457–462, doi:10.1016/j.proeps.2015.06.045.

42. Lee, S.; Choi, Y. Reviews of unmanned aerial vehicle (drone) technology trends and its applications in the mining industry. *Geosyst. Eng.* **2016**, *19*, 197–204, doi:10.1080/12269328.2016.1162115.
43. Park, S.; Choi, Y. Applications of Unmanned Aerial Vehicles in Mining from Exploration to Reclamation: A Review. *Minerals* **2020**, *10*, 663, doi:10.3390/min10080663.
44. Shahmoradi, J.; Talebi, E.; Roghanchi, P.; Hassanalian, M. A Comprehensive Review of Applications of Drone Technology in the Mining Industry. *Drones* **2020**, *4*, 34, doi:10.3390/drones4030034.
45. Allen, J. Short term spectral analysis, synthesis, and modification by discrete Fourier transform. *IEEE Trans. Acoust. Speech Signal Process.* **1977**, *25*, 235–238, doi:10.1109/tassp.1977.1162950.
46. Wyłomańska, A.; Żak, G.; Kruczek, P.; Zimroz, R. Application of tempered stable distribution for selection of optimal frequency band in gearbox local damage detection. *Appl. Acoust.* **2017**, *128*, 14–22, doi:10.1016/j.apacoust.2016.11.008.
47. Hebda-Sobkowicz, J.; Zimroz, R.; Wyłomanska, A. Selection of the informative frequency band in a bearing fault diagnosis in the presence of non-gaussian noise-Comparison of recently developed methods. *Appl. Sci.* **2020**, *10*, 2657, doi:10.3390/APP10082657.
48. Hebda-Sobkowicz, J.; Zimroz, R.; Pitera, M.; Wyłomańska, A. Informative frequency band selection in the presence of non-Gaussian noise—A novel approach based on the conditional variance statistic with application to bearing fault diagnosis. *Mech. Syst. Signal Process.* **2020**, *145*, doi:10.1016/j.ymssp.2020.106971.
49. Wodecki, J.; Michalak, A.; Zimroz, R. Optimal filter design with progressive genetic algorithm for local damage detection in rolling bearings. *Mech. Syst. Signal Process.* **2018**, *102*, 102–116, doi:10.1016/j.ymssp.2017.09.008.

Review

A Technology of Hydrocarbon Fluid Production Intensification by Productive Stratum Drainage Zone Reaming

Oleg Bazaluk [1], Orest Slabyi [2], Vasyl Vekeryk [2], Andrii Velychkovych [3], Liubomyr Ropyak [4] and Vasyl Lozynskyi [5,*]

1   Belt and Road Initiative Institute for Chinese-European Studies (BRIICES), Guangdong University of Petrochemical Technology, Maoming 525000, China; bazaluk@ukr.net
2   Department of Technical Mechanics, Ivano-Frankivsk National Technical University of Oil and Gas, 076019 Ivano-Frankivsk, Ukraine; burewisnyk@gmail.com (O.S.); vasyl_vekeryk@hotmail.com (V.V.)
3   Department of Construction and Civil Engineering, Ivano-Frankivsk National Technical University of Oil and Gas, 076019 Ivano-Frankivsk, Ukraine; a_velychkovych@ukr.net
4   Department of Computerized Engineering, Ivano-Frankivsk National Technical University of Oil and Gas, 076019 Ivano-Frankivsk, Ukraine; l_ropjak@ukr.net
5   Department of Mining Engineering and Education, Dnipro University of Technology, 49005 Dnipro, Ukraine
*   Correspondence: lvg.nmu@gmail.com

Abstract: The paper proposes a new technology for fluid production intensification, in particular hydrocarbons, which is implemented via significant increasing of the local wellbore diameter in the interval, where the productive stratum is present. The proposed technology improves the well productivity by increasing the filtration surface area and opening new channels for filtering fluids into the well. The innovative, technical idea is to drill large diameter circular recesses in planes perpendicular to the well axis. After that, the rock mass located between the circular recesses are destroyed by applying static or dynamic axial loads. The required value of the axial force is provided by the weight of the standard drilling tool. As a result of the study, the analytical relations to specify the admissible radius of circular recesses and admissible thickness of rock mass between two adjacent circular recesses from the condition of safe operation are obtained. The numerical analysis carried out for typical reservoir rocks substantiated the possibility of well diameter local reaming twenty times. A special tool for circular recess drilling is developed and the principle of its operation is described. The advantage of the proposed approaches is the low energy consumption for well diameter reaming. Our technology will have special economic expediency for the intensification of production from hydrodynamically imperfect wells and under the condition of fluid filtration according to the expressed nonlinear law.

Keywords: well reamer; rock mass; intensification of production; nonlinear law; completion of wells; innovative technology

Citation: Bazaluk, O.; Slabyi, O.; Vekeryk, V.; Velychkovych, A.; Ropyak, L.; Lozynskyi, V. A Technology of Hydrocarbon Fluid Production Intensification by Productive Stratum Drainage Zone Reaming. *Energies* **2021**, *14*, 3514. https://doi.org/10.3390/en14123514

Academic Editor: Sergey Zhironkin

Received: 19 May 2021
Accepted: 10 June 2021
Published: 13 June 2021

Publisher's Note: MDPI stays neutral with regard to jurisdictional claims in published maps and institutional affiliations.

1. Introduction

Today, the current problem of the energy industry is development of classical and new technologies (including environmentally attractive and energy-saving mining ones) to increase well productivity for oil/gas drilling [1,2]. New technological schemes for coal reserves working out with gasification are developing, which will additionally allow non-commercial and abandoned mine reserves to be used and the mining enterprises duration to be extended [3,4]. In general, the country's energy sector is the basis of the economy and industry, because any production requires energy resources [5–7]. Our study is involved in solving this problem.

A well flow rate increases if the well diameter increases for the interval of the productive stratum, the depression of the stratum decreases, as well as the pressure gradient of the well wall, and the drainage zone expands. Note that we are not talking about the classic reamer applications to increase wellbore but about increasing the diameter of the well more than five times in a clearly defined interval. It is possible to increase the diameter of the wellbore in the interval of the productive stratum at the end of well development immediately after drilling or during the overhaul of existing production or injection wells. Sometimes, liquid and gas extraction from the productive stratum is limited by the requirement not to exceed a certain value of the filtration rate, at which the intensive removal of sand into the well begins. If the production is limited by the abovementioned geological factor, the maximum allowable flow rate of the well is directly proportional to the radius of its bottomhole.

Based on the analysis of radial and spherical inflows of linear or nonlinear fluids to the well, the following conclusion was made.

The greater effect of well diameter change on the flow rate is associated with the more significant well deviation from the hydrodynamic perfection of its completion and the more expressed nonlinearity of the fluid filtration law. This is especially expressed at a significant flow rate of the well when there is a crisis of the linear law of filtration in the bottomhole zone. In this case, the predicted increase in the well flow rate can reach the value of $n^{2/3}$, where n—the multiplicity of the well diameter—increases [8–10].

The works [11,12] have carried out a thorough analysis of the expected effects of the well diameter increasing in the productive stratum interval. In particular, this makes it possible to solve the following tasks: the colmatted layer removal in the bottomhole zone to reduce the resistance of fluid inflow into the well; completion of existing and creation of new filtration channels especially typical for limestone collectors; and prevention of sand separation by reducing depression on the stratum and creating quality gravel filters.

Therefore, a significant increase of the well diameter in the productive stratum can provide a significant increase of well productivity. Thus, development of a technology for the significant increasing of the wellbore diameter of the productive stratum interval is an issue at stake for the oil and gas industry. This technology should be technically simple and implemented at a reasonable financial cost. The urgency of the task and the expected profitability of the results of its solution became the main motivators of our study.

To date, a number of devices have been designed to expand the diameter of the well in a certain interval. Usually, drilling tools operate in aggressive and abrasive environments under the action of intense loads, so their performance is subject to special requirements [13,14]. This issue is especially significant for tools used for difficult technological operations. In particular, a number of methods are applied to improve the performance of high-tech drilling tools. Design methods envisage drill pipe design modification [15], threaded joint improving [16–18], ensuring the accuracy of threads [19] and preventing their self-unscrewing [20], parametric optimization of cutters, supports and washing units of rock-destroying tools [21,22], etc. Among the technological methods are the rational reinforcement of steel parts [23], optimization of the formation of helical surfaces and surfaces of complex topology [24], and application of flexible [25,26] and functionally gradient [27–30] coatings. Operational methods to increase drilling tool performance include substantiation of drilling modes taking into account the force [31,32] and temperature interactions [33], ensuring thorough flushing of the bottomhole [34], and usage of vibration protection for drill strings [35,36]. Researchers pay special attention to effective ways to eliminate complications during drilling and improve well completion technologies [37,38]. The design features of drilling tools, specifics of operation conditions and contact interaction with the well cause vibrations and dynamic loads. General approaches to contact analysis of pipes and drilling tools with an elastic medium are presented in [39,40]. The most radical way to solve the problem of tool vibration protection is based on the use of special vibration protection devices: elastic couplings, drill shock absorbers and elastic spindles of downhole motors [41], and dynamic extinguishers and specialized dampers [42].

In general, modern equipment to ream well diameter can be categorized by design into two groups. The first group includes reamers with fixed PDC cutters [43] and reamers with retractable cutters [44,45], which are not able to increase the well diameter more than twice. Such devices are mainly used to remove the colmatted layer from the walls of the well, align the wellbore, increase the drilling speed of large diameter wells, etc. These tools are actively used for both marine and earth drilling, but their designs do not allow well diameter to be significantly reamed and this can significantly affect the flow rate.

The second group includes devices to ream the well diameter more than twice. Among the mechanical devices, this group include reamers with deflecting PDC cutters and reamers with deflecting cones on the legs [46,47]. Hypothetically, such devices can ream the well diameter several times for several paths. Non-mechanical ways to expand the wellbore for significant increasing of the well diameter include erosion of the rock using a rotary hydro-sandblasting head [48] or erosion of the rock by circulating fluid outside the casing [49]. A geomechanical model of rock mass, the behavior of which depends on the mining and geological conditions and mining parameters of the hydrocarbon production process, is presented in [50–52]. In studies [53,54] the experimental study of the thermal reaming of the borehole via axial plasmatron are discussed. Although the considered methods allow the well diameter to be significantly reamed, their application is associated with high energy consumption and low productivity for cases when it is necessary to increase the well diameter more than five times. In our opinion, the main gap in the known research is the inefficient use of axial force to implement the process of well diameter local reaming. The disadvantage of the considered non-mechanical methods of well reaming is the entire rock mass is destroyed by the directed flow. Accordingly, the application of such methods requires significant energy and time, and their efficiency depends on the mechanical properties of the rock and decreases nonlinearly with the increasing diameter of the recess. Devices for mechanical reaming of the well diameter destroy the rock mass via cutting and require a significant torque. It should be noted that the bottomhole assembly limits the value of torque that can be applied to the tool. Therefore, obtaining recesses of a large diameter is possible only when drilling in several passes. The drill string has a lot of weight and we are able to create a large axial load without any technical difficulties. In view of this, the idea arose to develop an approach that would allow the effective use of axial force to destroy rocks in the process of local reaming of the well diameter.

This study aims to develop technological and technical support for local increasing of well diameter (more than five times) in the productive stratum interval. To achieve this goal, the following tasks were set:

- specification of ideas and hypotheses of the study;
- development of formation stages of the expanded drainage zone of a productive stratum;
- specification of admissible dimensions of circular recesses and interval of their drilling from conditions of safety operation;
- testing the possibility of the proposed technology application based on the condition of strength of the formed reservoir rock mass; and
- development of the technical mean design to implement the technology of well diameter local reaming.

2. Materials and Methods

At the first stage of the research, we used the bibliosemantic method and content analysis to deeply study the problem and choose an effective way to solve it. The analysis of previous research based on scientific literature sources, electronic resources, as well as practical and production experience of the authors made it possible to select the necessary scientific data according to a certain logic, classify them, and specify relations links and relationships between them. As a result, the purpose and objectives of the study were formulated.

The main ideas and concepts of the innovative technology for well diameter reaming in the area of productive stratum were formulated in order to increase its flow rate using the methods of structural-logical and system analysis. It was determined that the well reaming process should be carried out in several stages, and well diameter increasing should be significant (more than five times) and carried out in a clearly defined interval of the well. The process of the technology development was organized to provide the main volume of the rock collapsing only under the influence of axial force, the magnitude of which is provided by the weight of the drilling tool.

Based on the methods of mechanics of deformable solids, the problem of the limit equilibrium of the rock mass formed by two circular recesses was formulated and solved. An analytical method for solving this problem has been developed to determine the stress state and evaluate the strength of the considered system. The reliability of the obtained results was confirmed via the validity of the geometric-linear formulation of the problem, strict implementations of mathematical methods tested in the literature for analytical research, and convergence of the results of partial (limit) cases with known results. In general, the possibility of applying the proposed technology for typical reservoir rocks is substantiated.

For numerical approbation of the results, some typical rocks were chosen: siltstone, sandstone, fine-grained limestone, and dolomite.

Next, we applied the so-called method of the basic unit (basic design). The method is based on the usage of a basic structure to be transformed into a machine for the desired purpose by attaching special equipment. The basic unit is a drilling tool, which is produced in series. Therefore, using the method of the basic unit (basic design), as well as the sequential compounding method, a special tool and special configurations of the drill string were developed for the practical implementation of the proposed technology.

Let us formulate the basic concepts of a new method of well diameter significant reaming in the area of productive stratum to increase the flow rate of the well. Well diameter reaming envisages several stages, which are schematically shown in Figure 1.

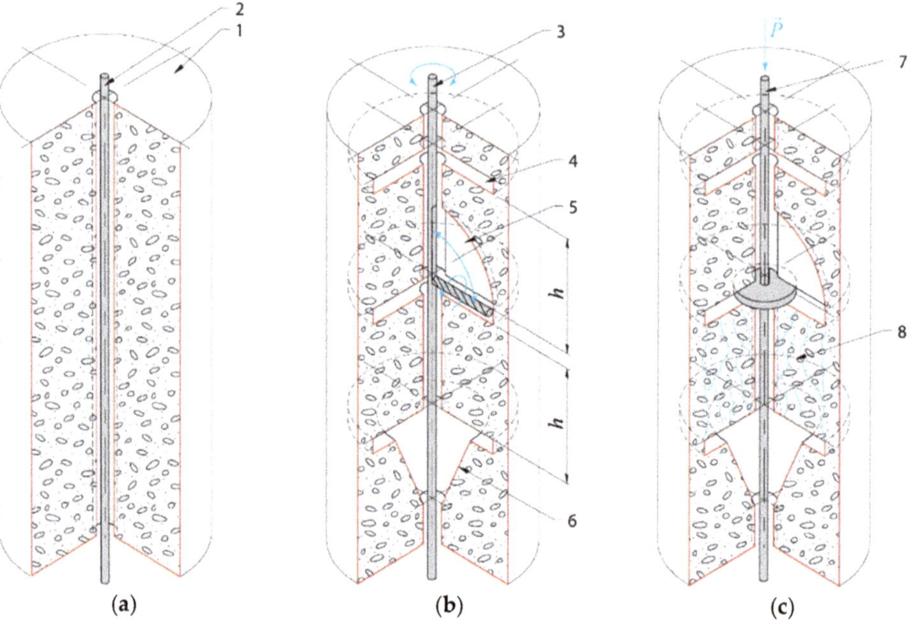

(a)   (b)   (c)

Figure 1. Stages of productive stratum expanded drainage zone formation: 1—rock, 2—drill string, 3—special bottomhole assembly for radial hole drilling in the well, 4—circular recess, 5—recess formed by drilling of the radial hole, 6—conical hole, 7—special bottomhole assembly for cantilever rock destruction and grinding, 8—zone of rock destruction by static and dynamic axial force.

The first stage is drilling of the well with the nominal radius of r in the productive stratum (Figure 1a). After that, a special bottomhole assembly is lowered into the well with a device with a bit that can be deflected. This bit drills a deep circular recess with the radius of R in the wellbore interval to be reamed with the predetermined step of h. The recess that is furthest from the bottomhole is drilled in the form of a cone (Figure 1b). Next, the bottomhole assembly with a jar or other device that can create a variable axial load and a device with retractable legs is lowered into the well. This assembly applies static and dynamic axial force to rock mass formed by two circular recesses. As a result, the rock mass is destroyed, crushed, and removed by washing liquid to the surface (Figure 1c).

3. Results

3.1. Specification of Circular Recess Allowable Dimensions and Interval of Its Drilling

The proposed technology envisaged that the rock mass between circular recesses should remain intact at the second stage of the operation (Figure 1b) and be guaranteed to be destroyed when applying axial force, which is technologically possible to apply at the third stage of the process (Figure 1c).

The allowable size of circular recesses and the interval of their drilling h (see Figure 1b) are specified by the strength of the rock mass between two adjacent recesses. We assume intuitively the following: too small a specified drilling interval or a too big radius of circular recesses can lead to premature destruction of the rock mass and an emergency situation (clamping the deviated bit by the rock mass). In addition, if the drilling interval is specified too big, the maximum applicable axial force P may not be sufficient to destroy the rock mass.

To find out whether we are right in the declared assumptions, let us consider the following problem.

Let us consider the rock mass in the form of a round plate with the thickness of h and the outer radius of R having a technological slot and a through hole with the radius of r (Figure 2). The plate is loaded with bulk forces, the intensity of which is determined by the specific weight of the rock, and the plate can be additionally loaded with axial force P.

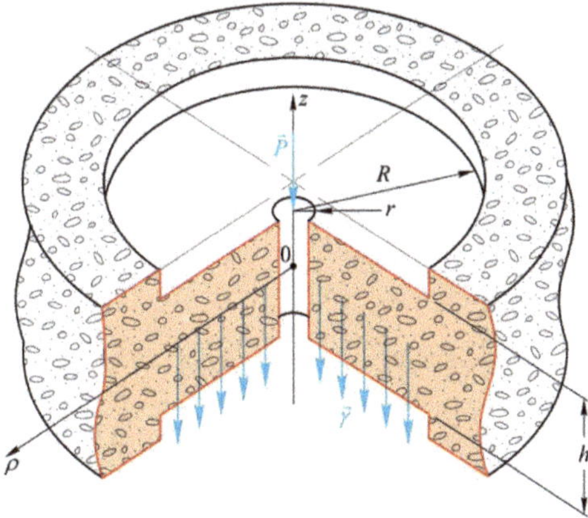

Figure 2. Calculation scheme of the rock mass.

The proposed technology assumes that the plate is thick. As the thick plate is not very susceptible to bending, we will take into account the shear of the rock when assessing the strength. Let us consider the limit equilibrium of the plate (Figure 2):

$$\int_s \tau\, ds - \int_V \gamma\, dV - P = 0 \tag{1}$$

After the transformations we obtain a relation for the tangential stress averaged over the thickness of the plate:

$$\tau = \frac{1}{2}\left(\frac{P}{\pi \rho h} + \gamma \rho\left[1 - \left(\frac{r}{\rho}\right)^2\right]\right), \ \rho \in [r, \ R],\tag{2}$$

where s is the area of the plate side surface of radius ρ and V is the volume of the plate.

Analysis of relation (1) indicates that large values of P near the through hole may have a boundary effect (the case when the tangential stresses in the hole are greater than the stresses in the clamping zone of the plate). We assume that the technical means of plate loading (a device with retractable legs) will eliminate the manifestation of this effect. Therefore, we neglect the influence of the described boundary effect and assume $\rho = R$ in expression (1) and obtain the rock mass strength condition:

$$\tau_{\max} = \frac{1}{2}\left(\frac{P}{\pi R h} + \gamma R\left[1 - \left(\frac{r}{R}\right)^2\right]\right) \le \tau_u,\tag{3}$$

where τ_u is the maximum resistance to rock displacement (determined by the strength passport of a particular rock).

Analysis of relation (2) shows that the boundary state of the plate loaded by bulk forces depends on the plate radius and the specific gravity of the reservoir rock. Therefore, in order to provide integrity of the plate formed by two adjacent circular recesses within drilling, it is necessary to meet the following conditions:

$$\frac{1}{2}\gamma R\left[1 - \left(\frac{r}{R}\right)^2\right] \le \frac{\tau_u}{\zeta\, k},\tag{4}$$

where ζ is the coefficient of operational conditions (depends on the coefficient of water saturation of the rock, the coefficient of porosity, thermal impact, etc.) and k is the coefficient of strength.

The thickness of the rock mass $h_{\max}$ between the circular recesses should be specified from the condition of its possible destruction under the applied axial force. Therefore, the maximum thickness of the rock mass $h_{\max}$ to be destroyed by the applied axial force P is specified by the formula:

$$\frac{1}{2}\gamma R\left[1 - \left(\frac{r}{R}\right)^2\right] \le \frac{\tau_u}{\zeta\, k},\tag{5}$$

where λ is the coefficient of axial force application dynamism (for static application of axial force $\lambda = 1$, for sudden application). If there is a necessity for percussion mechanisms or application of deep vibrators, the coefficient of dynamism can be determined using the methods presented in [55–57].

3.2. Specification of Circular Recess Allowable Dimensions and Interval of Its Drilling

The possibility of applying the proposed method of wellbore reaming is limited by the strength of the cantilever rocks and the possibility of their further destruction. To do this, we numerically estimated several typical types of reservoirs. Hydrocarbon collectors are rocks with pores, cavities, or systems of cracks and are able to retain and filter fluids (oil, gas, etc.). The vast majority of reservoir rocks are of sedimentary origin, but in practice, there is often a combination of different types of reservoirs with a predominance of one or another type. Four typical reservoir rocks were selected for the study; Table 1 presents their characteristics [58].

We assume that the rock mass with the following characteristics is formed as a result of drilling: $r = 0.2\ m$, $\zeta = 0.2$, $k = 1.5$.

According to (3) we obtain the maximum stresses, $\tau_{\max}$, in the reservoir rocks referring to the radius of the circular recesses (Figure 3). Assuming that the maximum stresses, $\tau_{\max i}$, for the selected rock are equal to the allowable stresses, $[\tau]_i$, we determine the allowable radius of the circular recesses $[R]_i$. For the selected rocks, the allowable radius of the recesses are as follows: siltstone $[R]_1 = 2.82$ m, sandstone $[R]_2 = 3.1$ m, fine-grained limestone $[R]_3 = 3.46$ m, and dolomite $[R]_4 = 2.18$ m.

Table 1. Characteristics of reservoir rocks.

Reservoir Rock	Siltstone	Sandstone	Limestone Fine-Grained	Dolomite
Index i	1	2	3	4
Specific weight of the rock, N/m^3	1.95×10^4	2.4×10^4	2.71×10^4	2.94×10^4
Maximum shear resistance, KPa	82	110	140	95
Allowable stresses, $[\tau]_i = \tau_{ui}/\zeta_i\, k_i$, KPa	54.7	73.3	93.3	63.3

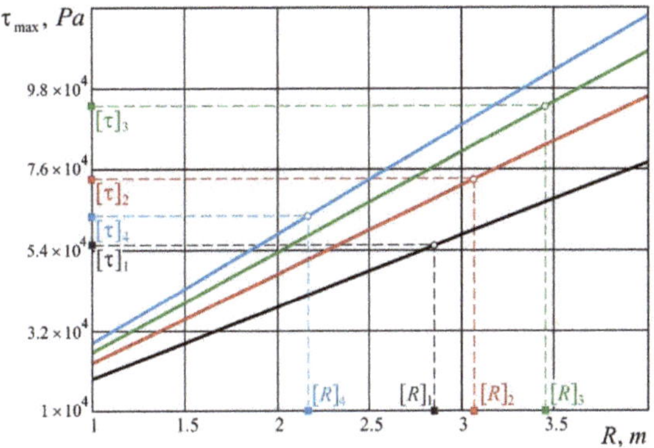

Figure 3. Determination of allowable drilling radii of circular recesses from the condition of rock mass strength: 1—siltstone, 2—sandstone, 3—fine-grained limestone, 4—dolomite.

Due to design and technological limitations of the device for circular recesses drilling, their radius does not exceed 2.5 m ($R \leq 2.5$ m). Therefore, the obtained results show that based on the technical capabilities of recess drilling, the condition of strength of the cantilever rock mass is met for most reservoir rocks.

To determine the maximum allowable thickness of the rock mass, which can be formed between two adjacent circular recesses, we use formula (4). Axial static loading of 1000 KN was statically applied to the formed rock mass by heavy weight drill pipes. Figure 4 shows the allowable thickness of the rock mass referring to the circular recess radius (abscissas of the curve break corresponds to the values of $[R]_i$). It is observed that the radius of the rock mass (if other conditions are constant) leads to a rapid decrease in the allowable thickness of the rock mass only at a certain interval. Outside this interval, the rate of decline of the allowable thickness decreases, stabilizes, and then stops altogether. The reliability of the obtained results is confirmed by the validity of the problem statement and the strict implementation of the mechanical and mathematical methods tested in the literature for analytical research [25,42].

In practice, the thickness h of the rock mass should be slightly less than the calculated value h_{max} to ensure guaranteed destruction of the rock and to neutralize the error of the bottomhole assembly position in the well. Under certain conditions, a certain step of drilling circular recesses h at their radius of $R = 1.7$–2.5 m makes it difficult to implement the proposed technology using only static axial force (for example, reservoir rock and $i = 3$ in Figure 4). Note that h_{max} linearly depends on the coefficient of dynamism λ. Therefore, if necessary, h_{max} can be adjusted to a larger range to provide rock mass destruction by applying a dynamic axial load.

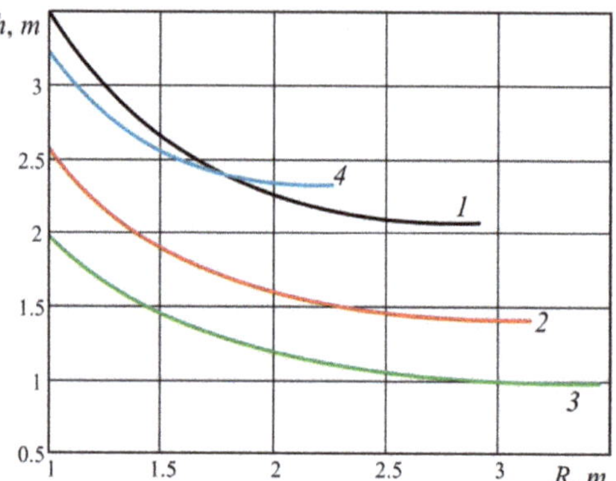

Figure 4. Determination of the circular recess drilling interval: 1—siltstone, 2—stone sandstone, 3—fine-grained limestone, 4—dolomite.

3.3. Technical Means for Local Reaming of the Well Diameter

To explain the technology of wellbore reaming, let us consider the schematic diagrams of special drilling equipment and describe the principle of its operation. Figure 5 schematically shows the special bottomhole assembly (BHA) used at the second and third stages of wellbore diameter reaming.

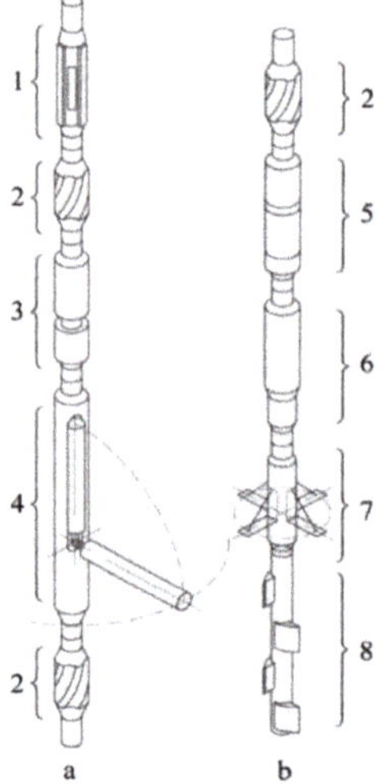

Figure 5. Schemes of bottomhole assemblies: (**a**) circular and conical recesses drilling and (**b**) rock mass destruction and grinding; 1—anchor, 2—centralizer, 3—turning gear, 4—special tool with deflecting bit, 5—shock absorber, 6—mechanical vibrogenerator (jar) of longitudinal oscillations, 7—blade raising device, 8—reamer with fixed blades.

The special BHA used for circular recess drilling (Figure 5a) consists of an anchor (1) of known design, designed to fix the drill string relative to the wellbore, two centralizers (2) used to orient the BHA relative to the axis of the well, a rotary device (3) with remote control, and a special device with a rotating bit (4).

The sequence to obtain circural recesses with this BHA is as follows:

- drilling of the well of nominal radius r;
- the special bottomhole assembly is lowered into the well to a given depth and fixed;

- washing liquid starts circulation, causes the rock-destroying tool rotation, and serves as a trigger for the remote control system on the circular recess drilling to start;
- the remote control system provides a control signal to the linear drive which through the rod (1), via the lever mechanism (2), deflects the rock-destroying tool at an angle of 90 degrees (Figure 6);
- the rotary device (3) rotates the device with the rotary bit around its axis by 360 degrees, and as a result, the circular recess is formed;
- the rock-destroying tool returns to its original position on the signal of the control system, and the BHA is fixed; and
- the BHA rises to a given height h, after which the whole process is repeated.

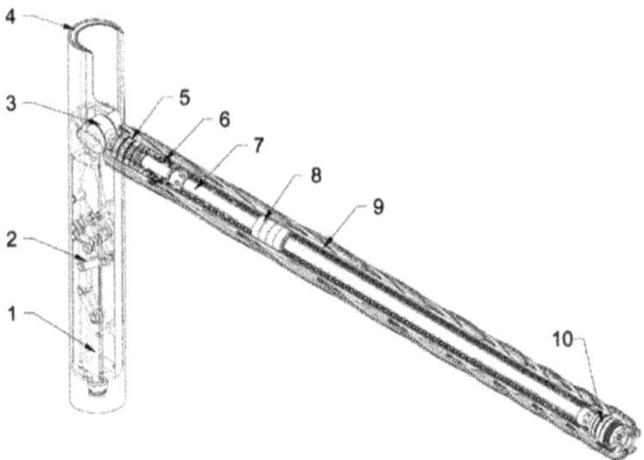

Figure 6. Structural diagram of the special tool with the rotating bit: 1—rod of the linear actuator, 2—lever mechanism, 3—hinge, 4—cavity between the housing and the casing for supplying the flushing fluid, 5—rolling support, 6—reducer, 7—shaft, 8—turbine sections, 9—body—drill bit; 10—rolling support.

After completing the drilling, the tool is removed from the well.

Figure 6 shows the special tool design with the rotating bit. It consists of a hollow bit (see number 10 in Figure 6) with cutters fixed on its surface driven through the gearbox 6 by a section of turbines (8) with movable stators and fixed rotors, or a section of a screw motor with a hollow rotor and torsion bar (not shown) in it. The turbine rotors are attached to the shaft (7) with rolling bearings (5) and (10) at the ends. One end of the shaft (7) is attached to the hinge (3), which can rotate 90 degrees via the lever mechanism (2) due to the movement of the rod of the linear actuator (1). The developed assembly of the special device confirms the possibility of its implementation in the section of the BHA with an overall size of 197 mm and more. Thus, the assembly scheme of the device for the tool with a diameter of 197 mm makes possible usage of the deflecting bit (10) with a diameter of 140–145 mm with the drive in the form of standard sections of turbine drills or sections of a propeller engine with a hollow rotor with a diameter of 85 mm. The maximum length of the bit (10) is determined based on the condition of the strength of the rotary mechanism (2) and the hinge (3), in particular, they should hold the effort to ensure the volumetric destruction of the rock by the bit. Based on the preliminary calculations, the length of the bit is limited to 1.75–2.5 m depending on the bit operation mode, rock mass, and bit cutter charactersitics. We used methods to assess the strength and rigidity of long structures, including pipe columns, pipelines, and special tools operation in the conditions of contact interaction with the elastic environment and at operation in difficult geotechnical conditions [59,60].

The BHA (Figure 5b) is lowered into the well after circular recess drilling destroys the rock mass between the circular recesses. It consists of a section of heavy weighted drill pipes which carries out its orientation relative to the axis of the well; one of the known drill shock absorbers (5) [61,62], which protect the drill string from vibration and control the

BHA dynamics [63]; the vibration generator or jar (6); the special device to retract legs (7), which allows independent rotation of the legs around the axis of the BHA; and the section of the eccentric well reamers (6).

The destruction of the rock mass using this bottomhole envisages the following sequence:

- the BHA is lowered into the well, which is positioned in one of the known ways so that the device to retract the legs is in front of the drilled circular recesses;
- the legs retract to the circular recess after the command;
- intensive circulation of the washing liquid begins with simultaneous rotation of the drill string;
- the axial load is applied to the rock mass, the static component of which is created by the weight of the of heavy weighted drill pipes—due to the operation of the vibrator. As a result, the main cracks appear in the rock and its destruction occurs; and
- to prevent trapping of the drill string and high-quality cleaning of the well, pieces of rock that fall into the space between the drill string and the walls of the well are further crushed using the section of eccentric expanders 8 (Figure 3) and are carried to the surface with flushing fluid.

After the destruction of one cantilever rock mass, further destruction of the following is carried out, after which the BHA is removed from the well.

4. Discussion

Well diameter increasing in the interval of the productive stratum leads to an increase in the flow rate of the well, reducing the depression on the stratum and the pressure gradient on the well wall. Increasing the contact area between the bottomhole and the reservoir area (increasing the filtration surface area, opening new channels for filtering fluids into the well, etc.) allows hydrocarbon fluid production to be intensified. It should be emphasized that this is not a classic reaming of the wellbore with traditional reamers, but well diameter increasing more than five times in a clearly defined interval. The greater effect of well diameter change on the flow rate is associated with the more significant well deviation from the hydrodynamic perfection of its completion and the more expressed nonlinearity of the fluid filtration law [8,9]. This is especially expressed at a significant flow rate of the well when there is a crisis of the linear law of filtration in the bottomhole zone. In this case, the predicted increase in the well flow rate can reach the value of $n^{2/3}$ where n—the multiplicity of the well diameter—increases. The procedure of well diameter increasing in the interval of the productive stratum can be carried out immediately after drilling or during the overhaul of existing production or injection wells.

We propose the innovative technology for local well diameter increasing (more than five times) in a clearly defined interval. The peculiarity of the proposed technology is a two-stage process of well reaming. At the first stage, large-diameter circular recesses are drilled in planes that are perpendicular to the well axis. The second stage is the destruction, grinding, and removal to the surface of the rocks located between the drilled circular recesses. Static and dynamic axial forces applied to the special bottomhole assembly are used for rock mass destruction. The sequence of technological operations for wellbore local increasing is presented.

The sequence of technological operations on the local increase in the diameter of the wellbore is presented. The designs of drill string configurations that should be used to implement the proposed technology are proposed:

- layout for drilling disk and conical recesses;
- layout for destruction and crushing of rock mass.
- a special tool with a rotating bit for drilling disk recesses has been developed, its design is presented and the principle of operation is described. The designs of bottomhole assembly for the proposed technology implementation have been presented:
- BHA for circular and conical recess drilling; and
- BHA for rock mass destruction.

A special tool for circular recess drilling and its design are presented, and the principle of its operation is described.

In order to substantiate the possibility of the technology implementation and for determination of its applicability limits, we formulated and obtained a solution to the problem of the limit equilibrium of the rock mass formed between two adjacent circular recesses. We obtained analytical relations to specify the admissible radius of the circular recesses (formula 3) and admissible thickness of the rock mass (formula 4) on the condition of the safe performance of planned technological processes. In general, the possibility of the proposed technology application for typical reservoir rocks is substantiated, in particular for siltstone, sandstone, fine-grained limestone, and dolomite. The carried out analysis showed that the technology can increase the local diameter of the well by twenty times. The tasks of the next stages of our research are field and bench tests of the special tool with the rotating bit and evaluation of the quality of sludge removal when drilling circular recesses of large diameter.

Today, the oil and gas industry are actively developing and improving the classical [64–66] and developing and testing new technologies to increase the productivity of wells [2,67,68]. Each of these technologies has its advantages and disadvantages, as well as economically feasible applications [1,66]. Among all the variety of approaches offered, we try to find our niche. The technology we have developed is technically simple and is implemented at a reasonable financial cost. To destroy the rock mass, our technology effectively uses the axial force provided by the weight of existing drilling tools so the technology is energy efficient. This favourably differs the proposed technology from others aimed at increasing the contact area between the bottomhole and the collector area. Our technology can be used immediately after drilling or in the process of overhauling a well.

5. Conclusions

Analysis of existing studies has shown that under certain geological conditions it is possible to achieve a significant intensification of hydrocarbon production by local reaming of the well diameter in the productive interval. The paper has proposed that the new technology can increase the local well diameter by more than five times in a clearly defined interval. The peculiarity of the proposed technology is a two-stage process of well reaming. At the first stage, large-diameter circular recesses are drilled in planes that are perpendicular to the well axis. The second stage is the destruction, grinding, and removal to the surface of the rocks located between the drilled circular recesses. Static and dynamic axial forces applied to the special bottomhole assembly are used for rock mass destruction.

To test the feasibility of the proposed technology, the authors have considered the problem of rock mass limit equilibrium formed between two adjacent circular recesses. Analytical relations were obtained to specify the admissible radius of the circular recess and admissible thickness of rock mass from the condition of safe operation. Assuming that the maximum stresses for the reservoir rock are equal to the allowable stresses, the maximum allowable values of the radii of the circular recesses were specified. In particular, for the selected reservoir rocks the allowable radius of the circular recess are as follows: 2.82 m for siltstone, 3.1 m for sandstone, 3.46 m for fine-grained limestone, and 2.18 m for dolomite. The thickness of the rock mass between the circular recesses (drilling interval) was specified from the condition of the possibility of rock destruction under the action of axial force. In particular, the maximum allowable drilling interval of circular recesses with a radius of 2 m with a static application of axial force of 100 kN ranges from 1.2 m for fine-grained limestone to 2.3 m for dolomite. The value of the calculated intervals can be adjusted largely in the case of using devices that generate the dynamic component of the axial force during rock mass destruction. In general, the possibility of applying the proposed technological operations for typical reservoir rocks is substantiated.

For the technical implementation of the proposed technology, special bottomhole assemblies have been developed. A special tool to drill circular recesses has also been developed, its design is presented, and the principle of operation is described. The study showed the possibility of its application for bottomhole assemblies with a radius of 195 mm and more and the possibility of drilling circular recesses with a diameter of up to 2.5 m.

Author Contributions: Conceptualization, O.S. and V.V.; methodology, O.S. and V.V.; software, O.S.; validation, L.R. and V.L.; formal analysis, A.V.; investigation, A.V.; resources, L.R.; data curation, A.V.; writing—original draft preparation, O.S. and A.V.; writing—review and editing, V.L.; visualization, L.R.; supervision, L.R.; project administration, O.B.; funding acquisition, O.B. All authors have read and agreed to the published version of the manuscript.

Funding: This study was carried out as part of the project "Belt and Road Initiative Institute for Chinese-European studies (BRIICES)" and was funded by the Guangdong University of Petrochemical Technology.

Institutional Review Board Statement: Not applicable.

Informed Consent Statement: Not applicable.

Data Availability Statement: Data are contained within the article.

Acknowledgments: The team of authors express their gratitude to the reviewers for valuable recommendations that have been taken into account to improve significantly the quality of this paper.

Conflicts of Interest: The authors declare no conflict of interest.

Nomenclature

h	thickness of the rock mass between two adjacent circular recesses;
R	radius of the circular recess;
r	radius of the through hole in the rock mass;
γ	specific weight of the rock;
P	axial force, which loads the rock mass;
$[R]$	the maximum allowable radius of the circular recess;
ρ	polar coordinate;
s	area of the side surface of the rock mass with the radius of ρ;
V	volume of rock mass;
τ	tangential stresses in the rock mass;
τ_{max}	maximum tangential stresses in the rock mass;
τ_u	ultimate resistance to rock displacement;
ζ	coefficient of operational conditions (depends on water saturation of the rock, the porosity coefficient, thermal impact);
k	safety factor;
λ	the coefficient of axial force application dynamics.

References

1. Al-Rbeawi, S.; Artun, E. Fishbone type horizontal wellbore completion: A study for pressure behavior, flow regimes, and productivity index. *J. Pet. Sci. Eng.* **2019**, *176*, 172–202. [CrossRef]
2. Abolhasanzadeh, A.; Reza Khaz'ali, A.; Hashemi, R.; Jazini, M. Experimental study of microbial enhanced oil recovery in oil-wet fractured porous media. *Oil Gas Sci. Technol. Rev. IFP Energ. Nouv.* **2020**, *75*, 73. [CrossRef]
3. Falshtynskyi, V.; Saik, P.; Lozynskyi, V.; Dychkovskyi, R.; Petlovanyi, M. Innovative aspects of underground coal gasification technology in mine conditions. *Min. Miner. Depos.* **2018**, *12*, 68–75. [CrossRef]
4. Lozynskyi, V.; Medianyk, V.; Saik, P.; Rysbekov, K.; Demydov, M. Multivariance solutions for designing new levels of coal mines. *Rud. Geol. Naft. Zb.* **2020**, *35*, 23–32. [CrossRef]
5. Tutak, M.; Brodny, J.; Szurgacz, D.; Sobik, L.; Zhironkin, S. The Impact of the Ventilation System on the Methane Release Hazard and Spontaneous Combustion of Coal in the Area of Exploitation—A Case Study. *Energies* **2020**, *13*, 4891. [CrossRef]
6. Sekerin, V.; Dudin, M.; Gorokhova, A.; Bank, S.; Bank, O. Mineral resources and national economic security: Current features. *Min. Miner. Depos.* **2019**, *13*, 72–79. [CrossRef]
7. Haiko, H.; Saik, P.; Lozynskyi, V. The Philosophy of Mining: Historical Aspect and Future Prospect. *Philos. Cosmol.* **2019**, *22*, 76–90. [CrossRef]
8. Ursul, A.; Ursul, T. Environmental Education for Sustainable Development. *Future Human Image* **2018**, *9*, 115–125. [CrossRef]
9. Xiang, H.; Valery Kadet, V. Modern methods of underground hydromechanics with applications to reservoir engineering. *J. Hydrodynam. B* **2016**, *28*, 937–946. [CrossRef]

10. Mufazalov, R.S. Tim's Theorem: A New Paradigm for Underground Hydrodynamics Part 2. *ROGTEC Oil Gas Mag.* **2019**, *58*, 64–80.
11. Poltavskaya, M.D.; Verzhbickij, V.V.; Gun'kina, T.A. The influence of increased well bore diameter on an increase in wells productivity. *Perm J. Pet. Min. Eng.* **2013**, *6*, 74–85.
12. Yanukyan, A.P.; Lushpeev, V.A.; Nagaeva, S.N.; Sorokin, P.M. Optimization of the Gas Wells Performance Indicators. *Int. J. Ocean. Oceanogr.* **2016**, *10*, 1–11.
13. Zhu, X.H. Analysis of reamer failure based on vibration analysis of the rock breaking in horizontal directional drilling. In *Handbook of Materials Failure Analysis with Case Studies from the Oil and Gas Industry*; Butterworth-Heinemann: Oxford, UK, 2016; pp. 199–214. [CrossRef]
14. Jing, J.; Lu, Y.; Zhu, X. Weight Distribution Characteristics During the Process of Hole Enlargement When Drilling. *Arab. J. Sci. Eng.* **2018**, *43*, 6445–6459. [CrossRef]
15. Vlasiy, O.; Mazurenko, V.; Ropyak, L.; Rogal, O. Improving the aluminum drill pipes stability by optimizing the shape of protector thickening. *EEJET* **2017**, *85*, 25–31. [CrossRef]
16. Shats'kyi, I.P.; Lyskanych, O.M.; Kornuta, V.A. Combined Deformation Conditions for Fatigue Damage Indicator and Well-Drilling Tool Joint. *Strength Mater.* **2016**, *48*, 469–472. [CrossRef]
17. Onysko, O.; Kopei, V.; Medvid, I.; Pituley, L.; Lukan, T. Influence of the Thread Profile Accuracy on Contact Pressure in Oil and Gas Pipes Connectors. *Lect. Notes Mech. Eng.* **2020**, 432–441. [CrossRef]
18. Pryhorovska, T.; Ropyak, L. Machining Error Influnce on Stress State of Conical Thread Joint Details. In Proceedings of the International Conference on Advanced Optoelectronics and Lasers, (CAOL), Sozopol, Bulgaria, 6–8 September 2019; pp. 493–497. [CrossRef]
19. Onysko, O.R.; Kopey, V.B.; Panchuk, V.G. Theoretical investigation of the tapered thread joint surface contact pressure in the dependence on the profile and the geometric parameters of the threading turning tool. *IOP Conf. Ser. Mater. Sci. Eng.* **2020**, *749*, 012007. [CrossRef]
20. Shatskyi, I.; Ropyak, L.; Velychkovych, A. Model of contact interaction in threaded joint equipped with spring-loaded collet. *Eng. Solid Mech.* **2020**, *8*, 301–312. [CrossRef]
21. Ropyak, L.Y.; Pryhorovska, T.O.; Levchuk, K.H. Analysis of Materials and Modern Technologies for PDC Drill Bit Manufacturing. *Prog. Phys. Met.* **2020**, *21*, 274–301. [CrossRef]
22. Pryhorovska, T.A.; Chaplinskiy, S.S. Finite element modeling of rock mass cutting by cutters for PDC drill bits. *Neftyanoe Khozyaystvo Oil Ind.* **2018**, *1*, 38–41. [CrossRef]
23. Ropyak, L.; Schuliar, I.; Bohachenko, O. Influence of technological parameters of centrifugal reinforcement upon quality indicators of parts. *East. Eur. J. Enterp. Technol.* **2016**, *1*, 53–62. [CrossRef]
24. Ropyak, L.Y.; Vytvytskyi, V.S.; Velychkovych, A.S.; Pryhorovska, T.O.; Shovkoplias, M.V. Study on grinding mode effect on external conical thread quality. *IOP Conf. Ser. Mater. Sci. Eng.* **2021**, *1018*, 012014. [CrossRef]
25. Shats'kyi, I.P.; Makoviichuk, M.V.; Shcherbii, A.B. Influence of a Flexible Coating on the Strength of a Shallow Cylindrical Shell with Longitudinal Crack. *J. Math. Sci.* **2019**, *238*, 165–173. [CrossRef]
26. Shatskyi, I.P.; Makoviichuk, M.V.; Shcherbii, A.B. Influence of Flexible Coating on the Limit Equilibrium of a Spherical Shell with Meridional Crack. *Mater. Sci.* **2020**, *55*, 484–491. [CrossRef]
27. Tarel'nik, V.B.; Konoplyanchenko, E.V.; Kosenko, P.V.; Martsinkovskii, V.S. Problems and Solutions in Renovation of the Rotors of Screw Compressors by Combined Technologies. *Chem. Petrol. Eng.* **2017**, *53*, 540–546. [CrossRef]
28. Shatskyi, I.P.; Ropyak, L.Y.; Makoviichuk, M.V. Strength optimization of a two-layer coating for the particular local loading conditions. *Strength Mater.* **2016**, *48*, 726–730. [CrossRef]
29. Tarelnyk, V.B.; Gaponova, O.P.; Konoplianchenko, Y.V.; Martsynkovskyy, V.S.; Tarelnyk, N.V.; Vasylenko, O.O. Improvement of Quality of the Surface Electroerosive Alloyed Layers by the Combined Coatings and the Surface Plastic Deformation. I. Features of Formation of the Combined Electroerosive Coatings on Special Steels and Alloys. *Metallofiz. Noveishie Tekhnol.* **2019**, *41*, 47–69. [CrossRef]
30. Shatskyi, I.P.; Perepichka, V.V.; Ropyak, L.Y. On the influence of facing on strength of solids with surface defects. *Metallofiz. Noveishie Tekhnol.* **2020**, *42*, 69–76. [CrossRef]
31. Grydzhuk, J.; Chudyk, I.; Velychkovych, A.; Andrusyak, A. Analytical estimation of inertial properties of the curved rotating section in a drill string. *East. Eur. J. Enterp. Technol.* **2019**, *7*, 6–14. [CrossRef]
32. Pryhorovska, T. Rock heterogeneity numerical simulation as a factor of drill bit instability. *Eng. Solid Mech.* **2018**, *6*, 315–330. [CrossRef]
33. Tatsiy, R.M.; Pazen, O.Y.; Vovk, S.Y.; Ropyak, L.Y.; Pryhorovska, T.O. Numerical study on heat transfer in multilayered structures of main geometric forms made of different materials. *J. Serb. Soc. Comput. Mech.* **2019**, *13*, 36–55. [CrossRef]
34. Panevnik, D.A.; Velichkovich, A.S. Assessment of the stressed state of the casing of the above-bit hydroelevator. *Neftyanoe Khozyaystvo Oil Ind.* **2017**, *1*, 70–73.
35. Dutkiewicz, M.; Gołębiowska, I.; Shatskyi, I.; Shopa, V.; Velychkovych, A. Some aspects of design and application of inertial dampers. *MATEC Web Conf.* **2018**, *178*, 06010. [CrossRef]
36. Velichkovich, A.S.; Velichkovich, S.V. Vibration-impact damper for controlling the dynamic drillstring conditions. *Chem. Pet. Eng.* **2001**, *37*, 213–215. [CrossRef]

37. Vytvytskyi, I.I.; Seniushkovych, M.V.; Shatskyi, I.P. Calculation of distance between elastic-rigid centralizers of casing. *Nauk. Visnyk Natsionalnoho Hirnychoho Universytetu* **2017**, *5*, 28–35.
38. Shatskyi, I.; Velychkovych, A.; Vytvytskyi, I.; Seniushkovych, M. Analytical models of contact interaction of casing centralizers with well wall. *Eng. Solid Mech.* **2019**, *7*, 355–366. [CrossRef]
39. Velychkovych, A.S.; Andrusyak, A.V.; Pryhorovska, T.O.; Ropyak, L.Y. Analytical model of oil pipeline overground transitions, laid in mountain areas. *Oil Gas Sci. Technol. Rev. IFP Energ. Nouv.* **2019**, *74*, 65. [CrossRef]
40. Shatskyi, I.; Perepichka, V. Problem of Dynamics of an Elastic Rod with Decreasing Function of Elastic-Plastic External Resistance. *Springer Proc. Math. Stat.* **2018**, *249*, 335–342. [CrossRef]
41. Velichkovich, A.; Dalyak, T.; Petryk, I. Slotted shell resilient elements for drilling shock absorbers. *Oil Gas Sci. Technol. Rev. IFP Energ. Nouv.* **2018**, *73*, 34. [CrossRef]
42. Velychkovych, A.; Petryk, I.; Ropyak, L. Analytical study of operational properties of a plate shock absorber of a sucker-rod string. *Shock Vib.* **2020**, *3292713*. [CrossRef]
43. Warren, T.M.; Sinor, L.A.; Dykstra, M.W. Simultaneous Drilling and Reaming with Fixed Blade Reamers. In Proceedings of the the SPE Annual Technical Conference and Exhibition, Dallas, TX, USA, 22–25 October 1995. [CrossRef]
44. Lunkad, S.F.; Grindhaug, G.; Hussain, S.; Walker, D.; Dhaher, K.S. Innovative Drilling Technologies Contribute in Implementing New Gas Lift Well Design on Statfjord Field, North Sea. In Proceedings of the SPE Artificial Lift Conference—Latin America and Caribbean, Salvador, Brazil, 27–28 May 2015. [CrossRef]
45. Murillo, R.; Santarini, F.; Hurtado, P.; Costa, R.; Diaz, M.A. New-Generation Underreamers Reduce Trips, Decrease Operating Time, and Eliminate Nonproductive Time. In Proceedings of the Offshore Technology Conference Asia, Kuala Lumpur, Malaysia, 20–23 March 2018. [CrossRef]
46. Moisyshyn, V.; Voyevidko, I.; Tokaruk, V. Design of bottom hole assemblies with two rock cutting tools for drilling wells of large diameter. *Min. Miner. Depos.* **2020**, *14*, 128–133. [CrossRef]
47. Lyons, W.C.; Plisga, G.J.; Lorenz, M.D. *Drilling and Well Completions, Standard Handbook of Petroleum and Natural Gas Engineering*; Gulf Professional Publishing: Houston, TX, USA, 2016; pp. 414–584. [CrossRef]
48. Dreus, A.Y.; Sudakov, A.K.; Kozhevnikov, A.A.; Vakhalin, Y.N. Study on thermal strength reduction of rock formation in the diamond core drilling process using pulse flushing mode. *Nauk. Visnyk Natsionalnoho Hirnychoho Universytetu* **2016**, 5–10. Available online: http://nvngu.in.ua/index.php/en/component/jdownloads/viewdownload/61/8504 (accessed on 25 May 2021).
49. Sarraf Shirazi, A.; Frigaard, I.A. Gravel packing: How does it work? *Phys. Fluids* **2020**, *32*, 053308. [CrossRef]
50. Dychkovskyi, R.; Shavarskyi, I.; Saik, P.; Lozynskyi, V.; Falshtynskyi, V.; Cabana, E. Research into stress-strain state of the rock mass condition in the process of the operation of double-unit longwalls. *Min. Miner. Depos.* **2020**, *14*, 85–94. [CrossRef]
51. Abdiev, A.R.; Mambetova, R.S.; Mambetov, S.A. Geomechanical assessment of Tyan-Shan's mountains structures for efficient mining and mine construction. *Gorn. Zhurnal* **2017**, *4*, 23–28. [CrossRef]
52. Bazaluk, O.; Sai, K.; Lozynskyi, V.; Petlovanyi, M.; Saik, P. Research into Dissociation Zones of Gas Hydrate Deposits with a Heterogeneous Structure in the Black Sea. *Energies* **2021**, *14*, 1345. [CrossRef]
53. Voloshyn, O.; Potapchuk, I.; Zhevzhyk, O.; Yemelianenko, V.; Zhovtonoha, M.; Sekar, M.; Dhunnoo, N. Experimental study of the thermal reaming of the borehole by axial plasmatron. *Min. Miner. Depos.* **2019**, *13*, 103–110. [CrossRef]
54. Voloshyn, O.; Potapchuk, I.; Zhevzhyk, O.; Yemelianenko, V.; Horiachkin, V.; Zhovtonoha, M.; Semenenko, Y.; Tatarko, L. Study of the plasma flow interaction with the borehole surface in the process of its thermal reaming. *Min. Miner. Depos.* **2018**, *12*, 28–35. [CrossRef]
55. Levchuk, K.G. Investigation of the vibration transfer process to a stuck drill string. *SOCAR Proc.* **2017**, *2*, 23–33. [CrossRef]
56. Levchuk, K.G. Diagnosing of a Freeze-In of Metal Drill Pipes by Their Stressedly-Deformed State in the Controlled Directional Bore Hole. *Metallofiz. Noveishie Tekhnol.* **2018**, *40*, 701–712. [CrossRef]
57. Moisyshyn, V.; Levchuk, K. Investigation on Releasing of a Stuck Drill String by Means of a Mechanical Jar. *Oil Gas Sci. Technol.* **2017**, *72*, 27. [CrossRef]
58. Ganat, T.A.A.O. *Fundamentals of Reservoir Rock Properties*; Springer: Cham, Switzerland, 2020. [CrossRef]
59. Shats'kyi, I.P.; Struk, A.B. Stressed state of pipeline in zones of soil local fracture. *Strength Mater.* **2009**, *41*, 548–553. [CrossRef]
60. Shatskyi, I.; Popadyuk, I.; Velychkovych, A. Hysteretic Properties of Shell Dampers. In *Dynamical Systems in Applications, Proceedings of the DSTA 2017, Łódź, Poland, 11–14 December 2017*; Awrejcewicz, J., Ed.; Springer: Cham, Switzerland, 2018; pp. 343–350. [CrossRef]
61. Velichkovich, A.S.; Popadyuk, I.I.; Shopa, V.M. Experimental study of shell flexible component for drilling vibration damping devices. *Chem. Pet. Eng.* **2011**, *46*, 518–524. [CrossRef]
62. Moisyshyn, V.M.; Slabyi, O.O. Creation of the Vibroprotective Device for Adjustment of Dynamics of a Column of Steel Drill Pipes and a Bit. *Metallofiz. Noveishie Tekhnol.* **2018**, *40*, 541–550. [CrossRef]
63. Slabyi, O. Studying the coupled axial and lateral oscillations of the drilling riser under conditions of irregular seaways. *East-Eur. J. Enterp. Technol.* **2018**, *3*, 27–33. [CrossRef]
64. Liew, M.S.; Danyaro, K.U.; Zawawi, N.A.W.A. A Comprehensive Guide to Different Fracturing Technologies: A Review. *Energies* **2020**, *13*, 3326. [CrossRef]
65. Basirat, R.; Goshtasbi, K.; Ahmadi, M. Geomechanical key parameters of the process of hydraulic fracturing propagation in fractured medium. *Oil Gas Sci. Technol. Rev. IFP Energ. Nouv.* **2019**, *74*, 58. [CrossRef]

66. Huang, Z.; Huang, Z. Review of Radial Jet Drilling and the key issues to be applied in new geo-energy exploitation. *Energy Procedia* **2019**, *158*, 5969–5974. [CrossRef]
67. Luo, X.; Gong, H.; Ziling, H.; Zhang, P.; He, L. Recent advances in applications of power ultrasound for petroleum industry. *Ultrason. Sonochem.* **2021**, *70*, 105337. [CrossRef]
68. Li, K.; Wang, D.; Jiang, S. Review on enhanced oil recovery by nanofluids. *Oil Gas Sci. Technol. Rev. IFP Energ. Nouv.* **2018**, *73*, 37. [CrossRef]

Article

Thermal Imaging Study to Determine the Operational Condition of a Conveyor Belt Drive System Structure

Dawid Szurgacz [1,2,*], Sergey Zhironkin [3,4,5], Stefan Vöth [6], Jiří Pokorný [7], A.J.S. (Sam) Spearing [8], Michal Cehlár [9], Marta Stempniak [10] and Leszek Sobik [11]

1 Center of Hydraulics DOH Ltd., 41-906 Bytom, Poland
2 Polska Grupa Górnicza S.A., ul. Powstańców 30, 40-039 Katowice, Poland
3 Department of Trade and Marketing, Siberian Federal University, 79 Svobodny Av., 660041 Krasnoyarsk, Russia; zhironkinsa@kuzstu.ru
4 Department of Open Pit Mining, T.F. Gorbachev Kuzbass State Technical University, 28 Vesennya St., 650000 Kemerovo, Russia
5 School of Core Engineering Education, National Research Tomsk Polytechnic University, 30 Lenina St., 634050 Tomsk, Russia
6 Technische Hochschule Georg Agricola (THGA), Westhoffstraβe 15, 44791 Bochum, Germany; stefan.voeth@thga.de
7 Faculty of Safety Engineering, VSB–Technical University of Ostrava, Lumírova 13/630, 700 30 Ostrava-Výškovice, Czech Republic; jiri.pokorny@vsb.cz
8 School of Mines, China University of Mining and Technology, 1 Daxue Road, Tongshan District, Xuzhou 221116, China; sam.spearing@curtin.edu.au
9 Institute of Earth Sources, Faculty of Mining, Ecology, Process Technologies and Geotechnology, Technical University of Košice, Letná 9, 042 00 Košice, Slovakia; michal.cehlar@tuke.sk
10 Faculty of Geoengineering, Mining and Geology, Wroclaw University of Science and Technology, Na Grobli 15, 50-421 Wroclaw, Poland; marta.stempniak5@wp.pl
11 KWK ROW Ruch Chwałowice, ul., 44-206 Rybnik, Poland; sobik55@poczta.fm
* Correspondence: dawidszurgacz@vp.pl

check for **updates**

Citation: Szurgacz, D.; Zhironkin, S.; Vöth, S.; Pokorný, J.; Spearing, A.J.S.; Cehlár, M.; Stempniak, M.; Sobik, L. Thermal Imaging Study to Determine the Operational Condition of a Conveyor Belt Drive System Structure. *Energies* **2021**, *14*, 3258. https://doi.org/10.3390/en14113258

Academic Editor: Nikolaos Koukouzas

Received: 13 April 2021
Accepted: 26 May 2021
Published: 2 June 2021

Publisher's Note: MDPI stays neutral with regard to jurisdictional claims in published maps and institutional affiliations.

Abstract: The paper discusses the results of a study carried out to determine the thermal condition of a conveyor power unit using a thermal imaging camera. The tests covered conveyors in the main haulage system carrying coal from a longwall. The measurements were taken with a thermal imaging diagnostic method which measures infrared radiation emitted by an object. This technology provides a means of assessing the imminence and severity of a possible failure or damage. The method is a non-contact measuring technique and offers great advantages in an underground mine. The thermograms were analysed by comparing the temperature distribution. An analysis of the operating time of the conveyors was also carried out and the causes of the thermal condition were determined. The main purpose of the research was to detect changes in thermal state during the operation of a belt conveyor that could indicate failure and permit early maintenance and eliminate the chance of a fire. The article also discusses the construction and principle of operation of a thermal imaging camera. The findings obtained from the research analysis on determining the thermal condition of the conveyor drive unit are a valuable source of information for the mine's maintenance service.

Keywords: thermal imaging; belt conveyor; diagnostics; underground mining; mechanical failure; preventative maintenance

1. Introduction

The discovery of infrared radiation gave rise to the science of thermography. It was discovered in 1800 by Friedrich Wilhelm Herschel, an English astronomer. In the second half of the 19th century, the following scientists, Kirchoff, Boltzman, Wiena and Planck, whose research laid the foundation for the development of thermal imaging, deepened the knowledge of thermal imaging. The first applications were for the military, where the first infrared indicators were built in the mid-20th century. In the 1960s, the first thermal imaging device appeared, which today is a thermal imaging camera [1–4].

Thermal imaging measures an object body whose temperature is higher than zero because it emits thermal radiation. This thermal radiation is the part of the electromagnetic spectrum; its wavelength falls between 760 and 1 mm. This radiation is detected and measured by the thermal imaging device in two different ways—when the thermal detector absorbs infrared radiation completely (of any wavelength) and when the photon detector reacts only to radiation of a specific wavelength. The detector of a thermal imaging camera enables the energy of infrared radiation to be changed into an electrical signal. In the individual signal processing modules, the signal is amplified, converted into digital form and converted into the temperature value of the individual points of the image matrix. This is how a map of the distribution (thermogram) of the temperature of the object under investigation is created [5–8].

The thermal imaging camera works on the principle of converting infrared radiation that can be emitted or reflected by an object, into an electrical signal and later into an image displayed on a computer monitor. The camera is composed of an optical system, an infrared radiation detector, electronic amplification, processing and a visualization path [9,10].

Belt conveyors are mechanical, hydraulic or pneumatic means of transport, they operate in continuous or cyclic motion. Their purpose is to transport the excavated material over often considerable distances, with varying conveying speeds, capacities and conveyor belt lines. In underground coal mines, they are the primary form of transport [11,12]. The drive systems used in mining can exclude or hinder diagnostic measurements. The results which are obtained by means of various measurements can be processed by dedicated software FLIR Tools [13]. Studies on the development of longwall conveyors are presented in works [14–22]—they are part of the innovative development of machinery and equipment [23–27].

The popularity of the thermal imaging method to assess the technical condition of belt conveyors in a mine has been increasing [28]. The first experimental studies using thermal imaging cameras were described in works [29–32], whose findings and the method developed contributed to minimising failures primarily in the mines of Polish State Mining and Metallurgical Combine (KGHM). Multiple diagnostic methods are recommended for costly machines and process lines [33–36]. Control testing can prevent the occurrence of fires, which are one of the most dangerous hazards in underground mines. Excavations in closed areas are subject to natural hazards, mainly methane [37–47] and fire [48]. They can cause serious damage to machinery and equipment, and even pose a threat to human health and life [49–51].

Based on the research carried out, the main causes of the thermal condition for the drive unit were defined as: belt slip in the drive, problems with optimal cooling of the drive, bearing friction, seizure of the brake system, seizure of the drive drums, and seizure of the pulleys. These causes are mainly generated in the contact zone: improper cooling of the drive unit, the drum coming into contact with the belt, or the pulley coming into contact with the belt.

The main objective of this study was to identify the thermal condition of conveyor belt component structures and to analyse the risk of critical temperature increases. The tests were focussed on the drive unit, specifically the engine, the braking system and the gearbox. In order to measure the actual temperature distribution occurring in the main haulage belt conveyors, it was necessary to analyse the operating time of the belt conveyors and determine the cause of any thermal anomaly. The analysis of the working time of the main haulage conveyors was related to one working day in this study. The results are presented in the form of measurement images. They were developed using dedicated software. The obtained characteristics for the thermal state are presented in the form of diagrams. This paper presents a real-life example of a thermal condition survey for a measuring unit using a thermal imaging camera.

2. Materials and Methods

The use of thermal imaging is a very important and useful research method because, as a method for object diagnosis, it allows fast, safe and also accurate measurements in even restrictive space [52–71]. In a deep mine environment, cameras can be used to work in smoky, dusty and dark environments. The use of the thermal imaging method in the mining industry offers a wide range of research opportunities in view of the heat production that takes place during the operation of all powered equipment. Factors such as ambient temperature, humidity, air velocity, air volume in the excavation and emissivity have a significant influence on the measurement results [72]. Using long-wave infrared radiation in the measurements, thermal radiation is recorded. The camera captures objects, people and high-temperature sources in limited or no visibility conditions [73–78].

Thermal imaging cameras use energy that increases as the temperature of an object increases, and can be obtained from any object whose temperature is above zero. The measurements result in a total temperature distribution over the background of the object, which can be seen by the colour variation in the measurement image. The advantages of thermal imaging cameras are that they are non-invasive and can locate faults invisible to the naked eye. The test with a thermal imaging camera is based on measuring the temperature from the external surface, where the temperature distribution is non-uniform [79]. In order to obtain the relevant quantities, an average is determined which forms the basis for fault finding as temperatures increase above the normal operating ones.

In industry, thermography is used to control technological processes and, more specifically, the thermal state in order to predict and prevent failures. The image taken by the thermal imaging camera reflects the temperature of the device under examination and other surfaces, allowing the technical condition to be assessed. Equipment such as power grids, main fan stations, boilers for district heating and conveyor belts, among others, are examined using thermal imaging. In order to be considered reliable, the measurement must be carried out over a longer period of time and operate to its specification, e.g., the conveyor belt must be loaded with excavated material [80–90].

2.1. Objective and Scope of the Study

The objective of this study was to identify the thermal condition of an operational belt conveyor drive unit in an underground coal mine. The following tasks were completed:

- tests and measurements on the conveyor drive unit,
- an analysis of operating times of conveyors,
- determination of the causes of the thermal condition for the construction of conveyors,
- an analysis of the results and recommendations.

2.2. Analysed Main Haulage Conveyors

The main haulage belt conveyors used in the study transported the excavated coal from the longwall. The longwall mining was carried out conventionally with roof caving. The longwall was equipped with a powered roof support, a double-drum shearer and a scraper conveyor. The length of the longwall is 238 m and the panel length is 480 m. The thickness of the seam is between 2.5 and 3.1 m, with a slope between 23° and 25°. The main haulage system from the longwall transports the excavated material to a 1000 m^3 silo located in the mining shaft area. The analysed haulage system consists of six belt conveyors with a total length of 1846 m. The parameters of the analysed main haulage are presented in Table 1 and their location in Figure 1.

Table 1. Technical parameters of the analysed main haulage system.

Number of the Conveyor	Type of the Conveyor	Power (kW)	Belt Width (m)	Belt Length (m)	Performance Maximum (t/h)
PT-1	Intermet-1200	2×250	1.2	480	1388
PT-2	Vacat-1400	3×315	1.4	420	1512
PT-3	Intermet-1200	2×160	1.2	80	1220
PT-4	Pioma-1200	2×250	1.2	140	1220
PT-5	Pioma-1200	2×250	1.2	260	1134
PT-II	Pioma-1400	2×250	1.4	410	1500
PT-I	Bogda-1400	2×132	1.4	56	1500

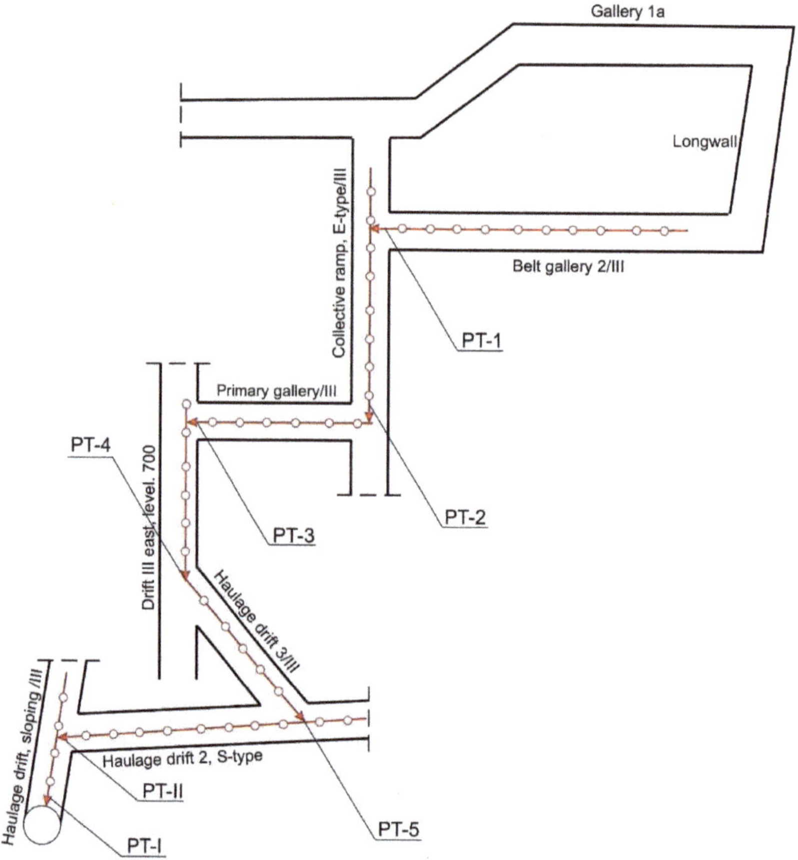

Figure 1. Layout of the conveyors of the main haulage system, (PT—a belt conveyor).

The conveyor routes are made of coils supported on lower trestles, which are spaced every 3 m and each has two Ø 133 mm pulleys (Figure 2b), they serve to guide the lower belt in a V arrangement with a constant inclination angle of 10° and variable advance (−2°, 0°, 2°). The upper band is guided along the triangular supports to form a trough with an angle of 35° (Figure 2a). Each of the side pulleys of the top support has an oblique 2° lead-out in the belt direction and a belt distance of 1.2 m.

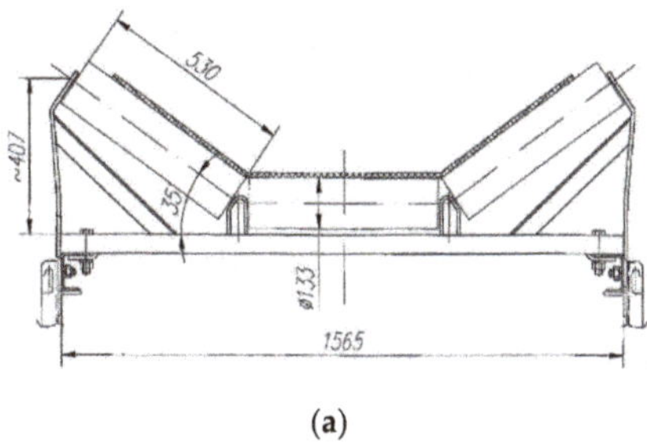
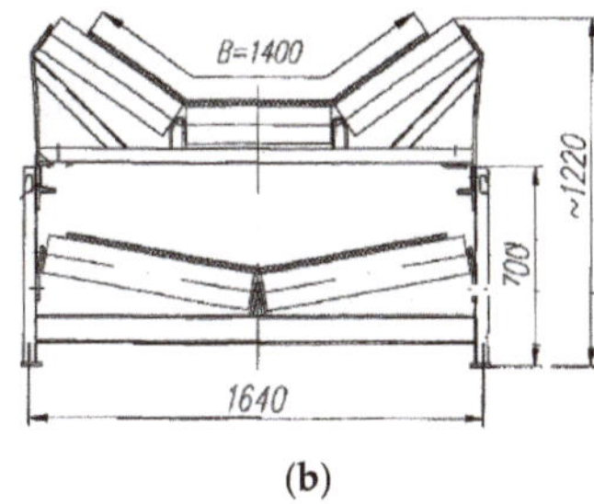

(a)   (b)

Figure 2. Support structure of the conveyor route: (**a**) The upper belt forms a trough with an angle of 35°; (**b**) View of the upper and lower belt routing.

2.3. Description of the Measuring Apparatus

A thermal imaging camera (shown in Figure 3a,b) works by processing infrared radiation that is emitted or reflected by objects. The resulting electrical signal is transformed into an image viewed on a monitor. The camera is built of an optical system, an infrared detector, a visualisation circuit and electronic amplification. The camera reads the measurement of any object with a temperature above zero without relative 0 °C, which is a source of infrared radiation, and its intensity depends on the temperature and surface features of a given object. The range of detection (sensing), recognition and observation identification depends mainly on three parameters: the viewing angle of the camera, thermal resolution and number of detectors in the array [91]. The devices shown in Figure 3 are equipped with a laser pointer that allows the temperature to be recorded at a specific point during the measurement from the object or location. A pyrometer (Figure 3c) is used for non-contact temperature measurement. It works by analysing the thermal radiation emitted by the objects as a whole.

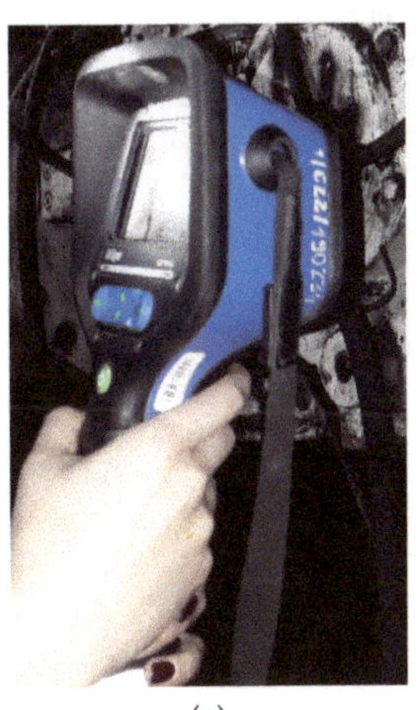

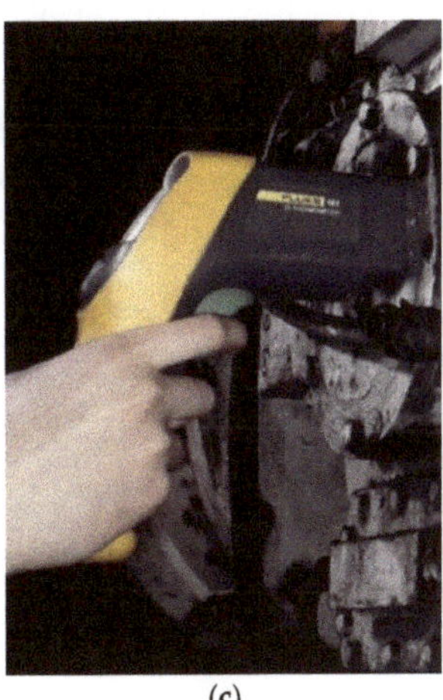

(a)   (b)   (c)

Figure 3. The measuring equipment used in the study of the thermal condition of the belt conveyor drive unit: (**a**) Dräger UFC 9000 thermal imaging karma; (**b**) FLIR i60 thermal imaging camera; (**c**) FLUKE 561 pyrometer.

2.4. Design of the Conveyor Drive Unit

The drive unit consists of a gearbox, clutch and motor connected via the coupling case. The conveyor drive drums are driven by drive units. The transmission of the take-off torque from the gearbox to the drums is affected by means of couplings. The drive unit is built on a drive drum module. The gearbox is attached to the drive body via an intermediate plate. Drive units consisting of motors and gearboxes require water cooling. Figure 4 shows an example of the construction of the drive unit.

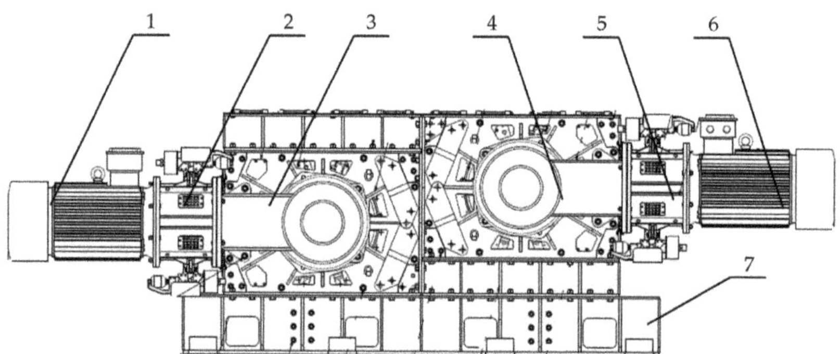

Figure 4. Conveyor drive unit, where; 1—engine, 2—left brake system (I), 3—left gearbox (I), 4—right gearbox (II), 5—right brake system (II), 6—engine, 7—foundation.

3. Results

It can be quite difficult to carry out measurements using a thermal imaging camera for this purpose in an underground mine. One of the main factors that influence the result is the prevailing dust in the excavation. The correct temperature range for the drive unit is influenced by the length of the route, the variable load, and the size of the drive drums. The main problem during the research was to obtain a suitable measuring distance. Conveyor drive components such as the motor, gearbox and braking system are built into a recess due to the dimensions of the workings. For major conveyor installations, a fixed thermal imaging device could be used and data sent to a central control room for continuous monitoring. Exceeding a threshold temperature, predetermined from field data, could trigger an alarm, for example. Maintenance personnel could then be sent to the unit to investigate the temperature anomaly and conduct preventative maintenance if needed.

The following parameters were introduced to minimise measurement interference: emissivity, humidity, ambient temperature, and the distance of the camera from the object. Each of these measurements was additionally determined using a pyrometer type device. It was not possible to place the measuring equipment on a tripod due to the dimensional constraints of the excavation. The measured air temperature at the drive locations varied between 22 and 30 °C. The measurements taken were sequential, with a frequency of 5 min. The entire measurement session for a single test object lasted approximately 60 min, resulting in 12 measurements.

3.1. Analysis of the Working Time of the Main Conveyors

The operating time of the main haulage conveyors depends on many factors, including the mine's operating system and planned daily tonnage. The main haulage unit under analysis operates on a five-shift system. This is characterised by four mining shifts and a fifth maintenance shift. A maintenance-related stoppage of the conveyors to perform necessary checks or repairs is made between 5:30 and 8:00 a.m. The working time of the main haulage conveyors was analysed by data collected from the ZEFIR system. This system performs the function of continuous supervision of mine operations, for the operational management, alerting, documentation and analysis of the production process. Figure 5 shows a visualisation of the working time of the analysed main haulage conveyors, and Table 2 presents a summary of the effective use of the working time of the analysed belt conveyors.

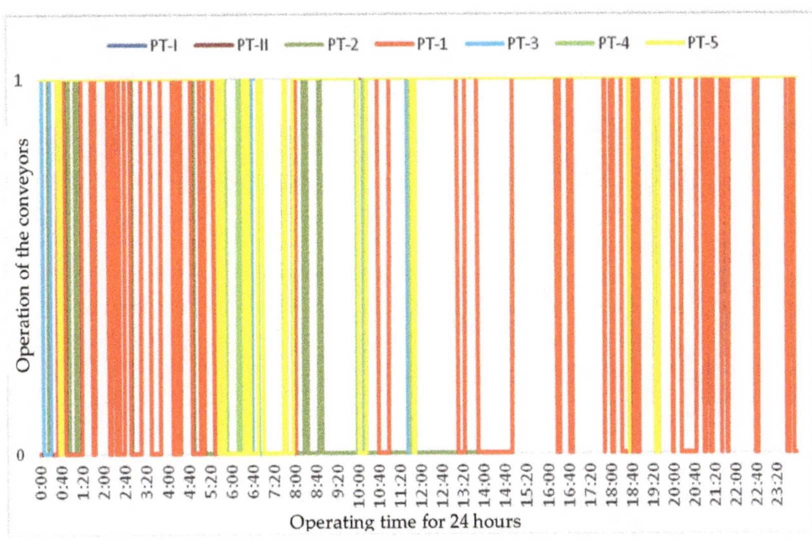

Figure 5. Graph for the operating time of the main haulage belt conveyors, where: 1—the operating time (it is in motion) of the conveyor belt, 0—the idle time (the machine is halted).

Table 2. Measuring the running time of the main haulage belt conveyors.

No. of the Conveyor	Type	Location	Working Time (min)	Stoppage Duration (min)
PT-I	1-400	Haulage drift/III	1300	140
PT-II	1-400	Haulage drift 2, S-type	1297	143
PT-2	1-400	Collective ramp, E-type/III	698	742
PT-1	1-200	Belt gallery 2/III	916	524
PT-3	1-200	Primary gallery/III	1282	158
PT-4	1-200	Drift III east, level 700	1290	150
PT-5	1-200	Haulage drift 3/III	1291	149

The diagram (Figure 5) illustrates the operation of the main haulage belt conveyors on a daily basis. It allows us to view the stoppages that have occurred during its daily operation. The belt conveyors whose running time measurement is shown in the above diagram are explained in Table 2.

3.2. Inspection of a Conveyor Drive Unit Using a Thermal Imaging Camera

Testing of the individual drive units of the main haulage conveyors was carried out two hours after start-up of the morning shift. All conveyors tested were loaded with excavated coal material. It was assumed that the temperature value should stabilize after this time from the start up. Obtaining a series of measurement images from a single conveyor drive during sixty minutes of operation allowed us to calculate the minimum and maximum temperatures for the drive unit. The unit consists of the motor, gearbox and brake system. The construction of the drive, including the drums, was omitted from measurements. The tested constructions of the drive system consisted of two drive units, one in the left-hand version (II drive) and one in the right-hand version (I drive).

Two thermal imaging cameras—FLIR 60 and Dräger UCF 9000—were used in the study. The measurement series taken on the equipment made it possible to locate the hottest areas within the drive structure. The following thermograms (Figures 6 and 7) show the recorded temperatures for selected drive units from the conveyors analysed.

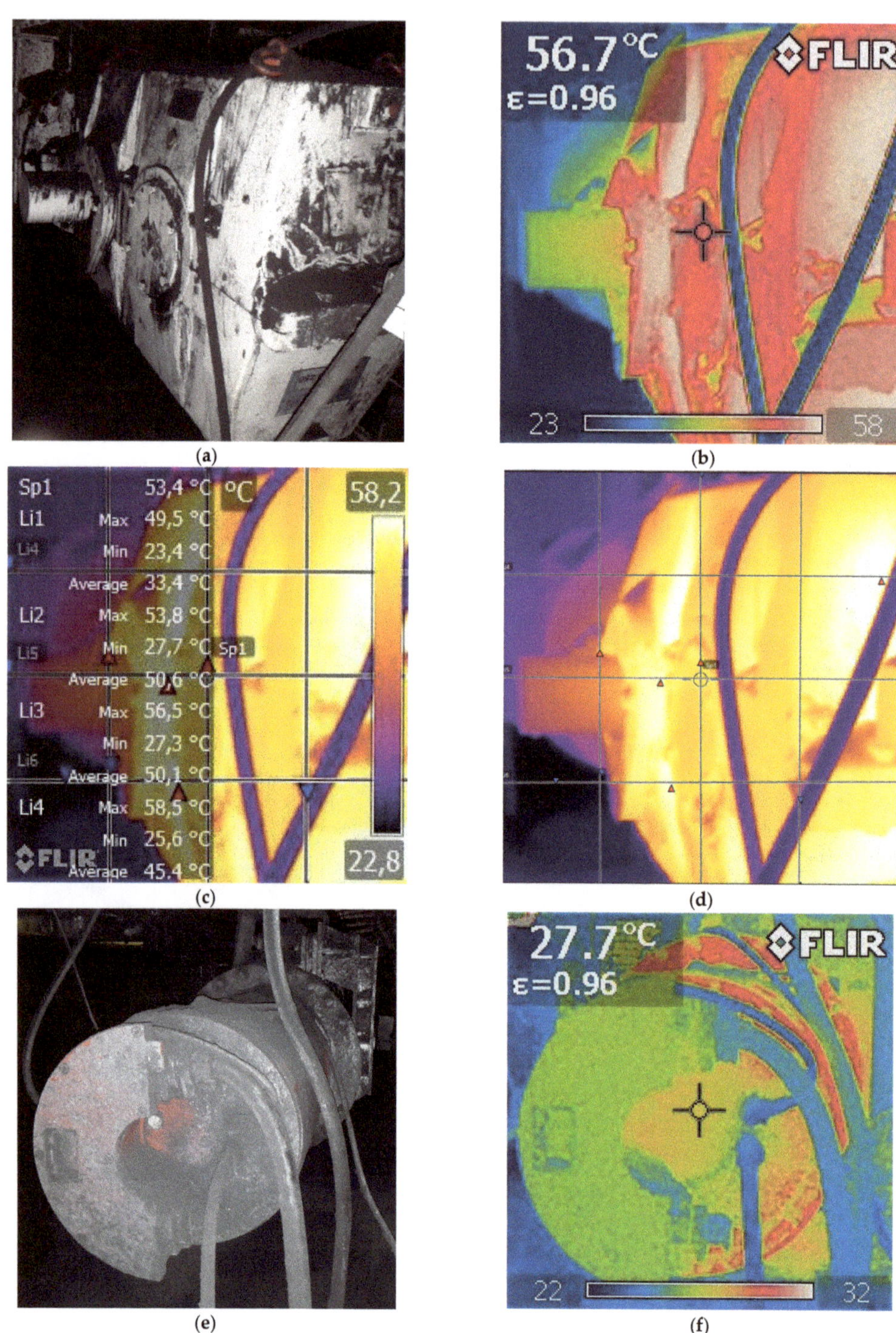

Figure 6. *Cont.*

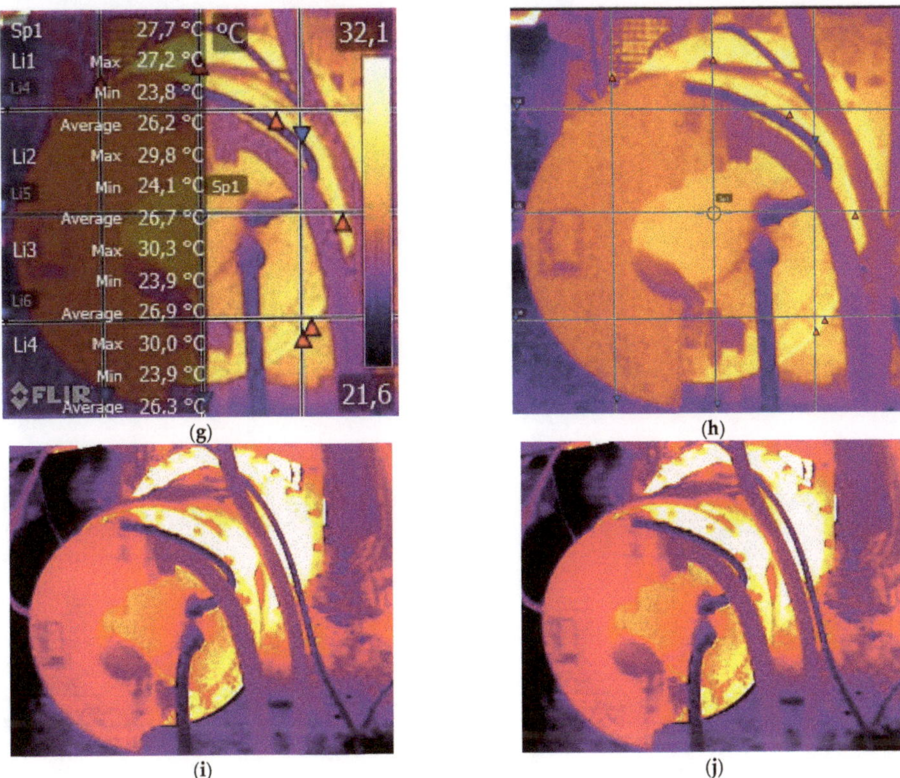

Figure 6. A view from a measurement session made with a thermal imaging camera together with processing in the software FLIR Tools for (II) drive unit of belt conveyor PT-4 from the analysed main haulage, where heat distribution is presented for: (**a**) photo of the tested right gearbox (II); (**b**) thermogram with visible thermal effect for the right gearbox (II); (**c,d**) thermogram showing the heat distribution for the right gearbox (II); (**e**) photo of the tested right engine (II), (**f**) thermogram with visible thermal effect for the right engine (II); (**g,h**) thermogram showing the heat distribution for the right engine (II); (**i,j**) thermogram showing the heat distribution taken with the Dräger UFC 9000 camera.

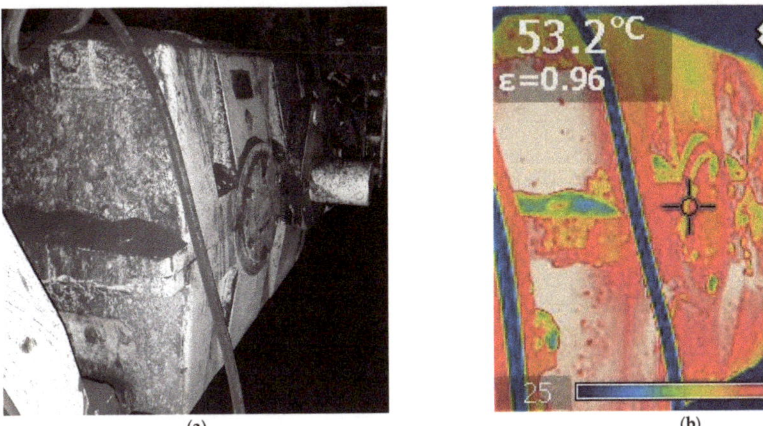

Figure 7. *Cont.*

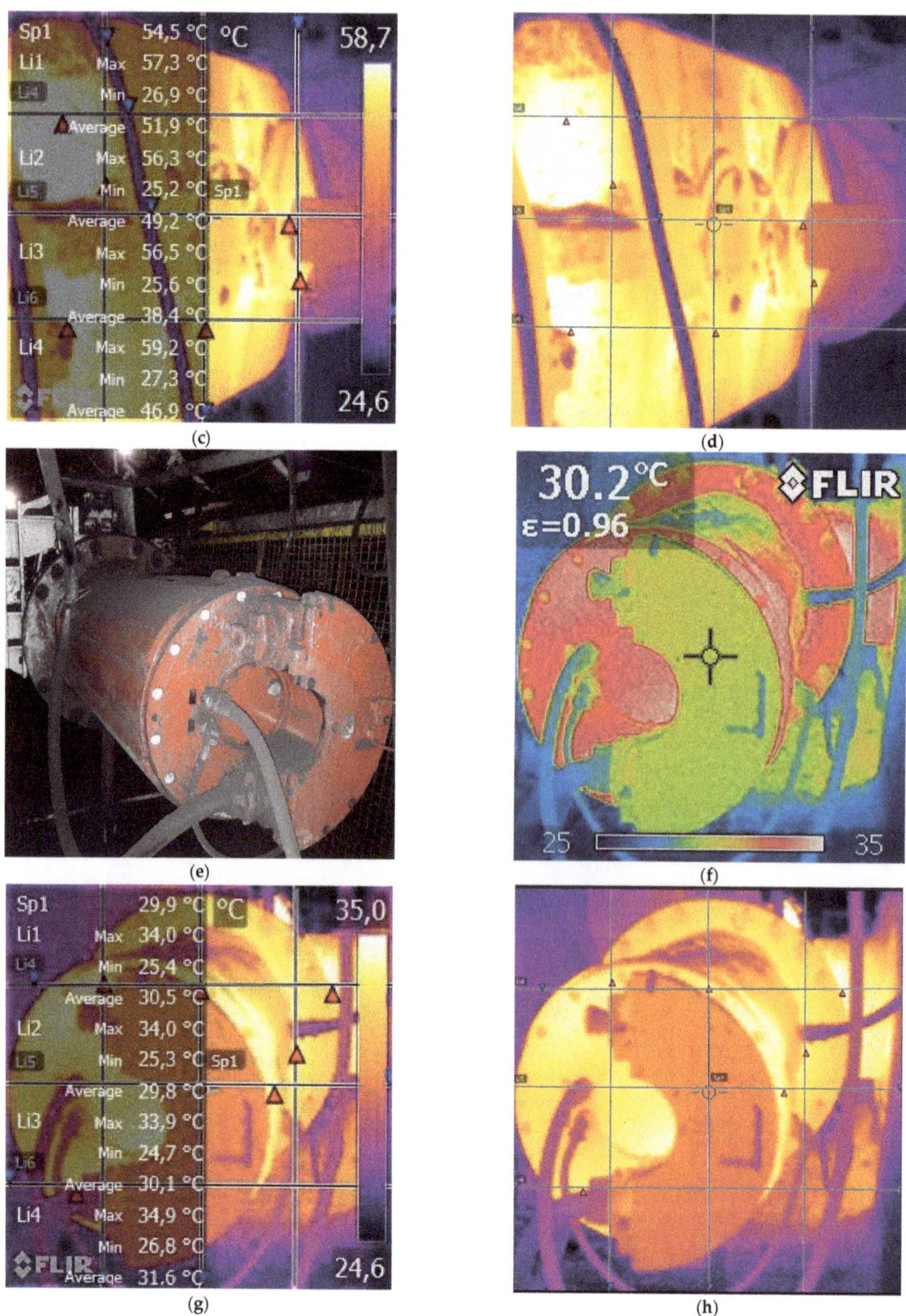

Figure 7. *Cont.*

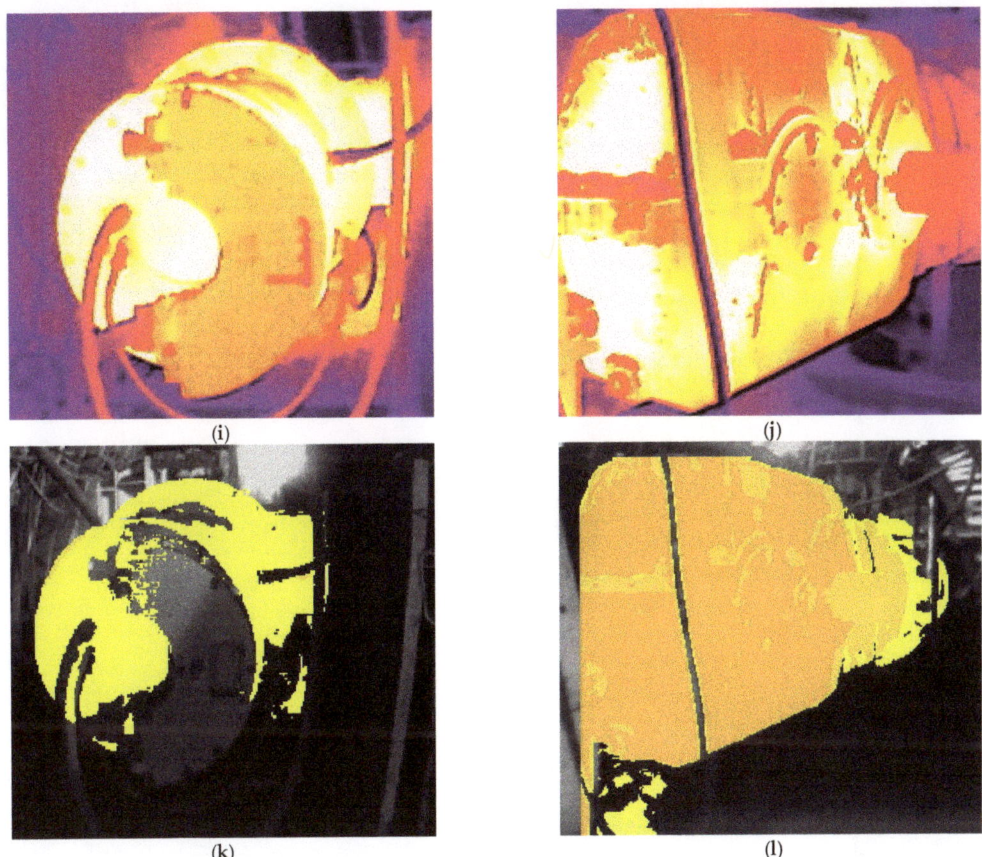

Figure 7. A view from a measurement session made with a thermal imaging camera together with processing in the software FLIR Tools for (I) a drive unit of a belt conveyor PT-4 from the main haulage analysed, where the heat distribution is presented for: (**a**) photo of the tested left gearbox (I); (**b**) thermogram with visible thermal effect for the left gearbox (I); (**c,d**) thermogram showing heat distribution for the left gearbox (I); (**e**) photo of the tested left motor (I); (**f**) thermogram showing heat distribution for the left motor (I); (**g,h**) thermogram showing the thermal effect for the left motor (I); (**i,j**) thermogram showing the heat distribution with the Dräger UFC 9000 camera; (**k**) temperature distribution for the motor in the excavation environment; (**l**) temperature distribution for the gearbox in the excavation environment.

The test results obtained (Figure 6) were analysed using a process based on special software FLIR Tools in order to determine the temperature distribution within the test object. The software FLIR Tools uses the function of creating a line to catch points for which a black triangle is used to mark maximum temperatures (Figures 6c,g and 7c,g,k,l). The blue triangles indicate the minimum temperatures. A series of temperature points for the gearbox and motor were thus obtained. This function greatly simplifies and shortens measurement times, as a single image provides information on the complete temperature distribution. On this basis, the average highest and lowest temperatures of the gearboxes, motors and braking systems of the belt conveyors studied were determined, which were used in the analysis of the thermal condition of the main haulage conveyor drive units.

3.3. Test Results

In the 60 min test interval, the graphs show minor decreases and increases in the minimum and maximum temperatures. There were no interruptions to the operation of the conveyor due to a temporary stoppage. The measured water temperature for engine and transmission cooling was around 21 °C. The water after passing through the cooling system at the discharge into the dewatering pipeline was, on average, 33 °C.

The obtained test results for the conveyor drive unit are presented in Table 3 below, showing the thermal state characteristics for: the entire gearbox, motor, and braking system. In the following graphs (Figures 8 and 9), the red colour indicates the values for the maximum temperature, while the blue colour indicates the values for the minimum temperature.

Table 3. Temperature measurement for belt conveyor drive unit PT-4.

Conveyor's Drive Unit	Minimum Temperature °C	Maximum Temperature °C	Amplitude Temperature °C
gear I	46.4	51.7	5.3
gear II	47.3	52.6	5.3
motor I	24.0	28.0	4.0
motor II	25.0	30.1	5.1
braking system I	27.2	32.0	4.8
braking system II	32.2	37.2	5.0

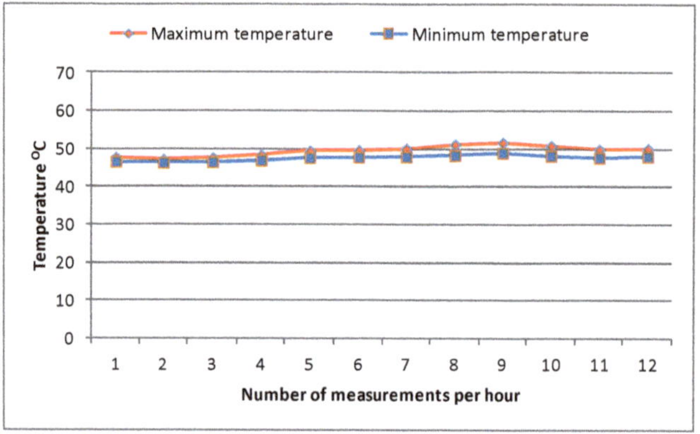

Figure 8. Temperature variation over 60 min for gear I of conveyor PT-4.

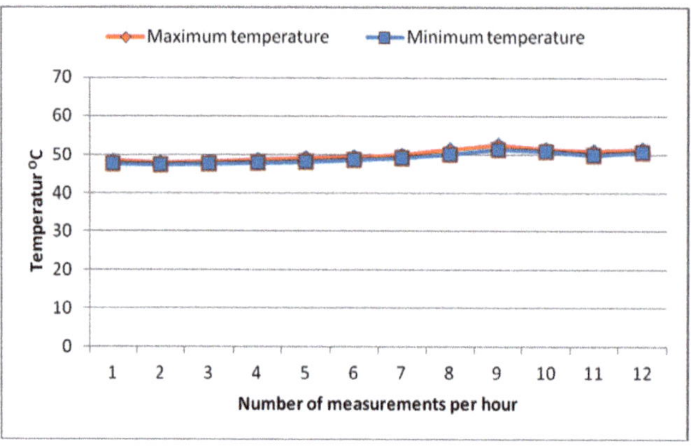

Figure 9. Temperature variation over 60 min for gear II of conveyor PT-4.

4. Discussion

The obtained characteristic curves of heat distribution in the form of thermograms (Figure 6) and graphs (Figures 7 and 8) constitute an evaluation of the thermal condition of the individual elements of the tested driving unit of the main haulage belt conveyor. The measurements made refer to stable (constant) operation. For all tested main haulage conveyors (Figure 1), the highest temperatures were recorded for gearbox II and had a significant effect on the maximum temperature of the entire drive. For the PT-2 conveyor drive unit located in the collective ramp on E/III, the lowest of the maximum temperatures at around 29.9–34.2 °C. was recorded. Higher temperatures (50.1–54.6 °C) obtained from the measurements were characteristic of the PT-1 conveyor drive located in the belt gallery. In contrast, the same type of conveyor PT-3 achieved temperatures of 42.8–47.6 °C. In further measurements, temperatures remained between 47.2 and 52.6 °C for the drive unit of the PT-4 conveyor located in the eastern drift III. The presented comparison of temperatures obtained for the conveyor drive unit (Figure 9) shows that the PT-5 conveyor drive has the highest temperatures, remaining at 62.7 °C, which means it exceeded the critical value of 60 °C. This may indicate intensive use of the conveyor. The analysed measurements for the PT-II conveyor drive were at a level of 48.3–51.0 °C, which indicates correct operation. For the PT-I conveyor, temperatures ranging from 38.0 to 43.3 °C were recorded.

A key element in the overall test procedure is to determine the values at which alarm states associated with temperatures exceeding 60 °C can be identified. The determination of these values is based on a statistical analysis of the results obtained for the main substitution tested. A summary of the measurements obtained (Figure 10) for the main drive units are shown, with the temperature warning levels marked by a red dashed line.

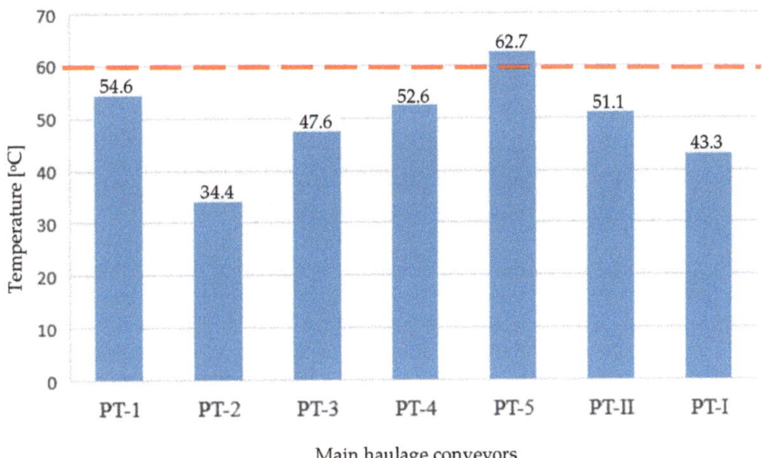

Figure 10. Summary of test results obtained for the drive units.

The summary of obtained measurements for the presented Figure 9, which illustrates all tested main haulage conveyors, can be the basis for searching for diagnostic changes related to improper operation. In the following discussion, based on the results obtained, the average temperature rise in relation to the ambient was determined [92]:

$$\Delta_T = \frac{(T_I - T_0) + (T_{II} - T_0)}{2}, (°C) \tag{1}$$

where:

Δ_T–average temperature rise °C,

T_I–temperature of the gear I °C,
T_{II}–temperature of the gear II °C,
T_0–ambient temperature °C.

Table 4 shows that the effects of ambient temperature for the analysed drive units of the main haulage conveyors are different, which significantly affects the average temperature rise.

Table 4. Data obtained from measurements carried out in the surroundings of the drive unit of the main haulage belt conveyors.

No. of the Conveyor	PT-1	PT-2	PT-3	PT-4	PT-5	PT-II	PT-I
Ambient temperature °C	23.0	26.0	24.0	24.0	27.0	24.0	23
Air temperature °C	26.6	22.4	18.6	18.2	18.8	19.2	22.4
Reflected temperature °C				22.0			
Air flow (m/s)	2.27	2.93	2.20	2.20	0.44	2.53	1.04
Air volume in excavation (m³)	0.47	0.78	0.64	0.64	0.11	0.67	0.16
Humidity (%)	85	77	79	85	80	74	79
Relative humidity (%)				50			
Vertical distance (m)				1.0			
Emissivity				0.96			
Average temperature rise Δ_t	29.10	9.40	20.55	28.15	33.35	25.70	17.65
Amount of measurements	42	97	43	52	43	56	79

5. Conclusions

Thermal imaging is characterised by a non-invasive research method. Conducting this type of research is quite difficult in an underground mine. The use of stationary thermal condition monitoring on the main haulage can be quite difficult due to the dimensions of the workings. The application of the thermal imaging method to the monitoring of industrial processes, including underground mining, has made it possible to assess, based on the test results obtained, the thermal condition of the belt conveyor drive units (Figure 9).

Thermal imaging technology makes it possible to solve many diagnostic problems easily and inexpensively by using a thermal imaging camera. The thermograms obtained (Figures 6 and 7) depict the surface of the object under investigation, i.e., the areas with the highest or lowest temperature.

Based on the research and analysis carried out, the main causes of the thermal condition of the conveyor drive unit were defined. These are the location and method of installation in the excavation. The tested units were located in a hollow of the excavation, which for ventilation reasons affects the ventilation and adequate heat discharge. In order to address these issues, the design of underground workings must consider those remarks which can significantly improve the operating conditions of the machine and the quality of work for people. The relatively early identification of these causes can have a direct impact on the operational reliability of the conveyor drive unit. The relatively early detection of a thermal condition based on temperature measurements with a thermal imaging camera contributes to minimising the probability of a fire.

The thermal imaging inspections carried out allowed the thermal condition of the power unit to be determined and identified the location of intense heat generation, which may indicate the beginnings of a fault condition. The thermal imaging measurements made it possible to diagnose the thermal condition of the drive unit without stopping the belt conveyor operation. The thermal state characteristics obtained for the drive unit under test determined whether a critical temperature occurs. A key element of the entire test procedure was to determine the values at which emergency states associated with temperature exceedances above 60 °C could be visualised.

Author Contributions: Conceptualization, D.S. and M.S.; methodology, D.S. and L.S.; software, D.S., and M.S.; validation, D.S., S.Z., J.P., S.V. and M.C.; formal analysis, D.S., S.Z. and A.J.S.S.; investigation, D.S.; resources, D.S.; data curation, D.S and S.Z.; writing of the original draft preparation, D.S. and M.S.; writing of review and editing, D.S. and A.J.S.S.; visualization, D.S and A.J.S.S.; supervision, D.S.; project administration, D.S., L.S. and M.S.; funding acquisition, D.S. and S.Z. All authors have read and agreed to the published version of the manuscript.

Funding: This research received no external funding.

Institutional Review Board Statement: The study was conducted according to the guidelines of the Declaration.

Informed Consent Statement: Not applicable.

Data Availability Statement: Not applicable.

Acknowledgments: The paper was developed within the framework of Engineer's Thesis called "Thermal imaging test for the thermal condition of conveyor belt structures". The research task carried out as part of the student internship was courtesy of KWK ROW Ruch Chwałowice.

Conflicts of Interest: The authors declare no conflict of interest.

References

1. Lucchi, E. Applications of the infrared thermography in the energy audit of buildings: A review. *Renew. And Sust. Ener. Rev.* **2018**, *82*, 3077–3090. [CrossRef]
2. Lucchi, E. Non-invasive method for investigating energy and envirnomental performances in existing buildings. In Proceedings of the PLEA Conference on Passive and Low Energy Architecture, Louvain-la-Neuve, Belgium, 13–15 July 2011.
3. Lucchi, E. Thermal transmittance of historical brick masonries: A comparison among standard data, analytical calculation procedures, and in situ heat flow meter measurements. *Energy Build.* **2017**, *134*, 171–184. [CrossRef]
4. Lucchi, E. Thermal transmittance of historical stone masonries: A comparison among standard, calculated and measured data. *Energy Build.* **2017**, *151*, 393–405. [CrossRef]
5. Minkina, W.; Klecha, D. Atmospheric transmission coefficient modeling in the infrared for termovision measurements. *J. Sen. Sen. Sys.* **2016**, *5*, 17–23. [CrossRef]
6. Mazur, D.; Herbut, E.; Walczak, J. Termowizja jako metoda diagnostyczna. *Rocz. Nauk. Zoot.* **2006**, *2*, 171–181.
7. Więcek, B.; De Mey, G. *Termowizja w Podczerwieni, Podstawy i Zastosowania*; Wydawnictwo PAK: Warsaw, Polnad, 2011.
8. Minkina, W.; Dudzik, S. *Infrared Thermography: Errors and Uncertainties*; John Wiley & Sons: Hoboken, NJ, USA, 2009.
9. Żabicki, D. Kamery termowizyjne w badaniach diagnostycznych. *Elektroinstalator* **2014**, *1*, 20–24.
10. Howell, K.; Dudek, K.; Soroko, M. Thermal camera performance and image analysis repeatability in equine thermography. *Infr. Phy. Technol.* **2020**, *110*, 103447. [CrossRef]
11. Więcek, B.; Pacholski, K.; Olbrycht, R.; Strąkowski, R.; Kałuża, M.; Borecki, M.; Wittchen, W. *Termografia i Spektrometria w Podczerwieni, Zastosowania Przemysłowe*; Wydawnictwo WNT: Warsaw, Poland, 2017.
12. Michalik, P.; Zajac, J. Use of Thermovision for Monitoring Temperature Conveyor Belt of Pipe Conveyor. *Appl. Mech. Mater.* **2014**, *68*, 238–242. [CrossRef]
13. Błażej, R. *Ocena Stanu Technicznego Taśm Przenośnikowych z Linkami Stalowymi*; Oficyna Wydawnicza Politechniki: Wrocławskiej, Poland, 2018.
14. Antoniak, J. *Przenośniki Taśmowe w Górnictwie Podziemnym i Odkrywkowym*; Wydawnictwo Politechniki Śląskiej: Gliwice, Poland, 2010.
15. Szrek, J.; Wodecki, J.; Błażej, R.; Zimroz, R. An Inspection Robot for Belt Conveyor Maintenance in Underground Mine—Infrared Thermography for Overheated Idlers Detection. *Appl. Sci.* **2020**, *10*, 4984. [CrossRef]
16. Bortnowski, P.; Gladysiewicz, L.; Krol, R.; Ozdoba, M. Tests of Belt Linear Speed for Identification of Frictional Contact Phenomena. *Sensors* **2020**, *20*, 5816. [CrossRef]
17. Kawalec, W.; Król, R.; Suchorab, N. Regenerative Belt Conveyor versus Haul Truck-Based Transport: Polish Open-Pit Mines Facing Sustainable Development Challenges. *Sustainability* **2020**, *12*, 9215. [CrossRef]
18. Kawalec, W.; Suchorab, N.; Konieczna-Fuławka, M.; Król, R. Specific Energy Consumption of a Belt Conveyor System in a Continuous Surface Mine. *Energies* **2020**, *13*, 5214. [CrossRef]
19. Bortnowski, P.; Gładysiewicz, L.; Król, R.; Ozdoba, M. Energy Efficiency Analysis of Copper Ore Ball Mill Drive Systems. *Energies* **2021**, *14*, 1786. [CrossRef]
20. Bajda, M.; Hardygóra, M. Analysis of Reasons for Reduced Strength of Multiply Conveyor Belt Splices. *Energies* **2021**, *14*, 1512. [CrossRef]
21. Lipinski, P.; Brzychczy, E.; Zimroz, R. Decision Tree-Based Classification for Planetary Gearboxes' Condition Monitoring with the Use of Vibration Data in Multidimensional Symptom Space. *Sensors* **2020**, *20*, 5979. [CrossRef] [PubMed]
22. Gąsior, K.; Urbańska, H.; Grzesiek, A.; Zimroz, R.; Wyłomańska, A. Identification, Decomposition and Segmentation of Impulsive Vibration Signals with Deterministic Components—A Sieving Screen Case Study. *Sensors* **2020**, *20*, 5648. [CrossRef]

23. Kozłowski, T.; Wodecki, J.; Zimroz, R.; Błażej, R.; Hardygóra, M. A Diagnostics of Conveyor Belt Splices. *Appl. Sci.* **2020**, *10*, 6259. [CrossRef]
24. Wodecki, J.; Góralczyk, M.; Krot, P.; Ziętek, B.; Szrek, J.; Worsa-Kozak, M.; Zimroz, R.; Śliwiński, P.; Czajkowski, A. Process Monitoring in Heavy Duty Drilling Rigs—Data Acquisition System and Cycle Identification Algorithms. *Energies* **2020**, *13*, 6748. [CrossRef]
25. Góralczyk, M.; Krot, P.; Zimroz, R.; Ogonowski, S. Increasing Energy Efficiency and Productivity of the Comminution Process in Tumbling Mills by Indirect Measurements of Internal Dynamics—An Overview. *Energies* **2020**, *13*, 6735. [CrossRef]
26. Hebda-Sobkowicz, J.; Zimroz, R.; Wyłomańska, A. Selection of the Informative Frequency Band in a Bearing Fault Diagnosis in the Presence of Non-Gaussian Noise—Comparison of Recently Developed Methods. *Appl. Sci.* **2020**, *10*, 2657. [CrossRef]
27. Schmidt, S.; Zimroz, R.; Chaari, F.; Heyns, P.S.; Haddar, M. A Simple Condition Monitoring Method for Gearboxes Operating in Impulsive Environments. *Sensors* **2020**, *20*, 2115. [CrossRef]
28. Ziętek, B.; Banasiewicz, A.; Zimroz, R.; Szrek, J.; Gola, S. A Portable Environmental Data-Monitoring System for Air Hazard Evaluation in Deep Underground Mines. *Energies* **2020**, *13*, 6331. [CrossRef]
29. Borkowski, P.J. Comminution of Copper Ores with the Use of a High-Pressure Water Jet. *Energies* **2020**, *13*, 6274. [CrossRef]
30. Zimroz, R. *Metody Adaptacyjne w Diagnostyce Układów Napędowych Maszyn Górniczych*; Oficyna Wydawnicza Politechniki: Wrocławskiej, Poland, 2010.
31. Błażej, R.; Sawicki, M.; Konieczna, M.; Kozłowski, T.; Kirjanów, A. Automatic analysis of themrograms as a means for estimating technical of a gear system. *Diagnostyka* **2016**, *2*, 43–48.
32. Sawicki, M.; Stefaniak, P.; Zimroz, R.; Błażej, R. Badania eksperymentalne stanu technicznego elementów przenośnika taśmowego z wykorzystaniem metody termowizyjnych, Interdyscyplinarne zagadnienia w górnictwie i geologii. *Interdyscyplinarne Zagadnienia w Górnictwie i Geologii* **2013**, 203–211. Available online: http://diagbelt.pwr.edu.pl/Publikacje/J.Z.W.G.I.G._tomIV_2013 .pdf (accessed on 28 May 2021).
33. Sawicki, M.; Stefaniak, P.; Zimroz, R.; Błażej, R.; Król, R. Wykorzystanie metod termowizyjnych do badania stanu technicznego układów napędowych przenośników taśmowych w O/ZG Polkowice-Sieroszowice. *Interdyscyplinarne Zagadnienia w Górnictwie i Geologii* **2014**, 177–183. Available online: http://labdiag.pwr.wroc.pl/radzim/papers/2014/Sawicki%20i%20inni_KDiMU\T1 \textquoteright14_vO.docx (accessed on 28 May 2021).
34. Zimroz, R.; Stefaniak, P.; Hardygóra, M. Wybrane zagadnienia diagnostyki procesów roboczych i stanu technicznego elementów maszyn górniczych. *Inżynieria Maszyn.* **2014**, *19*, 7–16.
35. Hulewicz, A. Diagnostyka termowizyjna w elektrotechnice. *Pozn. Univ. Technol. Acad. J. Electr. Eng.* **2017**, *89*, 259–269. [CrossRef]
36. Soroko, M. Termowizja w sporcie—Kontrola kontuzji siatkarzy. *Acta Bio Opit. Informa. Med.* **2010**, *1*, 46–47.
37. Wittchen, W.; Mazur, A. Metoda termowizji jako narzędzie pomocnicze w procesach symulacji numerycznej. *Prace IMŻ.* **2012**, *1*, 188–191.
38. Minkina, W.; Rutkowski, P.; Wild, W. Podstawy pomiarów termowizyjnych. Część II—Współczesne rozwiązania systemów termowizyjnych, błędy metody. *PAK* **2000**, *1*, 11–14.
39. Bartelus, W. *Diagnostyka Maszyn Górniczych: Górnictwo Odkrywkowe*; Wydawnictwo Śląsk: Katowice, Poland, 1998.
40. Tutak, M.; Brodny, J.; Szurgacz, D.; Sobik, L.; Zhironkin, S. The Impact of the Ventilation System on the Methane Release Hazard and Spontaneous Combustion of Coal in the Area of Exploitation—A Case Study. *Energies* **2020**, *13*, 4891. [CrossRef]
41. Tutak, M. The Influence of the Permeability of the Fractures Zone Around the Heading on the Concentration and Distribution of Methane. *Sustainability* **2020**, *12*, 16. [CrossRef]
42. Tutak, M.; Brodny, J. The Impact of the Strength of Roof Rocks on the Extent of the Zone with a High Risk of Spontaneous Coal Combustion for Fully Powered Longwalls Ventilated with the Y-Type System—A Case Study. *Appl. Sci.* **2019**, *9*, 5315. [CrossRef]
43. Tutak, M.; Brodny, J. Forecasting Methane Emissions from Hard Coal Mines Including the Methane Drainage Process. *Energies* **2019**, *12*, 3840. [CrossRef]
44. Tutak, M.; Brodny, J. Predicting Methane Concentration in Longwall Regions Using Artificial Neural Networks. *Int. J. Environ. Res. Public Health* **2019**, *16*, 1406. [CrossRef] [PubMed]
45. Tutak, M.; Brodny, J. Analysis of the Impact of Auxiliary Ventilation Equipment on the Distribution and Concentration of Methane in the Tailgate. *Energies* **2018**, *11*, 3076. [CrossRef]
46. Brodny, J.; Tutak, M. Analyzing Similarities between the European Union Countries in Terms of the Structure and Volume of Energy Production from Renewable Energy Sources. *Energies* **2020**, *13*, 913. [CrossRef]
47. Brodny, J.; Tutak, M. Analysing the Utilisation Effectiveness of Mining Machines Using Independent Data Acquisition Systems: A Case Study. *Energies* **2019**, *12*, 2505. [CrossRef]
48. Brodny, J.; Tutak, M. Exposure to Harmful Dusts on Fully Powered Longwall Coal Mines in Poland. *Int. J. Environ. Res. Public Health* **2018**, *15*, 1846. [CrossRef]
49. Szurgacz, D.; Tutak, M.; Brodny, J.; Sobik, L.; Zhironkina, O. The Method of Combating Coal Spontaneous Combustion Hazard in Goafs—A Case Study. *Energies* **2020**, *13*, 4538. [CrossRef]
50. Nowicki, J.; Hebda-Sobkowicz, J.; Zimroz, R.; Wylomanska, A. Local Defect Detection in Bearings in the Presence of Heavy-Tailed Noise and Spectral Overlapping of Informative and Non-Informative Impulses. *Sensors* **2020**, *20*, 6444. [CrossRef] [PubMed]

51. Hebda-Sobkowicz, J.; Gola, S.; Zimroz, R.; Wyłomańska, A. Identification and Statistical Analysis of Impulse-Like Patterns of Carbon Monoxide Variation in Deep Underground Mines Associated with the Blasting Procedure. *Sensors* **2019**, *19*, 2757. [CrossRef]

52. Polnik, B. Tests of a longwall shearer diagnostic system using infrared camera. *Meas. Auto. Monit.* **2015**, *6*, 249–251.

53. Chajda, J.; Poloszyk, S.; Różański, L. Termowizja w diagnostyce technicznej maszyn technologicznych. *Arch. Technol. Masz. Automat.* **1999**, *19*, 65–82.

54. Gabryś, R. Termografia w diagnostyce. *Elektroinstalator* **2017**, *4*, 28–32.

55. Kuczyński, K. Bezkontaktowa diagnostyka urządzeń. *Elektro Info* **2010**, *11*, 46–48.

56. Żabicki, D. Termowizja w diagnostyce maszyn i urządzeń elektroenergetycznych. *Elektroinstalator* **2016**, *3*, 48–51.

57. Erazo-Aux, J.; Loaiza-Correa, H.; Restrepo-Giron, A.D.; Ibarra-Castanedo, C.; Maldague, X. Thermal imaging dataset from composite material academic samples inspected by pulsed thermography. *Data Brief* **2020**, *32*, 106313. [CrossRef] [PubMed]

58. Paramasivam, B. Investigation on the effects of damping over the temperature distribution on internal turning bar using Infrared fusion thermal imager analysis via SmartView software. *Measurement* **2020**, *162*, 107938. [CrossRef]

59. Tattersall, G.L.; Danner, R.M.; Chaves, J.A.; Levesque, D.L. Activity analysis of thermal imaging videos using a difference imaging approach. *J. Ther. Biol.* **2020**, *91*, 102611. [CrossRef]

60. Carvalho, R.; Nascimento, R.; D'Angelo, T.; Delabrida, S.; Bianchi, A.G.C.; Oliveira, R.A.R.; Azpúrua, H.; Garcia, L.G.U. A UAV-Based Framework for Semi-Automated Thermographic Inspection of Belt Conveyors in the Mining Industry. *Sensors* **2020**, *20*, 2243. [CrossRef]

61. Ghorbanzadeh, O.; Valizadeh Kamran, K.; Blaschke, T.; Aryal, J.; Naboureh, A.; Einali, J.; Bian, J. Spatial Prediction of Wildfire Susceptibility Using Field Survey GPS Data and Machine Learning Approaches. *Fire* **2019**, *2*, 43. [CrossRef]

62. Oishi, Y.; Ishida, H.; Nakajima, T.Y.; Nakamura, R.; Matsunaga, T. The Impact of Different Support Vectors on GOSAT-2 CAI-2 L2 Cloud Discrimination. *Remote Sens.* **2017**, *9*, 1236. [CrossRef]

63. Ozotta, O.; Gerla, P.J. Mapping Groundwater Seepage in a Fen Using Thermal Imaging. *Geosciences* **2021**, *11*, 29. [CrossRef]

64. Resendiz-Ochoa, E.; Saucedo-Dorantes, J.J.; Benitez-Rangel, J.P.; Osornio-Rios, R.A.; Morales-Hernandez, L.A. Novel Methodology for Condition Monitoring of Gear Wear Using Supervised Learning and Infrared Thermography. *Appl. Sci.* **2020**, *10*, 506. [CrossRef]

65. Milovanović, B.; Banjad Pečur, I. Review of Active IR Thermography for Detection and Characterization of Defects in Reinforced Concrete. *J. Imaging* **2016**, *2*, 11. [CrossRef]

66. Clark, M.R.; McCann, D.M.; Forde, M.C. Application of infrared thermography to the non-destructive testing of concrete and masonry bridge. *NDT E Int.* **2003**, *36*, 265–275. [CrossRef]

67. Meola, C.; Boccardi, S.; Carlomagno, G.M. An Excursus on Infrared Thermography Imaging. *J. Imaging* **2016**, *2*, 36. [CrossRef]

68. De Finis, R.; Palumbo, D.; Galietti, U. Mechanical Behaviour of Stainless Steels under Dynamic Loading: An Investigation with Thermal Methods. *J. Imaging* **2016**, *2*, 32. [CrossRef]

69. Robinson, J.B.; Shearing, P.R.; Brett, D.J.L. Thermal Imaging of Electrochemical Power Systems: A Review. *J. Imaging* **2016**, *2*, 2. [CrossRef]

70. Steen, K.A.; Villa-Henriksen, A.; Therkildsen, O.R.; Green, O. Automatic Detection of Animals in Mowing Operations Using Thermal Cameras. *Sensors* **2012**, *12*, 7587–7597. [CrossRef]

71. Eisele, A.; Lau, I.; Hewson, R.; Carter, D.; Wheaton, B.; Ong, C.; Cudahy, T.J.; Chabrillat, S.; Kaufmann, H. Applicability of the Thermal Infrared Spectral Region for the Prediction of Soil Properties Across Semi-Arid Agricultural Landscapes. *Remote Sens.* **2012**, *4*, 3265–3286. [CrossRef]

72. Malicki, W.; Miedziński, B. Przydatność termowizji do diagnostyki technicznej urządzeń energomechanicznych w zakładach górniczych. *Mechan. Automat. Górn.* **2009**, *7*, 34–38.

73. Fedorczak-Cisak, M.; Radziszewska-Zielina, E.; Orlik-Kożdoń, B.; Steidl, T.; Tatara, T. Analysis of the Thermal Retrofitting Potential of the External Walls of Podhale's Historical Timber Buildings in the Aspect of the Non-Deterioration of Their Technical Condition. *Energies* **2020**, *13*, 4610. [CrossRef]

74. Kasprzyk-Kucewicz, T.; Szurko, A.; Stanek, A.; Sieroń, K.; Morawiec, T.; Cholewka, A. Usefulness in Developing an Optimal Training Program and Distinguishing between Performance Levels of the Athlete's Body by Using of Thermal Imaging. *Int. J. Environ. Res. Public Health* **2020**, *17*, 5698. [CrossRef] [PubMed]

75. Siekański, P.; Paśko, S.; Malowany, K.; Malesa, M. Online Correction of the Mutual Miscalibration of Multimodal VIS–IR Sensors and 3D Data on a UAV Platform for Surveillance Applications. *Remote Sens.* **2019**, *11*, 2469. [CrossRef]

76. Maj, M.; Ubysz, A.; Hammadeh, H.; Askifi, F. Non-Destructive Testing of Technical Conditions of RC Industrial Tall Chimneys Subjected to High Temperature. *Materials* **2019**, *12*, 2027. [CrossRef] [PubMed]

77. Noszczyk, P.; Nowak, H. Inverse Contrast in Non-Destructive Materials Research by Using Active Thermography. *Materials* **2019**, *12*, 835. [CrossRef] [PubMed]

78. Galla, S. A Thermographic Measurement Approach to Assess Supercapacitor Electrical Performances. *Appl. Sci.* **2017**, *7*, 1247. [CrossRef]

79. Wernik, J. Investigation of Heat Loss from the Finned Housing of the Electric Motor of a Vacuum Pump. *Appl. Sci.* **2017**, *7*, 1214. [CrossRef]

80. Nieoczym, A.; Longwic, R.; Lotko, W. Wstępna diagnostyka pojazdu z wykorzystaniem kamery termowizyjnej. *Eksploatacja Testy* **2017**, *6*, 981–984.
81. Cheaito, R.; Gorham, C.S.; Misra, A.; Hattar, K.; Hopkins, P.E. Thermal conductivity measurements via time-domain thermoreflectance for the characterization of radiation induced damage. *J. Mater. Res.* **2015**, *30*, 1403–1412. [CrossRef]
82. Dular, M.; Coutier-Delgosha, O. Thermodynamic effects during growth and collapse of a single cavitation bubble. *J. Fluid Mecha.* **2013**, *736*, 44–66. [CrossRef]
83. Eriksson, P.; Andersson, J.Y.; Stemme, G. Thermal characterization of surface-micromachined silicon nitride membranes for thermal infrared detectors. *J. Microele. Syst.* **1997**, *6*, 55–61. [CrossRef]
84. Krešák, J.; Peterka, P.; Kropuch, S.; Novák, L. Measurement of tight in steel ropes by a mean of thermovision. *Measurement* **2014**, *50*, 93–98. [CrossRef]
85. Suszyński, Z.; Świta, R.; Łoś, J.; Zarzycka, M.B.; Kaleniecka, A.; Zarzycki, P.K. Fast assessment of planar chromatographic layers quality using pulse thermovision metod. *J. Chromatogr. A* **2014**, *1373*, 211–215. [CrossRef]
86. Nadolny, K.; Plichta, J. Comparative method of thermovision temperature measurement in single-pass internal cylindrical grinding. *Arch. Civ. Mech. Eng.* **2006**, *4*, 67–74. [CrossRef]
87. Ivanov, G.V.; Ivanov, V.G. Temperature and emissivity determination of small-size long-range object's using staring Thermovision Cameras. *Infrared Phys. Technol.* **2013**, *60*, 161–165. [CrossRef]
88. Sun, X.; Xu, H.; He, M.; Zhang, F. Experimental investigation of the occurrence of rockburst in a rock specimen through infrared thermography and acoustic emission. *Int. J. Rock Mech. Min. Sci.* **2017**, *93*, 250–259. [CrossRef]
89. He, M.; Jia, X.; Gong, W.; Faramarzi, L. Physical modeling of an underground roadway excavation in vertically stratified rock using infrared thermography. *Int. J. Rock Mech. Min. Sci.* **2010**, *47*, 1212–1221. [CrossRef]
90. Chan, W.T.; Sim, K.S.; Tso, C.P. Application of optical character recognition in thermal image processing. *Infrared Phys. Technol.* **2011**, *54*, 353–366. [CrossRef]
91. Madura, H.; Sosnowski, T.; Bieszczad, G. Termowizyjne kamery obserwacyjne—Budowa, zastosowanie i krajowe możliwości realizacji. *Przegl. Electrotech.* **2014**, *9*, 5–8.
92. Król, R. *Metody Badań i Doboru Elementów Przenośnika Taśmowego z Uwzględnieniem Losowo Zmiennej Strugi Urobku*; Wydział Geoinżynierii, Górnictwa i Geologii Politechniki: Wrocławskiej, Poland, 2018.

Article

A Multiple Criteria Decision Making Method to Weight the Sustainability Criteria of System Selection for Surface Mining

Michał Patyk [1], **Przemysław Bodziony** [1] and **Zbigniew Krysa** [2,*]

1   Faculty of Mining and Geoengineering, AGH University of Science and Technology, al. Mickiewicza 30, 30-059 Kraków, Poland; mpatyk@agh.edu.pl (M.P.); przembo@agh.edu.pl (P.B.)
2   Faculty of Geoengineering, Mining and Geology, Wroclaw University of Technology, Wybrzeże Wyspiańskiego 27, 50-370 Wrocław, Poland
*   Correspondence: zbigniew.krysa@pwr.edu.pl

Abstract: Selection and assessment of mining equipment used in open pit rock mines relies chiefly on estimates of overall exploitation cost. The rational arrangement of mining equipment and systems comprising loading machines, haul trucks and crushing plants should be preceded by a thorough analysis of technical and economic aspects, such as investment outlays and the costs of further exploitation, which largely determine the costs of mining operations and the deposit value. Additionally, the operational parameters of the mining equipment ought to be considered. In this study, a universal set of evaluation criteria has been developed, and an evaluation method has been applied for the selection of surface mining equipment and the processing system to be operated in specific mining conditions, defined by the user. The objective of this study is to develop and apply the new methodology of multi-criteria selection of open pit rock mining equipment based on multiple criteria decision-making (MCDM) procedures, to enable the optimization of loading, handling and crushing processes. The methodology, underpinned by the principles of MCDM, provides the dedicated ranking procedures, including the ELECTRE III. The applied methodology allows the alternative options (variants) to be ranked accordingly. Ultimately, a more universal methodology is developed, applicable in other surface mines where geological and mining conditions are similar. It may prove particularly useful in selection and performance assessment of mining equipment and process line configurations in mining of low-quality rock deposits. Therefore, we undertook to develop universal criteria and applications for the selection and performance assessment of process machines for surface mines, taking into account environmental aspects as well as deposit quality.

Keywords: surface mining; mining equipment; multiple criteria decision making (MCDM); ELECTRE III

Citation: Patyk, M.; Bodziony, P.; Krysa, Z. A Multiple Criteria Decision Making Method to Weight the Sustainability Criteria of System Selection for Surface Mining. *Energies* **2021**, *14*, 3066. https://doi.org/10.3390/en14113066

Academic Editor: Sergey Zhironkin

Received: 28 March 2021
Accepted: 20 May 2021
Published: 25 May 2021

1. Introduction

The profitability of rock production mining is closely linked to the individual components of the machine's working process. The rational selection of mining system configurations including vehicles and loading and processing machines should be based on a thorough technological and economic analysis that factors in mining conditions. The selection of machines should take into consideration all key elements specific for mining process [1]. Stevanovic et al. [2] recognize that geological conditions are essential to assessing the se-lection of a mining system and consider them the most important among 6 other criteria. Of particular importance in selection of the process line system is the exact number, type and operating capacity of machinery and equipment. Any over- or underestimation may adversely and irreversibly impact on the net value of the entire mining project, hence the quest for reliable methods for selecting machines based on precise and exact criteria [3]. Choosing the optimal mining equipment is challenging problems and depends on many criteria and the applied method of selecting machines: deterministic where fleet size is the function of production requirements and more complex taking into account uncertainty based on approximation algorithms and stochastic models [4]. Voronov et al. [5] indicate that the optimal selection of machines is a formidable task, requiring vast amounts of input data especially in the case of mixed equipment fleets. Selection of mining equipment has received a great deal of attention, and the problem has grown in importance lately as the quality of deposits to be mined has deteriorated. Samanta et al. [6] based on the case of mining equipment selection notice that large numbers of factors can lead to unrealistic outcome or difficulty in making criteria comparison. In real life situations, the fact that information is unavailable or incomplete may result in data (attributes) being not deterministic but fuzzy-imprecise [3].

Exploitation of low-quality deposits is a vitally important and topical issue. Low quality deposits or deposit sections, containing various kinds of inclusions (e.g., karstic formations), are, in current mining practice, either not exploited or hauled to rock dumping sites for selective dumping, leaving the possibility of their exploitation at a later stage. However, this process does not contribute to the sustainable exploitation of the deposit but helps postpone the problem of re-exploitation.

Therefore, in the case of low-quality deposits, it is necessary to strive for sustainable exploitation and the best possible use of the entire deposit in parallel with the target deposit, which may contribute to increasing extraction and extending mine life [7–9]. In addition, the use of such deposits or parts of deposits reduces the amount of dumped material and the size of the dumps, which translates into lower waste generation and long-term sustainable management of the deposit [10,11], especially since the exploitation of low-quality deposits may be profitable [12]. The mining sector is responsible for GHG emissions from the operating mining equipment and from electricity generation still reliant on fossil fuels. The global demand for minerals is increasing steadily and the process routes in mineral extraction require larger amounts of energy to extract and process minerals from low-quality deposits, and hence the emission levels are rising [13].

Selection of mining equipment is a complex multi-criteria decision problem. The main objective of this study is to develop new interdisciplinary criteria taking into account both tangible and intangible aspects, and to explore potential applications of the Multiple-Criteria Decision-Making (MCDM) tool, such as to facilitate the optimal selection of environment friendly equipment that should satisfy the involved decision makers. Identification of relevant factors is the crucial step in decision making and optimization process, particularly in the case of low-quality deposits, when the quality criterion determines the actual price which may be close to the production costs. In this economic perspective the problem of low-quality deposits is similar to exploitation of small deposits, specifically to the mining of minerals for modern technology (rare earth elements) [14]. The fact that the output is relatively small is of major importance to decision-makers especially in the context of capital expenditures, a large proportion of which are the costs of mining equipment.

2. Materials and Methods

This article explores the feasibility of using three variants of mining equipment layout for the exploitation of low-quality deposits. Analyses carried out in the mine have shown that it is possible to separate karst fraction from the rest of the deposit material by means of preliminary crushing. The karst is located mainly in the 0–40 mm fraction and constitutes about 20% of the analyzed part of the deposit. This means that about 20% is not suitable for further use in the production process and must be transported to the dump. However, around 80% of the material can be used. The analyses showed that it is possible to separate karst fraction from the rest of the deposit material by preliminary crushing.

A set of criteria having relevance to evaluation and selection of open pit mining equipment are presented below. Relying on the MCDM approach, they can be effectively used to solve discrete multiple criteria decision problems, such as ranking problems. The group of 8 criteria and is subdivided into 4 subsets, covering the environmental, economic, technological aspects as well as mining operations and reliability. In consideration of those aspects, we undertook to develop universal criteria and applications for the selec-tion and assessment of process machines for surface mines, including low-quality deposits, sustainability and environmental issues. The applications of MCDM in solving issues of sustainable development are currently important topics of research and were described by Siksnelyte, et al. [15], Kumar et al. [16] and Trojanowska and Nęcka [17]. For this reason, environmental and energy consumption aspects are taken into account when defining the criteria that describe the decision model.

The main focus is on applications of Multiple-Criteria Decision-Making (MCDM) tools, including one of the related MCDM methods—ELECTRE III. A method is proposed for ranking the set of alternatives to be evaluated basing on multiple and conflicting crite-ria, organized in a hierarchical structure. This hierarchy allows the decision-maker to identify various intermediate sub-problems to be addressed. Consequently, the analysis of the criteria is carried out according to the subsets defined in the hierarchy and following the precedence relations principle in the bottom-up approach. To effectively handle such hierarchical structures, the application of the extended ELECTRE-III is recommended. According to Hokkanen et al. [18] ELECTRE III is a most effective outranking methodology as it uses thresholds for modelling of imprecise data. Thus, alternative solutions (alterna-tives) applicable to surface mining machine systems can be sorted and ranked accordingly. The proposed methods are universally applicable to other types of mines [19].

They rely on a number of criteria allowing the user's preference options to be stated (i.e., strict, weak, low or indifferent) such that the value of a membership function should be derived. The ELECTRE III, based on a fuzzy outranking relationship, was developed to handle ranking problem, such as evaluation of the mining equipment layout in the specified mining conditions encountered in open pit mines where the deposits are of poor quality.

This study showcases the applications of ELECTRE III, which are more widespread than those of the simpler ELECTRE methods, to be used when handling similar problems, especially in the area of mining engineering, mechanical engineering, manufacturing systems and logistics and supply chains. Sitorus et al. [20] carried out a thorough overview on publications in the field of MCDM methods and their applications to mining and mineral processing and established that by 2018 the use of MCDM was reported in 90 studies whilst 24 out of them had relevance to mining equipment. The author of one paper [1] reported on the use of ELECTRE method. Most of the MCDM solutions in the sub-field of equipment selection were implemented using the AHP or hybrid method. Sousa et al. [21] used the ELECTRE I and PROMETHEE II method to select haul trucks to transport the ore in one of the Brazilian mines. The combination of the AHP and ELECTRE methods presents Stojanovic et al. [22] showing the application of the choice of operating technology for coal deposit. The AHP is used to handle the selection problem, the PROMETHEE effectively deals with ranking questions whilst ELECTRE III is able to handle both the ranking and sorting problems [23]. Nevertheless, as Hodget showed on the example of chemical manufacturing process, MCDM method may lead to different or no results for the same input data [24]. To conclude, our findings can be summarized recalling a quotation from Figueira et al. [25] who declared that research on ELECTRE III methods is far from being a dead field. It is just the opposite: the method is evolving and gains popularity, extending to new areas of expertise, supported by methodological and theoretical developments and dedicated user-friendly software implementations.

The step-by-step procedures recalled in this case study is outlined. The ELECTRE III method allows us to rank the finite set A of variants, evaluated according to a consistent set of criteria [26–28]. The variants are ranked according to the outranking relationship, denoted by S. Variant a is assumed to outrank variant b, which is denoted by aSb, as long as the available information on the decision maker's preferences, effectiveness (quality) of the variant evaluation and on the specific aspects of the problem provides sufficient evidence to prove that variant a is at least as good as b, unless there are valid reasons to reject this assumption [29,30]. The calculation procedure in the ELECTRE III method involves the following four stages:

- Construction of the decision maker's model of preferences with regard to particular criteria, including the definition of the respective weights and threshold values.
- Derivation of the valued outranking relationship—*S*.
- Variant ranking according to outranking relationships.
- Final ranking of variants.

The ELECTRE III method allows the decision-maker to express four states of preference when comparing variants a and b:

- Indifference, denoted by *aIb*.
- Weak preference for variant *a* over variant *b*, denoted by *aQb*.
- Strong preference for variant *a* over variant *b*, denoted by *aPb*.
- Incomparability of variants *a* and *b*, denoted by *aJb*.

Three thresholds are set to model these states of preference, determined separately for each j-criterion: threshold of indifference—q_j, threshold of preference—p_j and veto threshold—v_j. In addition, the relative importance of individual criteria is expressed by their respective weighting factors—k_j.

The outranking relationship evaluates the degree of credibility that a should be at least as good as *b*, which is denoted by the concordance index $c(a,b)$, whilst the discordance condition with respect to the relationship $S(a,b)$ is expressed by the discordance index $D_j(a,b)$ (Figure 1).

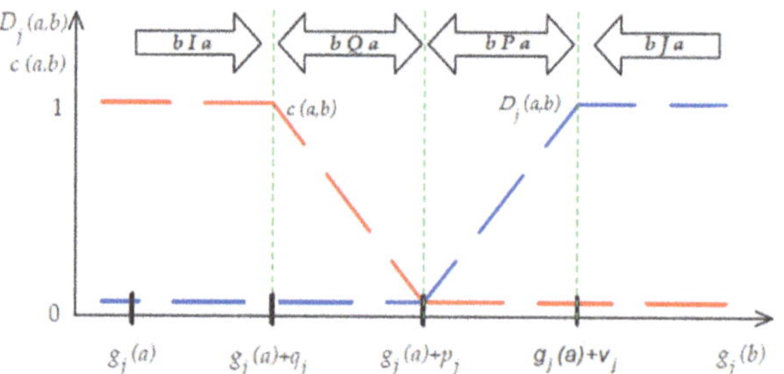

Figure 1. The four-state preference model (indifference *I*, weak preference *Q*, strong preference *P*, incomparability *J*) constructed on the basis of threshold values q_j, p_j, v_j [31].

An obvious advantage of using valued outranking relationships is that they are largely insensitive to parameter changes, both arbitrary and necessary ones.

The final ranking of variants in the ELECTRE III method is based on joining the descending and ascending distillations. The result is one complete preorder, in the form of a graph, which is the final ranking of the analyzed variants, showing the relations between them. The ELECTRE III method allows the conflicting criteria to be incorporated into a single analytical procedure. Even though the ELECTRE III method may appear fairly complicated, dedicated software is available to facilitate the procedure [32].

3. Results

The following aspects were taken into consideration when constructing the family of criteria and completing the set of alternatives, and in selection of the ranking method and implementations of the user preference models:

- Complexity and intricate nature o of evaluation criteria.
- Implementation of several decision-makers' preference models, the decision-makers acting as independent experts.
- Preference models giving the relative weight of each criterion, as well as the relationship between the weak and strong preference, and indifference between the alternatives being evaluated.

- Uncertainty on the part of decision makers as to whether the analyzed variants and preference thresholds should be regarded as incomparable.
- The significance of modelling of the decision-making processes, requiring a repeated reliability analysis.

3.1. Designing Decision Variants

In the case of karstic formations, ungraded material (barren rock) accounts for 20% of the limestone deposit. The term "deposit quality 80" used in our paper means that the limestone content in the intermediate product is 80%, assuming that the rock extracted from the deposit should be preliminarily crushed to remove the karst fractions with the grain size of 0–40 mm from the final product. The grain size in the fraction that is to be further processed (intermediate product) exceeds 40 mm [33].

In karstic formations, 20% of the limestone deposit under study comprises ungraded material (barren rock), which is unfit for use further in the process. In our study, we used the term "deposit quality 80", meaning that the intermediate product consists in 80% of limestone, and assuming that the rock extracted from the deposit should be preliminarily crushed in order to remove the karst fraction with a particular size distribution of 0–40 mm from the final material. The fraction that is fit for use further in the process (intermediate product) has a particular size distribution of more than 40 mm [33].

Three Decision Variants of the processing systems to be deployed have been proposed, making use of machines currently operated in the mine. The primary crushing operations will take place either in an electric-powered mobile crusher (MC) or in a diesel powered or stationary crusher (SC2), depending on the decision variant. The excavated material can be loaded onto the mobile crusher (MC) by a hydraulic excavator (E), whilst wheeled loaders (WL) will be used to load it onto the haulage trucks (HT). Depending on the decision variant, trucks will be deployed to:

- Transport the excavated to the primary crusher (SC2 or MC).
- Transport excavated material to Aggregate Mining Plant (AMP).
- Transport the extracted rock to the external dump (ED).

In the decision variant no. 2 belt conveyor (BC) was also used for the transport of intermediate product between the excavation site and Aggregate Mining Plant (AMP).

Schematics of all decision variants were presented at the Figures 2–4 [33].

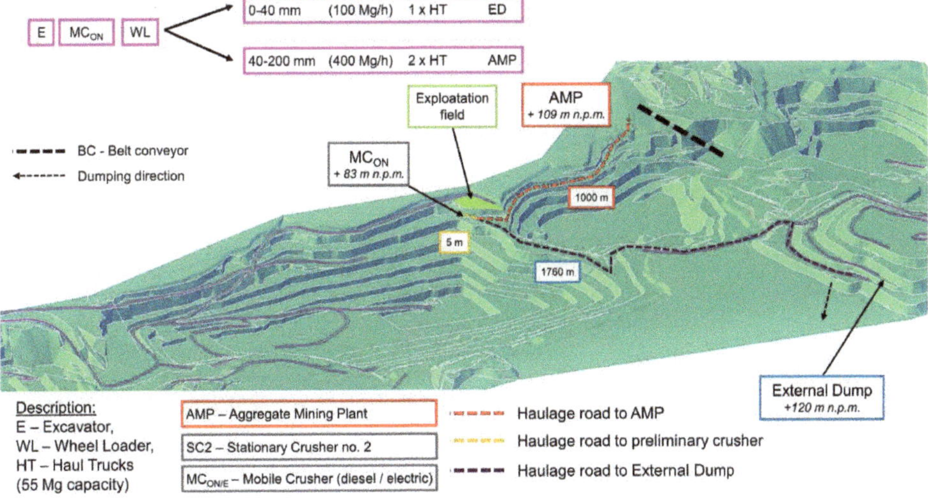

Figure 2. Primary crusher locations in the analyzed mine—Decision Variants no. 1.

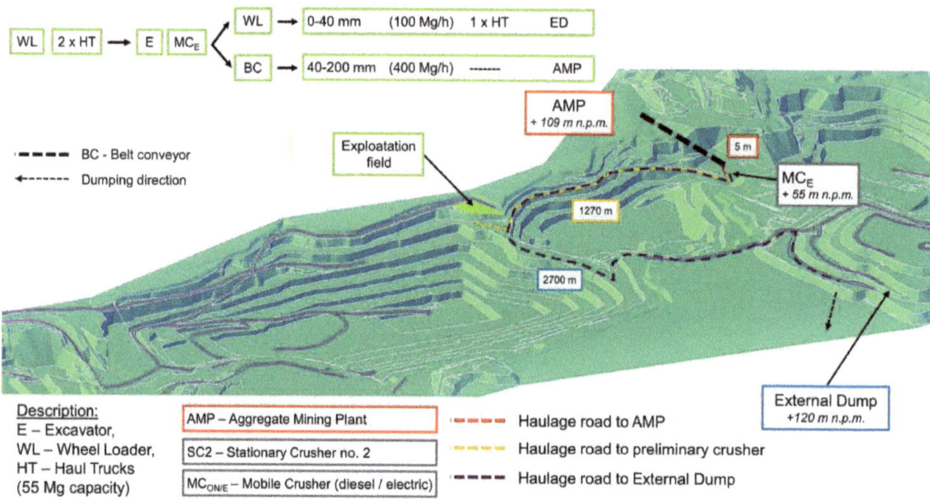

Figure 3. Primary crusher locations in the analyzed mine—Decision Variants no. 2.

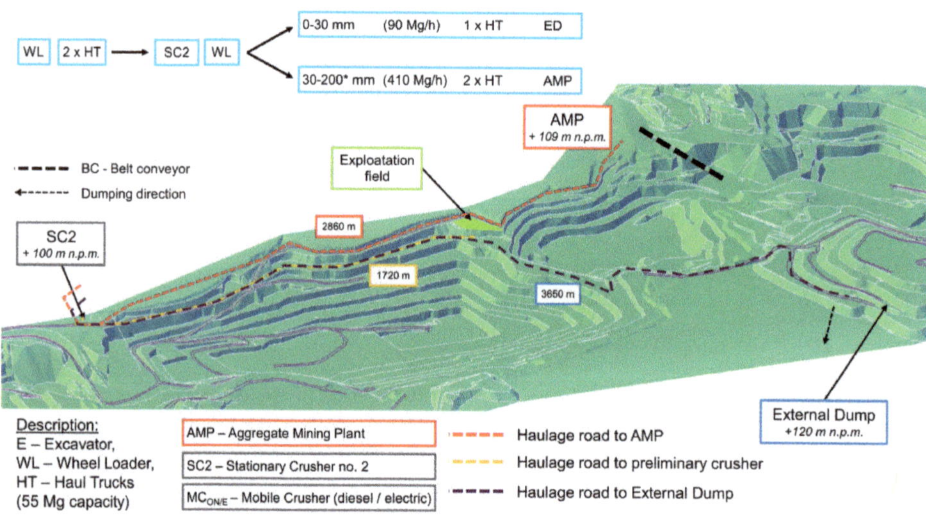

Figure 4. Primary crusher locations in the analyzed mine—Decision Variants no. 3.

The main difference between decision variant no. 1 and no. 2 is requirements for the movement and transport mobile crusher (MC) to the different points in the mine. This is necessary because of the blasting methods in this type of mining (see Figures 2 and 3).

In all variants, mining operations continued for two working shifts, the number of working hours per one shift is assumed to be 7 (total working hours).

3.2. Criteria for Selection and Evaluation of Mining Equipment in Open Pit Rock Mines

The family of criteria having relevance to evaluation and selection of the mining equipment are summarized below. A comprehensive analysis of the decision-making problem relies on a set of consistent criteria, covering the engineering, economic, environmental and operational aspects as well as reliability issues.

- K1—length of transport routes.

- K2—machine fleet size.
- K3—reliability index.
- K4—distance of the crushing unit from residential buildings.
- K5—energy consumption by the mining equipment.
- K6—CO emissions from the mining equipment.
- K7—size of external dump.
- K8—process-related costs.

3.2.1. Technology-Related Criteria

K1—Length of transport routes—a minimized quantifiable criterion that determines the total number of kilometers covered by vehicles, including haulage trucks and a belt conveyor, during two working shifts, expressed in (km);

K2—Machine fleet size—a minimized quantifiable criterion that determines the total number of machines making up the fleet (decision alternative), expressed in (pcs) (Table 1);

Table 1. Number of machines in each variant [12].

Machine Type	Name	W1	W2	W3
haul truck	Komatsu HD 465	3	3	5
excavator	CAT 34	1	1	-
wheel loader	Komatsu WA600 HL	1	2	2
mobile crusher diesel	Powerscreen Premiertrak 1180	1	-	-
mobile crusher electric	Powerscreen Premiertrak 1180	-	1	-
stationary crusher	-	-	-	1
belt conveyor	-	-	1	1
total	-	6	8	8

3.2.2. Exploitation and Reliability Criterion

K3—Reliability index K_{gt}—a maximized criterion expressing the probability that a given machine or system should be operational and running at a specified time. It has to be obtained for each type of equipment that make up the mining machine system, expressed as the mean value for all types of machines (haulage trucks, loaders and processing machines). The reliability index is expressed in (%).

$$K_{gt} = \frac{\sum_{i=1}^{n} t_i^{(j)}}{\sum_{i=1}^{n} t_i^{(j)} + \sum_{i=1}^{n} t_i^{(n)}} \tag{1}$$

where: $t_i^{(j)}$—effective working time in i-th day of operation (h), $t_i^{(n)}$—time required for repairs and maintenance, including:

$$\sum_{i=1}^{n} t_i^{(n)} = t_{er} + t_{es} + t_{wd} \tag{2}$$

t_{er}—effective repair downtime (h), t_{es}—repairs done by external services (h), t_{wd}—workshop downtime (h).

A comprehensive analysis of a decision-making problem requires a consistent family of criteria covering technological, environmental and economic aspects.

3.2.3. Environmental Criteria

K4—Distance from the crushing unit to residential buildings—a maximized quantifiable criterion that determines the distance of the preliminary crushing site (crusher's location) to the nearest residential buildings, expressed in meters, constituting the basis for establishing the nuisance zone associated with noise, vibration and dust emissions.

K5—Energy/power consumption by the mining equipment—a minimized quantifiable criterion that determines the total energy expenditure by mining and processing machines, expressed in (MJ/day). It allows for identification of those machines and systems that cause the lowest degree environmental nuisance due to the lowest energy consumption during two working shifts. Fuel consumption during the ride by haulage trucks depends on the length and configuration of haulage roads.

K6—CO emissions from the mining equipment—a minimized quantifiable criterion that determines the total amount of CO emissions by a machine or system, expressed in (g/kWh). It enables the identification of mining machines and systems that produce the least environmental nuisance in terms of car-bon monoxide emissions—i.e., which mining machine or system is responsible for the CO lowest emission during two working shifts, based on telemetry data (Table 2). CO emissions are considered instead of CO_2 or NOx since a common benchmark is required to compare emissions from different sources. Direct emissions produced by the machine were considered in this study, without analyzing the emissions related to the value chain e.g., as in the case of electric systems.

Table 2. Fuel and energy consumption and CO emissions by mining equipment [12,34–37].

Machine Type/Name	Power Installed (kW)	Av. Consumption of Energy Carriers	CO Emissions (g/kWh)
excavator/CAT 349	317	39 l/h	6.5
wheel loader/Komatsu WA600 HL	393	40 l/h	3.5
haul trucks/Komatsu HD 465	551	38 l/h	3.5
mobile crusher plant (diesel)/Powerscreen Premiertrak 1180	205	28 l/h	5.0
mobile crusher plant (electric)/Powerscreen Premiertrak 1180	185	148 kW/h	zero emission
belt conveyor	180	144 kW/h	zero emission
preliminary crushing + rectangular screen stationary Crusher no. 2 + stationary plant	1300	1043 kW/h	zero emission

K7—Size of external dump—a minimized quantifiable criterion that determines the amount of waste rock dumped during one working shift, expressed in (Mg). It quantifies the amount of material dumped within the area of the mine, which is a major determinant of the actual size of the dumping site, embracing additional environmental fees and transport costs, as well as dumping costs.

3.2.4. Economic Criteria

K8—Process-related costs—a minimized quantifiable economic criterion related to the aggregation of all significant costs borne to obtain the intermediate product and ungraded material, i.e., all the constituent costs generated by the mining equipment during a working shift, expressed in (EUR/day).

For the purposes of the ELECTRE III method, a matrix of criteria values was constructed (Table 3). In order to assess and select an optimum mining machine system, we conducted in-depth interviews with three experts—a senior mine manager (E1) (Table 4, Figure 5a), a mine maintenance manager (E2) (Table 5, Figure 5b), an academic expert whose area of research is surface mining (E3) (Table 6, Figure 5c) and a local government official expert on environment (E4) (Table 7, Figure 5d). Accordingly, each expert was asked to define the relative importance of all the criteria and the preference levels and to include them in the decision thresholds q_i—indifference, p_i—preference, v_i—veto.

Table 3. Values of the specified common family of criteria (criteria evaluation matrix)—final values of the criteria for each analyzed variant (mining machines and systems).

Criteria	K1 Length of Transport Routes	K2 Machine Fleet Size	K3 Reliability Index	K4 Distance of the Crushing Unit	K5 Energy Consumption	K6 CO Emissions	K7 Size of External Dump	K8 Process-Related Costs
unit	km/day	pcs.	%	m	MJ/day	g/kWh	Mg	EUR/day
preference	min	min	max	max	min	min	min	min
W1	454	6	93.1	2192	103,194	90.70	1400	732
W2	570	8	94.0	1190	125,925	95.67	1400	991
W3	2181	8	93.6	1975	170,885	125.84	1260	1304

Table 4. Decision-maker's preference model—Expert 1—Mine Manager, Chief Executive Officer.

No.	Preference Direction	Indifference Threshold	Preference Threshold	Veto Threshold	Relative Importance
K1	min	25	50	100	8
K2	min	0	1	2	4
K3	max	0.5	0.75	1	5
K4	max	100	200	500	9
K5	min	1000	15,000	20,000	6
K6	min	2	6	25	6
K7	min	30	600	1000	7
K8	min	120	240	600	8

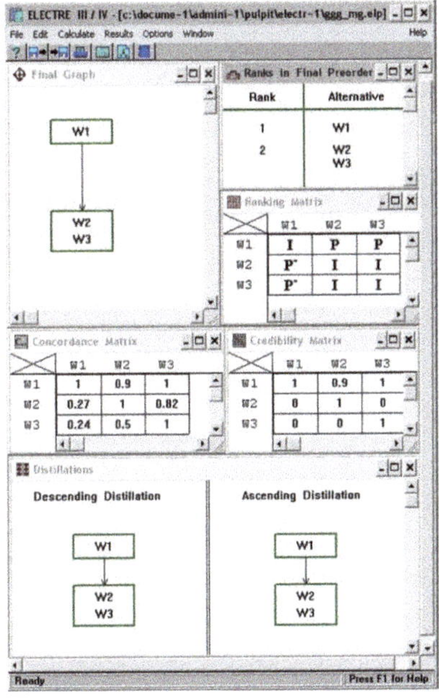

(a)

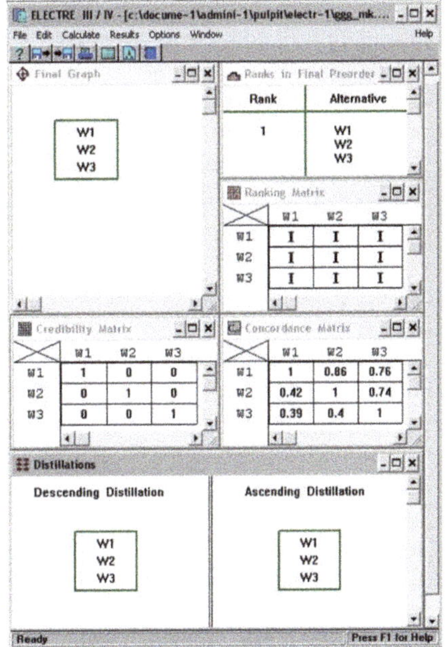

(b)

Figure 5. *Cont.*

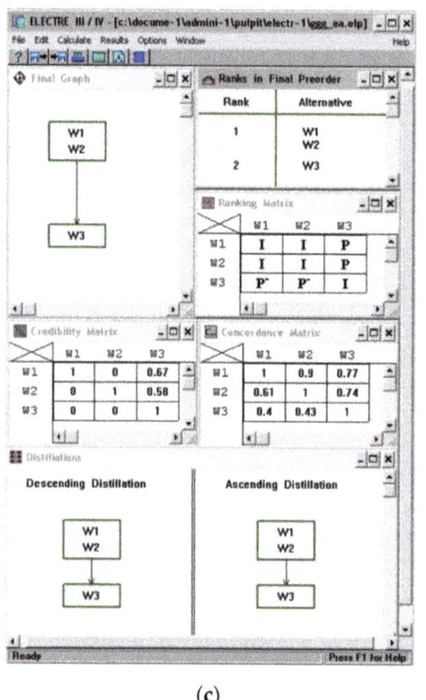

(c)

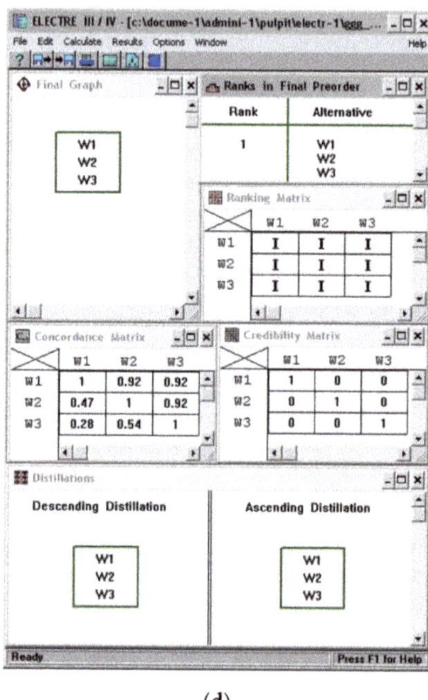

(d)

Figure 5. Final rankings, outranking, credibility and concordance matrices for the experts' preference models: (**a**) E1, (**b**) E2, (**c**) E3, (**d**) E4.

Table 5. Decision-maker's preference model—Expert 2—Mine Manager—Mining Operations Manager.

No.	Preference Direction	Indifference Threshold	Preference Threshold	Veto Threshold	Relative Importance
K1	min	200	500	1000	7
K2	min	1	2	4	7
K3	max	0.1	0.3	0.5	7
K4	max	200	400	650	8
K5	min	2500	20,000	50,000	5
K6	min	10	15	20	2
K7	min	60	100	200	5
K8	min	30	60	120	9

Table 6. Decision-maker's preference model—Expert 3—academic expert.

No.	Preference Direction	Indifference Threshold	Preference Threshold	Veto Threshold	Relative Importance
K1	min	100	300	500	5
K2	min	1	3	5	3
K3	max	0.2	0.5	0.7	5
K4	max	1300	500	1000	5
K5	min	3000	10,000	30,000	4
K6	min	5	10	20	4
K7	min	50	100	150	5
K8	min	180	360	720	9

Table 7. Decision-maker's preference model—Expert 4—local government official expert on environment.

No.	Preference Direction	Indifference Threshold	Preference Threshold	Veto Threshold	Relative Importance
K1	min	25	50	100	8
K2	min	1	2	4	6
K3	max	0.05	0.35	0.75	4
K4	max	300	600	1200	8
K5	min	3750	7500	15,000	4
K6	min	5	10	25	9
K7	min	30	60	120	8
K8	min	60	120	240	2

Preference levels stated by decision-makers are collated in Tables 4–7. Preference directions of each of the criteria (min or max value) are included, indicating minimization or maximization of the value of the criterion.

The values of the relative weights of the criteria, accepted in accordance with the preferences of the decision-maker, are presented in Table 8.

Table 8. Relative and average importance of the criteria considered by experts (E1–E4).

No.	Preference Direction	E1	E2	E3	E4	Average
K1	min	8	7	5	8	7
K2	min	4	7	3	6	5
K3	max	5	7	4	4	5.25
K4	max	9	8	5	8	7.5
K5	min	6	5	4	4	4.75
K6	min	6	2	4	9	5.25
K7	min	7	5	5	8	6.25
K8	min	8	9	9	2	7

These models of decision-makers' preferences were constructed independent of each other, without consultations between experts. In addition, an expert data analyst was called in to identify, explain and create an underlying mathematical model with regard to respective fields of expertise and to select all methods that should enable the solution of the given decision problem. Accordingly, four final rankings were generated, for each of the adopted preference models, respectively (Figure 5).

The final order for expert E1 (Mine Manager and Chief Executive Officer) indicates that for the preference model in question, alternative W1 outranks all other alternatives and is therefore the most advantageous solution. Alternatives W2 and W3 are indistinguishable from each other. For expert E2 (Mining Operations Manager) and E4 (local government official expert on environment), all variants are equally advantageous and, at the same time, indistinguishable. The final order for expert E3 (academic expert) indicates that alternatives W1 and W2 are optimum and, at the same time, indistinguishable and that they outrank alternative W3.

MCDM methods consider the preferences of decision-makers, they will present a set of subjective solutions. In this case, one can easily observe the areas that the decision-makers considered most relevant, depending on the functions they performed. Thus, the relative importance defined by experts E1 and E2, was associated mostly with economic and technical and operational parameters which directly affect the production costs and the financial performance of the mine. As regards the decision-maker E3, apart from the cost-related criterion, which was classified as the most important one, relative importance was evenly distributed among the remaining criteria at an average level. Decision-maker E4 focused chiefly on negative environmental impacts of the mining plant on its surroundings and it becoming a source of nuisance.

The presented multi-aspect analysis of the issue enables us to find a compromise solution and to identify the most important criteria, both in terms of mine performance and its environmental impacts. Accordingly, the most advantageous variant was selected (W1) and the least favorable variant W3 was firmly rejected.

4. Discussion

This study explores the application of a multiple-criteria decision-making (MCDM) method, or more specifically, one of the available tools within this approach—ELECTRE III—to solving problems that involves the selection and assessment of mining machinery under specific rock mining conditions. The presented ELECTRE III method proves useful in solving problems related to the assessment of mining equipment performance, and the final results obtained indicate the desired directions of operation.

The results of all analyses show that alternative W1 has an advantage over all other alternatives. Alternative W1 meets nearly all key criteria emphasized by experts in the field on surface mining, academic experts and local government official expert on environment. Moreover, this alternative outranks all other alternatives in terms of process-related costs—a criterion that has been given the highest relative importance by E1, E2 and E3 decision-makers. The criteria with the highest level of importance, according to most decision-makers, were as follows: K4—Distance of the crushing unit from residential buildings (7.5 points), K1—Length of transport routes (7.0 points) and K8—Process-related costs (7.0 points). It is also important to note the high rank of alternative W1 compared to other criteria and its relation to other alternatives in expert assessments.

5. Conclusions

The set of criteria adopted to evaluate and select the mining equipment are largely those applicable to all types of selection problems. Hence a comprehensive analysis of a decision-making problem uses a set of task-specific criteria, taking into account technological, economic, environmental operational aspects as well as reliability.

The proposed method could be universally applicable in a number of mines, especially when the fleet of machines is to be replaced. This method allows the experts to conduct a rigorous analysis of the issue and gain information basing on results the outputs obtained at intermediate levels.

In addition, ELECTRE III offers the decision-maker an opportunity to define the local preference model in each node of the hierarchy, according to their objectives and taking into account the specific feature of the sub-problem. Furthermore, the analysis has clearly demonstrated that an improved layout and configuration of machine systems will result in lower costs, both in terms of capital expenditure and, last but not least, also operating costs.

Despite the specificity related to reduced profitability of low-quality deposits exploitation, the implementation of the MCDM presented no major difficulties, which proves its universality.

Author Contributions: Conceptualization, M.P. and P.B.; methodology, P.B.; software, P.B.; validation, M.P., P.B. and Z.K.; formal analysis, Z.K.; investigation, M.P. and P.B.; resources, M.P.; data curation, M.P.; writing—original draft preparation, P.B., M.P. and Z.K.; writing—review and editing, Z.K.; visualization, M.P.; supervision, Z.K. All authors have read and agreed to the published version of the manuscript.

Funding: This research was funded by AGH University of Science and Technology, Faculty of Mining and Geoengineering; subsidy number: 16.16.100.215.

Institutional Review Board Statement: Not applicable.

Informed Consent Statement: Not applicable.

Data Availability Statement: The data can be accessed upon request any of the authors.

Conflicts of Interest: The authors declare no conflict of interest.

References

1. Bodziony, P.; Kasztelewicz, Z.; Sawicki, P. The Problem of Multiple Criteria Selection of the Surface Mining Haul Trucks. *Arch. Min. Sci.* **2016**, *61*, 223–243. [CrossRef]
2. Stevanović, D.; Lekić, M.; Kržanović, D.; Ristović, I. Application of MCDA in selection of different mining methods and solutions. *Adv. Sci. Technol. Res. J.* **2018**, *12*, 171–180. [CrossRef]
3. Bazzazi, A.A.; Osanloo, M.; Karimi, B. A new fuzzy multi criteria decision making model for open pit mines equipment selection. *Asia-Pac. J. Oper. Res.* **2011**, *28*, 279–300. [CrossRef]
4. Burt, C.N.; Caccetta, L. Equipment Selection for Surface Mining: A Review. *Interfaces* **2014**, *44*, 143–162. [CrossRef]
5. Voronov, Y.; Voronov, A.; Voronov, A. Quality criterion of the loading and transport system operation at open-pit mines. *E3S Web Conf.* **2019**, *105*. [CrossRef]
6. Samanta, B.; Sarkar, B.; Mukherjee, S.K. Selection of Opencast Mining Equipment by Multi-Criteria Decision-Making Process. *Min. Technol.* **2002**, *111*, 136–142. [CrossRef]
7. Haldar, S.K. Chapter 12—Elements of Mining, Mineral Exploration (Second Edition). *Princ. Appl.* **2018**, 229–258.
8. Petlovanyi, M.V.; Lozynskyi, V.H.; Saik, P.B.; Sai, K.S. Modern experience of low-coal seams underground mining in Ukraine. *Int. J. Min. Sci. Technol.* **2018**, *28*, 917–923. [CrossRef]
9. Pimentel, B.S.; Gonzalez, E.S.; Barbosa, G.N.O. Decision-support models for sustainable mining networks: Fundamentals and challenges. *J. Clean. Prod.* **2016**, *112*, 2145–2157. [CrossRef]
10. Capasso, I.; Lirer, S.; Flora, A.; Ferone, C.; Cioffi, R.; Caputo, D.; Liguori, B. Reuse of mining waste as aggregates in fly ash-based geopolymers. *J. Clean. Prod.* **2019**, *220*, 65–73. [CrossRef]
11. Segura-Salazar, J.; Tavares, L.M. Sustainability in the Minerals Industry: Seeking a Consensus on Its Meaning. *Sustainability* **2018**, *10*, 1429. [CrossRef]
12. Patyk, M.; Bodziony, P.; Kasztelewicz, Z. Analysis of quarrying equipment operating cost structure. *J. Pol. Miner. Eng. Soc.* **2019**, *21*, 311–318.
13. Mudd, G.M. The Environmental sustainability of mining in Australia: Key mega-trends and looming constraints. *Resour. Policy* **2010**, *35*, 98–115. [CrossRef]
14. Moore, K.R.; Whyte, N.; Roberts, D.; Allwod, J.; Leal Ayala, D.; Bertrand, G.; Bloodworth, A.J. The re-direction of small deposit mining: Technological solutions for raw materials supply security in a whole systems context. *Resour. Conserv. Recycl. X* **2020**, *7*. [CrossRef]
15. Siksnelyte-Butkiene, I.; Zavadskas, E.K.; Streimikiene, D. Multi-Criteria Decision-Making (MCDM) for the Assessment of Renewable Energy Technologies in a Household: A Review. *Energies* **2020**, *13*, 1164. [CrossRef]
16. Kumar, A.; Sah, B.; Singh, A.R.; Deng, Y.; He, X.; Kumar, P.; Bansal, R.C. A review of multi criteria decision making (MCDM) towards sustainable renewable energy development. *Renew. Sustain. Energy Rev.* **2017**, *69*, 596–609. [CrossRef]
17. Trojanowska, M.; Nęcka, K. Selection of the Multiple-Criiater Decision-Making Method for Evaluation of Sustainable Energy Development: A Case Study of Poland. *Energies* **2020**, *13*, 6321. [CrossRef]
18. Hokkanen, J.; Salminen, P. ELECTRE III and IV Decision Aids in an Environmental Problem. *J. Multi-Criteria Decis. Anal.* **1997**, *6*, 215–226. [CrossRef]
19. Mutlu, M.; Sari, M. Kullanimi multi-criteria decision making methods and use of in mining industry. *Sci. Min. J.* **2017**, *4*, 181–196.
20. Sitorus, F.; Cilliers, J.; Brito-Parada, R. Multi-criteria decision making for the choice problem in mining and mineral processing: Applications and trends. *Expert Syst. Appl.* **2019**, *121*, 393–417. [CrossRef]
21. De Sousa, W.T., Jr.; Souza, M.J.F.; Cabral, I.E.; Diniz, M.E. Multi-Criteria Decision Aid methodology applied to highway truck selection at a mining company. *Rem. Rev. Esc. Minas* **2014**, *67*, 285–290. [CrossRef]
22. Stojanovic, C.; Bogdanovic, D.; Urosevic, S. Selection of the optimal technology for surface mining by multi-criteria analysis. *Kuwait J. Sci.* **2015**, *42*, 170–190.
23. Vujic, S.; Hudej, M.; Miljanovic, I. Results of the promethee method application in selecting the technological system at the majdan III open pit mine. *Arch. Min. Sci.* **2013**, *58*, 213–219.
24. Hodgett, R.E. Comparison of Multi-Criteria Decision-Making Methods for Equipment Selection. *Int. J. Adv. Manuf. Technol.* **2016**, *85*, 1145–1157. [CrossRef]
25. Figueira, J.R.; Greco, S.; Roy, B.; Slowinski, R. An overview of ELECTRE methods and their recent extensions. *J. Multi-Criteria Decis. Anal.* **2013**, *20*, 61–85. [CrossRef]
26. Roy, B. *The Outranking Approach and the Foundations of ELECTRE Methods. Readings in Multiple Criteria Decision Aid. Bana e Costa C.A.*; Springer: Berlin, Germany, 1990; pp. 155–183.
27. Roy, B. Decision Aid and Decision Making. *Eur. J. Oper. Res.* **1990**, *45*, 324–331. [CrossRef]
28. Roy, B. *Wielokryterialne Wspomaganie Decyzji*; Wydawnictwo Naukowo-Techniczne: Warszawa, Poland, 1990.
29. Vincke, P. *Multicriteria Decision–Aid*; John Wiley & Sons: Chichester, UK, 1992.
30. Vincke, P. Outranking Approach. In *Multicriteria Decision Making*; Gal, T., Stewart, T.J., Hanne, T., Eds.; International Series in Operations Research & Management Science; ISOR: Boston, MA, USA, 1999; Volume 21, pp. 305–333.
31. Sawicki, P. Wielokryterialny dobór operatora usług logistycznych. *Logistyka* **2001**, *4*, 59–66.
32. Govindan, K.; Brandt Jepsen, M. ELECTRE: A comprehensive literature review on methodologies and applications. *Eur. J. Oper. Res.* **2016**, *250*, 1–29. [CrossRef]

33. Bodziony, P.; Patyk, M.; Kasztelewicz, Z. Multi-criteria decision making for the choice equipment in mining with application of AHP method. New trends in production engineering: Monograph. *Soc. Sci.* **2019**, 48–59.
34. *Caterpillar Performance Handbook*; 48th ed.. Caterpillar Inc.: Peoria, IL, USA, 2018. Available online: https://wheelercat.com/wp-content/uploads/2018/07/SEBD0351_ED48.pdf (accessed on 1 March 2021).
35. *Specification @ application Handbook*; Edition 31. Komatsu Ltd.: Komatsu, Japan, 2013. Available online: https://www.directminingservices.com/wp-content/uploads/2011/05/Edition31.pdf (accessed on 1 March 2021).
36. Powerscreen 1180 Premiertrak Jaw Crusher Specification—Rev 6. 01/01/2015. Available online: http://csrpower.com.my/details/Powerscreen%201180%20Premiertrak.pdf (accessed on 1 March 2021).
37. *EMEP/EEA Air Pollutant Emission Inventory Guidebook 2019*; EEA Report; EEA: Luxembourg, 2019; Volume 13.

 *energies*

Article

Energy Efficiency Analysis of Copper Ore Ball Mill Drive Systems

Piotr Bortnowski, Lech Gładysiewicz, Robert Król and **Maksymilian Ozdoba** *

Department of Mining and Geodesy, Faculty of Geoengineering, Mining and Geology, Wroclaw University of Science and Technology, 50-421 Wroclaw, Poland; piotr.bortnowski@pwr.edu.pl (P.B.); lech.gladysiewicz@pwr.edu.pl (L.G.); robert.krol@pwr.edu.pl (R.K.)
* Correspondence: maksymilian.ozdoba@pwr.edu.pl

Abstract: Milling is among the most energy-consuming technological stages of copper ore processing. It is performed in mills, which are machines of high rotational masses. The start of a mill filled to capacity requires appropriate solutions that mitigate the overloading. One method for increasing the energy efficiency of ball mills is to optimize their drive systems. This article looks at two variants of drive systems with efficiencies higher than the already existing solutions. The first variant is a low-speed synchronous motor with permanent magnets without a gearbox, and the second variant is an asynchronous high-efficiency motor with a gearbox and a fluid coupling. The energy performance analysis of the three solutions was based on the average energy consumption indicator per mass unit of the milled material and on the energy consumption per hour. The investigations required models of the drive systems and analyses with the use of the Monte Carlo methods. The highest energy efficiency is observed in the case of the solution based on the permanent magnet motor. However, the drive system with the high-speed motor offers a gentle start-up possibility owing to the fluid coupling.

Keywords: fluid coupling; ball mill; electric motor; drive system; grinding; energetic efficiency

check for updates

Citation: Bortnowski, P.; Gładysiewicz, L.; Król, R.; Ozdoba, M. Energy Efficiency Analysis of Copper Ore Ball Mill Drive Systems. *Energies* **2021**, *14*, 1786. https://doi.org/10.3390/en14061786

Academic Editor: Sergey Zhironkin

Received: 26 February 2021
Accepted: 19 March 2021
Published: 23 March 2021

Publisher's Note: MDPI stays neutral with regard to jurisdictional claims in published maps and institutional affiliations.

1. Introduction

Ore beneficiation and processing plants are among the biggest industrial facilities with respect to energy consumption. Therefore, research in this sector has been naturally focused on maximizing the energy efficiency. Energy efficiency is defined as an optimal (profitable) usage of energy for the purposes of current production or services. As the majority of this industrial sector uses highly-emissive sources of electric energy, the energy efficiency has in this case become an important point on the European Union's agenda [1]. A policy based on limiting the energy consumption is also an effective method for improving the competitiveness of an industrial plant and for maintaining constant technological progress, while reducing CO_2 emissions to the environment [2]. The first stage in the processing of raw materials (i.e., comminution) is the most energy-consuming, and accounts for up to 4% of global electric energy consumption [3–6]. In the United States, the milling process is estimated to account for 0.5% of the primary energy consumption, 3.8% of the total electric energy consumption, and 40% of the total energy consumption in the mining industry [7]. In Poland, the annual electric energy consumption from copper ore processing plants is approximately 2.5 TWh [8]. Primary ore comminution is performed with jaw and hammer crushers. Ore is then transferred to ball mills, in which it is first subjected to coarse grinding, and in the second phase (i.e., fine grinding) it is treated with grinding bodies (rods or cylpebs), which are introduced into the mill. Specific energy consumption during the milling process depends on the geomechanical parameters of the raw material and on the design parameters of the mill [9,10]. The energy demand indicator increases exponentially together with the decrease of the required grain size [11]. The cost of the energy used in comminution represents $50 \div 60\%$ of the entire ore processing costs [12]. The above reason explains the importance of the research into new solutions for improving the energy efficiency of the comminution process.

2. Directions for Ore Processing Optimization

The literature mentions a number of examples of optimization-focused activities based on modeling the phenomena that occur during the milling process. Numerical simulation methods allow detailed analyses of the behavior of the material in the mill. For example, a previous publication [13] described a simulated model of a tumbling ball mill that allowed a number of operating recommendations to be formulated in order to increase the milling efficiency. In another example of the application of advanced numerical techniques, the authors of a previous study [14] used the ANSYS Fluent (manufacturer Ansys, Inc., Canonsburg, PA, USA) package to analyze a model of multiphase flow in a ball mill. The results of their research allowed a precise prediction of the flow with the free surface of the slurry in the mill.

Other optimization methods are based on the modernization of the milling process or on modifications of the mill design. If an adequate mill type is chosen for each milling stage already while designing the entire technological system, the total energy efficiency of the milling process is significantly influenced. Ball mills are applicable only in the primary milling. The milling efficiency and the related energy demand render impossible the use of ball mills for fine milling [15,16]. The number and the type of internal mill components (i.e., the type of the inner lining, the type and number of the diaphragms) are important for increasing the efficiency of the milling process, and this in turn translates directly into increased energy efficiency [17]. Mills with drum diameters greater than 5.5 m show lower energy efficiency at an optimal fill level of 40% [18]. The grinding media—their type and size—are also of significance [19]. The ratio between the size of the grinding media and the size of the comminuted particles is especially important. Research performed on this aspect demonstrates that the size of the particles in the mill feed significantly influences the energy efficiency of the comminution process, and by using an optimal crusher at the beginning of the technological system, this efficiency may be increased by as much as 10%. Wet milling significantly increases the energy consumption [20]. Specific energy demand by a mill depends on its efficiency, which is directly related to the speed of the mill. Optimal mill speed is typically 65 $\div$ 75% of its critical speed, at which the milling process is no longer observed. In industrial conditions, mill speed can be regulated, for example, with the use of systems based on programmable logic controllers (PLCs) [21]. In the future, an important aspect of energy optimization may involve the concept of a single technological system incorporating devices that allow automation, as well as the processing and exchanging of data (Industry 4.0) for the purpose of monitoring the working conditions and controlling the parameters of the mill.

In another approach, the energy efficiency of milling may be improved by adjusting the parameters of the comminuted material with the use of chemical additives or water. Chemical additives may, however, have a negative environmental impact [22].

Energy savings may also be found in the ball mill drive systems. In the scale of a processing plant, the electric motors of these drive systems account for 95% of the total electric energy demand [23]. Therefore, the type of the motor used has a significant influence on the efficiency of the drive system. One of the solutions aimed at improving this efficiency is to use synchronous motors induced with permanent magnets. Motors of this type have a very high efficiency [24–26], and therefore they were used, in a low-speed version, in the modernized copper ore processing plant [27–29]. However, their operation entails a number of problems. Although low-speed motors eliminate the need to use a gearbox in the drive system, their start-up is rapid, which can damage its internal components and cause the overloading of the power supply line [30].

3. Analysis of Variant Ball Mill Drive Systems

The basic element of a ball mill is the drum, in which the milling process takes place (Figure 1). The length of the drum in the analyzed mill (without the lining) is 3.6 m, and the internal diameter is 3.4 m. The mass of the drum without the grinding media is 84 Mg. The drum is set in rotational motion by driving a ring gear mounted on the drive side. The ring gear is engaged with a pinion gear. In versions with low-speed motors, the pinion is set on a shaft directly connected to the motor, and in versions with high-speed motors the pinion is set on an output shaft of the gearbox. The nominal mill efficiency, at optimum speed, is 80 Mg/h. The dimensions and the high mass of the rotating elements cause the drive system to be overloaded when the mill filled with copper ore is started. For this reason, the mill is stopped only in the case of a failure or planned maintenance work. The aim is to minimize the potential number of start-ups, as the mill at a standstill must be later started under full load. In the case of a prolonged downtime, the moist ore inside the mill hardens. During start-up, the mass is thus unevenly distributed with respect to the rotational axis of the mill. This condition is similar to the state of unbalance. This phenomenon causes unfavorable dynamic reactions in the mill bearings and excessive loads in other components of the drive system. Such conditions facilitate increased friction in the system, as well as noise emissions. They also increase the risk of damage to the motor itself, its overheating, or overloading of the power supply line [31–33].

Figure 1. Ball mill of the primary milling stage with the visible gear ring.

3.1. Traditional Drive System with SAS Motors

Designed already at the end of the twentieth century, the investigated drive systems of the ball mills installed in the copper ore processing plant were provided with SAS motors (Figure 2). The motors are synchronized in steady motion and powered from a 6 kV grid. Depending on the mill type, the power range of the SAS motors was 400 kW to 1120 kW. Low-power units work with 187.5 RPM, while higher-power units work with 166.6 RPM. Owing to this solution, the ball mill drive system does not have a gearbox [34].

(a)

(b)

Figure 2. Ball mill drive system with low-speed SAS motor without intermediate gearbox: (**a**) prior to modernization, (**b**) after modernization on the right [34].

The nominal efficiency of the SAS motors, as declared by the manufacturer, is between 90% and 93%. Because they were manufactured in the 1970s, their efficiency after several decades in operation is estimated to have been reduced to 85%. Their compact design makes SAS motors particularly prone to the influence of an adverse, aggressive working environment. The most common operating problems experienced in motors of this type include rotor seizure or insulation damage due to moisture from the wet ore. The renovation cost of a single motor exceeds 25,000 USD.

Measurements of the current parameters with the use of the HIOKI loggers (HIOKI 9625, HIOKI E.E. CORPORATION, Nagano, Japan) show that the power consumed by the SAS motor in steady motion varies around a mean value of 602 kW (Figure 3). The rated power of this motor is 630 kW.

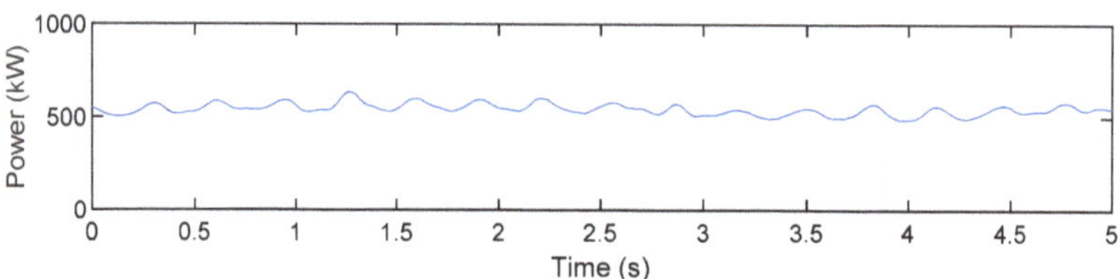

Figure 3. Active power of the SAS 630 kW motor during normal operation with filled ball mill.

Because of low efficiency and increasing maintenance costs, a decision was made to look for new solutions that would improve the energy efficiency [34].

3.2. The Drive System with a Low-Speed Motor Type LSPMSM SMH-1732T

One of the solutions aimed at increasing the energy efficiency of the milling process is to use new, energy-saving motors. The prototype drive system is based on a low-speed synchronous LSPMSM SMH-1732T motor (Figure 4), induced with permanent magnets. The design of the motor allows it to be directly connected to the power grid. The electric machine works in a three-phase system, with a supply voltage of 6 kV. The nominal power of the motor is 630 kW, and its most important advantage lies in its high efficiency, which is up to 97% [28,35]. The prototype LSPMSM motor replaced the SAS motor in one of the mills [36].

Figure 4. New ball mill drive system with the prototype energy-saving SMH motor [37].

As in the case of the SAS motor, the new drive system does not have a gearbox or a starter. The start of the charged ball mill is very violent, as shown in the graphs from the industrial HIOKI logger (Figure 5).

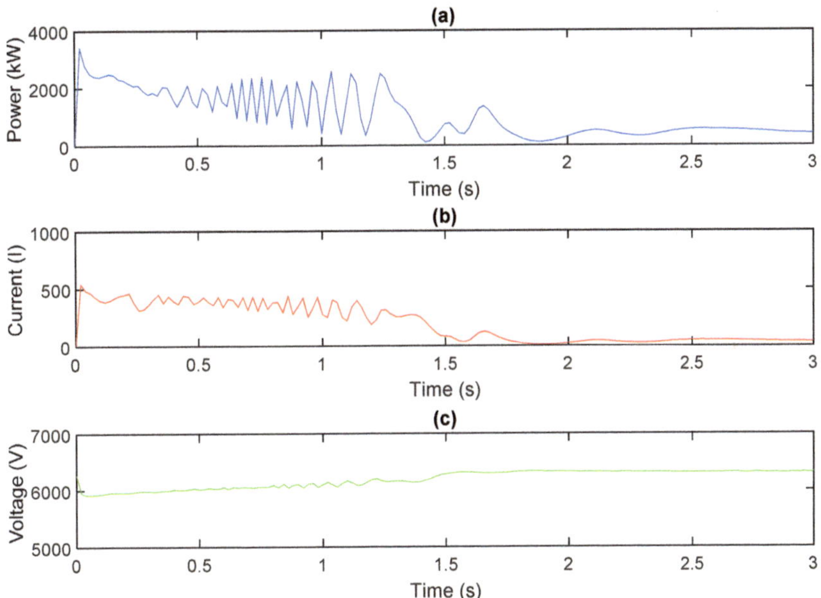

Figure 5. Start-up of the LSPMSM SMH-1732T motor with a filled ball mill after two hours of downtime: (**a**) active power, (**b**) current intensity, (**c**) voltage.

The start-up of the new motor lasts for approximately 2 s (Figure 5). Immediately after the motor starts, its instantaneous active power reaches the peak value of 3500 kW, being more than 5-fold in excess of its nominal power (Figure 5a). The start-up also entails significant changes of current intensity in the windings of the motor (Figure 5b). The highest current values of 550 A are observed in the initial phase of the start-up. This phenomenon is accompanied by a voltage drop at approximately 6% of the voltage in the power supply line (Figure 5c). High instantaneous values of the start-up current and of the active power, with a simultaneous voltage drop, pose a considerable risk and may lead to an overload in the power supply line [28]. The start-up of the motor is obstructed due to not only high mass moments of the mill inertia but also the braking torque from the permanent magnets within the entire range of revolutions, which is clearly seen in the graph representing the mechanical characteristic of the motor (Figure 6). The torque generated on the motor shaft is the result of the torque from the cage and of the braking torque from the permanent magnets (Figure 6a).

After a violent start-up of the motor, the recorded active power becomes stable at 430 kW (70% of the nominal motor power) (Figure 7). As in the case of the SAS motor, the active power was measured for a loaded mill operating at the nominal efficiency of 80 Mg/h. In this case, the fluctuations of the active power in steady motion (Figure 7) are decidedly lower than in the case of the SAS motor (Figure 3).

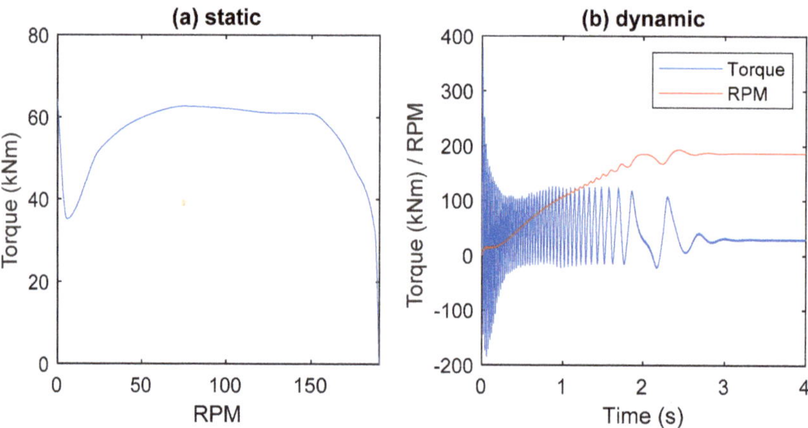

Figure 6. Mechanical characteristic of the LSPMSM SMH-1732T motor: (**a**) static, (**b**) dynamic [38].

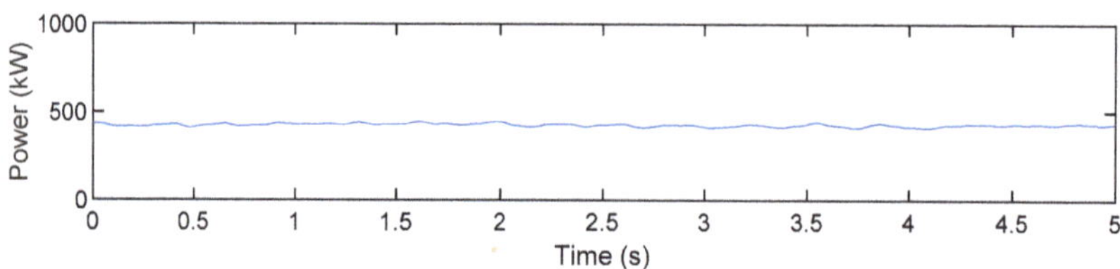

Figure 7. Active power of the LSPMSM SMH 1732T 630 kW motor during normal operation with filled ball mill.

One of the options for eliminating the problems related to the start-up process may be to implement a frequency converter in the power supply system [39,40], but a solution dedicated to this type of motors has not been developed yet.

3.3. Drive System with Asynchronous High-Efficiency Motor, with Gearbox, and with Fluid Coupling

Fluid coupling systems may be an alternative to expensive drive systems with frequency converters. Dynamic analyses demonstrated clearly that a fluid coupling may not be used in combination with a low-speed motor. The torque transmitted by a fluid coupling depends on the angular velocity of the input shaft and on the diameter of the shaft [40,41]. Thus, no coupling design exists that would enable transferring a torque of 32.1 kNm (the nominal torque of the SMH 630 kW motor) at 187.5 RPM (nominal revolutions of the SMH motor). The commercially available fluid couplings are used for revolutions above 700 RPM, with the torque increasing exponentially together with the increasing revolutions of the coupling [42]. Therefore, the use of fluid couplings is not possible in the above-described drive systems, both with the low-speed SAS and with the new SMH motors. For this reason, detailed consideration was given to the concept of a new drive system comprising a high-speed motor in a high-efficiency class (at least IE3), a gearbox, and a fluid coupling. The schematic diagram of the system is shown in Figure 8.

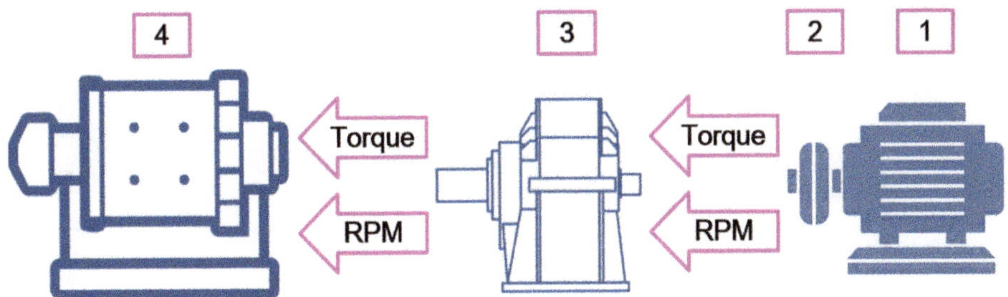

Figure 8. Schematic diagram of the drive system allowing the use of a fluid coupling for mill start-up: 1—asynchronous motor 630 kW 1000 RPM, 2—fluid coupling, 3—gearbox, 4—ball mill.

The electric motor used in this system had the same power as the motors used in the previous versions. The nominal revolutions of the motor are 1000 RPM. Therefore, an additional gearbox (3), with a 1:5 ratio, was required. The intermediate gearbox ensures optimum mill revolutions. The fluid coupling (2) was installed between the motor and the gearbox, on the high-speed shaft. Based on dynamic calculations, the selected coupling was TVV 866 with rotor diameter equal to 978 mm. Couplings of this type (TVV) are equipped with a delay chamber, which stores part of the fluid from the working circuit during downtime. A momentary decrease of the amount of fluid in the working circuit allows for more gentle torque behavior during the start-up. With time, the working fluid returns to the working circuit of the coupling, enabling it to transfer the nominal torque. The analysis of the new drive system was based on dynamic simulations of starting a mill fully charged with copper ore (Figure 9).

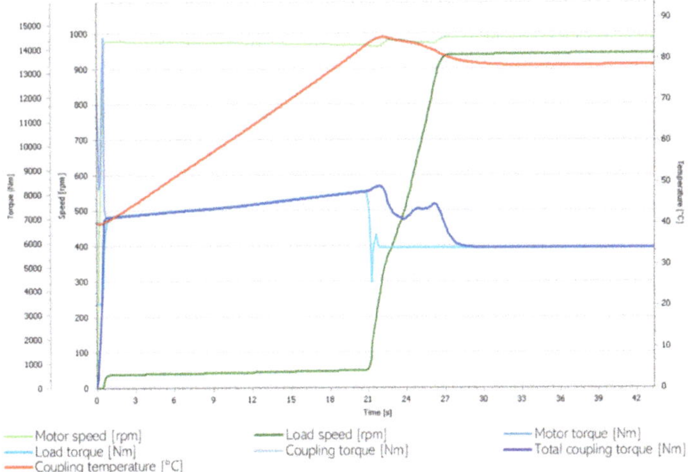

Figure 9. Dynamic simulation of the ball mill start-up process with the use of the TVV 866 fluid coupling.

The mill with the new drive system reaches the intended revolutions over a time of 27 s, and revolutions increase gently. This time duration can be thus assumed to be the mill start-up time proper. The final moment of the start-up process can be identified as the moment in which the motor starts to balance the anti-torque of the mill (power transmission). Both the entire torque transferred by the fluid coupling and the motor revolutions are stabilized. The working fluid used in the coupling is mineral oil, which reaches a maximum temperature of 85.1 °C after 21 s from the start-up. The working fluid temperature becomes stable at 79 °C after approximately 30 s. The coupling transfers a nominal torque of 5800 Nm after 29 s and thus works in the medium load range. The nominal coupling slip in such conditions is 3.2%. In order to verify the thermal conditions, an additional simulation of two mill start-up processes was performed at short time intervals. The simulation results are shown in the graph below (Figure 10).

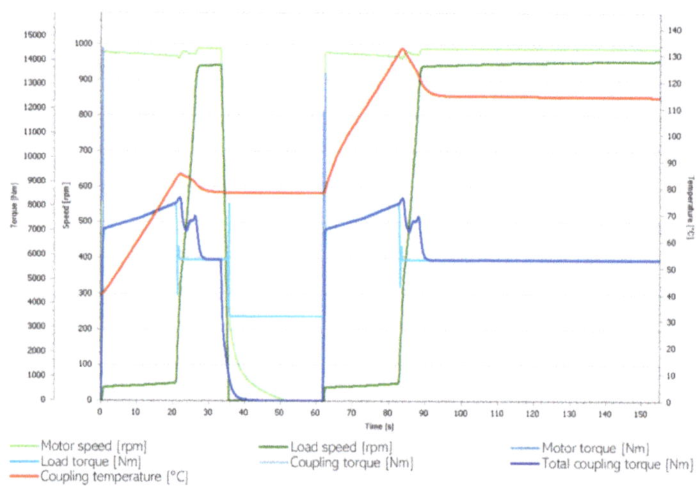

Figure 10. Dynamic simulation of two ball mill start-up processes with the use of the TVV 866 fluid coupling.

During the start-up, the energy transmission inside the coupling is the highest and results in an increased oil temperature. Performing two start-up processes over a short time period allows the assessment of both the maximum temperature increases and the cooling efficiency. Upon the second start-up, the maximum temperature of the working fluid momentarily reaches approximately 130 °C and then falls. The identified working temperatures of the coupling remain within the average range of loads and are thus safe for the drive system. The simulations were made for an engine with a rated power of 630 kW, operating with an effective power of 600 kW. The moment of inertia reduced to the motor shaft is 90.5 kgm^2 (the load torque reduced to the motor shaft is approx. 4600 Nm). The initial temperature was 40 °C.

4. Energy Efficiency Evaluation of the Analyzed Drive Systems

The simulations confirm that the application of a fluid coupling in the new drive system allows control of the ball mill start-up process. However, a complete analysis of each drive variant should also importantly involve the energy aspect. A reliable evaluation of the accuracy of the selected option must include an analysis and comparison of the energy efficiency of the drive systems. The comparison was based on two simple indicators. The first is the average electric energy consumption of the motor operating continuously for one hour. The second is the specific electric energy demand of the ball mill (i.e., the so-called specific energy consumption). This indicator defines the amount of electric energy required to process 1 Mg of copper ore. For this reason, it is strongly correlated with the first determined indicator. Note should be taken, however, that it does not allow for the quality-related parameters of the feed material and of the product (i.e., for the grain size composition).

This analysis was based on the measured electric energy consumption for the SAS and the SMH motors. With the known distributions of the ball mill operating parameters (efficiency and active power), the Monte Carlo method can be used to simulate the work areas in the analyzed variants. The simulation allows the expected energy consumption variations to be represented in relation to the efficiency and enables an evaluation of the influence of these two parameters on the specific energy consumption. Both the instantaneous power of the drive system and the instantaneous efficiency of the ball mill were considered as input parameters for the simulation model. The model also allowed for the relationships between these parameters. The variability of the instantaneous drive system power was described on the basis of the measurement data. Table 1 shows the assumed variability parameters for the active power of the SAS motor (Variant 1) and of the SMH motor (Variant 2). The new system with the fluid coupling (Variant 3), which was not tested in actual operation, was described on the basis of the results obtained from the simulated start-up. In Variant 3, the motor with the nominal power of 630 kW was assumed to work at an average power of 600 kW.

Table 1. Statistical description of the active power values (in kW) used in the simulation.

No.	Parameter	Variant 1 (SAS)	Variant 2 (SMH)	Variant 3 (New Drive)
1	Average	601.77	429.75	600.04
2	Standard error	0.57	0.45	0.08
3	Standard deviation	29.68	7.74	1.43
4	Variance	880.93	59.84	2.04
5	Kurtosis	0.03	−0.20	−1.12
6	Skewness	0.50	−0.22	0.09

An important aspect of the simulation was to develop such a technological model of the operating ball mill that would allow for the energy consumption variability as a function of the production variability. The production variability allows for the random character of the feed stream (the ore fed to the mill) and for the variability of the output stream from the mill. Based on the measurement results, an assumption was made that the variability of the mill efficiency is ±5% of its nominal efficiency of 80 Mg/h. Figure 11 is a bar chart showing the probability distribution of ball mill efficiencies.

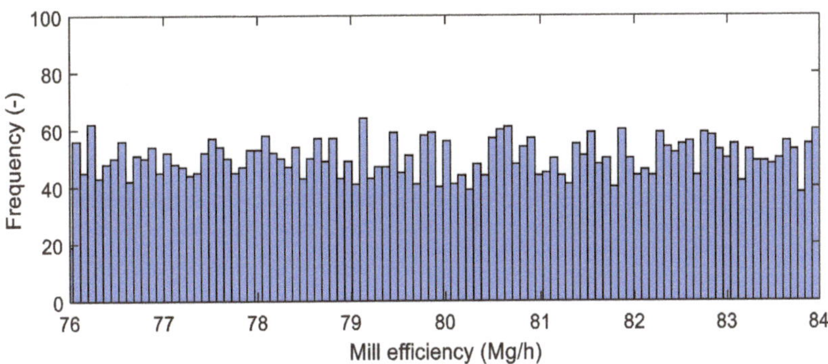

Figure 11. Bar chart of ball mill efficiencies (average value 80.02, standard error 0.033, standard deviation 2.32).

Figure 12 is a schematic diagram of the simulation algorithm. The output of the model provides statistically represented energy-consumption indicators.

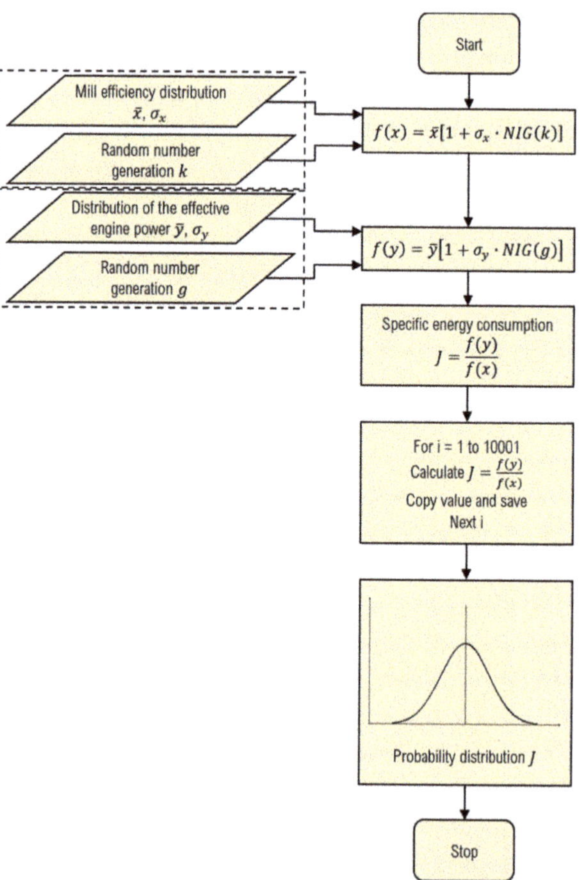

Figure 12. Algorithm of the Monte Carlo simulation.

The proper part of the Monte Carlo simulation consisted of 10,000 simulations, which served to calculate the variation range of the energy-consumption indicators. The random character of the process results in the formation of concentrated event areas (Figure 13a). Based on the statistical analysis, average points (Figure 13b) and specific energy consumption distributions (Figure 14) were identified for each drive variant.

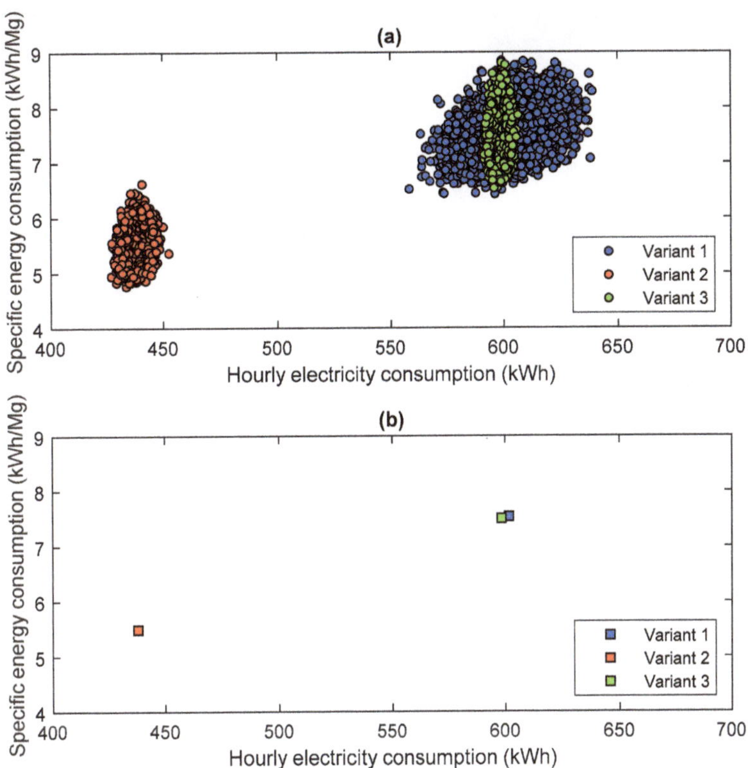

Figure 13. Monte Carlo simulation results—relationship between the specific energy consumption of the mill and the hourly electric energy consumption by a particular drive: (**a**) work areas of the selected drive systems, (**b**) averaged work points for Variant 1 (601.71 kWh; 7.53 kWh/Mg), Variant 2 (438.14 kWh; 5.49 kWh/Mg), and Variant 3 (598 46 kWh; 7.50 kWh/Mg).

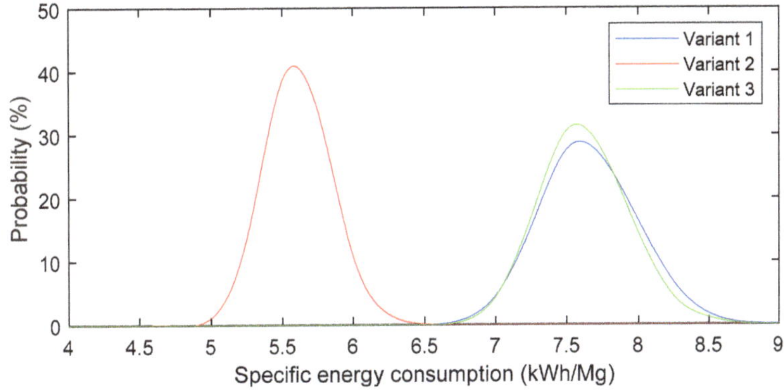

Figure 14. Probability distribution of specific energy consumption for the analyzed drive systems.

Tables 2 and 3 show statistical parameters characterizing the obtained indicators and values of simulation errors: MAE (mean absolute error) and RMSE (root mean squared error).

Table 2. Statistical description and errors of hourly electric energy consumption by each of the analyzed drives, obtained in Monte Carlo simulations. MAE: mean absolute error; RMSE: root mean squared error.

No.	Parameter	Variant 1 (SAS)	Variant 2 (SMH)	Variant 3 (New Drive)
1	Average	601.709	438.141	598.457
2	Standard deviation	10.823	3.038	1.993
3	Variance	117.143	9.231	3.974
4	Kurtosis	−0.012	0.042	0.041
5	Skewness	0.012	−0.011	0.034
6	MAE	8.642	2.408	1.585
7	RMSE	10.823	3.038	1.993

Table 3. Statistical description and errors of specific electric energy for each of the analyzed drives, obtained in Monte Carlo simulations.

No.	Parameter	Variant 1 (SAS)	Variant 2 (SMH)	Variant 3 (New Drive)
1	Average	7.530	5.489	7.496
2	Standard deviation	0.340	0.230	0.310
3	Variance	0.116	0.053	0.096
4	Kurtosis	0.030	0.129	0.101
5	Skewness	0.224	0.237	0.256
6	MAE	0.272	0.183	0.247
7	RMSE	0.340	0.230	0.310

5. Conclusions

The choice of a drive system for a ball mill used in copper ore comminution is largely dictated by the energy-related criteria. The energy efficiency of the three analyzed drives was identified in Monte Carlo simulations, which allow predictions of the most probable energy-consumption indicators. The research demonstrated that the old drive systems based on the SAS low-speed motors, which have been in operation for many years, have an average hourly electric energy consumption of 601.71 kWh and a specific energy consumption of 7.53 kWh/Mg. The replacement of the SAS motors with the new synchronous SMH motors induced with permanent magnets results in significant reductions of both the hourly energy consumption (to the level of 438.1 kWh) and the specific energy consumption (to the level of 5.49 kWh/Mg). The reduction in the energy consumption of the milling process is thus as much as 27%. The electricity demand indicators obtained by the simulation are in line with theoretical forecasts and measured values [12].

Solutions ensuring the gentle start-up of a mill filled to capacity are not available in the case of drive systems with low-speed SAS and SMH motors. The only solution seems to consist of pairing such motors with frequency converters. However, such solutions are not commercially available yet. An effective solution, which mitigates the problems related to the start-up process of a ball mill, can be provided by implementing asynchronous high-speed, high-efficiency motors. Based on the results of dynamic calculations, the new drive system design was paired with the TVV 866 fluid coupling. The simulated start-up process of the mill equipped with the new coupling demonstrated that this solution indeed allows for more gentle start-up behavior. As a result, the duration of the start-up process can be extended from 2 s (for the currently used drives) to 27 s. The analysis demonstrated that despite the two additional elements that affect the overall efficiency of the power transmission system (the gearbox and the fluid coupling), the energy-consumption indicators (the average, hourly electric energy consumption, and the specific energy consumption) remain slightly lower than in the traditional solution based on the SAS motor. The energy efficiency of this solution is much less advantageous than in the case of the SMH motor with permanent magnets. However, if compared with the traditional drives based on SAS motors (i.e., the drives to be replaced by the new solution), the average reduction in energy consumption is 0.5%. Taking into perspective the scale of energy demand from the ore processing plant, the above difference allows an average daily savings of 78 kWh of electric energy to be obtained in the operation of a single mill. On an annual basis, this results in one mill providing more than 23 MWh of electric energy savings.

Taking into account only the energy-related criteria, the drive system with the SMH asynchronous permanent magnet motor seems to be the best solution. However, the choice of the drive system should also be dictated by economic criteria. If the operation of the drives with the SMH motors shows that the costs of their repairs due to overloads at start-up exceed the energy savings, then the solution with the fluid coupling may prove more advantageous.

Author Contributions: Conceptualization, methodology, software, and validation: M.O. and P.B.; writing—review, editing, and supervision: R.K. and L.G.; project administration: R.K. and L.G.; final text prepared by M.O., P.B., L.G. and R.K. All authors have read and agreed to the published version of the manuscript.

Funding: The research work was co-founded with the research subsidy of the Polish Ministry of Science and Higher Education granted for 2021.

Institutional Review Board Statement: Not applicable.

Data Availability Statement: The data presented in this study are available on request from the corresponding author.

Conflicts of Interest: The authors declare no conflict of interest.

References

1. Malinauskaite, J.; Jouhara, H.; Ahmad, L.; Milani, M.; Montorsi, L.; Venturelli, M. Energy efficiency in industry: EU and national policies in Italy and the UK. *Energy* **2019**, *172*, 255–269. [CrossRef]
2. Giridhar Kini, P.; Bansal, C.R. Energy Efficiency in Industrial Utilities. *Energy Manag. Syst.* **2011**. [CrossRef]
3. Kumar, A. Technomine, Mining Technology. Available online: http://technology.infomine.com/reviews/comminution/welcome.asp?view=full (accessed on 4 November 2020).
4. Jankovic, A.; Valery, W.; La Rosa, D. *Fine Grinding in the Australian Mining Industry*; Universiti Sains Malaysia: Nibong Tebal, Malaysia, 2008.
5. Jeswiet, J.; Szekeres, A. Energy Consumption in Mining Comminution. *Procedia CIRP* **2016**, *48*, 140–145. [CrossRef]
6. DOE. *Comminution and Energy Consumption: Report of the Committee on Comminution and Energy Consumption*; U.S. Department of Energy: Washington, DC, USA, 1981.
7. De Bakker, J. Energy Use of Fine Grinding in Mineral Processing. *Metall. Mater. Trans. E* **2014**, *1*, 8–19. [CrossRef]
8. Wojciechowski, B. Nie tylko wydobywamy. *Energetyka Cieplna i Zawodowa* **2015**, *2*, 20–23.
9. Rumpf, H. Problems of scientific development in particle technology, looked upon from a practical point of view. *Powder Technol.* **1977**, *18*, 3–17. [CrossRef]
10. Sidor, J. Directions in development of mills for raw materials and mineral binders grinding. *Mater. Ceram.* **2016**, *68*, 61–69.
11. Tamblyn, R.J. Analysis of Energy Requirements in Stirred Media Mills. Ph.D. Thesis, University of Birmingham, Birmingham, UK, 2009.
12. Krawczykowski, D.; Gawenda, T.; Foszcz, D. Comparison of real and theoretically estimated energy consumption for ball grinder. *Min. Geoengin.* **2006**, *30*, 79–90.
13. Wang, M.H.; Yang, R.Y.; Yu, A.B. DEM investigation of energy distribution and particle breakage in tumbling ball mills. *Powder Technol.* **2012**, *223*, 83–91. [CrossRef]
14. Mayank, K.; Malahe, M.; Govender, I.; Mangadoddy, N. Coupled DEM-CFD Model to Predict the Tumbling Mill Dynamics. *Procedia IUTAM* **2015**, *15*, 139–149. [CrossRef]
15. Marchal, G. Industrial experience with clinker grinding in the HOROMILL. In Proceedings of the IEEE Cement Industry Technical Conference (Paper), Hershey, PA, USA, 20–24 April 1997; pp. 195–211.
16. Tesema, G.; Worrell, E. Energy efficiency improvement potentials for the cement industry in Ethiopia. *Energy* **2015**, *93*, 2042–2052. [CrossRef]
17. Telichenko, V.I.; Sharapov, R.R.; Lozovaya, S.Y.; Skel, V.I. Analysis of the efficiency of the grinding process in closed circuit ball mills. In Proceedings of the MATEC Web of Conferences, EDP Sciences, Amsterdam, The Netherlands, 23–25 March 2016; Volume 86, p. 4040.
18. Rowland, C.A.; Erickson, M.T. Large Ball Mill Scale-Up Factors to Be Studied Relative to Grinding Efficiency. *Miner. Metall. Process.* **1984**, *1*, 165–172. [CrossRef]

19. Fuerstenau, D.W.; Lutch, J.J.; De, A. The effect of ball size on the energy efficiency of hybrid high-pressure roll mill/ball mill grinding. *Powder Technol.* **1999**, *105*, 199–204. [CrossRef]
20. Zheng, J.; Harris, C.C.; Somasundaran, P. A study on grinding and energy input in stirred media mills. *Powder Technol.* **1996**, *86*, 171–178. [CrossRef]
21. Costea, C.; Silaghi, H.; Silaghi, M.; Silaghi, P. Mill speed control using programmable logic controllers. In Proceedings of the International Conference on Mathematical Methods and Computational Techniques in Electrical Engineering, Timişoara, Romania, 21–23 October 2010; World Scientific and Engineering Academy and Society (WSEAS): Stevens Point, WI, USA, 2010; pp. 26–30, ISBN 978-960-6766-60-2.
22. El-Shall, H.; Somasundaran, P. Physico-chemical aspects of grinding: A review of use of additives. *Powder Technol.* **1984**, *38*, 275–293. [CrossRef]
23. Pacholski, E.; Iskierski, L. The analisis of influence of start-up mode of drive motors to electric network parameters. *Electr. Mach. Trans. J.* **2013**, *2*, 99.
24. Bianchi, N.; Bolognani, S.; Frare, P. Design criteria for high-efficiency SPM synchronous motors. *IEEE Trans. Energy Convers.* **2006**, *21*, 396–404. [CrossRef]
25. Lu, X.; Iyer, K.L.V.; Mukherjee, K.; Kar, N.C. Development of a novel magnetic circuit model for design of premium efficiency three-phase line start permanent magnet machines with improved starting performance. *IEEE Trans. Magn.* **2013**, *49*, 3965–3968. [CrossRef]
26. Stoia, D.; Chirilă, O.; Cernat, M.; Hameyer, K.; Ban, D. The behaviour of the LSPMSM in asynchronous operation. In Proceedings of the EPE-PEMC 2010—14th International Power Electronics and Motion Control Conference, Ohrid, North Macedonia, 6–8 September 2010; pp. T4–T45.
27. Mayer, C.B.; Johnson, R.M.; Gray, D.J. Solution of a Serious Repetitive Vibration Problem on a 4500-hp Single-Pinion Synchronous Motor Ball Mill Drive. *IEEE Trans. Ind. Appl.* **1985**, *IA-21*, 1039–1046. [CrossRef]
28. Zawilak, T.; Zawilak, J. Energy-efficient motor in ball mill applcation. *Drives Control* **2017**, *19*, 68–72.
29. Gao, X.; Wang, X.; Wei, Z. Design and control of high-capacity and lowspeed doubly fed start-up permanent magnet synchronous motor. *IET Electr. Power Appl.* **2018**, *12*, 1350–1356. [CrossRef]
30. Mirošević, M.; Maljković, Z. Effect of sudden change load on isolated electrical grid. In Proceedings of the Electrical Systems for Aircraft, Railway and Ship Propulsion, ESARS, Aachen, Germany, 3–5 March 2015; pp. 1–4.
31. McElveen, R.F.; Toney, M.K. Starting high-inertia loads. *IEEE Trans. Ind. Appl.* **2001**, *37*, 137–144. [CrossRef]
32. Leng, S.; Ul Haque, A.R.N.M.R.; Perera, N.; Knight, A.; Salmon, J. Soft Start and Voltage Control of Induction Motors Using Floating Capacitor H-Bridge Converters. *IEEE Trans. Ind. Appl.* **2016**, *52*, 3115–3123. [CrossRef]
33. Zenginobuz, G.; Çadirci, I.; Ermiş, M.; Barlak, C. Soft starting of large induction motors at constant current with minimized starting torque pulsations. *IEEE Trans. Ind. Appl.* **2001**, *37*, 1334–1347. [CrossRef]
34. Pacholski, E.; Leśnik, M. Sas motors exploitation in kghm polska miedz s.a. division of concentrator. Experience, problems, priority activities. *Electr. Mach. Trans. J.* **2012**, *1*, 153–157.
35. Zawilak, T. Synchronous motors excited by permanent magnets in high power drives. *Przegląd Elektrotechniczny* **2017**, *1*, 175–178. [CrossRef]
36. Kisielewski, P.; Leśnik, M.; Pacholski, E.; Zawilak, J.; Zawilak, T. Design, construction and testing of prototypes series of large synchronous motors with permanent magnets. *Electr. Mach. Trans. J.* **2016**, *3*, 191–195.
37. KGHM Polska Miedź S.A. *Promotional Materials*; KGHM Polska Miedź S.A.: Lubin, Poland, 2020.
38. Start, L.; Magnet, P.; Motor, S.; Ball, I.N. Line start permanent magnet synchronous motor in ball mill application. *Electr. Mach. Trans. J.* **2016**, *2016*, 169–173.
39. Ni, R.; Xu, D.; Blaabjerg, F.; Wang, G.; Li, B.; Lu, K. Synchronous switching of non-line-start permanent magnet synchronous machines between inverter and grid drives. In Proceedings of the 2016 IEEE Energy Conversion Congress and Exposition (ECCE), Milwaukee, WI, USA, 18–22 September 2016; Volume 31, pp. 3717–3727. [CrossRef]
40. Brun, K.; Meyenberg, C.; Thorp, J. Hydrodynamic torque converters for oil & gas compression and pumping applications: Basc principles, performance characteristics and applications. In Proceedings of the Asia Turbomachinery & Pump Symposium, Singapore, 22–25 February 2016; Turbomachinery Laboratories, Texas A&M Engineering Experiment Station: College Station, TX, USA, 2016; p. 15.
41. Heisler, H. Hydrokinetic fluid couplings and torque converters. In *Advanced Vehicle Technology*; Elsevier: Amsterdam, The Netherlands, 2002; pp. 98–116. ISBN 9780750651318.
42. Huitenga, H.; Mitra, N.K. Improving startup behavior of fluid couplings through modification of runner geometry: Part I—Fluid flow analysis and proposed improvement. *J. Fluids Eng. Trans. ASME* **2000**, *122*, 689–693. [CrossRef]

 *energies*

Article

Analysis of Reasons for Reduced Strength of Multiply Conveyor Belt Splices

Mirosław Bajda *[iD] and **Monika Hardygóra**

Faculty of Geoengineering, Mining and Geology, Wroclaw University of Science and Technology, Na Grobli 15 St., 50-421 Wrocław, Poland; monika.hardygora@pwr.edu.pl
* Correspondence: miroslaw.bajda@pwr.edu.pl; Tel.: +48-604-095-374

Abstract: Belt conveyors are used for the transportation of bulk materials in a number of different branches of industry, especially in mining and power industries or in shipping ports. The main component of a belt conveyor is its belt, which serves both as a support for the transported material along the conveyor route and as an element in the drive transmission system. Being crucial to the effective and reliable operation of the conveyor, the belt is also its most expensive and the least durable element. A conveyor belt comprises a core, covers and edges. A multiply textile belt, in which the core is constructed of synthetic fibers such as polyamide, polyester or aramid, is the oldest and still the most commonly used conveyor belt type. The plies are joined with a thin layer of rubber or another material (usually the material is the same as the material used in the covers), which provides the required delamination strength to the belt and allows the plies to move relative to each other as the belt is bent. Belts are installed on the conveyors in a closed loop in order to join belt sections, whose number and length depend on the length and type of the belt conveyor. Belts are joined with each other in a splicing procedure. The cutting of the belt core causes belt splices to be prone to concentrated stresses. The discontinued core also causes the belt to be the weakest element in a conveyor belt loop. The article presents the results of strength parameter tests that were performed on laboratory and industrial splices and indicated the reasons for the reduced strength of conveyor belt splices. Splice strength is reduced mainly due to incorrect preparation of the spliced surfaces and to different mechanical parameters of the spliced belts.

Keywords: textile conveyor belts; multiply conveyor belts; multiply belt splices; laboratory splice tests

 check for updates

Citation: Bajda, M.; Hardygóra, M. Analysis of Reasons for Reduced Strength of Multiply Conveyor Belt Splices. *Energies* **2021**, *14*, 1512. https://doi.org/10.3390/en14051512

Academic Editor: Sergey Zhironkin

Received: 14 February 2021
Accepted: 7 March 2021
Published: 9 March 2021

Publisher's Note: MDPI stays neutral with regard to jurisdictional claims in published maps and institutional affiliations.

1. Introduction

Transportation systems consisting of belt conveyors are considered to be the most efficient and effective solution for transporting large amounts of bulk materials [1]. The most expensive and the least durable element of a belt conveyor is its conveyor belt, as it directly contacts the transported material and is therefore prone to such damage as punctures, longitudinal cuts and tears. It serves to support and move the transported material along the conveyor [2,3]. In actual operating conditions, the conveyor belt suffers from impacts caused by the transported material. These impacts are observed in locations where the transported material is fed to the conveyor, frequently leading to belt damage. In many cases, the belt is damaged beyond further use [4]. Such damage causes the entrepreneur to suffer financial losses due to the need to replace the damaged belt section and, as a result, to also make new splices.

Another function of the belt is to transfer longitudinal forces required to overcome resistance to motion. The belt comprises a core, which is expected to transfer loads. The core is protected by covers and edges.

Multiply textile belts are the oldest type of conveyor belts, and they are still commonly used. This type of belt was patented by Thomas Robinson and first used in 1891 in a magnetite mine in New Jersey, USA. Textile cores in multiply belts are presently most typically manufactured of 100% synthetic fibers, such as polyamide, polyester or aramid. The textile plies in the core are spliced with a thin layer of rubber or another material (usually the material is the same as the material used in the covers), which provides the required delamination strength to the belt and allows the plies to move relative to each other as the belt is bent on the drive, tail and take-up pulleys [5–8]. This type of conveyor belt is most widely employed in underground mines, power plants, cement plants, harbors, etc., as well as in other locations where materials are transported with the use of the belt. conveyors. The splice in such type of a belt is a layer-based structure with a complex distribution of stresses due to disturbed belt structure, which results from discontinuing the textile plies in the belt core. Splices, which allow shorter belt sections to be joined into a loop having a length corresponding to the length of the conveyor, are an underestimated element of the belt conveyor [9–11].

In underground bituminous coal mines, the length of belt sections is limited by the size of the excavations and of the transport shaft. A belt on a reel has a certain volume, and therefore, in underground mines, belts are typically transported to the installation location in sections no longer than 150 m. This is not a problem in surface mines. Belts used in such mines are transported on special double reels that allow the transportation of continuous belt sections with no splices and lengths reaching 700 m. The weight and size of the reels with belt sections of such lengths cause problems when the belt is transported on public roads from the manufacturer to the place of installation. These problems are worth overcoming, however, as the belt loop has very few splices.

Efforts towards installing belt sections of lengths reaching technical limits make splices the weakest link in the belt loop [12]. The smaller their number, the greater the reliability of the entire loop. From the perspective of its reliability, a conveyor belt loop is a system of spliced belt sections arranged in series, and the strength of splices installed in belts operated in mines rarely reaches the full strength of a new belt. Splices are thus areas most prone to developing discontinuities in the belt loop. In order to avoid splice breaks, a number of different splice monitoring methods are implemented. Splices are monitored on occasions when belt loop inspections are performed (for both the sections and the splices) by the maintenance crew or with the help of computer-aided digital image analysis [13–15]. Another implemented solution involves measuring changes in the lengths of distances between special magnets installed in the belt [16]. Research has also been performed into automatic splice inspections in magnetic systems [15].

Splice strength is affected by a number of factors [17,18]. The most important of these factors include the splicing method and the choice of splicing materials. The above have a decisive influence on splice fatigue life. Another important factor is the quality of the installed splice, which depends on the proper geometry of splices [19–21], which should be adjusted to the belt's design and operating conditions, as well as on observing best practices in the field of splicing technology. The pressure to reduce conveyor downtime (i.e., to avoid production-related losses) and harsh conditions in underground mines has a negative influence on the static and dynamic strength of splices. The above fact has been confirmed in numerous tests performed by Laboratorium Transportu Tasmowego (Belt Conveying Laboratory, further: LTT) as part of research works and expert opinions requested by conveyor belt producers and users. Such research has been continued for over 25 years [22–24], and during that time more than 300 belt splices were tested for numerous companies from Poland and abroad. The results of this research became an impulse for more detailed works regarding the values and distributions of stresses in splice bonds [25]. This research project resulted in improved reliability of cold-vulcanized splices, their increased service life, and a lower cost of their installation. These effects were obtained by identifying the properties of the conveyor belts and of the splicing materials which have a significant influence on the stress values in the adhesive bonds and thus on splice service life.

Based on research performed for the mining industry, reduced splice strength was identified to result from defective splicing procedures and materials. Figure 1 shows the percentage shares for the reasons behind the lowered strength of splices in multiply conveyor belts.

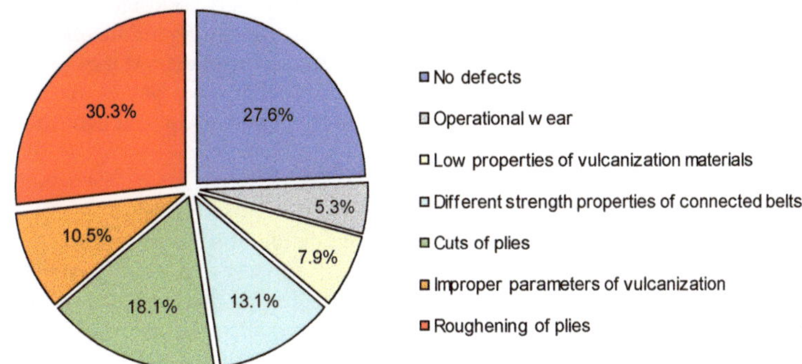

Figure 1. Percentage share for the reasons behind the lowered strength of splices in multiply conveyor belts [25].

Although 27.6% of splices were found not to show any faults regarding the installation technique, they did show lowered strength. An additional 13.1% of splices demonstrated lowered strength as a result of attempts to splice belts having different strength properties. Splice strength decrease caused by inappropriate belt selection is a significant problem to be avoided. While the selection of materials appropriate for a particular type of belt seems a relatively easy task in light of the research results available in the field, the selection of appropriate belts is practically beyond the capacity of the splicing specialist. The belts to be spliced are selected by splicing companies only on the basis of the belt width, its nominal tensile strength and the number of plies in the core. It is typically impossible to select belts that would be produced by the same manufacturer and thus be made of identical materials. As conveyor belt manufacturers use various materials (e.g., various textile plies), some strength parameters of conveyor belts having an identical nominal strength and the number of plies may significantly differ. The results of tests performed in companies that employ belt conveyors to transport bulk materials demonstrate that these differences lead to reduced belt loop strength in the location of the splice.

The strength of a single conveyor belt splice determines the strength of the entire belt loop installed on the conveyor [26], and therefore splicing technology is an issue of key importance. Tests of splices for industrial applications performed by LTT demonstrate clearly that improper technology used in preparing textile plies for splicing may be a reason behind splice strength reductions by as much as several tens of percent. The results of laboratory tests presented in this article indicate the causes of this type of fault.

The article presents the results of laboratory tests into the strength parameters of splices in multiply conveyor belts with various strength characteristics. The presented test results also include the results for splices installed on belts damaged due to improper pre-splicing preparation of the surfaces of textile plies. The analysis of the results leaves no doubt that the two factors have a negative influence on splice strength by lowering splice quality.

2. Theoretical Background

The measurement of splice breaking force allows the identification of belt strength loss due to the introduction of a splice, for example during its installation on the belt conveyor. This simple test provides information of key importance for the belt user. It also demonstrates that splices in multiply textile belts are still a major limitation to taking full advantage of the belt strength. Lowered splice strength is primarily caused by a number of mistakes made during the splicing procedure (Figure 1). Another reason may lie in an inappropriate selection of the spliced belts, i.e., in splicing belt sections that have different mechanical properties. These mistakes can be eliminated from the splicing procedure by increasing the quality-related demands placed on the splicing technicians. Avoiding splice strength reductions due to improperly selected belts appears to be a greater challenge. Belts offered by manufacturers using different materials and production technologies may show significantly different properties even if they are selected on the basis of identical nameplate tensile strengths and number of plies in the core. This is the object of the theoretical considerations presented below and confirmed further in this article with the results of laboratory tests.

2.1. The Mathematical Model

The mathematical modeling of splices in multiply conveyor belts was an issue addressed inter alia in [27,28]. In view of the above publications, the influence of various mechanical properties of splices on the stress distribution observed in the splice was analyzed for a conveyor belt model based on the following assumptions:

- the textile plies in the belt core are considered to be elastic elements subject to Hooke's law,
- due to the low elastic modulus value of rubber, which is incomparably smaller than in the case of the plies, normal stresses in the layers of rubber between the plies and in the belt covers were ignored,
- the stresses in the textile plies are balanced in the belt cross-section, which is located at a finite distance from the failure location,
- in the area where load disturbances are observed, the layer between the plies is subject to non-dilatational strains,
- the rubber between the plies is treated as a linear elastic body subject to Hooke's law.

Further assumptions included:

- the values of longitudinal elasticity moduli for textile plies in the spliced belts are different,
- the differences in elasticity moduli values between individual plies of the same belt are negligibly small in comparison to the difference between the spliced belts,
- when tensioning the undamaged belt core, strain in individual plies is equal.

With the above assumptions, the stresses which occur in textile plies of a belt core, in any longitudinal cross-section of the belt, are shown in Figure 2.

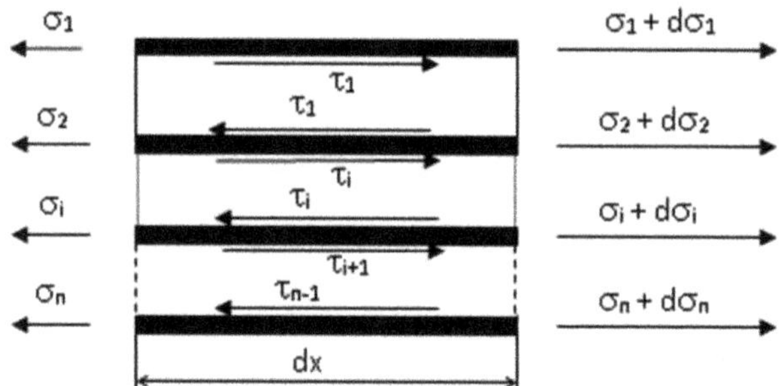

Figure 2. Stresses in an elementary length of a belt core cross-section, τ—shear stresses, σ—tensile stresses.

Considering the balance conditions of forces acting on the elements of individual plies in the cross-section of the belt element with a length "dx" leads to the following balance equations:

- the case for an ith ply, where $1 \leq i < n$, and n is the number of plies in the belt, is described by Equation (1),

$$\sigma_i + \tau_i dx = \tau_{(i-1)} dx + \sigma_i + d\sigma_i \tag{1}$$

- the case for an nth ply is described by Equation (2),

$$\sigma_n = \tau_{(n-1)} dx + \sigma_n + d\sigma_n \tag{2}$$

The plies adjacent to each other in the cross-section of the belt core are joined with a layer of inter-ply rubber (or a layer of adhesive rubber). Consideration for the relationship between the extent of deformation in the layer of inter-ply rubber and the shear stresses acting on this layer, represented in Equation (3),

$$\tau_i = \frac{G}{d} \Delta U \tag{3}$$

allows the formulation of differential Equations (4)–(6), which describe the stresses in individual plies:

- for the first ply ($i = 1$),

$$\frac{d^2 \sigma_1}{dx^2} = \frac{G}{d} \left[\frac{1}{E_2} \sigma_2 - \frac{1}{E_1} \sigma_1 \right] \tag{4}$$

- for plies from the second one ($i = 2$) to the penultimate one ($i = n - 1$),

$$\frac{d^2 \sigma_i}{dx^2} = \frac{G}{d} \left[\frac{1}{E_{(i+1)}} \sigma_{(i+1)} - \frac{2}{E_i} \sigma_i + \frac{1}{E_{(i-1)}} \sigma_{(i-1)} \right] \tag{5}$$

- for the last, external ply in the cross-section ($i = n$),

$$\frac{d^2 \sigma_n}{dx^2} = \frac{G}{d} \left[\frac{1}{E_{(n-1)}} \sigma_{(n-1)} - \frac{1}{E_n} \sigma_n \right] \tag{6}$$

The symbol G used in Equation (3) is the transverse elasticity modulus of the inter-ply rubber, d is the thickness of the inter-ply rubber and ΔU is the displacement of the adjacent plies. The symbol E_i used in the differential equations is the elasticity modulus of the ith ply.

The above relationships form a system of "n" differential equations (where "n" is the number of plies in the belt) describing the stresses in the plies of a belt subjected to uniaxial tension. Moreover, the balance Equation (7) holds for any transverse cross-section of the belt and replaces any of the above equations:

$$\sum_{i=1}^{n} \sigma_i = const \tag{7}$$

In Equation (7), σ_i is the stress in the ith ply, and n represents the number of plies in the belt.

With identical strain in the plies outside the zone affected by the cut, the values of stresses in these plies depends on the elastic modulus values in individual plies and is described by Equation (8), in which σ_T is the stress in the conveyor belt:

$$\sigma_i = \sigma_T \cdot \frac{E_i}{\sum_{i=1}^{n} E_i} \tag{8}$$

The above mathematical model of a splice in a multiply belt is a system of "n" second-order ordinary differential equations, in which "n" is the number of plies in the conveyor

213

belt. This allows analyses of stress distributions in the spliced area of a multiply conveyor belt, in which individual plies were discontinued at the contact points of individual steps. The assumed boundary conditions for the solution are defined by the characteristic geometrical and material parameters of the splice, such as the location of the failure in the cross-section of the core, the length of the step or the mechanical properties of the belt and of the joining materials. The application of this model in the calculations of stress distributions in the splice requires an instantiation of the number of equations, the values of the required parameters and the initial conditions. The calculations consist in numerically solving a system of differential equations by substituting the derivatives with differential quotients and by solving the resulting system of linear algebraic equations.

2.2. Calculation Results

The stress distributions in the spliced area of a four-ply belt were calculated for the following cases:

- the mechanical properties of the splices in the belts were uniform,
- the mechanical properties of the splices in the belts were not uniform.

A series of calculations was performed for various assumed difference values of longitudinal elasticity moduli for the plies of the spliced belts. Due to the splice symmetry, the calculations needed to be performed only for two cases of the failure location in the cross-section of the core: when the cut is made in the external ply (the beginning of the first and the end of the second step), and when the cut is made in a ply located directly under the external ply (the contact point between the first and the second step, and between the second and the third step). The results of the calculations are shown in Table 1. The following constant values were assumed in the calculations: $E = 2000$ kN/m, $G = 750$ kN/m^2, $d = 0.001$ m.

Table 1. Calculated values of stress concentrations.

Difference between Elasticity Moduli for the Plies $E_1 < E_2$	Maximum Stress Concentration in Comparison to:							
	Stresses in Splices of Identical Belts				Stresses in Undamaged Belt			
	σ_1/σ_0	σ_2/σ_0	σ_3/σ_0	σ_4/σ_0	σ_1/σ_{01}	σ_2/σ_{02}	σ_3/σ_{03}	σ_4/σ_{04}
at the end of step 1								
50%	0.83	-	1.09	1.08	1.24	-	1.59	1.12
40%	0.86	-	1.07	1.06	1.28	-	1.56	1.1
30%	0.89	-	1.05	1.05	1.33	-	1.53	1.09
20%	0.93	-	1.03	1.03	1.39	-	1.5	1.07
10%	0.96	-	1.02	1.02	1.43	-	1.49	1.06
at the beginning of step 2								
50%	0.79	-	1.11	1.1	1.18	-	1.62	1.14
40%	0.83	-	1.09	1.08	1.24	-	1.59	1.12
30%	0.85	-	1.08	1.07	1.27	-	1.58	1.11
20%	0.9	-	1.05	1.05	1.35	-	1.53	1.09
10%	0.94	-	1.03	1.03	1.4	-	1.5	1.07
at the end of step 2								
50%	0.88	0.89	-	1.22	0.92	1.3	-	1.82
40%	0.9	0.91	-	1.19	0.94	1.33	-	1.78
30%	0.92	0.94	-	1.14	0.96	1.37	-	1.7
20%	0.94	0.96	-	1.1	0.98	1.4	-	1.64
10%	0.97	0.99	-	1.04	1.01	1.45	-	1.55
at the beginning of step 3								
50%	0.88	0.89	-	1.24	0.92	1.28	-	1.85
40%	0.89	0.91	-	1.2	0.93	1.33	-	1.79
30%	0.91	0.93	-	1.16	0.95	1.36	-	1.73
20%	0.94	0.95	-	1.11	0.98	1.39	-	1.66
10%	0.96	0.96	-	1.07	0.99	1.42	-	1.6

The theoretically calculated values of maximum stress concentrations in the splices of conveyor belts with plies having different mechanical properties indicate a superposition of the disturbances in load distributions that can be observed in the zone affected by the failure and that result from the location of the damaged splice and from the influence of the discontinued core structure.

3. Materials and Methods

3.1. Problem Formulation

The strength of a splice in a multiply conveyor belt is always smaller than the strength of the belt itself [22]. This fact results from a number of factors, the most important of which include:

- the number of textile plies in the belt core,
- the length of the splice step,
- adhesive properties of the bonding materials,
- strength properties of the textile plies,
- strength properties of the rubber between the plies and of the adhesives,
- the splicing technology.

The design of the splice in a multiply belt necessitates that the cross-section in the splice step contacts will have one ply less than in the spliced belt.

Therefore, the belt strength in those cross-sections is reduced by the value of $1/n$, where n is the number of plies in the belt. The reduction of belt strength is thus in reverse proportion to the number of plies. The actual strength loss is even greater as a notch phenomenon occurs due to ply discontinuity in the splice. The loss is further increased by an uneven distribution of shear stresses in the adhesive bond and by stress concentration at splice contact points. Figure 3 shows an example of stress distribution (identified in tests performed at LTT) [25] in the adhesive bond of two splices having an identical nominal strength of 1000 kN/m and an identical number of textile plies. The step length in those splices was 250 mm, and the total splice length was 3 × 250 mm = 750 mm.

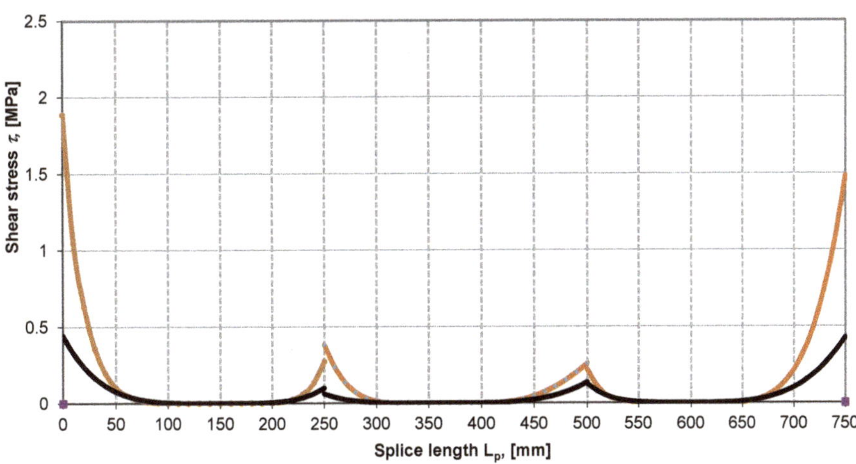

Figure 3. Shear stress distribution τ along the total splice length L_p. Splice in belt type EP 1000/4 (black), splice in belt type P 1000/4 (orange) [25].

The splices differed in the material used for belt plies. The results indicated in orange were obtained from a P-type belt with polyamide core, and the results in black from an EP-type belt with polyamide-polyester core. The highest shear stresses were recorded on the contact points of the outer steps (L_p = 0 and 750 mm). These are several times higher than the stresses recorded on the contact points of the middle steps (L_p = 250 and 500 mm).

Research indicated that the stress concentration factor k_τ, see Equation (9), has values between 2.6 and 3.6:

$$k_\tau = \frac{\tau_{max}}{\tau_{sr}} \qquad (9)$$

where τ_{max}—maximum shear stress in ply contact points; τ_{sr}—mean shear stress in the splice.

Research performed by Hardygóra [29] demonstrated that tensile stresses in belt plies also have an uneven distribution, and the stress concentration factor of tensile stresses in the plies located in the vicinity of ply contact points in splice k_σ (see Equation (10)) has values between 1.7 and 1.9:

$$k_\sigma = \frac{\sigma_{max}}{\sigma_{sr}} \qquad (10)$$

where σ_{max}—maximum tensile stress in the ply; σ_{sr}—mean tensile stress in belt plies outside the splice.

The development of shear and tensile stresses may lead to the destruction of the splice in a multiply belt due to:

- the coming apart of the plies, if the shear strength of the adhesive bond is exceeded,
- the breaking of textile plies in the belt core, if their tensile strength is exceeded.

Static tensile strength tests of splices demonstrate that the latter is the most common case of splice destruction. On the other hand, the plies in the splice come apart practically only in the case when the adhesive bond has faults due to mistakes in the performed splicing procedure.

3.2. Conveyor Belt Splice Specimens

The object of the research was splices in multiply conveyor belts with polyamide-polyester (type EP, E—polyester, P—polyamide) or polyamide (type P) textile carcass. The splices were installed with the use of the heat curing method, in which the required adhesion force between the individual elements of the splice was obtained owing to appropriate vulcanization temperature and pressure in the vulcanization press. The splices were built according to the schematic drawing in Figure 4.

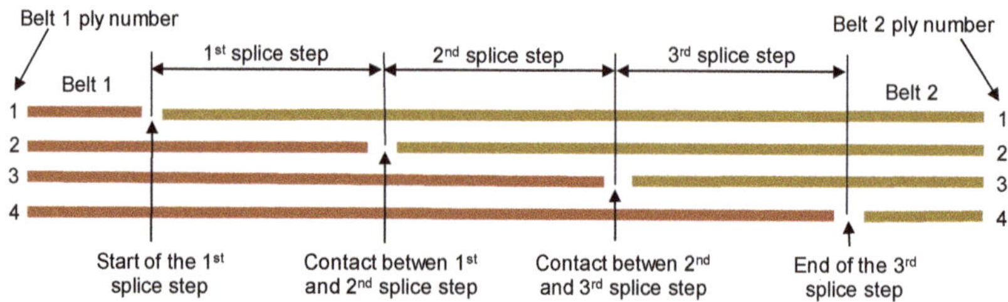

Figure 4. Schematic drawing of a splice in a 4-ply belt.

The drawing shows a splice installed in a belt with four textile plies in the core. It is a three-step lap-type splice. This means that the number of steps in the splice is smaller by one than the number of plies in the belt core. The splices indicated in Table 2 with numbers 6 and 9 were installed as a strap type. This fact means that unlike in the case of lap-type splices, the number of steps in the latter type of splice is equal to the number of plies in the core.

Table 2. Results of tensile strength tests for industrial splices.

Splice No.	Belt Type	Splice Tensile Strength		Splice Strength Loss
		Required [kN/m]	Measured [kN/m]	[%]
1	EP 1800/5	1224	663	46
2	EP 1800/5	1224	653	47
3	GTP 1250/3	708	530	25
4	P 630/3	357	220	38
5	GTP 2000/4	1275	897	30
6	GTP 2000/4	1700 [1]	1375	19
7	GTP 1800/4	1147	888	23
8	GTP 1800/4	1147	800	30
9	GTP 2500/5	21,251	1814	15
10	GTP 1250/3	708	576	19
11	GTP 1400/4	892	592	34
12	GTP 2000/4	1265	700 [1]	45

[1] Strap-type splices.

3.2.1. Laboratory Splices

The test specimens were divided into two groups. The first group contained laboratory splices, prepared at LTT. The splices were used to investigate how ply roughening influences their strength. Prior to splicing, the canvas had to be appropriately and precisely prepared. Ensuring that the fabric is carefully cleaned of residual rubber and that the test results are repeatable required appropriate conditions to be provided. Figure 5 shows the belt fabric prepared for splicing.

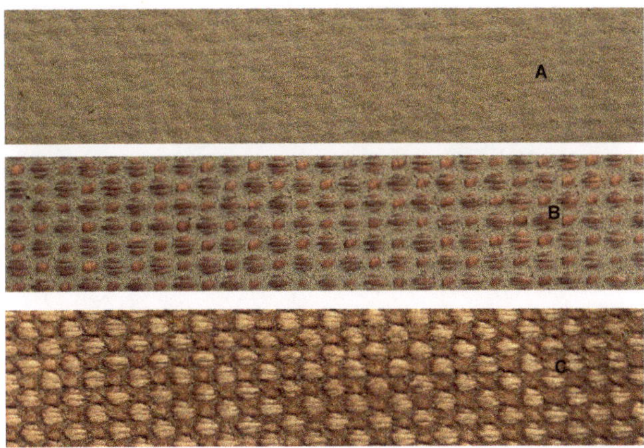

Figure 5. Textile plies: (**A**)—unroughened, (**B**)—roughened friction rubber, (**C**)—roughened fabric.

In order to identify how different mechanical belt properties influence splice strength, tests were performed also on splices prepared in laboratory conditions.

The tests were performed on four-ply belts with a textile EP core having a nominal tensile strength of 2000 kN/m. The belts were obtained from three different manufacturers. They differed in the material used for belt plies (P or EP). The tested belts had never been used on a conveyor. They were supplied by the manufacturers in 10 m long sections. They were given the following symbols: Belt 1, Belt 2 and Belt 3. Prior to performing the tests, belt parameters were checked for compliance with ISO 14890 [30]. The belts were found to comply with the standard.

3.2.2. Industrial Splices

The second group of test specimens consisted of industrial splices made by conveyor belt operators. The splices had been made in belts operated in underground mines as well as above the ground surface. They had been cut from the belt loop and supplied to LTT in order to test their strength parameters. The specimens included splices installed both in ordinary belts (no. 1, 2 and 4 in Table 1) and in fire-resistant belts, dedicated to working in an environment prone to the risk of gas (methane) or coal dust explosion. These splices are indicated in Table 1 with the symbol GTP (acronym from Polish "Górnicze Taśmy Przenośnikowe"—mining conveyor belts). The belts had different tensile strengths: 630, 1250, 1400, 1800, 2000 and 2500 kN/m. The number of textile plies was between 3 and 5. The differences reflect the fact that the belts represent a wide range of applications.

3.3. Methods of Experimental Research—Tests of Conveyor Belts and Their Splices

The described methods here for testing the strength parameters of conveyor belts and their splices are standardized [31–34]. The standards provide precise instructions on the number and preparation of test specimens, as well as on the test conditions. As a result, different laboratories can perform conveyor belt tests in an identical manner, and the obtained results are comparable.

This article presents the results of tests performed at the Belt Conveying Laboratory, Wroclaw University of Science and Technology (LTT).

LTT has almost thirty years of experience in laboratory tests of conveyor belts, splices, rubber, fabric, rubber compounds and plastics. The laboratory also holds the Research Laboratory Accreditation Certificate No. AB 710 awarded by Polish Center for Accreditation [35].

LTT cooperates with the industry and carries out research and scientific projects for the industry. It also provides consultancy on conveyor belts and their splices at the stages of designing and operating conveyor transportation systems. It offers verifications of the parameters of new conveyor belts (as an independent party) and the monitoring of belt parameter changes during belt operation. The laboratory issues expert opinions on the quality of splices and uses the results of these tests to indicate the reasons for reduced splice strength.

3.3.1. Splice Strength Tests

Splice tensile strength tests were performed according to a test method described in the PN-C-94147 standard [31]. This test method consists of placing a full-length splice in the jaws of the testing machine and in subsequently tensioning it at a constant speed of 100 mm/min until core rupture. During the test, a record is made of both the tensile force and the corresponding splice elongation. The splice specimens were from 200 to 400 mm in width.

Splice strength tests were performed in the two testing machines shown in Figure 6. The machine on the right (yellow color) is type ZP-40, capable of testing splice specimens up to 4000 mm in length and 200 mm in width. The splices may be stretched with a force of up to 400 kN. In order to meet the demands resulting from a trend to increase belt nominal strength, the ZP-40 test rig at LTT has been upgraded. Due to its design limitations, the old rig only allowed testing belt specimens having a nominal strength of up to 3200 kN. Therefore, in 2019, the laboratory was equipped with the ZP-100 splice tensile testing machine (Figure 6 on the right). The new test rig allows tests of belt splices having a nominal strength of up to 7800 kN. The possibility to use a tensile force exceeding 1000 kN enabled tests of splice specimens having a width of up to 500 mm. The ZP-100 machine is currently the only testing machine of this size in Poland to allow strength tests of full-length (8000 mm) splices.

Figure 6. Test stands.

The results of splice strength tests were verified against the required strength defined in relationship (11), as per [31]:

$$R_z = 0.85 \cdot R_r \cdot \frac{n-1}{n} \quad \text{kN/m} \tag{11}$$

where R_r—actual belt tensile strength, kN/m; n—number of plies in the belt core; 0.85—factor allowing for stress concentration at step contact points in areas where the ply is discontinued.

The splice strength is compared to the actual belt strength. If the actual belt strength cannot be identified, then the splice strength is compared to the nominal belt strength provided by the manufacturer. The final result is in the form of an arithmetic mean from tests performed on three specimens.

3.3.2. Delamination Strength Tests of the Adhesive Bond

The delamination strength of the adhesive bond in the splice (in the longitudinal direction) was tested following the method described in ISO 252:2018 [32]. This test method consists of delaminating the textile plies. The delamination process was performed at the speed of 100 mm/min along a minimum length of 100 mm. During the test, the delaminating force was recorded in time. The test specimens were 25 mm in width and a minimum of 300 mm in length. The tests were performed on the INSTRON 4467 testing machine. Figure 7 shows the belt specimen during the delamination test.

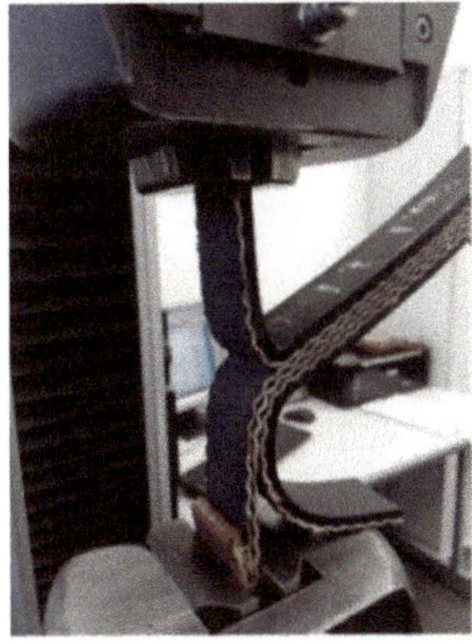

Figure 7. Delamination between ply 1 and ply 2.

Delamination strength tests of the adhesive bond were performed for each splice step. In the case of 4-ply belts, three splices were delaminated: on the first splice step (between ply 1 and ply 2), on the second splice step (between ply 2 and ply 3) and on the third splice step (between ply 3 and ply 4).

The adhesive strength W between the plies in the longitudinal direction can be calculated from Equation (12):

$$W = \frac{F_m}{b} \quad kN/m \tag{12}$$

where W—adhesion strength, in [kN/m]; F_m—mean delaminating force, in [kN], as a median; b—nominal specimen width, in [mm].

The final result is provided in the form of an arithmetic mean value obtained from three measurements on each splice step.

3.3.3. Testing the Shear Strength of the Adhesive Bond

The shear strength of the adhesive bond is tested in accordance with the method described in PN-C-94147 [31]. Figure 8 is a schematic view of the specimen prepared for shear strength tests of the adhesive bond.

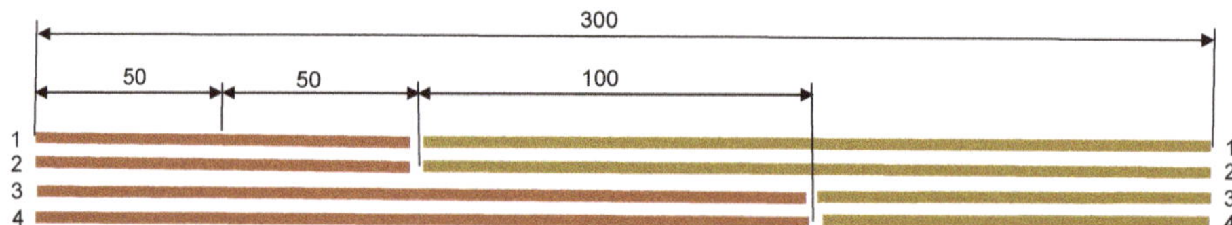

Figure 8. Schematic view of the specimen prepared for splice shear strength tests.

Test specimens were 30 mm × 300 mm in size and have cuts in their two upper plies and two bottom plies. The cuts were 100 mm in length. The specimens were placed in the clamps of the testing machine and tensioned with a speed of (100 ± 10) mm/min until the specimen is broken.

Figure 9 shows the specimen placed in the clamps of the INSTRON 4467 testing machine.

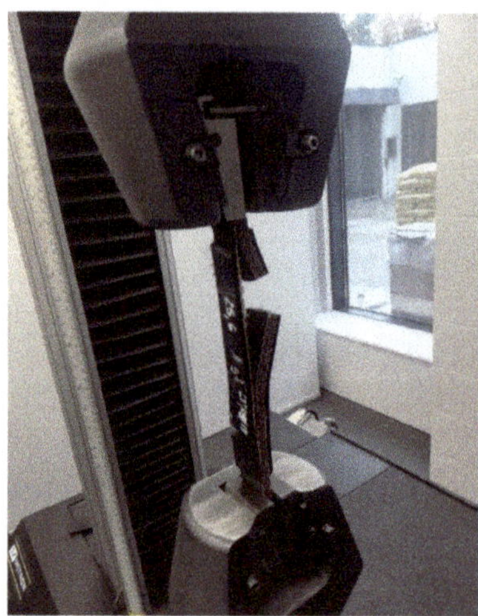

Figure 9. Schematic view of the specimen prepared for splice shear strength tests.

Shear strength of the adhesive bond τ should be calculated following Equation (13):

$$\tau = \frac{P}{b \cdot l} \quad kN/m^2 \tag{13}$$

where P—breaking force, in [kN]; l—length of the layer subjected to shear, in [m]; b—specimen width, in [m].

The final result is provided in the form of an arithmetic mean value from five measurements.

3.3.4. Belt Strength Tests

Belt tensile strength was tested following the method described in ISO 283:2016-01 [33]. The belt specimen having the shape of an oar with the clamp part of an appropriate length and width is placed in the clamps of the testing machine. The length of the specimen and the width of the clamp part depend on the strength of the belt. A 100 mm long measurement basis is marked on the specimen in order to measure belt elongation. The testing machine is equipped with a video extensometer, which measures belt elongation during the tensioning of the belt. The test results provided by the machine are then interpreted in appropriate software and subsequently stored and displayed as a graphic representation on the computer screen located on the machine. After the specimen is placed in the clamps of the testing machine, it is tensioned at a constant speed of 100 mm/min until it breaks or the core strength is significantly reduced. During the test, a record is made of both the tensile force and the belt elongation. The tests were performed on the LabTest 6.100 testing machine. It was calibrated in an accuracy class of 0.5 (while the standard requires calibration in an accuracy class of 1), and it was approved by the Polish Central Office of Measures. Figure 10 shows the testing machine with the belt specimen secured in the clamps and with the video extensometer.

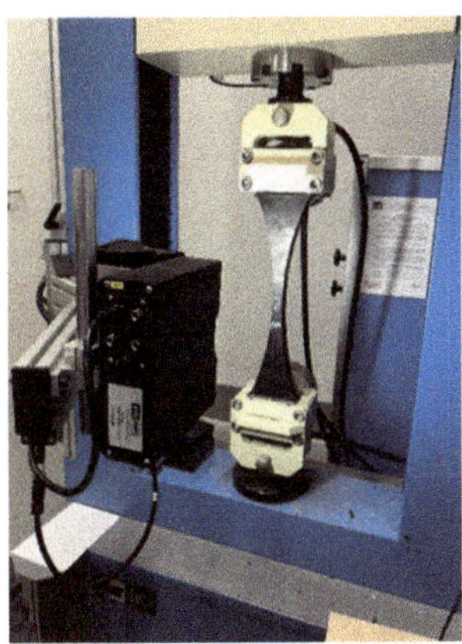

Figure 10. The LabTest 6.100 testing machine.

The belt strength R test results were calculated from relationship (14):

$$R = \frac{F_{max}}{b \; kN/m} \tag{14}$$

where F_{max}—maximum value of the tensile force, kN; n—specimen width at its most narrow part, m.

As a standard, the tests are performed on three specimens. If the belt breaks in an area beyond the measurement basis, the test result is rejected and an additional specimen is tested.

3.3.5. Tests of Belt Modulus of Elasticity

Tests of the modulus of elasticity for textile core belts were performed with the use of a standard method described in ISO 9856:2005 [34]. The method requires rectangular belt specimens, 500 mm in length and 50 mm in width and cut along the length of the belt. The measurement basis for marking the specimen elongation is required to be 100 mm. The belt specimens are subjected to sinusoidal tensile loads within the range from 2% to 10% of the belt nominal strength at a frequency of 0.1 Hz. The analysis is performed on an elastic hysteresis from 200 load cycles (Figure 11). After 200 load cycles, the force–elongation curve is used to read the values of plastic elongation Δl_p and of elastic elongation Δl_e for the tested specimen. The modulus of elasticity E is defined as the relationship between the increase of stress ΔF and the increase of belt elastic strain ε_{elast} (Equations (15) and (16)):

$$E = \frac{\Delta F}{\varepsilon_{elast}} \quad \text{N/mm} \tag{15}$$

$$\varepsilon_{elast} = \frac{\Delta l_e}{l_o} \cdot 100\% \tag{16}$$

where ε_{elast}—belt elastic strain, %; l_o—initial length of the reference section, mm.

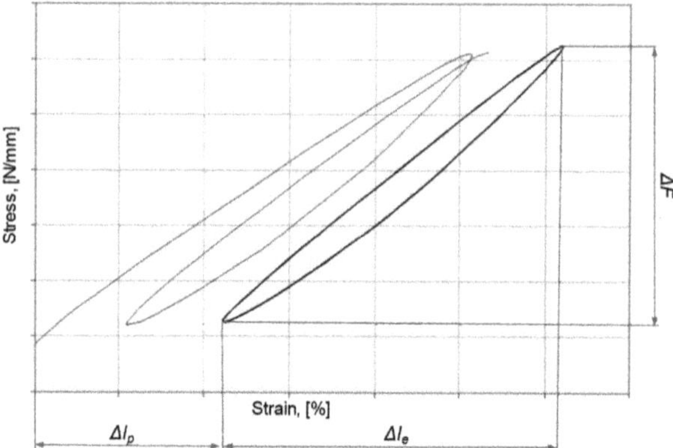

Figure 11. The 1st and the 200th hysteresis loop for a belt specimen.

Figure 11 shows examples of the test results for belt modulus of elasticity obtained with the standard method [36]. The hysteresis loops indicated in the graph are the 1st and the 200th loop for a selected belt specimen.

The tests of the modulus of elasticity were performed on the Zwick/Roell Amsler HC 25 machine for dynamic tests up to 25 kN, equipped with an extensometer mounted on the specimen for measuring its elongation.

4. Results and Discussion

4.1. The Influence of Ply Roughening on the Strength Parameters of Splices

Currently, the majority of splices tested by LTT for the industry partners meet the required values of strength parameters. The splices demonstrate the tensile strength R_z, which is described with relationship (11).

However, some of the splices do not have the required strength. Table 2 presents the results of tensile strength tests performed for several selected splices that did not demonstrate the minimum tensile strength and failed as a result of breaks in the belt plies.

Investigations of the reasons for the lowered splice strength presented in Table 1 led to the conclusion that in each case, the spliced surfaces of the textile belts had been roughened in a manner that affected their structure. Ply surfaces are prepared directly before they are covered with glue in order to clean them from the residual friction rubber or to level any rubber surface irregularities. This operation should be performed with much care so as not to damage the ply structure. Plies not covered by friction rubber are not recommended for roughening. As shown in an example in Figure 12, this recommendation had not been observed in the cases discussed here.

Figure 12. Roughened ply in the splice on belt type EP 1800/5.

The splice partially shown in Figure 12 (splice 1 in Table 2) failed upon reaching a strength of 663 kN, which is just 54% of its required strength. The splice failed gradually. Already at the value of 400–450 kN/m, the spliced plies started to break. The reason for the plies breaking at such low loads was found to lie in the fact of the surface roughening. In this case, the roughened plies not only broke at the splice contact points but also became locally fractured in various locations on their surfaces. The structure of the textile plies had been damaged due to excessive roughening. The damaged fibers in the cords are clearly visible after the ply is removed (see Figure 13). The splice partially shown in Figure 13 (splice 2 in Table 2) failed at a load 47% lower than required. As in the case of the previous splice, excessively roughened plies were observed.

Figure 13. Roughened ply in the splice on belt type EP 1800/5.

In order to find the degree at which ply roughening decreased the strength of the belt itself, comparative tensile strength tests were performed for belt specimens obtained both from the area of the splice and from outside of this area. The specimens from the splice area were cut to avoid any ply contact locations. Table 3 contains the test results for three selected splices, referred to in Table 2 by numbers 1, 11 and 12.

The results presented in Table 3 indicate that the reason behind significantly lowered splice strength lies in the reduced belt strength in the splice area. As roughening is performed only for the two bonded plies in each splice step, and the results from Table 3 describe the simultaneous breaking of all belt plies, the strength loss in the roughened plies should be assumed to be greater than the results obtained during the belt tests.

Table 3. Test results of splice strength.

No.	Belt Type	Splice Tensile Strength, [kN/m]		Belt Strength Loss in the Splice Area, [%]
		outside the Splice	in the Splice	
1	EP 1800/5	1802	1243	31
11	GTP 1400/4	1395	746	46
12	GTP 2000/4	1985	1392	30

In order to investigate the influence of ply roughening on ply strength, tests were performed on a conveyor belt type EP 2000/4. This is a four-ply belt with a textile, polyamide-polyester core having a nominal tensile strength of 2000 kN/m. The belt was delaminated between the second and the third ply, obtaining two-ply cores referenced with symbol EP 1000/2. The 1800 mm long belt was divided into three sections of equal lengths, designated with letters A, B and C. Section A was not subjected to any preparation. Section B was roughened on both sides to remove the friction rubber until the ply fabric was reached, with much attention not to damage the plies. Section C was roughened on both sides to remove the friction rubber and subsequently the ply fabric was further roughened until the surface reached uniform roughness. Figure 5 shows the surfaces prepared in accordance with the above descriptions.

Specimens for tensile strength tests were then cut from each of the belt sections. The results of these tests are shown in Table 4.

Table 4. Results of strength tests for specimens of belts A, B and C.

Belt Symbol	Specimen Designation	Belt Tensile Strength	
		[kN/m]	[%]
EP 1000/2	A	1135	100
EP 1000/2	B	1073	94
EP 1000/2	C	519	46

Specimens A were not roughened, and their strength was thus assumed at 100%. The strength of specimens B decreased by approximately 6% despite efforts not to damage the ply fabric. The strength of specimens C decreased by approximately 46% with respect to the value recorded for specimens A. It should be noted, however, that the evaluation of the degree of roughening is subjective and therefore the results may be in a wide range of values.

In order to identify the influence of ply roughening on splice strength properties, the tests were performed on three different belts type 2000/4 provided by three different manufacturers. The tests consisted in delaminating the belts between the second and the third ply and in subsequently preparing the delaminated surfaces using the method described above (see Figure 5). The plies were then spliced again using the hot vulcanization method. Mean results of the shear strength and the delamination strength of the splice specimens are shown in Table 5.

The results of delamination and shear strength tests were compared with the same parameters, which had been identified for the belts used in the above splices. The obtained results allow a definite conclusion that the best strength properties were observed in splices in which the plies were not roughened (splice A), and the worst strength properties in splices in which ply surfaces are excessively roughened until the fabric became rough and the fibers are damaged (splice C).

The comparison of the results of shear and delamination strength tests for the adhesive bond between the splice with roughened plies (splice C) and the splice with no ply roughening (splice A) indicates clearly that roughening significantly affected the strength test results. The tested shear strength for splice C was only at 21% to 63% of the values obtained in the case of splice A. The tested delamination strength for splice C was also much lower than the results for splice A, reaching only 55–77% of the strength observed in the unroughened splices.

Table 5. Splice shear and delamination strength test results.

No.	Belt Type	Specimen Designation	Shear Strength		Delamination Strength	
			[MPa]	[%]	[N/mm]	[%]
1	EP 2000/4	Belt No. 1	3.08	100	8.4	100
2		Splice A	2.69	87	7.8	93
3		Splice C	2.07	67	5.0	59
4	GPM 2000/4	Belt No. 2	3.75	100	10.8	100
5		Splice A	2.68	71	10.9	101
6		Splice B	2.71	72	4.3	40
7		Splice C	1.58	42	2.3	21
8	GPM 2000/4	Belt No. 3	4.23	100	10.7	100
9		Splice A	3.05	72	6.6	62
10		Splice B	2.67	63	5.3	49
11		Splice C	1.80	42	4.1	38

4.2. Different Properties of Spliced Belts Influencing Splice Strength

As the relationship between the elongations and stresses observed at conveyor belt tensioning is non-linear, the consideration of the elongation–stress relationship as a linear relationship requires defining the elastic modulus for particular load limits and for a particular minimum load. The value of the modulus was calculated in the full range of the set belt loads. The value of the longitudinal elastic modulus for textile plies was calculated with the use of software and on the basis of the closest to linear part of the stress–strain curve (in the set range). The calculated values of longitudinal elasticity moduli for plies in various load ranges are shown in Table 6.

Table 6. Values of longitudinal elasticity moduli for the belt cores in various load ranges.

Load Range, [kN]	Percentage of Nominal Load, [%]	Longitudinal Elasticity Modulus of Belt Core, E_p [kN/m]·10^3		
		Belt No. 1	Belt No. 2	Belt No. 3
2–4	10–20	2.6	0.8	1.1
4–6	20–40	2.0	0.8	1.0
6–8	30–40	1.1	1.1	1.5
8–10	40–50	1.3	1.3	1.8
10–12	50–60	1.8	1.6	1.9
12–14	60–70	2.2	1.7	2.7
14–16	70–80	2.4	1.9	2.6
16–18	80–90	3.0	2.2	2.5
18–20	90–100	3.7	2.1	3.0
10–20	50–100	2.7	1.9	2.3
14–20	70–100	2.8	2.0	2.3

Selected conveyor belts were used to prepare six splices for strength tests. Figure 14 shows the shape and the dimensions of the splices. The splices were hot-vulcanized and prepared by joining the belts in various combinations: Belt 1 with Belt 1 (1-1), Belt 2 with Belt 2 (2-2), Belt 3 with Belt 3 (3-3), Belt 1 with Belt 2 (1-2), Belt 1 with Belt 3 (1-3) and Belt 2 with Belt 3 (2-3).

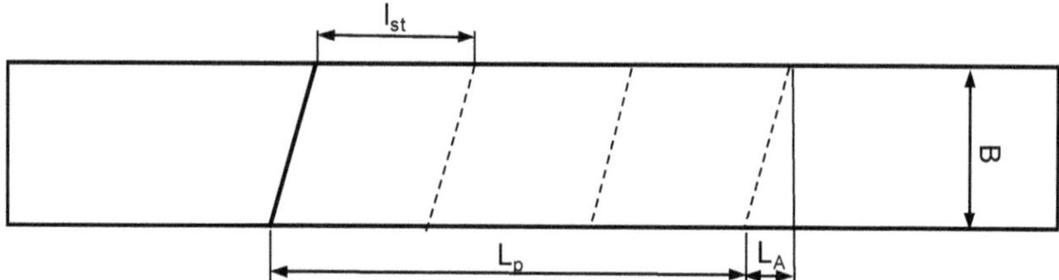

Figure 14. The shape and the dimensions of the tested splices: lst—length of the splice step; 250 mm, L_p—splice length, 750 mm; L_A—bias length, 0.3·B = 210 mm; B—belt width, 700 mm.

The measured splice strength values are presented in Table 7. The obtained splice strength was compared against the strength required as per PN/C-94147 [31] and against the nominal strength of the spliced belts (which was 1000 kN/m). In the case when different belts were spliced, a comparison was also made with the strength of the splice in the same belt. The values presented in the table, in the cells where the rows representing splices of different belts (1-2, 1-3, 2-3) cross with the columns representing splices of identical belts (1-1, 2-2, 3-3), define the relationship (expressed as percentage) between the strength measured for splices of different belts and the strength measured for splices of identical belts (assumed as 100%).

Table 7. Values of longitudinal elasticity moduli for the belt cores in various load ranges.

Splice Symbol (Numbers of Spliced Belts)	Mean Splice Rupture Strength, [kN/m]	Splice Strength vs. the Required Value, [%]	Splice Strength vs. the Strength Recorded for Splices of Identical Belts, [%]			Splice Strength vs. the Nominal Belt Strength, [%]
			1-1	2-2	3-3	
1-1	670	105	-	-	-	67
2-2	635	100	-	-	-	64
3-3	780	122	-	-	-	78
1-2	468	73	70	74	-	47
1-3	605	95	90	-	78	61
2-3	613	96	-	97	79	61

The tests indicate that splices of belts whose plies have different mechanical properties (different longitudinal modulus of elasticity) show lower strength than splices of identical belts. The strength of splices between different belts, referred to as 1-2, 1-3 and 2-3, was 70–97% of the strength recorded for splices 1-1, 2-2, 3-3. The difference between the values of elastic moduli for plies in different belts is observed across the entire range of the carried loads (see Table 6).

The difference between the mechanical properties of the belt with polyester-polyamide core (EP) and the mechanical properties of the belt with polyamide core (P) results from the different mechanical properties of the two materials. The elastic modulus value is low for polyamide and high for polyester. The selection of belts with identical strength and number of plies but made of materials with significantly different mechanical properties was a conscious choice dictated by the intention to obtain a clear demonstration of the influence of this configuration on the splice strength.

The difference between the elastic moduli observed in the case of belts with polyamide core is caused by the fact that the manufacturers of conveyor belts use different materials in the plies and different manufacturing technologies. As a result, apparently identical belts having the same nominal tensile strength, number of plies and width show different mechanical properties, and this leads to the lowered splice strengths demonstrated in the tests. While the difference between the polyester-polyamide (EP) belt and the polyamide (P) belt is obvious to splicing professionals, there is practically no method that would allow belts of identical nameplate parameters to be properly selected without strength tests (excluding a scenario in which the belts are provided by the same manufacturer). The splice strength is additionally reduced (with respect to splices of identical belts) due to increased concentrations of tensile stresses in the plies and to increased shearing stresses in the rubber (adhesive) layer between the plies.

5. Conclusions

The main results and conclusions can be summarized as follows:

1. The laboratory tests of belts and splices, which were performed in both laboratory conditions and in actual mining plants, confirmed the conclusions of the theoretical considerations pointing to differences between the mechanical properties of spliced belts as one of the reasons behind reduced splice strength.
2. In multiply conveyor belt splices, stresses concentrate both in the plies and in the adhesive bond located in the cross-sections of the ply contact points, leading to reduced splice strength. However, tests of some splices made in industrial installations show strengths lower than expected from the losses due to stress concentrations and reduced number of plies in the splice. Investigations allowed an observation that the reason behind such cases lies in the inappropriate preparation of belt plies in the process of splicing.
3. As the plies are cleaned and roughened with excessive intensity in order to remove residual friction rubber, the ply fabric becomes exposed and the belt strength is reduced, in effect leading to reduced splice strength. The strength loss may reach up to several tens of percent. The roughening of plies in vulcanized splices also lowers their shear and delamination strength.
4. The tests demonstrated that the highest strength parameters were observed in the case of splices in which the plies were not cleaned from the friction rubber, and therefore this procedure should be performed only if required in order to level irregularities or reduce the thickness of the friction rubber. The roughening procedure cannot cause the fabric of the plies to become exposed, as this inevitably damages the fabric.
5. The research results also indicate that the observed decrease in the strength of the same type of belts provided by different manufacturers is due to different mechanical properties of the plies, and in this case, to different elastic moduli. Knowledge about the belt modulus of elasticity is therefore crucial for rationally designing belt splices, analyzing their dynamics and improving the effective usage of the belt, which is the most expensive and the most important element of a belt conveyor.
6. Tests of selected conveyor belts indicated that plies of the same type show different mechanical properties if provided by different manufacturers. This fact is due to the use of different materials and belt manufacturing technologies.
7. The laboratory tests of full-length splices demonstrated that properly made splices of identical belts meet or even exceed the strength requirements offered in current standards.
8. The laboratory tests of the belts and splices confirmed the conclusions of belt splice tests performed in the mining plants and pointed to the difference between the mechanical properties of the spliced belts as one of the reasons behind reduced splice strength.

Author Contributions: Conceptualization, M.H.; methodology, M.H.; software, M.B.; validation, M.H.; formal analysis, M.H.; investigation, M.H., M.B.; resources, M.H., M.B.; data curation, M.H.; writing—original draft preparation, M.B.; writing—review and editing, M.H., M.B.; visualization, M.B.; supervision, M.H.; project administration, M.H. All authors have read and agreed to the published version of the manuscript.

Funding: This research was co-financed by the Polish Ministry of Science and Higher Education for 2020.

Institutional Review Board Statement: Not applicable.

Informed Consent Statement: Not applicable.

Data Availability Statement: The data presented in this study are available on request from the corresponding author.

Conflicts of Interest: The authors declare no conflict of interest.

References

1. Kawalec, W.; Suchorab, N.; Konieczna-Fuławka, M.; Król, R. Specific Energy Consumption of a Belt Conveyor System in a Continuous Surface Mine. *Energies* **2020**, *13*, 5214. [CrossRef]
2. Mu, Y.; Yao, T.; Jia, H.; Yu, X.; Zhao, B.; Zhang, X.; Ni, C.; Du, L. Optimal scheduling method for belt conveyor system in coal mine considering silo virtual energy storage. *Appl. Energy* **2020**, *275*, 115368. [CrossRef]
3. Zimroz, R.; Krol, R. Failure analysis of belt conveyor systems for condition monitoring purposes. *Min. Sci.* **2009**, *128*, 36, 255–270.
4. Ambriško, Ľ.; Marasová, D. Experimental Research of Rubber Composites Subjected to Impact Loading. *Appl. Sci.* **2020**, *10*, 8384. [CrossRef]
5. Andrejiova, M.; Grincova, A.; Marasova, D. Analysis of tensile properties of worn fabric conveyor belts with renovated cover and with the different carcass type. *Eksploatacja i Niezawodność—Maint. Reliab.* **2020**, *22*, 472–481. [CrossRef]
6. Marasova, D.; Ambrisko, L.; Andrejiova, M. Examination of the process of damaging the top covering layer of a conveyor belt applying the FEM. *Measurement* **2017**, *112*, 47–52. [CrossRef]
7. Molnar, W.; Nugent, S.; Lindroos, M.; Apostol, M.; Varga, M. Ballistic and numerical simulation of impacting goods on conveyor belt rubber. *Polym. Test.* **2015**, *42*, 1–7. [CrossRef]
8. Rudawska, A.; Madleňák, R.; Madleňáková, L.; Droździel, P. Investigation of the effect of operational factors on conveyor belt mechanical properties. *Appl. Sci.* **2020**, *10*, 4201. [CrossRef]
9. Bajda, M.; Blazej, R.; Jurdziak, L.; Hardygóra, M. Impact of differences in the durability of vulcanized and adhesive joints on the operating costs of conveyor belts in underground mines. Scientific Papers of the Institute of Mineral and Energy Economy. *Pol. Acad. Sci.* **2017**, *99*, 71–88.
10. Doroszuk, B.; Król, R. Conveyor belt wear caused by material acceleration in transfer stations. *Min. Sci.* **2019**, *26*, 189–201. [CrossRef]
11. Jurdziak, L. The conveyor belt wear index and its application in belts replacement policy. In Proceedings of the Ninth International Symposium on Mine Planning and Equipment Selection, Athens, Greece, 6–9 November 2000; Balkema, A.A.: Rotterdam, The Netherlands; pp. 589–594.
12. Temerzhanov, A.; Stolpovskikh, I.; Sladkowski, A. Analysis of reliability parameters of conveyor belt joints. *Transp. Probl.* **2012**, *7*, 107–112.
13. Bancroft, B.; Ch, F.; Pilarski, T. Belt Vision System for Monitoring Mechanical Splices. In Proceedings of the Conference materials: Longwall USA International Exhibition and Conference, Pittsburgh, PA, USA, 3–5 June 2003.
14. U.S. Department of Energy's. Effective Conveyor Belt Inspection for Improving Mining Productivity. U.S. Department of Energy Office of Energy Efficiency and Renewable Energy. 2004. Available online: http://www.nrec.ri.cmu.edu/projects/belt_inspection/tech/effectconvey.pdf (accessed on 11 May 2017).
15. Kozłowski, T.; Wodecki, J.; Zimroz, R.; Błazej, R.; Hardygóra, M. Diagnostics of Conveyor Belt Splices. *Appl. Sci.* **2020**, *10*, 6259. [CrossRef]
16. Mazurkiewicz, D. Monitoring the condition of adhesive-sealed belt conveyors in operation. *Eksploat. Niezawodn. Maint. Reliab.* **2005**, *3*, 41–49.
17. Bajda, M.; Błażej, R.; Hardygóra, M. Impact of selected parameters on the fatigue strength of splices on multiply textile conveyor belts. *IOP Conf. Ser. Earth Environ. Sci.* **2016**. [CrossRef]
18. Błażej, R.; Bajda, M.; Hardygóra, M. Monitoring creep and stress relaxation in splices on multiply textile rubber conveyor belts. *Acta Montan. Slovaca* **2017**, *22*, 116–125.
19. Heitzmann, P.; Fröbose, T.; Wakatsuki, A.; Overmeyer, L. Optimization of textile conveyor belt splices using Finite Element Method (FEM). *Logist. J.* **2016**. [CrossRef]
20. Long, X.; Li, X.; Long, H. Analysis of influence of multiple steel cords on splice strength. *J. Adhes. Sci. Technol.* **2018**, *32*, 2753–2763. [CrossRef]
21. Xianguo, L.; Xinyu, L.; Zhenqian, S.; Changyun, M. Analysis of Strength Factors of Steel Cord Conveyor Belt Splices Based on the FEM. *Adv. Mater. Sci. Eng.* **2019**. [CrossRef]
22. Hardygóra, M.; Bajda, M.; Błażej, R. Laboratory testing of conveyor textile belt joints used in underground mines. *Min. Sci.* **2015**, *22*, 161–169. [CrossRef]
23. Kirjanów, A. Analysis of the results of the strength tests of finger splices. *Min. Sci.* **2015**, *22*, 31–37. [CrossRef]
24. Wozniak, D. Laboratory tests of indentation rolling resistance of conveyor belts. *Measurement (London)* **2020**, *150*, 107065. [CrossRef]
25. Bajda, M.; Błażej, R.; Hardygóra, M.; Jurdziak, L. Project NCBiR no PBS3/A2/17/2015. In *Joints of Multi-Ply Conveyor Belts with Increased Functional Durability (Złącza Wieloprzekładkowych Taśm Przenośnikowych o Zwiększonej Trwałości Eksploatacyjnej)*; Wrocław University of Science and Technology: Wroclaw, Poland, 2018; in press. (In Polish)

26. Bajda, M.; Blazej, R.; Jurdziak, L. Partial replacements of conveyor belt loop analysis with regard to its reliability. In Proceedings of the 17th International Multidisciplinary Scientific GeoConference SGEM, Albena, Bulgaria, 27 June–6 July 2017; pp. 645–652. [CrossRef]

27. Madziarz, M. *Wpływ Konstrukcji i Technologii Wykonywania Połączeń Tkaninowych Taśm Przenośnikowych na Ich Wytrzymałość (Influence of the Splice Design and the Splicing Technology in Textile Conveyor Belts on Their Strength)*; Wrocław University of Science and Technology, doctoral dissertation: Wroclaw, Poland, 1998. (In Polish)

28. Błażej, R. Wpływ Właściwości Mechanicznych Rdzenia Taśm Przenośnikowych Tkaninowo-Gumowych na Wytrzymałość ich Połączeń (The Influence of the Mechanical Properties of the Core of Rubber-Textile Conveyor Belts on Splice Strength). Ph.D. Dissertation, Wrocław University of Science and Technology, Wroclaw, Poland, 2001. (In Polish)

29. Research project no 8T12A 03521. *Optimization of the Construction of Multi-Ply Conveyor Belt Splices*; Wrocław University of Science and Technology, Mining Institute: Wroclaw, Poland, 2004.

30. International Standard ISO 14890. *Conveyor Belts—Specification for Rubber or Plastics Covered Conveyor Belts of Textile Construction for General Use*; European Committee for Standarization: Brussels, Belgium, 2013.

31. Polish Standard PN-C-94147. Rubber products. In *Guiding Principles for Conveyor Belts Jointing by Vulcanization*; Polski Komitet Normalizacyjny PKN: Warsaw, Poland, 1997.

32. International Standard ISO 252. Conveyor belts. Adhesion between constitutive elements. In *Test Methods*; European Committee for Standarization: Brussels, Belgium, 2018.

33. International Standard ISO 283. Textile conveyor belts. Full thickness tensile strength, elongation at break and elongation at the reference load. In *Test Method*; European Committee for Standarization: Brussels, Belgium, 2016.

34. International Standard ISO 9856. Conveyor belts. In *Determination of Elastic and Permanent Elongation and Calculation of Elastic Modulus*; European Committee for Standarization: Brussels, Belgium, 2005.

35. The official website of the Belt Conveying Laboratory. Available online: www.ltt.pwr.edu.pl (accessed on 25 August 2020).

36. Furmanik, K.; Pracik, M. Experimental identification of rheological parameters standard model of belt conveyor. *Mechanics* **2011**, *2*, 25–33.

Article

A Step-by-Step Procedure for Tests and Assessment of the Automatic Operation of a Powered Roof Support

Dawid Szurgacz [1], Sergey Zhironkin [2,3,4,*], Michal Cehlár [5], Stefan Vöth [6], Sam Spearing [7] and Ma Liqiang [7]

1   Center of Hydraulics DOH Ltd., 41-906 Bytom, Poland; dawidszurgacz@doh.com.pl
2   Department of Trade and Marketing, Siberian Federal University, 79 Svobodny av., 660041 Krasnoyarsk, Russia
3   Department of Open Pit Mining, T.F. Gorbachev Kuzbass State Technical University, 28 Vesennya St., 650000 Kemerovo, Russia
4   School of Core Engineering Education, National Research Tomsk Polytechnic University, 30 Lenina St., 634050 Tomsk, Russia
5   Faculty of Mining, Ecology, Process Technologies and Geotechnology, Technical University of Košice, Letná 9, 042 00 Košice, Slovakia; michal.cehlar@tuke.sk
6   Technische Hochschule Georg Agricola (THGA), Westhoffstraße 15, 44791 Bochum, Germany; Stefan.Voeth@thga.de
7   School of Mines, China University of Mining and Technology, 1 Daxue Road, Tongshan District, Xuzhou 221116, China; sam.spearing@curtin.edu.au (S.S.); ckma@cumt.edu.cn (M.L.)
*   Correspondence: zhironkinsa@kuzstu.ru

Abstract: A powered longwall mining system comprises three basic machines: a shearer, a scraper (longwall) conveyor, and a powered roof support. The powered roof support as a component of a longwall complex has two functions. It protects the working from roof rocks that fall to the area where the machines and people work and transports the machines and devices in the longwall as the mining operation proceeds further into the seam by means of hydraulic actuators that are adequately connected to the powered support. The actuators are controlled by a hydraulic or electro-hydraulic system. The tests and analyses presented in the developed procedure are oriented towards the possibility of introducing automatic control, without the participation of an operator. This is important for the exploitation of seams that are deposited at great depths. The primary objective was to develop a comprehensive methodology for testing and evaluating the possibility of using the system under operating conditions. The conclusions based on the analysis presented are a valuable source of information for the designers in terms of increasing the efficiency of the operation of the system and improving occupational safety. The authors have proposed a procedure for testing and evaluation to introduce an automatic control system into the operating conditions. The procedure combines four areas. Tests and analyses were carried out in order to determine the extent to which the system could be potentially used in the future. The presented solution includes certification and executive documentation.

Keywords: safety function; electro-hydraulic control system; powered roof support

Citation: Szurgacz, D.; Zhironkin, S.; Cehlár, M.; Vöth, S.; Spearing, S.; Liqiang, M. A Step-by-Step Procedure for Tests and Assessment of the Automatic Operation of a Powered Roof Support. *Energies* **2021**, *14*, 697. https://doi.org/10.3390/en14030697

Received: 11 January 2021
Accepted: 26 January 2021
Published: 29 January 2021

Publisher's Note: MDPI stays neutral with regard to jurisdictional claims in published maps and institutional affiliations.

1. Introduction

The development of the underground coal mining industry is boosted by the need to improve the efficiency of the mining process, at the same time, maintaining the required level of safety and protecting the environment. Machines and devices with innovative designs that automate the process of mining are the key elements required to improve the mining process. One of the basic directions of development in the world mining industry is the automation of a longwall unit. Today, manufactured longwall shearers have improved mining efficiency with a face feed speed of almost 10 m/min. This means that the sections of the powered roof support must be moved within a maximum of 9 s for sections that are 1.5 m wide, and 10.5 s for sections with a width of 1.75 m. It is not possible to obtain such performance parameters using traditional, manual control of the powered roof support. These requirements and the fact that the design of a support section is complex means that a hydraulic system needs partial or full automation to achieve large hydraulic fluid flows [1–6]. The automation process reduces the total number of operators controlling the support sections and makes it possible to remove them from the immediate vicinity of the support that is being moved. This has a positive effect on safety by removing personnel from the active face.

The procedure developed based on research and analysis includes four stages. These are essential for manufacturers of longwall roof supports and producers of control systems, as it accurately and easily shows what the sequence of design, research, and production work should be. From a scientific point of view, the third stage is the most important one, as the results obtained, either positive or negative, make it possible to analyse whether the assumptions made in Stage 1 coincide with the design made in Stage 2. As a result of the research on prototypes and the acceptance of prototype modifications, documentation is developed, as presented in Stage 4. The procedure takes into account all possible safety requirements and the possibility of effective preparation of automatic control. The nature of the presented concept is open, as it takes into account the possibility of introducing specific conditions in which the seam will be mined, and the selection system used. The procedure can be modified to fit the research and development of automatic control devices and future requirements.

The depth of underground hard coal mining has been constantly increasing in the past years. This, in turn, increases natural hazards such as exogenous fire, seismicity, and methane hazards [7,8]. The longwall system in terms of these hazards challenges manufacturers, researchers, and mining companies to ensure an appropriate level of occupational safety. Currently, a considerable involvement of researchers in solving problems related to methane hazards [8–15], endogenous fires [7], and seismicity [16] can be observed worldwide. Therefore, global coal mining is at the forefront of the desired energy resource alongside oil and gas [17–19]. Coal mining in an underground mine is carried out using a longwall system that uses such machines as a shearer that mines the coal, a scraper conveyor for haulage, and a powered roof support for the longwall system. They operate together and form a longwall complex (Figure 1).

Monitoring the machinery and equipment in the longwall complex and mining and geological conditions largely contributes to the maintenance of sustainable development of mining in the world and improvement of work safety [20–28]. Recently, there has been extensive activity in the development of research on the application of new technologies in mining thanks to Industry 4.0 [29–32]. The selection of technical parameters of machines and equipment is important due to the arduous conditions in which they will work. It is possible to evaluate correctly selected machines through active monitoring of their operating parameters and analysis of operating conditions [33–39]. One of the main sources of hazards causing accidents at work in mining plants is the technical infrastructure and, in particular, machinery and technical equipment. In 2019, the rate of accident and mortality associated with technical hazards caused by machines and equipment was approximately 50% of all events [40]. To ensure safe use and proper hazard assessment, machine and control equipment designers place the greatest emphasis on structural safety aspects. The related requirements are defined in standards [41–46] harmonised with the Machinery Directive [47]. However, due to technical progress, such standards do not exist for most of the machine control systems currently developed. Therefore, in order to ensure compliance with the essential requirements of the Directive, a package of technical standards on functional safety issues was created; it includes the concept of safety assessment of machinery [41,43].

The mining and geological conditions in which the powered roof support system is to be used can be divided into dynamic impacts of the rock mass on the longwall excavation where the roof support system works. This phenomenon is more dangerous than mining and geological conditions in which the rock mass does not tend to tremble.

Sudden clamping of the excavation means a static displacement that overloads the longwall complex that is supposed to be supported by the powered roof support. The load that impacts powered roof supports results from the movement of rock masses at a certain speed. The main task of powered longwall supports is to transfer loads resulting from changing mining and geological conditions, allowing for deformation of the longwall excavation to a minimum degree [48].

The powered roof support is a hydraulic machine powered by high pressure with an oil–water concentrate as an energy carrier. Spragging of the support section of the powered support for the required height of the excavation and its support is conducted by hydraulic legs. They constitute a structural connection between the canopy and floor base. As a result of the impact of the rock mass, the hydraulic legs carry the load depending on the conditions under which the powered support operates. Their protection is ensured by placing a safety valve in the hydraulic system [49]. It is vital to carry out an economic analysis of the effective protection of the powered roof support together with all machines of the longwall complex. In the past, several studies on the automation of the entire mining process were conducted [50–54]. The selection of a safety valve for the hydraulic system of the leg of the powered roof support working in the conditions where tremors often occur is key as it affects the safety of machines and people [55,56].

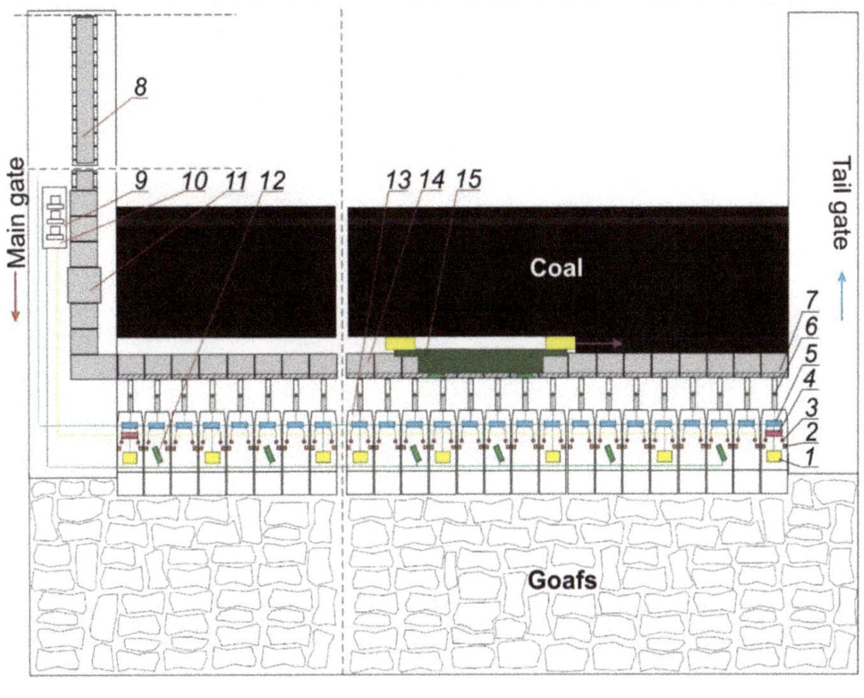

(a)

Figure 1. *Cont.*

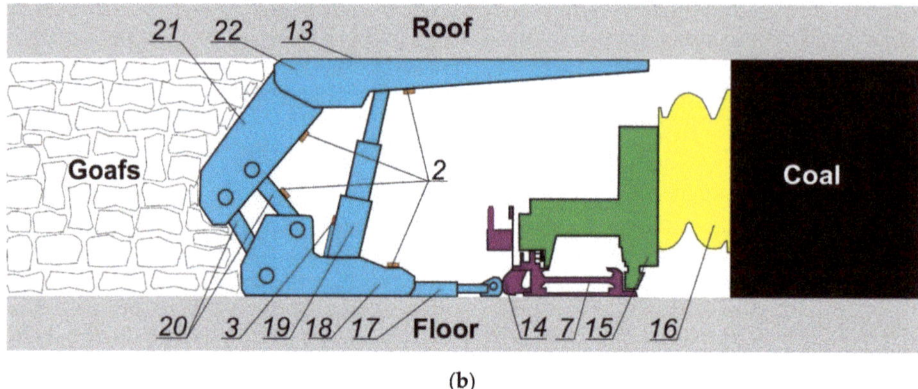

(**b**)

Figure 1. Arrangement of machines and equipment in a longwall complex: (**a**) cross-section of the longwall and (**b**) longitudinal section of the longwall, where 1—power supply, 2—inclinometer, 3—pressure sensor, 4—central controller, 5—controller with an executive block, 6—route sensor, 7—longwall scraper conveyor, 8—belt conveyor, 9—shearer control station with closed circuit television, Closed Circuit Television (CCTV) monitors and monitoring of operating parameters of the powered roof support, 10—scraper conveyor, 11—crusher, 12—CCTV camera, 13—powered roof support, 14—sliding system connection with a scraper conveyor, 15—shearer, 16—cutting unit, 17—sliding system, 18—floor base, 19—hydraulic leg, 20—lemniscate system, 21—shear support, and 22—roof.

2. Materials and Methods

Coal mining in an underground mine is based on a longwall complex that uses the powered roof support. In most cases, a hydraulic system based on manual controls is used to control the roof support. This is mainly determined by the economic efficiency of the company. It is reasonable to work towards reducing the number of workers near working machines to improve work safety. It is also important to measure the number of failures of machines and equipment carrying out technological processes related to hard coal mining.

The development of longwall complex automation technology in recent years has been directed towards the identification of shearer operation parameters based on the monitoring. A device that plays a crucial role here is a sensor that is applied in various systems designed to monitor the parameters of a longwall complex [57–59]. It is intended to improve occupational safety. With the development of intelligent computing technologies, it has become possible to develop intelligent algorithms related to neural networks, fuzzy logic, hybrid methods [58–77]. This formed the foundation for the development of an automatic control system for a powered roof support in a longwall complex.

2.1. System Design and Development

The powered roof support section, with hydraulic legs as its main elements, directly affects the required protection of the excavation. Vital functions are performed by other hydraulic cylinders including advancint the armoured face conveyor and in the powered roof support, such as the conveyor slide or the support of the canopy. Monitoring the pressure in the legs and determining the geometry of the section position in the excavation is an essential factor in the exploitation of longwall complexes. The possibility of automating specific sequences of work of particular elements of the support and controlling their implementation allows limiting the presence of the miners in the excavation. Equally important is the fact that the section of the roof support is guided by the profile of the excavation by means of appropriate cooperation of the section of the roof support with a scraper conveyor. Based on the presented concept, a vision of the system was designed together with a visualisation of the operating parameters, which is shown in Figure 2. A diagram of the actual control system of the powered system as a test demonstrator is shown in Figure 3.

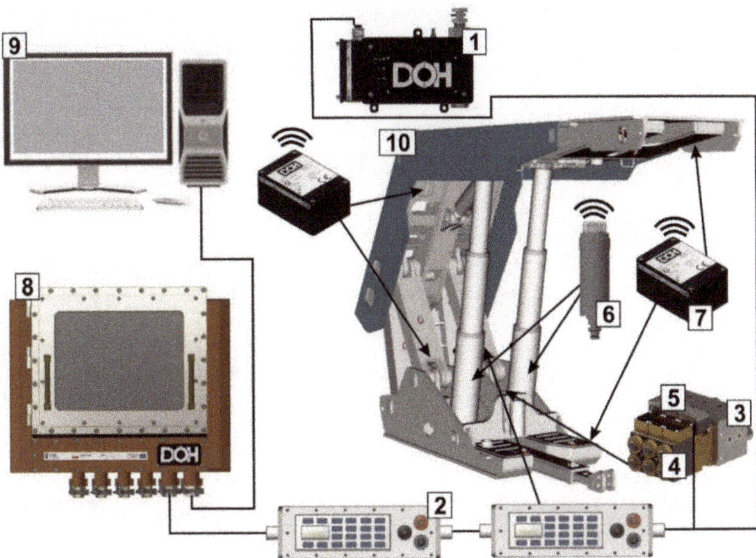

Figure 2. The concept of the devices that are part of the system; 1—intrinsically safe power supply, 2—controller, 3—electro-hydraulic executive block with control inserts, 4—solenoid valve, 5—control bar, 6–pressure sensor, 7—inclinometer, 8—underground computer, 9—surface computer, and 10— powered roof support section.

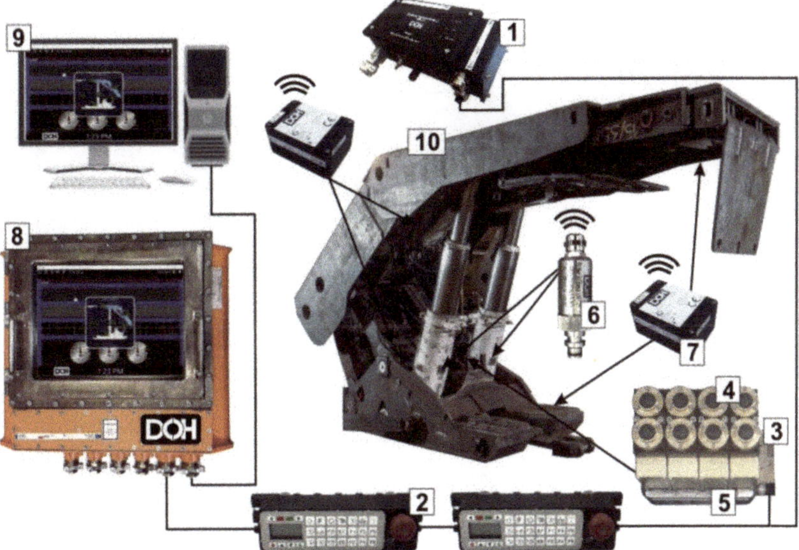

Figure 3. Prototype devices that are part of the system; 1—intrinsically safe power supply, 2— controller, 3—electro-hydraulic executive block with control inserts, 4—solenoid valve, 5—control bar, 6—pressure sensor, 7—inclinometer, 8—underground computer, 9—surface computer, and 10—powered roof support section.

2.2. The Requirements for the Design Procedure

The concept of the software and database structure was based on the assumptions of the analysis of the control system operation possibilities. A special criterion is to determine the areas which will constitute a visualisation of the working parameters. The information gathered in the database will be used to report on the production process of coal mining. The concept for the development of computer software architecture is based on the research and operating experience of the authors. Based on the theoretical analyses made, the assumptions of software architecture modules and databases were developed. Due to the gaps and limitations caused by the computing power of a computer operating in an underground mine, the focus was put on collecting relevant information such as pressure measurement in powered roof supports and their position geometry. One of the elements of creating a database is to test whether the proposed structure and data resource correspond to the requirements set at the design stage. The designed software architecture together with the database will constitute the information base of the control system.

The architecture diagram for the software for the computer that will operate on the surface and one that will operate underground are shown in Figures 4 and 5.

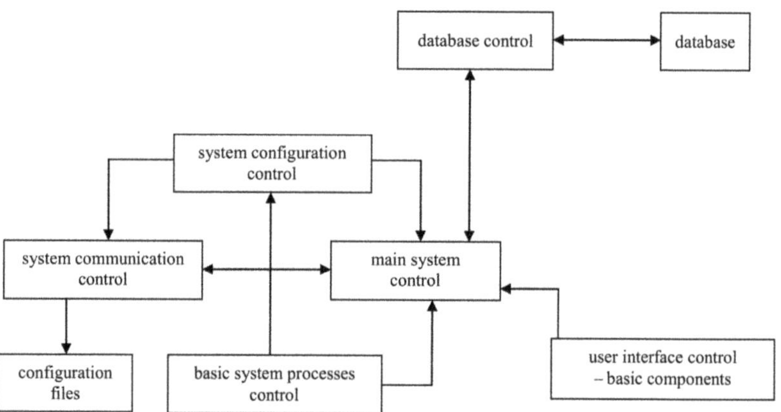

Figure 4. Design of the underground computer together with the database.

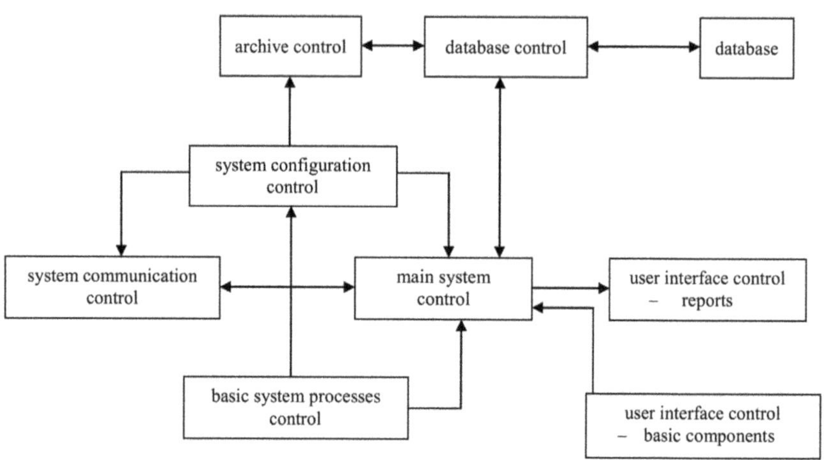

Figure 5. Design of the surface computer together with the database.

3. The Concept of Safety and Hazard Assessment

The electro-hydraulic system designed to control the powered roof support should perform its intended control functions even under damage or disturbance conditions in a predictable manner and with specified reliability. Measures taken to minimise the likelihood of such damage or disturbances and their extent depends on the level of hazard associated with the specific control function.

Research on the evaluation of the reliability of the implemented safety function for the electro-hydraulic control system of the powered roof support is aimed at reducing the risk of hazardous events to an acceptable level. The hazard reduction can be carried out based on standards [41–46], to optimally achieve the protection of the system. In the process of designing and constructing the control system, the risks were analysed, and measures were taken to protect the operator from the existing hazards. The hazard assessment is the result of logical steps, allowing for systematic analysis. The system or the machine must be designed and constructed taking into account the results of the hazard assessment. The first step of the assessment is to identify system components (equipment) and later the sources of the hazard for all activities of the operation. The authors considered only those risk factors that have a fundamental impact on the assessment of the effects of a threat when estimating the hazard [20]. Four levels of hazard were adopted for research based on own experience.

(1) Unacceptable—hazard reduction is necessary, otherwise, the system cannot be authorised;
(2) Undesirable—hazard is only acceptable if the expenditure involved in reducing it is clearly higher than the effects achieved or if hazard reduction is not achievable;
(3) Acceptable—the hazard is only acceptable if the expenditure involved in reducing it is significantly higher than the effects achieved;
(4) Negligible—further investment in hazard reduction is not acceptable.

An individual assessment based on good practice and experience was adopted for the risk assessment. This method uses predefined value ranges and descriptive measures such as:

S_1—negligible (the risk is very rare, the chance of an event occurring is low, a probability of no more than 10%),
S_2—minor (light injuries, mild occupational disease, 11–30% probability),
S_3—serious (serious injury to one or more persons or death of one person, probability 31–60%),
S_4—severe (many people die, probability 61–80%),
S_5—catastrophic (very many deaths and practically total destruction of the system, a probability above 81%).

In general, it must be assumed that the hazard is a combination of the intensity of the occurrence of safety h and its consequences S:

$$R = h \cdot S, \tag{1}$$

The total hazard associated with the use of the system consists of several hazards and therefore, the total hazard can be assumed as follows:

$$R = \sum_{i=1}^{n} h_i \cdot S_i \tag{2}$$

where h_i—the intensity of the i-th hazard and S_i—the consequences of the i-th hazard.
The probability of the i-th hazard can be determined.

$$p_i = \frac{h_i}{\sum_{i=1}^{n} h_i} \tag{3}$$

Expected impact per unit of time:

$$E_{(s)} = \sum_{i=1}^{n} S_i \cdot p_i \qquad (4)$$

Additionally, as a result

$$R = E_{(s)} \cdot \sum_{i=1}^{n} h_i \qquad (5)$$

The result of the assessment is presented in Table 1.

The parameter determining the level of hazard is primarily the severity of the damage that may occur as a result of failure to perform the safety function once it has been recalled. The consequences of failure to perform the exemplary control functions of the electro-hydraulic powered roof support are almost always serious for the health and life of those operating the machine and those around it. The analysis of the required performance level (PL) is based on the identification of the tolerable hazard function [45].

When defining the level of safety assurance PL, we refer to it as the ability to perform safety functions under the expected conditions. There are five discreet levels of safety assurance marked as:

a—indicates the lowest probability of failure,
b—means the average probability of damage,
c—means a good probability of damage,
d—means a high probability of damage,
e—means the highest probability of damage.

Determination of the Safety Integrity Level (SIL) is based on the estimation of the value of hazard and the ability of the control system to reduce hazards [41]. As already mentioned above, the safety function can be set up in two different ways. The determined safety level of SIL and PL for the tested control system was also compared based on the probability of accidental equipment failure. Table 1 presents the results of the analysis of the safety function.

Table 1. The result of the analysis designed to determine the level of the safety function.

Safety Integrity Level (SIL)	Probability of Dangerous Damage per Hour	Safety Performance Level (PL)	Hazard Assessment Method R
N/A	$\geq 10^{-5}$ to $<10^{-4}$	a	S_1
SIL 1	$\geq 3 \times 10^{-6}$ to $<10^{-5}$	b	S_2
SIL 1	$\geq 10^{-6}$ to $<3 \times 10^{-6}$	c	S_3
SIL 2	$\mathbf{\geq 10^{-7}}$ **to** $\mathbf{<10^{-6}}$	d	S_4
SIL 3	$\geq 10^{-8}$ to $<10^{-7}$	e	S_5

Table 1 presents how to assesses the hazard and its reduction and determines the Safety Performance Level PL and the Safety Integrity Level SIL. In the analysis of a given safety function, PL and SIL2 levels were obtained, and the estimated hazard is assumed at S_2. The analysed electro-hydraulic control system performing the safety function is characterised by a safety level not worse than that required.

Research on the Development of the Electro-Hydraulic Control System

The control system together with the mechanical and hydraulic system is crucial and often determines the efficiency of the roof support. Therefore, it is well justified that scientific activity regarding the control system should be boosted. The search for a system that would allow for automatic operation without the need to position workers in the longwall is intensifying. The use of digital data—both for the roof support and the whole complex—shows that the demand for smart control has been continuously increasing.

Currently, it is not possible to eliminate the miners from the longwall. This is also not to be expected in the coming years. The main reason is that there are no measures that would precisely specify the mining and geological conditions of each longwall. However, this does not limit the development of control systems, for which the requirements are increasing and will grow. The objective is to focus on works outside the mining area as much as possible. Here, the control systems are fundamental. Research in this area mainly includes work on the development of an electro-hydraulic control system. A testing station was designed with a virtual controller built in. The station is used to define the number of control system parameters (Figure 6).

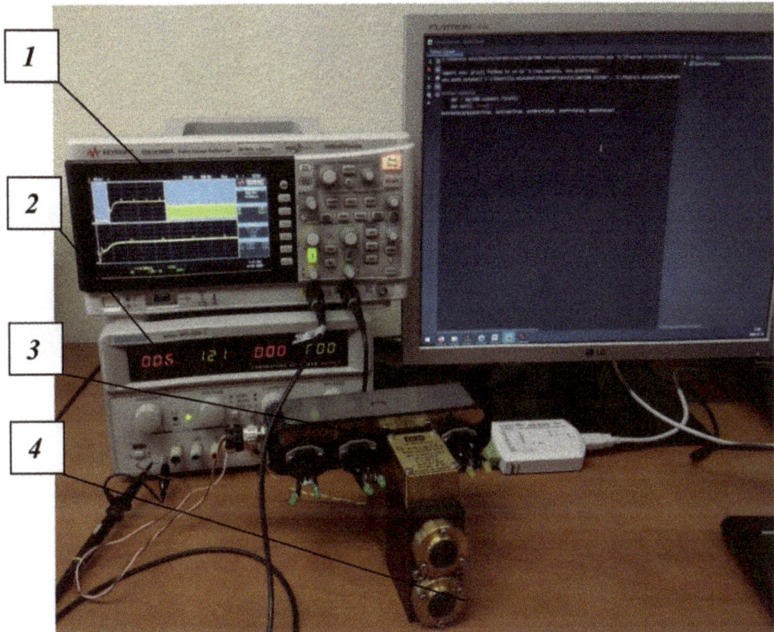

Figure 6. Test monitoring equipment, where 1—oscilloscope, 2—laboratory power supply, 3—controller module, and 4—solenoid valve.

The monitoring scope of the system is wide. It was used to determine the times for implementing the basic functions of the support such as the spreading and withdrawing of sections (extensions/slides of hydraulic legs mounted between floor bases and canopies). It measured the times of switching on the PWM signal and full signal supply for the basic functions of the powered roof support. The operating parameters of the control system obtained for individual phases of the section operation are shown in Figure 7.

The obtained results clearly indicate that the tested electro-hydraulic system has very low response times. This provides fast and reliable performance of the roof support. Such parameters make it possible to prepare both a comparative and sensitivity analysis of the control system and its components. Next, the results are used to adapt the control parameters to the mining and geological conditions of an individual longwall. It can, therefore, be assumed that virtual testing techniques applied for the systems that are already in use, new ones, or even prototypes can potentially facilitate their improvement. At the same time, the conditions to which these systems will be subject in real conditions are impossible to achieve in laboratory conditions.

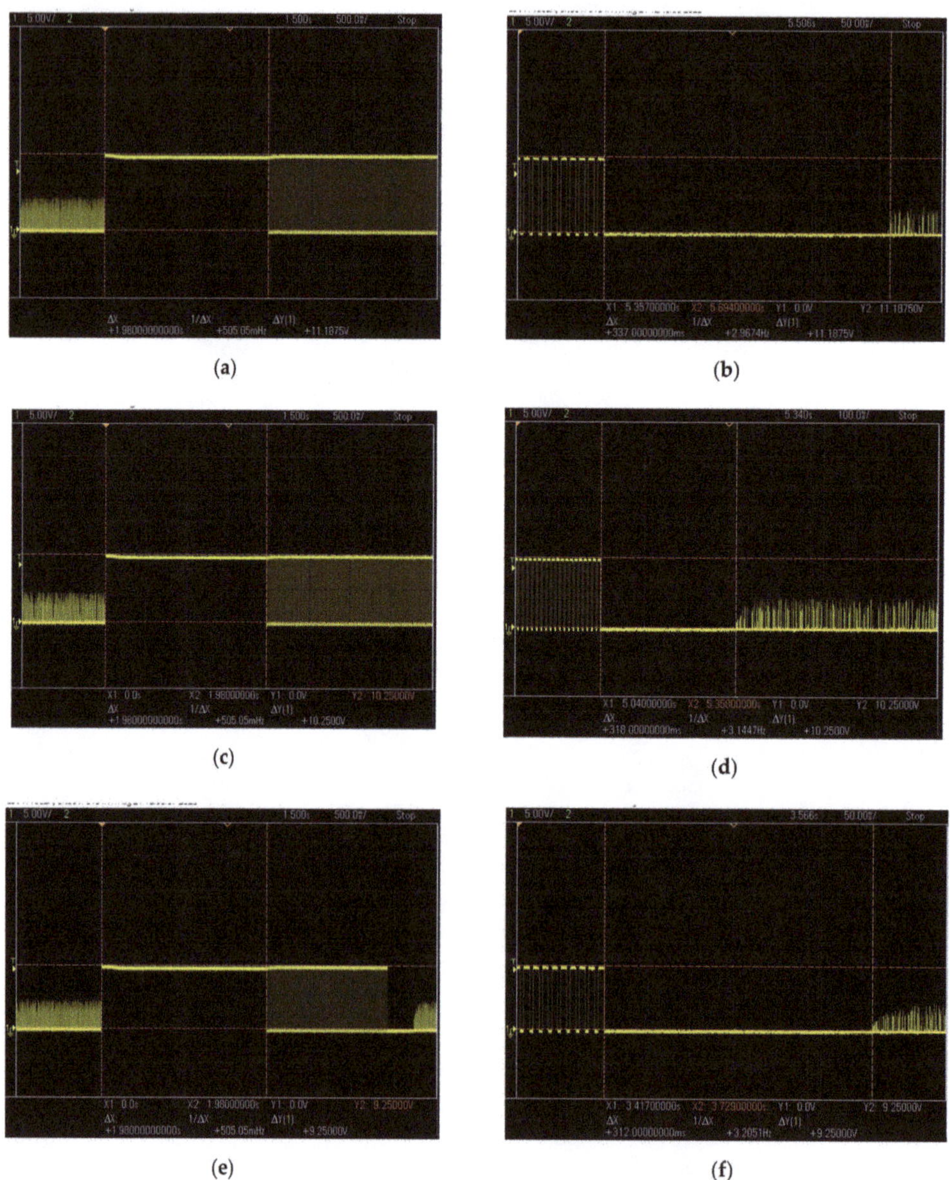

Figure 7. Switch-on time and delay of the Pulse Width Modulation (PWM) and full PWM signal for the function of sliding up and down the legs: (**a**) supply voltage of 12 V for switching on time ΔX = 1.98 s, (**b**) supply voltage of 12 V for delay time. (**c**) Supply voltage of 11 V for switching on time ΔX = 1.98 s, (**d**) supply voltage of 11 V for delay time, (**e**) supply voltage of 10 V for start-up time ΔX = 1.98 s, and (**f**) supply voltage of 10 V for delay time.

4. The Concept of Developing a Procedure as a Basis for the Introduction of an Automatic Control System

The development of a comprehensive design procedure, together with the procedure for conducting tests and the required safety assessment, is important in terms of adapting the system to the roof support structure and the conditions under which the system will operate. The procedure in Figure 8 includes four stages. The results should be used by manufacturers of roof supports and control systems, as it presents the order of design, research, and production. The third stage, from a scientific point of view, is the most important one, as the positive or negative results obtained make it possible to analyse whether the assumptions made in Stage 1 coincide with the design in Stage 2. As a result of research into prototypes and the acceptance of prototype amendments. The production documentation is then developed as outlined in Step 4. The authors of the study included all possible points for safety and efficient use.

Analysis of the electro-hydraulic control system together with the visualisation of the operation parameters of the powered roof support in the conditions of adverse effects of the rock mass shows that the construction of the powered roof support must be adapted properly. It is important to prepare the structure of the roof support so the control system and the sensors can be mounted properly to achieve the right operating parameters and ensure the necessary safety level. The highest level of functional safety for the components comprising the control and visualisation system requires the use of all testing methods. In particular, this concerns the safety of the miners.

Identification of potential hazards associated with the emergence of hazards resulting from the operating conditions of the control system is important. The mining and geological conditions in which longwall coal mining is carried out are one of the most dangerous. Step 2 includes a detailed safety assessment, which refers to the hazard assessment of the control system of the powered roof support designed for the operators. The hazard assessment will check whether the level achieved can be considered acceptable based on the possible occurrence of the hazard and the probability of its occurrence. The hazard reduction takes into account the hazard associated with the malfunctioning of the control system and must be designed in such a way that:

—it ensures the safety and prevents emergencies;
—defects in the computer hardware and software of the control system would not lead to dangerous situations;
—they are resistant to the loads resulting from their intended use and to the impact of dangerous situations;
—logic errors do not lead to dangerous situations;
—predictable human errors do not lead to dangerous situations.

Figure 8 presents a guide for engineers who design control systems. The manual is used following its established components, namely, when starting to work on a new system, the concept must be considered first, and then, design, research, and production must be considered. The guide (Figure 8) intends to organise and reduce engineers' working time. This procedure consists of four stages. In the first stage, the concept for the designed prototype of the system was taken into account together with the functional safety analysis. The second stage covered design aspects with the development of the control algorithm and software. The third stage includes tests. The research team designed bench and underground tests. The aim of this research is the functional ocean of the designed system. The last stage four is the introduction of the system into serial production. Here, manufacturing drawings, instructions for use, and a declaration of conformity are included.

The nature of the presented concept (Figure 8) is open, as it takes into account the possibility of introducing specific conditions in which the seam will be mined, and the selection system used. The procedure can be modified to fit the research and development of automatic control devices and future requirements.

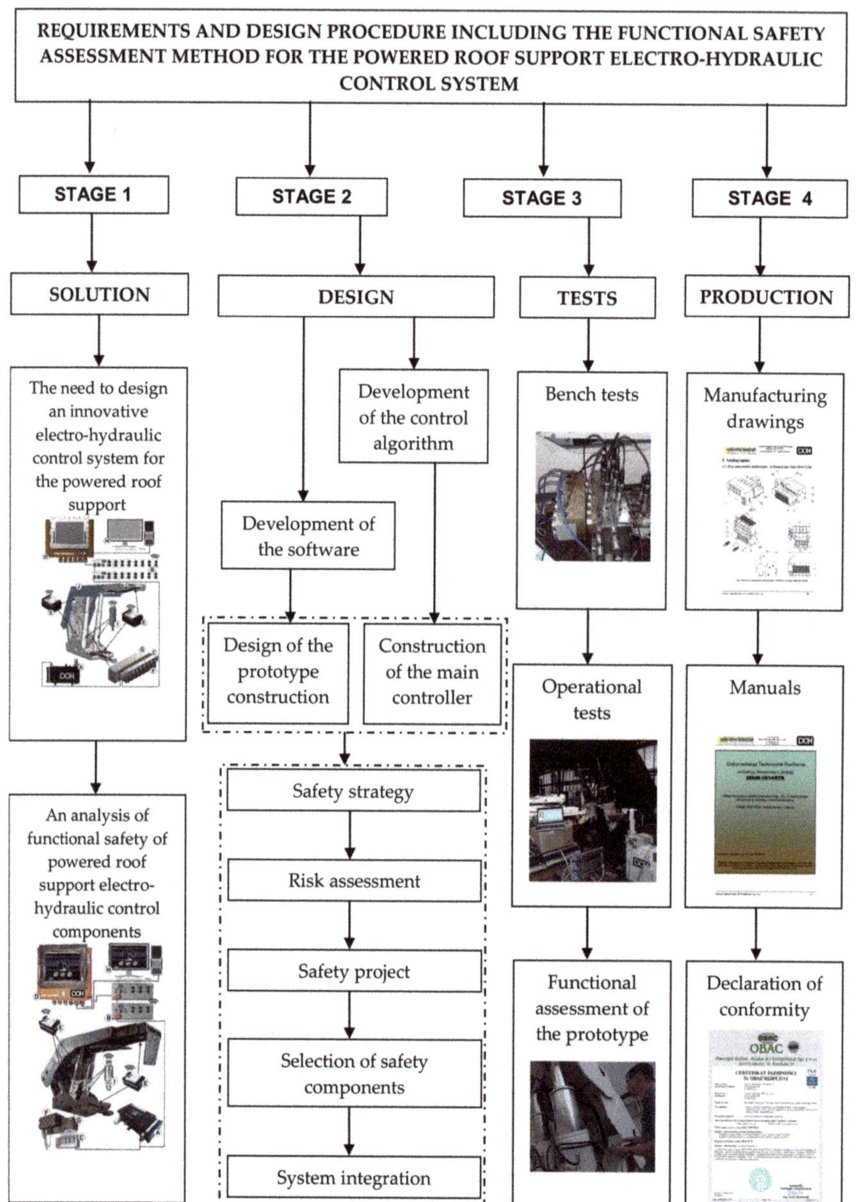

Figure 8. The procedure together with the procedure for testing and evaluation for the introduction of an electro-hydraulic control system and a visualisation system for underground mine operation.

The conditions in which the mining roof support operates are random. Consequently, the procedure must be used for every new mining longwall individually. The range of application should be adapted to calculated possible mining and geological conditions of a given area. Identified potential loads and requirements for the powered roof should be treated as a foundation of further research and calculations. Possible structural changes and other systems should be preceded by a thorough assessment of the condition of these conditions.

The procedure itself is open and flexible. This is crucial as this makes it possible to modify it depending on the conditions, requirements, and needs, as well as research capabilities. This applies to both manufacturers and users of the roof support.

The contractors who order a powered roof support should be aware that they must adapt the parameters and features for each longwall. They are responsible for the proper selection of the system that would fit the conditions in a given area. The safety of the miners is the most important. The support is the most expensive machine of the powered roof support complex, and its safe use impacts the economic efficiency of the entire mining operation process.

5. Conclusions

It is important to prepare the roof support in such a way that after it is installed in the longwall and launched, the automatic control systems take over the functions previously performed by the operator. Appropriate selection of the support section for mining conditions, equipping it with a set of sensors and a prepared intelligent control system, can together potentially fulfil expectations. The authors attempt to identify the main problems that need to be carefully considered, complemented by the knowledge gained from the research, and then to define the criteria. Based on the analysis of many tests carried out, the conditions (Figure 8) to be met for the powered roof support and its components have been clarified, which will allow preparing the structure of the support to meet the requirements necessary in terms of the automatic control system.

Based on the tests and analyses carried out for the resulting prototype of the electro-hydraulic control system, a detailed analysis of the safety function must be carried out before it is put into service as presented in Section 3 (Results). This analysis takes into account the required level of hazard and determines the level of safety integrity and the level of safety assurance with a probability of dangerous damage per hour. Based on the analysis carried out, this is a satisfactory level. This is of particular importance for the entire longwall complex.

Based on the design and research work on the development of an electro-hydraulic control system for the powered roof support to increase the efficiency of the entire complex, requirements and design procedures were developed together with the procedure for functional safety assessment.

The area of research and analysis in the third stage has provided new knowledge of equipment design. The developed system can have a huge impact on the efficiency of the work of the powered roof support and thus the entire longwall complex. In the future, this may result in a reduction in the number of workers during coal mining. So far, studies and analyses have not included such a comprehensive approach concerning increasing safety. The presented material expands the existing knowledge of control systems and visualisation of work parameters of the powered roof support and will be an important element in the process of effective use of machines and equipment in the longwall complex.

Author Contributions: Conceptualisation, D.S. and S.Z.; methodology, D.S. and S.S.; validation, D.S., S.Z., S.V., M.L., and M.C.; formal analysis, D.S., S.Z. and S.S.; investigation, D.S.; resources, D.S.; data curation, D.S. and S.Z.; writing of the original draft preparation, D.S.; writing of review and editing, D.S. and S.S.; visualisation, S.S.; supervision, D.S.; project administration, D.S. and S.Z.; funding acquisition, D.S. and S.Z. All authors have read and agreed to the published version of the manuscript.

Funding: The work was carried out within the project "Innovative electro-hydraulic control system for powered roof support" No. POIR.01.01.01-00-1129/15. The Operational Programme Smart Growth 2014–2020 carried out by the National Centre for Research and Development.

Institutional Review Board Statement: Not applicable.

Informed Consent Statement: Not applicable.

Data Availability Statement: Not applicable.

Conflicts of Interest: The authors declare no conflict of interest.

References

1. Fink, A.; Beikirch, H. MineLoc—Personnel Tracking System for Longwall Coal Mining Sites. *IFAC PapersOnLine* **2015**, *48*, 215–221. [CrossRef]
2. Si, L.; Wang, Z.; Liu, X.; Tan, C.; Liu, Z.; Xu, J. Identification of Shearer Cutting Patterns Using Vibration Signals Based on a Least Squares Support Vector Machine with an Improved Fruit Fly Optimization Algorithm. *Sensors* **2016**, *16*, 90. [CrossRef]
3. Si, L.; Wang, Z.B.; Tan, C.; Liu, X.H. A novel approach for coal seam terrain prediction through information fusion of improved D-S evidence theory and neutral network. *Measurement* **2014**, *54*, 140–151. [CrossRef]
4. Janik, M.; Kuska, J.; Świeczak, P.; Wojtas, M.; Fitowski, K. Zastosowanie nowoczesnych rozwiązań do zasilania sekcji obudowy zmechanizowanych w kompleksie ścianowym w kopalni„Ziemowit", ze szczególnym uwzględnieniem sterowań elektrohydraulicznych i wizualizacji parametrów pracy urządzeń. *Nap. Sterow.* **2011**, *13*, 104–108.
5. Kasprusz, A.; Mikuła, S.; Wojtas, M. Sterowanie elektrohydrauliczne DOH-matic do automatyzacji pracy obudowy zmechanizowanej. *Wiadomości Górnicze* **2013**, *5*, 275–282.
6. Peng, S.S.; Du, F.; Cheng, J.; Li, Y. Automation in U.S. longwall coal mining: A state-of-the art review. *Int. J. Min. Sci. Technol.* **2019**, *29*, 151–159.
7. Sobik, L.; Brodny, J.; Buyalich, G.; Strelnikov, P. Analysis of methane hazard in longwall working equipped with a powered longwall complex. *E3S Web Conf.* **2020**, *174*, 01011. [CrossRef]
8. Tutak, M.; Brodny, J. Predicting Methane Concentration in Longwall Regions Using Artificial Neural Networks. *Int. J. Environ. Res. Public Health* **2019**, *16*, 1406. [CrossRef]
9. Tutak, M. The Influence of the Permeability of the Fractures Zone Around the Heading on the Concentration and Distribution of Methane. *Sustainability* **2020**, *12*, 16. [CrossRef]
10. Tutak, M.; Brodny, J. The Impact of the Strength of Roof Rocks on the Extent of the Zone with a High Risk of Spontaneous Coal Combustion for Fully Powered Longwalls Ventilated with the Y-Type System—A Case Study. *Appl. Sci.* **2019**, *9*, 5315.
11. Tutak, M.; Brodny, J. Forecasting Methane Emissions from Hard Coal Mines Including the Methane Drainage Process. *Energies* **2019**, *12*, 3840. [CrossRef]
12. Tutak, M.; Brodny, J. Analysis of the Impact of Auxiliary Ventilation Equipment on the Distribution and Concentration of Methane in the Tailgate. *Energies* **2018**, *11*, 3076.
13. Brodny, J.; Tutak, M. Analyzing Similarities between the European Union Countries in Terms of the Structure and Volume of Energy Production from Renewable Energy Sources. *Energies* **2020**, *13*, 913. [CrossRef]
14. Brodny, J.; Tutak, M. Analysing the Utilisation Effectiveness of Mining Machines Using Independent Data Acquisition Systems: A Case Study. *Energies* **2019**, *12*, 2505. [CrossRef]
15. Brodny, J.; Tutak, M. Exposure to Harmful Dusts on Fully Powered Longwall Coal Mines in Poland. *Int. J. Environ. Res. Public Health* **2018**, *15*, 1846. [CrossRef]
16. Brodny, J. Tests of friction joints in mining yielding supports under dynamic load. *Arch. Min. Sci.* **2011**, *56*, 303–318.
17. BP Statistical Review of Word Energy 2019/68th edition. Available online: https://www.bp.com (accessed on 10 May 2020).
18. International Energy Agency: Coal Information 2019: Overview. Available online: https://www.iea.org (accessed on 22 May 2020).
19. *Raport: Górnictwo Węgla Kamiennego w Polsce 2018*; Instytut Gospodarki Surowcami Mineralnymi i Energią PAN: Kraków, Poland, 2019; Available online: https://www.min-pan.krakow.pl (accessed on 23 May 2020).
20. Białoń, A.; Pawlik, M. Problemy bezpieczeństwa i ryzyka na przykładzie urządzeń sterowania ruchem kolejowym. *Probl. Kolejnictwa* **2014**, *163*, 25–41.
21. Strawiński, T. Projektowanie funkcji bezpieczeństwa z wykorzystaniem podsystemu transmisji danych bezpieczeństwa. *Nap. Sterow.* **2014**, *10*, 118–122.
22. Yuan, Y.; Tu, S.; Zhang, X.; Li, B. Dynamic Effect and Control of Key Strata Break of Immediate Roof in Full Mechanized Mining with Large Mining Height. *Shock Vib.* **2015**, *2015*, 651818.
23. Lu, S.; Liu, S.; Wan, Z.; Cheng, J.; Yang, Z.; Shi, P. Dynamic Damage Mechanism of Coal Wall in Deep Longwall Face. *Adv. Civ. Eng.* **2019**, *2019*, 3105017.
24. Fan, G.; Zhang, D.; Wang, X. Mechanism of Roof Shock in Longwall Coal Mining under Surface Gully. *Shock Vib.* **2015**, *2015*, 803071. [CrossRef]
25. Bai, Q.; Tu, S. A General Review on Longwall Mining-Induced Fractures in Near-Face Regions. *Geofluids* **2019**, *2019*, 3089292. [CrossRef]
26. Zhai, J.; Liu, D.; Li, G.; Wang, F. Floor Failure Evolution Mechanism for a Fully Mechanized Longwall Mining Face above a Confined Aquifer. *Adv. Civ. Eng.* **2019**, *2019*, 8036928. [CrossRef]
27. He, J.; Dou, L.M.; Cai, W.; Lei Li, Z.; Ding, Y.L. In Situ Test Study of Characteristics of Coal Mining Dynamic Load. *Shock Vib.* **2015**, *2015*, 121053. [CrossRef]
28. Xiao, Z.; Liu, J.; Gu, S.; Liu, M.; Zhao, F.; Wang, Y.; Ou, C.; Zhen, M. A Control Method of Rock Burst for Dynamic Roadway Floor in Deep Mining Mine. *Shock Vib.* **2019**, *2019*, 7938491. [CrossRef]

29. Szrek, J.; Wodecki, J.; Błażej, R.; Zimroz, R. An Inspection Robot for Belt Conveyor Maintenance in Underground Mine—Infrared Thermography for Overheated Idlers Detection. *Appl. Sci.* **2020**, *10*, 4984.

30. Kozłowski, T.; Wodecki, J.; Zimroz, R.; Błażej, R.; Hardygóra, M. A Diagnostics of Conveyor Belt Splices. *Appl. Sci.* **2020**, *10*, 6259. [CrossRef]

31. Schmidt, S.; Zimroz, R.; Chaari, F.; Heyns, P.S.; Haddar, M. A Simple Condition Monitoring Method for Gearboxes Operating in Impulsive Environments. *Sensors* **2020**, *20*, 2115.

32. Hebda-Sobkowicz, J.; Zimroz, R.; Wyłomańska, A. Selection of the Informative Frequency Band in a Bearing Fault Diagnosis in the Presence of Non-Gaussian Noise—Comparison of Recently Developed Methods. *Appl. Sci.* **2020**, *10*, 2657. [CrossRef]

33. Tian, Q.; Guo, W. Reconfiguration of manufacturing supply chains considering outsourcing decisions and supply chain risks. *J. Manuf. Syst.* **2019**, *52*, 217–226.

34. Tuptuk, N.; Hailes, S. Security of smart manufacturing systems. *J. Manuf. Syst.* **2018**, *47*, 93–106. [CrossRef]

35. Panfilova, O.P.; Velikanov, V.S.; Usov, I.G.; Matsko, E.Y.; Kutlubaev, I.M. Calculation of Life on Functional Parts in the Structure of Mining Machines. *J. Min. Sci.* **2018**, *54*, 218–225. [CrossRef]

36. Golushko, S.K.; Cheido, G.P.; Shakirov, R.A.; Shakirov, S.R. Multi-Functional Mine Shaft Alarm System. *J. Manuf. Syst.* **2018**, *54*, 173–179. [CrossRef]

37. Lukichev, S.V.; Nagovitsyn, O.V. Modeling Objects and Processes within a Mining Technology as a Framework for a System Approach to Solve Mining Problems. *J. Manuf. Syst.* **2018**, *54*, 1041–1049. [CrossRef]

38. Kushal, K.S.; Nanda, M.; Jayanthi, J. Architecture Level Safety Analyses for Safety-Critical Systems. *Int. J. Aerosp. Eng.* **2017**, *2017*, 6143727. [CrossRef]

39. Wang, C.; Tu, S. Selection of an Appropriate Mechanized Mining Technical Process for Thin Coal Seam Mining. *Math. Probl. Eng.* **2015**, *2015*, 893232. [CrossRef]

40. Raport: Ocena Stany Bezpieczeństwa Pracy, Ratownictwa Górniczego Oraz Bezpieczeństwa Powszechnego w Związku z Działalnością Górniczo-geologiczną w 2019 Roku. Wyższy Urząd Górnicz. Katowice. 2019. Available online: https://www.wug.gov.pl (accessed on 20 May 2020).

41. *PN-EN 62061:2008P+A1:2013–6E Safety of Machinery—Functional Safety of Safety-related Electrical, Electronic and Programmable Electronic Control Systems*; Polski Komitet Normalizacyjny: Warszawa, Poland, 2005.

42. *PN-EN 61784–1:2011E Industrial Communication Networks—Profiles—Part 1: Field Buses Profiles*; Polski Komitet Normalizacyjny: Warszawa, Poland, 2011.

43. *PN-EN 61508 Functional Safety of Electrical/Electronic/Programmable Electronic Safety Related Systems*; NSAI Standard: Dublin, Ireland, 2010.

44. *PN-EN 61784-1:2011E Industrial Communication Networks—Profiles—Part 3: Functional Safety Fieldbuses—General Rules and Profile Definitions*; Polski Komitet Normalizacyjny: Warszawa, Poland, 2011.

45. *PN-EN ISO 13849-1:2008 Safety of Machinery. Safety Related Parts of Control Systems. Part 1: General Principles for Design*; Polski Komitet Normalizacyjny: Warszawa, Poland, 2008.

46. *PN-EN 62061:2008/A2:2016 Safety of Machinery. Functional Safety of Safety-related Electrical, Electronic and Programmable Electronic Control Systems*; Polski Komitet Normalizacyjny: Warszawa, Poland, 2016.

47. Directive 2006/42/EC: Machinery (Amending Directive 95/16/EC). Available online: https://eur-lex.europa.eu/legal-content/PL/LSU/?uri=CELEX:32006L0042 (accessed on 20 May 2020).

48. Szurgacz, D.; Brodny, J. Adapting the Powered Roof Support to Diverse Mining and Geological Conditions. *Energies* **2020**, *13*, 405.

49. Buyalich, G.; Byakov, M.; Buyalich, K.; Shtenin, E. Development of Powered Support Hydraulic Legs with Improved Performance. *E3S Web Conf.* **2019**, *105*, 03025. [CrossRef]

50. Holm, M.; Beitler, S.; Arndt, T.; Mozar, A.; Junker, M.; Bohn, C. Concept of Shield-Data-Based Horizon Control for Longwall Coal Mining Automation. In Proceedings of the 16th IFAC Symposium on Automation in Mining Mineral Processing, San Diego, CA, USA, 25–28 August 2013; pp. 98–103.

51. Ghose, A.K. Technology Vision 2050 for Sustainable Mining. In *Proceedings of the 6th International Conference on Mining Science and Technology*; Procedia Earth and Planetary Science: Amsterdam, The Netherlands, 2009; Volume 1, pp. 2–6.

52. Yuesen, Y.; Peng, D.; Yajun, X.; Liming, P.; Xiaojie, W. Research on Electro-hydraulic Control System for Hydraulic Support at Coal Mine. In *Proceedings of the 6th International Conference on Mining Science and Technology*; Procedia Earth and Planetary Science: Amsterdam, The Netherlands, 2009; Volume 1, pp. 1549–1553.

53. Xiang-en, C.; Lei, W. Research of Automation Integrated Monitoring System for the Fully Mechanized Coal Face. In *Proceedings of the Second International Conference on Mining Engineering and Metallurgical Technology*; Procedia Earth and Planetary Science: Amsterdam, The Netherlands, 2011; Volume 1, pp. 171–176.

54. Szurgacz, D. Electrohydraulic control systems for powered roof supports in hazardous conditions of mining tremors. *J. Sustain. Min.* **2015**, *14*, 157–163. [CrossRef]

55. Stoiński, K.; Mika, M. Dynamics of Hydraulic Leg of Powered Longwall Support. *J. Min. Sci.* **2003**, *39*, 72–77. [CrossRef]

56. Gil, J.; Kołodziej, M.; Szurgacz, D.; Stoiński, K. Introduction of standardization of powered roof supports to increase production efficiency of Polska Grupa Górnicza, S.A. *Min. Inf. Autom. Electr. Eng.* **2019**, *56*, 33–38. [CrossRef]

57. Bessinger, S.L.; Neison, M.G. Remnant roof coal thickness measurement with passive gamma ray instruments in coal mine. *IEEE Trans. Ind. Appl.* **1993**, *29*, 562–565. [CrossRef]

58. Chufo, R.L.; Johnson, W.J. A radar coal thickness sensor. *IEEE Trans. Ind. Appl.* **1993**, *29*, 834–840. [CrossRef]
59. Markham, J.R.; Solomon, P.R.; Best, P.E. An FT-IR based instrument for measuring spectral emittance of material at high temperature. *Rev. Sci. Instrum.* **1990**, *61*, 3700–3708. [CrossRef]
60. Aydin, I.; Karakose, M.; Akin, E. A multi-objective artificial immune algorithm for parameter optimization in support vector machine. *Appl. Soft Comput.* **2011**, *11*, 120–129. [CrossRef]
61. Ma, D.Y.; Liang, D.Y.; Zhao, X.S.; Guan, R.C.; Shi, X.H. Multi-BP expert system for fault diagnosis of power system. *Eng. Appl. Artif. Intell.* **2013**, *26*, 937–944. [CrossRef]
62. Seera, M.; Lim, C.P. Online motor fault detection and diagnosis using a hybrid FMM-CART model. *IEEE Trans. Neural Netw. Learn. Syst.* **2014**, *25*, 806–812. [CrossRef]
63. Bangalore, P.; Tjernberg, L.B. An artificial neural network approach for early fault detection of gearbox bearings. *IEEE Trans. Smart Grid* **2015**, *6*, 980–987.
64. Wang, T.; Chen, J.; Zhou, Y.; Snoussi, H. Online least squares one-class support vector machines-based abnormal visual event detection. *Sensors* **2013**, *13*, 17130–17155. [CrossRef]
65. Heng, A.; Zhang, S.; Tan, A.C.C.; Mathew, J. Rotating machinery prognostics: State of the art, challenges and opportunities. *Mech. Syst. Signal Process.* **2009**, *23*, 724–739. [CrossRef]
66. Winston, D.P.; Saravanan, M. Single parameter fault identification technique for DC motor through wavelet analysis and fuzzy logic. *J. Electr. Eng. Technol.* **2013**, *8*, 1049–1055. [CrossRef]
67. Li, B.; Li, D.Y.; Zhang, Z.J.; Yang, S.M.; Wang, F. Slope stability analysis based on quantum-behaved particle swarm optimization and least squares support vector machine. *Appl. Math. Model.* **2015**, *39*, 5253–5264. [CrossRef]
68. El-Baz, A.H. Hybrid intelligent system-based rough set and ensemble classifier for breast cancer diagnosis. *Neural. Comput. Appl.* **2015**, *26*, 437–446. [CrossRef]
69. Cervantes, J.; Li, X.O.; Yu, W. Imbalanced data classification via support vector machines and genetic algorithms. *Connect. Sci.* **2014**, *26*, 335–348. [CrossRef]
70. Li, K.; Chen, P.; Wang, S.M. An intelligent diagnosis method for rotating machinery using least squares mapping and a fuzzy neural network. *Sensors* **2012**, *12*, 5919–5939. [CrossRef] [PubMed]
71. Ayrulu-Erdem, B.; Barshan, B. Leg motion classification with artificial neural networks using wavelet-based features of gyroscope signals. *Sensors* **2011**, *11*, 1721–1723. [CrossRef] [PubMed]
72. Zhang, X.L.; Chen, W.; Wang, B.J.; Chen, X.F. Intelligent fault diagnosis of rotating machinery using support vector machine with ant colony algorithm for synchronous feature selection and parameter optimization. *Neurocomputing* **2015**, *167*, 260–279. [CrossRef]
73. Harish, N.; Mandal, S.; Rao, S.; Patil, S.G. Particle Swarm Optimization based support vector machine for damage level prediction of non-reshaped berm breakwater. *Appl. Soft Comput.* **2015**, *27*, 313–321. [CrossRef]
74. Silva, D.A.; Silva, J.P.; Neto, A.R.R. Novel approaches using evolutionary computation for sparse least square support vector machines. *Neurocomputing* **2015**, *168*, 908–916. [CrossRef]
75. Mani, G.; Jerome, J. Intuitionistic fuzzy expert system based fault diagnosis using dissolved gas analysis for power transformer. *J. Electr. Eng. Technol.* **2014**, *9*, 2058–2064. [CrossRef]
76. Ji, J.; Zhang, C.S.; Kodikara, J.; Yang, S. Prediction of stress concentration factor of corrosion pits on buried pipes by least squares support vector machine. *Eng. Fail. Anal.* **2015**, *55*, 131–138. [CrossRef]
77. Elbisy, M.S. Sea wave parameters prediction by support vector machine using a genetic algorithm. *J. Coast. Res.* **2015**, *31*, 892–899. [CrossRef]

Review

Increasing Energy Efficiency and Productivity of the Communition Process in Tumbling Mills by Indirect Measurements of Internal Dynamics—An Overview

Mateusz Góralczyk [1,*], **Pavlo Krot** [1,*], **Radosław Zimroz** [1] **and Szymon Ogonowski** [2]

1   Faculty of Geoengineering, Mining and Geology, Wroclaw University of Science and Technology, Na Grobli 15, 50-421 Wrocław, Poland; radoslaw.zimroz@pwr.edu.pl
2   Faculty of Automatic Control, Electronics and Computer Science, Silesian University of Technology, Akademicka 16, 44-100 Gliwice, Poland; szymon.ogonowski@polsl.pl
*   Correspondence: mateusz.goralczyk@pwr.edu.pl (M.G.); pavlo.krot@pwr.edu.pl (P.K.)

Received: 15 November 2020; Accepted: 10 December 2020; Published: 21 December 2020

Abstract: Tumbling mills have been widely implemented in many industrial sectors for the grinding of bulk materials. They have been used for decades in the production of fines and in the final stages of ore comminution, where optimal levels for the enrichment particles' sizes are obtained. Even though these ubiquitous machines of relatively simple construction have been subjected to extensive studies, the industry still struggles with very low energy efficiency of the comminution process. Moreover, obtaining an optimal size for the grinding product particles is crucial for the effectiveness of the following processes and waste production reduction. New, innovative processing methods and machines are being developed to tackle the problem; however, tumbling mills are still most commonly used in all ranges of the industry. Since heavy equipment retrofitting is the most costly approach, process optimization with dedicated models and control systems is the most preferable solution for energy consumption reduction. While the classic technological measurements in mineral processing are well adopted by the industry, nowadays research focuses on new methods of the mill's internal dynamics analysis and control. This paper presents a retrospective overview of the existing models of internal load motion, an overview of the innovations in process control, and some recent research and industrial approaches from the energy consumption reduction point of view.

Keywords: ball mills; energy efficiency; internal dynamics; DEM model; resonance oscillations; measurements

---

1. Introduction

Up to 70% of the costs of materials size-reduction operations fall on the rock particle's size reduction from 30–50 mm to 20–50 microns [1,2]. This operation is performed by the most common type of milling comminution equipment—ball milling. The design of ball mills was described back in the 18th century—and the principle of their operation has not changed; the only significant changes are the increased diameter and some new solutions for the mill's drives introduced. Energy consumption in ball and rod mills reaches up to 10–20 kWh per ton of rock. According to the study prepared by the U.S. Department of Energy [3] the biggest potential for energy savings in all of the energy-intensive ore and coal mining-related operations is to be found in grinding optimization (see Figure 1). To increase efficiency, designers of mining equipment tried to improve all the elements of ball mills, such as the geared mechanical drives, bearings, their lubrication systems, and electric motors. Moreover, efforts have been made to optimize the size of the grinding bodies, control the filling level of the drum with the material, and stabilize the rotation speed. However, the main problem of dramatically low energy efficiency is still unresolved—about 30% of grinding bodies are not involved in the dynamical process—they remain in a boundary dead zone. A large portion of the impacts of grinding bodies appears on the internal liners. There are several approaches to the activated material and grinding bodies' motion inside the mill.

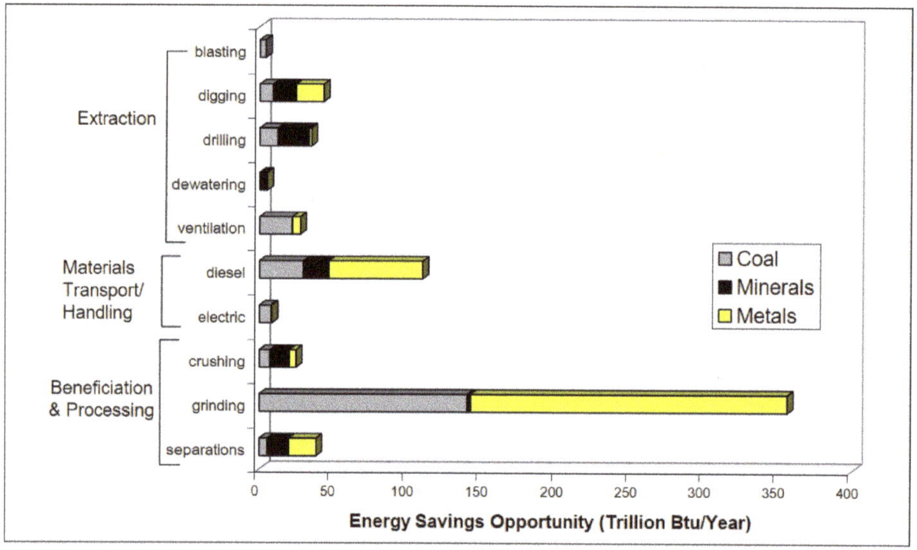

Figure 1. Energy saving potential for various energy-intensive processes in mining [3]. 1 Btu (British thermal unit) ≈ 1055 Joule.

First, new designs of energy-saving mills have been introduced in the market. For example, vertical (tower) Vertimill (Metso) and horizontal Isamill (Glencore Technology), which use, respectively, gravitation forces or several subsequent rotating disks inside to intensify the milling process. However, ball mills of the traditional design process the main portion of bulk materials in the world.

The next solution is the installation of so-called lifters, situated along the mill shell over the perimeter to promote higher trajectories of the balls falling down and to better mix the media treated. However, these additional elements quickly deteriorate due to the intensive wear and shock impacts on them. Hence, they may have an effect only within a short period of time after replacement.

A promising approach is to utilize the dynamic phenomena inside the mill. Namely, when the central part of bulk media moves in oscillatory (synchronous or resonance) mode against the mill shell, and the rest is passive part of the load. This approach is more attractive from the viewpoint of implementation because it does not require any additional modernization of mechanical equipment. Only the instrumentation for signals' measurement and processing, in combination with process parameters' control—based on existing automation systems—is needed in such cases. However, such control needs in-depth knowledge of intra-mill load dynamics and its stages estimation by the different channels—electric motors current or power, vibration, and acoustic emission inside the equipment, or spatial sounds and mechanical torques of multi-motor drives. The realization of such an approach can only be based on advanced instrumentation—including wireless sensors, and dynamical models (both analytical and discrete elements) and signal processing techniques. These methods should account for the gradual changing of treated material properties (input variations and reduced-by-time-fractions) simultaneously with the wear of grinding balls and internal protective liners of the mill itself. Additionally, the influences of such intentionally produced oscillations on the torsional dynamics of mill drives need estimation.

This paper intends to represent a comprehensive retrospective overview of existing methods and recent trends in ball mills modeling and control aimed at increasing their energy efficiency and productivity by indirect measurements of internal dynamics with signals of different physical nature.

2. State-of-the-Art in Mill Control

Being the main equipment used for fine-grinding in the raw materials industry and an element of the most energy-consuming stage in mineral processing, responsible for roughly half of the mining companies' energy consumption [4,5], tumbling mills were subjected to legitimate and intensive studies in the search for improving the efficiency of ore processing. Depending on the industry's nature, dedicated open and close grinding circuits are designed to perform comminution in a dry or wet environment [6,7]. Depending on the stage of the process and the requirement of the product grain size distribution (coarse, fine, or ultra-fine grinding) the specific energy rises exponentially with the final grains' size reduction [8]. In many applications, comminution is performed in order to achieve the desired degree of minerals liberation, that is, the percentage of the valuable mineral in the free form in relation to the valuable injected in the gangue (locked form) [9]. Thus, improved control over the comminution process has significant influences on the concentration processes, such as flotation [10,11] and bio-leaching [12]. Obtaining high efficiency from the energy and material consumption points of view, has a significant potential to contribute to decreasing the environmental impact of this very burdensome process. Due to the difficulty of balancing the large ore particle supply with the optimal feed rate, the autogenous (AG) and semi-autogenous (SAG) mills operate in a meta-stable state, which creates a demand for accurate and real-time assessments of a mill's load and its behavior [13].

Circuit retrofitting is the most difficult and costly approach to the grinding process optimization; thus, after initial circuit design, further optimization is achieved through the process control. In most cases it requires, however, process parameters' measurement for the feedback approach or accurate process models for the feed-forward approach [14]. Taking into account the complexity and the dynamics of the close grinding circuits (most popular in the mineral processing industry) the most suitable solution would be to measure or model the mill's output itself. The necessity of robustness and complexity of multivariate nonlinear predictive control of SAG ball mills is underlined in [15].

Direct observation of internal load motion is practically impossible; however, the knowledge about its trajectory plays the main role in the optimization of the disintegration of bulk material into smaller particles. Therefore, many scientists and engineers devoted their efforts to developing and improving indirect methods of studying the differences in operating conditions of ball mills. Some of the flagship examples of these types of measurements are the passive inertial measurements on the surface of a SAG mill, which have been used by Campbell et al. [16] to study the operating conditions of the machine, namely, to specify variations of the volumetric filling and motion characteristics. The vibrations of the machine's shell taking place as a result of the collisions of the grinding media' and ground ore with each other and with liners have been proven to provide informative data for condition surveillance when subjected to appropriate signal processing techniques. The usage of polar contour plots and spectrogram analysis has been presented as appropriate for deriving signals' characteristics that respond well to changes in operating conditions, such as frequency band power. Such indirect methods can be based on various signals—on vibration signature: [17–27]; acoustic data: [28–31]; using both—acoustic and vibration data together: [24,32–36]. The weight of the mill's charge has been also successfully evaluated by measuring strain changes in the mill's shell, together with obtaining some information about the dynamic behavior of grinding media [37]. Another possibility for monitoring the mill filling level is the measurement of the motor: power draw [38], torque [39] and other signals from the motor control units [18,24,40]. Since the engine's electrical signals' characteristics are dependent on the process of elevation of the particles inside the drum—their collisions and impacts [41]—they allow one to observe the intra-mill material dynamic behavior as well [42]. Another important set of issues to be taken into consideration includes the accelerated wear of lifters and liners, and the destruction of discharge grates emerging when the feed rate deviates from a desirable value. The identification of the regions where direct shell impact takes place, leading to the damage of liners and accelerated wear, by means of vibration data analysis, has been demonstrated in [43].

Physical variables describing the performance of ball mill have been successfully measured with an instrumented ball, equipped with a data storage module, a power supply, communications electronics, and inertial sensors connected to a small micro-controller [44–46].

A wide range of scientific research in that area led to various industrial applications of the automatic stabilization of the mill's motor power and the design of technical solutions for the control of its maximum level. Such solutions are becoming the standard in the raw materials industry worldwide, representing trends of Industry 4.0. Some of the most popular industrial solutions are StarCS from Mintek [47,48], MilSense from Outotek [49], and LoadIQ from FLSmidth [50]. Such systems usually measure a set of technological parameters, e.g., feed mass or volume flow rate, density, particle size distribution (PSD), and power drawn by the mill's engines. Most recently, on top of classic direct measurements, industrial solutions use indirect measurements based on vibration, strain, and acoustic or vision signals. Such systems allow for a 1–2% increase of efficiency, without reducing specific energy consumption. Taking into account that the largest mills have electrical power of about 20–30 MW, even a tenth of a percent reduction in energy consumption gives tremendous annual savings for plants usually having several mills.

Circuit control systems are being implemented widely in order to control the feed rate and run the mill at the optimum efficiency level, which means, depending on the specific objective function: maximizing output, minimizing energy consumption, or providing an accurate particle size distribution for the further stages of mineral processing. Examples of such systems are: Grinding Circuit Control (GCC) solutions, used in Canadian processing plants [51] (e.g., Strathcona Mill, Raglan Mill, Eland Mill), MillVis system [52] developed by AMEplus and KGHM Polska Miedź S.A. for all of the Polish milling sections, where rod-primary-mills and ball-mills for regrinding are used in Divisions of Concentrators (O/ZWR Lubin, O/ZWR Polkowice, O/ZWR Rudna), or an intelligent optimal-setting control (IOSC) applied in the Chinese iron ore concentration plants' grinding circuits [53]. While the systems operating in Canadian mines focus on accurate observation of the ore stream in multiple points of the circuit and reacting to the variations—in order to provide the stability of the milling process, the Polish process control and optimization system includes inertial and acoustic measurements aimed at the mill itself. More precisely speaking, in the case of MillVis, monitoring of the mill's performance and its technical state is based on inertial data, acquired by accelerometers distributed on the machine's shell—to diagnose liners and lifters; acoustic signal recorded in the direct neighborhood of the machine—for indirect assessment of grinding media and rotational speed; and additionally—video recording of the feed—to control feeds' granulation variability and lithological compound [54]. In the case of the Chinese processing plant, the developed approach includes a loop controller using case-based reasoning and a soft sensor for particle size distribution control based on a neural network, together with the fuzzy inference adjusting method [55]. The goal function of the IOSC is to provide an optimal rate of production, maintaining appropriate particle size distribution for further enrichment processes.

Increased control over the comminution process may be beneficial since it allows one to take advantage of the phenomenon occurring in the grinding chamber, which can improve the efficiency of the particles' size reduction. It was proven in the laboratory and industrial-scale studies of the internal mechanics of tumbling mills, that there is a resonant oscillation mode of the central part of the mill's charge, occurring at a certain value of the feed and rotational speed [56].

Such an oscillation of the material in the low-frequency range (1–3 Hz), if maintained, may lead to an increase of the mill's efficiency by 6–8% and a decrease of its energy consumption by 8–10% as it was discovered by industrial investigations in [57]. Some studies conducted on laboratory mills with different types of building materials give available energy savings up to 50% [58].

It was found in the experimental research that the resonant mode of oscillation can be preserved by changing the load factor in all the types of ball mills: ball mills using grinding media, semi-autogenous, and autogenous mills. Moreover, the granulometric characteristics of intra-chamber fill can affect the self-oscillatory effect, and thus the power intensity of the milling process [59]. Since many mills nowadays are still equipped with synchronous AC motors with not modifiable drive speeds, control of the operating mode (maintaining the resonance of intra-chamber material) in their case is possible only through changing the mill's filling level, or the slurry density—by supplying different amounts of process water to the mill's chamber.

Considering a complex system 'ore mill—magnetic separator', it is found in [60] that mill filling level with ore can be determined by the sign of the first derivative signal of the active power of the electric motor of the magnetic separator by the active power signal of the mill motor. For wet, autogenous grinding mills positive sign means under-loading, while for ball mills this parameter has the opposite meaning.

For many years most motors driving SAG, AG and ball mills were of fixed speed. As more accurate and faster controllers emerged, together with a decrease of costs and dimensions of the hardware, control of drive's operation became possible on an industrial scale. In order to optimize the material flow rate, decrease power draw, maintain maximized impact zone of cascading material, control the breakage rate function (in case of SAG mills), and to increase the availability of the comminution machines, more and more often mining companies decide to implement tumbling mills driven by the engines with modifiable speeds. There are two main solutions to make operating at variable rotation speed possible for tumbling mills [61]: cycloconverters—for the ones of high power and low speed, and multilevel voltage source inverters—appropriate for the mills demanding less power, operating at higher speeds.

The original method of ball mill control was proposed by the authors of the patent [62]. Their SmartMill uses a magnetic field created by electric magnets installed on the mill shell to keep the grinding media coupled with balls and prevent them from slipping in the "dead zone". In addition, the magnetic field helps to direct the grinding media to an optimal trajectory, resulting in increased drop height and impact energy. Researchers also carried out mathematical modeling of processes in a mill with electromagnets, which made it possible to determine the range of the optimal number of electromagnets located in each section of the mill and reduce the time of their activation (less energy consumption). The SmartMill technology will be the most energy-efficient for grinding magnetic ores due to the direct effect of the magnetic field on the material. This technology—as declared by authors—can reduce energy consumption up to 50% that has also been achieved in the new types of mills like Vertimill (Metso) [63] and Isamill (Glencore Technology) [64].

3. Methods of Measurements and Optimization of Tumbling Mills

The measurements aimed at the evaluation of operating conditions and optimization of comminution, mentioned in previous section, are listed in Tables 1 and 2 below. They have been divided according to the signal measured and the main criteria of optimization. The possible sources of informative data to be acquired on a tumbling mill or in its direct neighborhood, which were used by scientists and constitute an input for the process control systems used in the raw materials industry are acoustic emission in the surrounding of the mill, vibrations measured on the shell (possibly other parts of the machine), digital records of the output ore stream, current and other signals to be acquired from the motor control unit. A scheme showing the methods of measurement and the variables to be adjusted on the basis of multi-channel data acquisition is presented in Figure 2. Mill's performance, monitored based on vision, acoustic, inertial data and motor signals can be optimized by the input of grinding media, changing of the slurry density (by adding process water), adjusting of the speed (expressed as a % of the critical speed) or increasing/decreasing of the mill's load. All the mentioned adjustments influence the throughput and PSD of the product. Some of the crucial process parameters with their corresponding informative signals are presented in the Table 3.

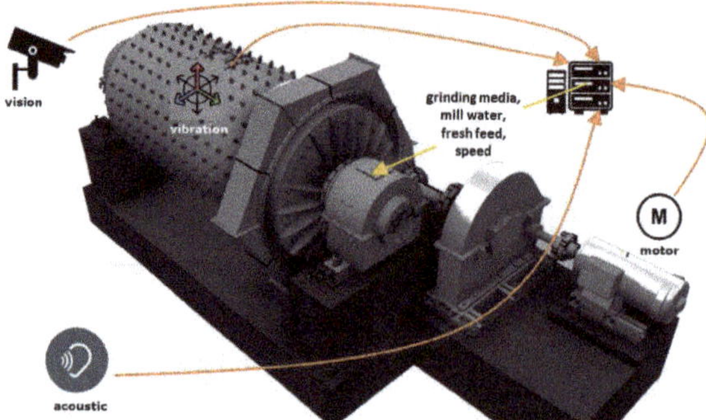

Figure 2. Schematic representation of the measured parameters and the parameters to be manipulated in order to obtain appropriate performance of the mill.

Following Table 1, one can notice that the majority of studies are based on vibration and acoustic emission measurements, which seem to be caused by three main reasons:

- Data acquisition methods and instrumentation for such signals are well proven in the industry;
- The price of the equipment is relatively low;
- Probably most importantly—it does not require direct contact of the sensor with the processed, highly abrasive material.

Acoustic emission technique is preferable for shock impacts sensing on the mill shell but requires wireless communication to record data. The same telemetry circuits are needed for strain gauges implementation on the rotating shafts of mechanical drives [65]. Therefore, it is important the measurement of angular backlashes in gear coupling [66] and to include in dynamical models of ball mills their drive-lines as the systems with non-linear parameters of stiffness [67].

Table 1. Examples of scientific papers contributing to the optimization of comminution in tumbling mills by signal monitored.

Signal Monitored	Work
Vibration	[16–27,32–37,68] (with use of strain transducer), [43,51] (estimated based on torque), [54]
Acoustic Emission	[24,28–36,51,54]
Power Draw	[38,51,69]
Instrumented Ball	[44–46]
Vision	[46,51,54]
Other	[24]—motor current, [39]—torque, [40]—bearing pressure and motor vector control, [42]—torque signal estimated on the basis of angular position and motor electric signals, [55]—product particle size (PPS) soft sensor

The multi-motor drives of heavy ball mills with open couplings of the peripheral tire and pinion gears are subjected to intensive wear and excessive angular and radial clearances. The different kinds of dynamical processes occurring in the drives of ball mills are investigated in [70,71]. One of the effects occurring in such kind of drives are out-of-phase torsional vibrations in parallel lines [72] having high amplitudes, which may interfere with internal load dynamics and significantly affect mill speed control. The most difficult mode for mill drives is to start under load especially for synchronous AC motors. Control methods of soft start are implemented for such cases [73].

Although electric motor parameters are quite easy to register and use for process stability estimation in the existing automation systems of industrial plants [74], a surprisingly small number of works in the domain of ball mills is discovered. This is most likely related to the simplicity of the signal processing methods used, which are not allowing to recognize multivariate correlations in material properties in the motor current data, which is reacting only to the integral load inside the mill.

Therefore, as it follows from Tables 2 and 3, the main efforts are undertaken in bulk media properties detection and related working conditions of the mill. Only a few studies are noted on the wear diagnostics of the grinding balls and protective liners, although this is very important for mills maintenance and balls replacement planning. Method of liners diagnostics is proposed in [69] based on analysis of infra-low frequencies (up to 0.01 Hz) of components in the active power spectrum. Another way to detect the wear of internal protective liners is to analyze self-excited torsional vibrations at the natural frequencies of the drive-line of the mill [75].

Table 2. Examples of scientific papers contributing to the optimization of comminution in tumbling mills by different criteria.

Purpose of Measurement	Work
Power draw reduction	[32]
Wear reduction	[31,43,54]
Particle size distribution optimization	[25,28,29,32,51,54,55]
Detection of characteristic working conditions	[17,23,27,51]
Fill level/behaviour	[16,18–22,26,30,31,34,37,39,40,42,44,45,68]
Other	[32]—pulp temperature, [51,55]—throughput maximization, [46]—DEM tuning, [69]—wear diagnostics of liners

Table 3. Examples of milling process parameters to be assessed by the measurement of particular signals.

Signal Measured/Process Parameters	Fill Level	Balls Wear	Liners Wear	Particle Size Distribution
Acoustic	✓		✓	✓
Vibration	✓		✓	✓
Vision				✓
Power draw	✓	✓	✓	
Torque	✓			
Other signals from electric motor	✓			

Application of the above-mentioned methods can be well described based on the MillVis system example [52]. As described in [54] the system uses vibration, acoustic, and vision measurements. Depending on the application (first or second stage grinding) it uses also other technological parameters like pulp density or average particle size. The latter are, however, accessible for mixed product streams from several mills sections and are used for global optimization purposes rather than individual mills performance improvement.

The system measures the mill's vibrations with wireless DataLogger using accelerometers installed on the mill's shell (see Figure 3A). Sensors and DataLogger rotate together with the mill, so wireless data transmission is required and device energy consumption optimization, since it operates using battery packages, rotating with the mill shell as well. According to the kinetics of grinding [76]—the vibration amplitude will rise while more of the kinetic energy of the grinding medium (e.g., steel balls) is transferred directly to the mill shell. The amplitude will be maximal when the balls hit the mill shell and the lowest when the balls sink in the processed material. Naturally, desired levels of the vibration amplitude will vary depending on the actual sensors position following from the rotation of the mill.

Thus, quite important is to correlate the amplitude measurements with the mill position. It is quite straightforward using the vibration signal from the individual sensor rotating together with the mill shell and low-pass filtering in frequency domain [77].

Figure 3B shows exemplary data from two full convolutions of the mill represented by blue and red parts of the chart. Black parts represent the final stages of the previous convolution and initial stages of the following convolution. The yellow sine-wave-shaped chart is the filtration result, representing the position of one of the sensors rotating with the mill shell. Binding the angular position of the sensor with the vibration signal values allows one to perform easy-to-understand signal analysis using e.g., polar plots.

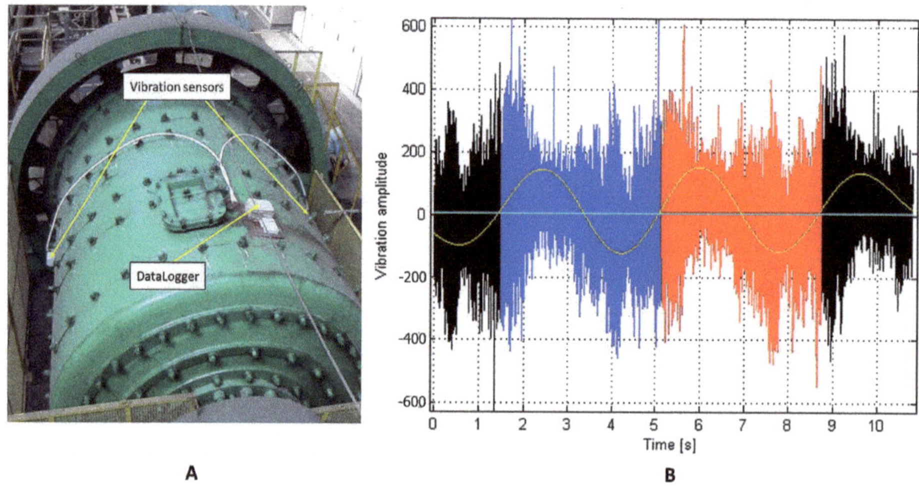

Figure 3. Location of the vibration sensors on the mill's shell (**A**) and time series of the ball mill vibration measurements (black, blue, and red representing separate full convolutions) and a rotation signal reconstructed from the original (**B**).

Figure 4 shows the comparison of the vibration signal for rod and ball mills. Each point on the polar plot represents the vibration amplitude (represented as a distance from the plot center) registered by the sensor in the given position during the mill rotation (represented by the angle value). One can easily observe the maximum amplitude angle changes with the process parameters variation (e.g., pulp density, throughput, grinding media charge).

The difference in the maximum amplitude angle for two compared mills is obvious and follows from different grinding media behaviour inside the mill's working chamber. The visual representation of the mill's behavior on the polar plots is useful for the operators to determine the state of the grinding process and the technical condition of the mill itself. Constant analysis of the above-mentioned parameters allows predictive maintenance and increases machinery availability.

Vibration signal amplitude analysis is only one of the system components. Values of technological process parameters, together with vibration signals' other parameters (e.g., dedicated indexes calculated at the frequency domain) and vision system measurements for the feed particles size distribution estimation are finally used by the dedicated software in the supervisory control layer to calculate optimal control set-points hints for the operator [78].

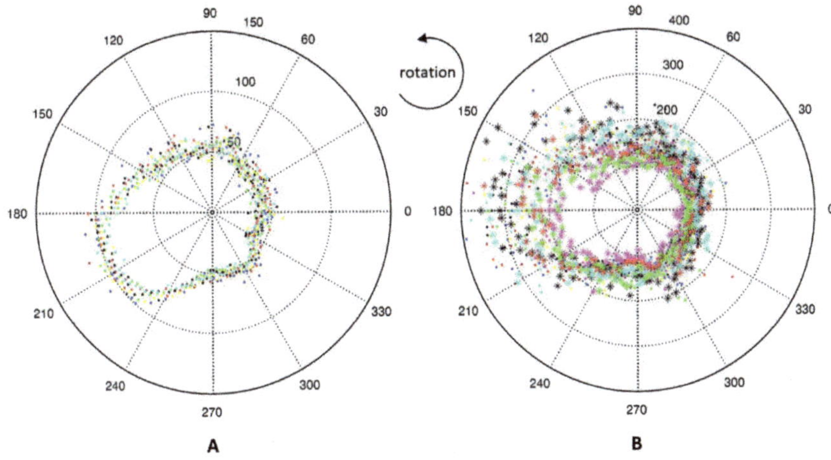

Figure 4. Ball mill (**A**) and rod mill (**B**) vibration signals presented on polar plots.

4. Analytical Models of Grinding Process Dynamics

4.1. Mathematical Modelling

Scientists tried to develop methods for controlling the process of grinding in ball mills with mathematical modeling methods. Basic semi-empirical relationships between mechanical work needed in the comminution process and the size of particles subjected to grinding, are the well known Kick's, Bond's and, Von Rittinger's equations applying to different thresholds of grain sizes [79].

- Kick's equation—for $d > 50$ mm

$$W_K = c_K(\ln(d_S) - \ln(d_G))$$ (1)

- Bond's equation—for 50 mm $> d > 0.05$ mm

$$W_B = c_B\left(\frac{1}{\sqrt{d_G}} - \frac{1}{\sqrt{d_S}}\right)$$ (2)

- Von Rittinger's equation—for $d < 0.05$ mm

$$W_R = c_R\left(\frac{1}{d_G} - \frac{1}{d_S}\right)$$ (3)

where W_K, W_B, W_R is the grinding work in kWh/t, c_K, c_B, c_R are the work indexes, usually given in kWh/t, d_S denotes pre-ground particle size, and d_G—the size of the particle after comminution (the values of d_S and d_G are usually taken as d_{80} from a granulometric curve, which describes the grain's size, below which the cumulative sum of smaller ones gives 80 percent of material's mass). However, it should be noted that energy consumption in the comminution processes still has not been theorized to a satisfactory degree [80].

In order to define different modes of operation depending on the geometry and rotational speed of a tumbling mill, the Froude number—a dimensionless ratio can be used to characterize bed dynamics and motion of the material during comminution [81,82]:

$$Fr = \frac{\omega^2 R}{g}$$ (4)

where ω is the rotational speed in revolutions per second, R—the radius of rotation in meters and g—the gravitational acceleration in (m/s^2).

Figure 5A shows experimentally determined regions of different motion modes of steel balls inside the laboratory drum, which is visually represented in Figure 6. The vertical axis corresponds to the number of particles converted to the height of bulk media h counting from the bottom of the drum.

To account for the nonlinear nature of the phenomenon and friction-induced self-oscillations, a Froude pendulum dynamics model was proposed in [83]. It contributed some theoretical considerations on friction dynamics, having strong potential to be a foundation for the analysis of energy-transferring mechanisms from the working tool (mill shell) to the treated media (bulk internal load)—including ore mills. The author represented internal load motion as the parametric oscillations using the Mathieu differential equation with periodically changing parameters, solutions of which are comprehensively represented as an Ince–Strutt diagram. The fragment of this diagram is shown below—Figure 5B, enabling the identification of stable regions in the mentioned model. A detailed formulation of the analytical functions for stable (shown shaded) and unstable regions boundaries on the Ince–Strutt diagram can be found in [84]. Briefly—a_0, b_2 and a_1, b_1 are boundary lines separating unstable (I) and stable (II) regions of the oscillations for π and 2π periodic solutions respectively; N_1, N_2, N_3—the transition from unstable (N_1, N_2) to stable (N_3) oscillations by the decreasing angular velocity of the rotation. The horizontal axis on this diagram corresponds to the excitation parameter and the vertical axis encompasses the relation of frequencies.

The shaded regions (II) determine the synchronization or out-of-synchronization in other regions (I) of the natural frequency and parametric excitation. The source of parametric oscillations can be any variable affecting the natural frequency: the moment of inertia, friction coefficient, filling level (mass of the internal load), etc. If these variables do not change, synchronization control can be achieved by the angular speed of mill rotation in case of regulated electric drives availability.

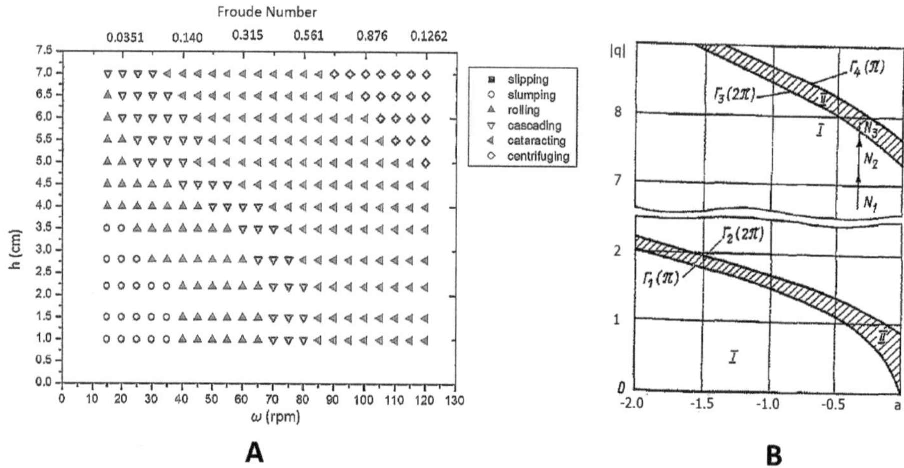

Figure 5. Characterization of steel granular materials' motion modes for given drum geometries and rotational speeds (**A**) [85] and the fragment of the Ince–Strutt diagram for the Froude pendulum-based dynamics model (**B**) [83].

The optimal modes of ore milling are controlled by the processing of the electric motor current or power signal. Conditions of synchronization are estimated based on the signal after band-pass filtering and high-resolution spectral density obtained online. Crossing the unstable regions by internal load motion is detected by an increase in spectrum amplitudes at the frequencies pre-calculated by the dynamical model.

It is known from the general theory of Mathieu equations in application to the physical systems that viscous damping—proportional to the momentary value of the angular velocity of contacting bodies motion (ore load and mill shell), can only narrow the regions of instability but not eliminate the unlimited increase of the amplitude of parametric oscillations. In other words, for the same relation of natural to exciting frequencies, e.g., for principal resonance in Figure 5B, higher amplitudes of periodical disturbance are required, but when its certain level is achieved, the amplitude of oscillations can develop to unlimited values. However, in real systems like ball mills, some non-linearity in energy dissipation always exists, and this is the reason for parametric oscillation attenuation.

Based on this approach, the mechanism of friction-induced oscillations is proposed in ball ore mills—including their journal bearings diagnostics and different types of other industrial machines [57], e.g., steel rolling mills [86–88], where the effect of synchronization is also observed, like in ball mills.

Regardless of applied models for frictional interaction investigation, this dynamic phenomenon has shown the following common features observed in many experimental studies [89]:

- Quick relative displacements observed at low speeds of mutual sliding of surfaces disappear with increasing speed;
- The amplitude and frequency of the separate displacements depend on the sliding speed, the mass of the moving body and the rigidity of the system;
- The first displacement of contacted bodies is much larger than the subsequent ones.

The appearance of abrupt displacements is explained either by the presence of a negative (falling) section of the friction force (moment) characteristic by sliding velocity or by an increase in the static friction force, which depends on the duration of the motionless contact.

Aiming to form an accurate mathematical description of the mill behavior, compromising between extreme operational states (Figure 6), a semi-phenomenological model extending the torque-arm-based equations [90], taking into account centrifugal mode of the load occurring at high rotational speed values, was developed by [91].

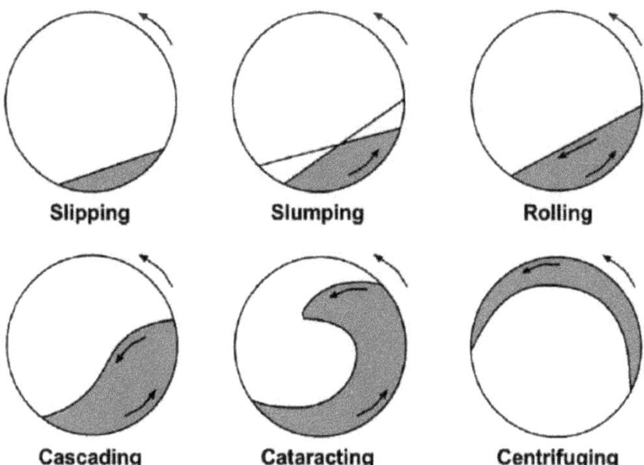

Figure 6. Internal load motion in the ball mill under different operational modes [85].

The author has shown that in the case of high-speed mills the Bond model is inadequate for describing their behavior and highlighted the need for investigation of the liners' design and model parameters. Models allowing for a better description of the material flow, including the influence of the mill's operating conditions and physical parameters of the ground ore, were presented by [92]. The authors used a geometric approach to investigate particles breakage process, obtaining equations describing the grinding media's motion and the slippage relationships of mill's charge In [93] mathematical models linking external signals (bearing pressure, power draw, and acoustic emission) with internal parameters of the mill, that can be utilized for optimization of the comminution process were presented.

A specific energy model and power consumption mathematical models derived with the use of the operational data from 4 industrial grinding circuits in Chilean Cu beneficiation plants were proposed by [94]. Despite having the potential to be used for predictive optimization of comminution processes, the model—referring to the authors' words—can be useful for the SAG mills' design purposes. In addition to the standard design variables (size of the mills, grinding media charge level, the concentration of solids in feed, rotation speed—expressed as a percentage of critical speed), an additional granulometric variable was included to account for PSD.

4.2. Discrete Element Method

The recent rapid increase of computational power allowed researchers and technicians to use numerical methods that reflect the granular nature of the feed better, applying discontinuous simulations. The most widely utilized technique of such type is a discrete element method (DEM) [95]. Discrete numerical models were already found to be valid for the study of granular assemblies' dynamics in 1979 [96]; however, for more than a decade DEM's practical applications remained restrained to 2D analyses because even when the problem was simplified to two dimensions, it still took "hours of super-computer time" to solve it [97]. Despite the fact that the computational abilities of that day's supercomputers were incomparable even to those of current smartphone devices, the authors of [98] have already managed to obtain some valuable information about the energy distribution, which occurs when ball-shape-represented particles collide. In a further study, the same authors found a potential of DEM to be employed for predictions of a real mill's power draw [97]. The method was later extended to three-dimensions and applied to assess the mill's power draw in a broad range of working conditions. Moreover, it could correctly model the motion of grinding media, what was confirmed by comparison with mill power consumption's torque-based measurement and with video recordings of the charge's behavior [99]. In [100] previously developed models were compared, to show their usefulness in describing different issues related to the charge movement. The model proposed in [101] was presented as an appropriate one for the definition of charge's outer boundary behavior [102]—was found a valuable contribution, properly describing the movement of individual balls, whereas the advantage of the one presented in [103] was the ability of a quick charge profile's computation. In [104] some constructive criticism about the definitions of significant parameters describing the behavior of intra-mill material has appeared. What is more, the authors presented how, and with which parameters as an input, the severity and incidence of impacting can be calculated, which is undoubtedly valuable from the mill's design point of view. In a work devoted to the modeling of mill's power draw with use of 3D discrete element method [105], the author proposes a method for decreasing the computational demand by assigning the mass of finer fractions, which require shorter timesteps, to the next coarser fraction classes. In this way, due to the increased timestep of the calculation, its speed was successfully reduced, allowing the modeling of the power consumption with the use of DEM—similarly accurately as in the case of empirical models however with higher flexibility. Another important contribution to the development of DEM as a tool for optimization and description of the comminution process in ball mills was the introduction of Smoothed Particle Hydrodynamics (SPH) [106,107]. It significantly improved the modeling of mill charge's dynamic behavior, enabling additional predictions related to the fluid slurry flow, as a result providing more realistic simulations. Examples of such simulations are presented in Figures 7 and 8 below.

There is a significant simplification used in DEM modeling, however, which is the representation of intra-mill material in the form of spheres. Most of the raw materials subjected to comminution comprise irregular particles, the interactions of which can not be accurately described when they are not represented with realistic, complex shapes. Some researchers extended DEM models, replacing circular particle representations in 2D with:

- Polygons [108];
- Ellipses [109].

In the case of 3D models, they used:

- Polyhedrons [110,111];
- Ellipsoids [112];
- Super quadratic functions [113].

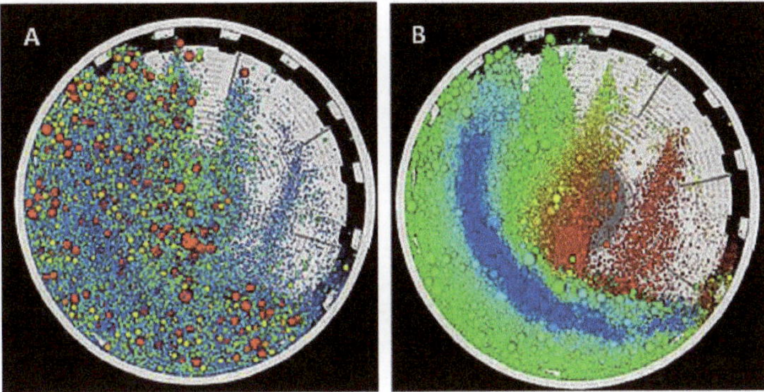

Figure 7. An example of typical mill flow pattern simulated with use of DEM. On the left-hand side (**A**) the particles are colored by diameter, dark blue being the finest and red the coarsest. On the right-hand side (**B**) the particles are colored by speed—blue being the slowest and red the fastest [106].

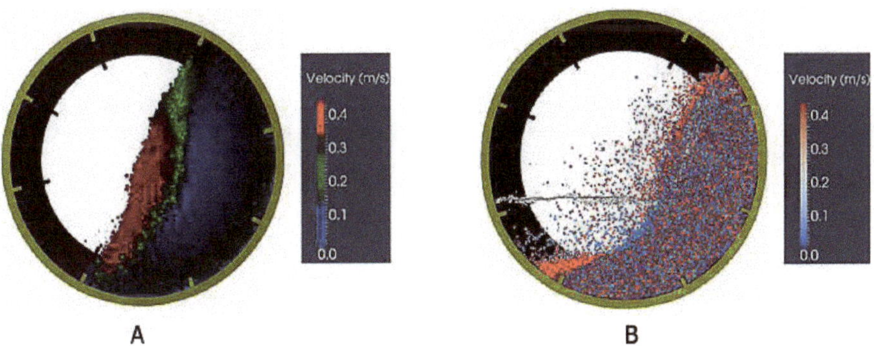

Figure 8. An example of a 3D DEM simulation of particle flow, based on (**A**) the only-solid-phase approach (basic DEM) and (**B**) combining DEM with SPH to take into account both phases: solid and liquid [107].

As an example of a compromise between the realistic representation of ground material particles and the computational complexity of the used model, the so-called "stacked spheres approach" or "glued spheres approach" can be given. In the models based on such an approach, a real particle is represented using particles of various sizes, which can overlap to create a surface resembling natural material [114]. An exhaustive review of DEM modeling of non-spherical particle systems can be found in [115].

The above-mentioned methods undoubtedly have advantages in describing the material's dynamic behavior in a more realistic way; however, some significant problems appear for the detection of contact between particles. Moreover, the computational power in the case of industrial scale-models limits their practical implementation.

5. Discussion and Conclusions

To provide optimal control of the comminution process under non-stationary properties of input raw materials, advanced measurement techniques are necessary. Despite the advanced design of modern ball mills' automation systems, the existing controllers can only stabilize the deviations at a certain level of motor power without a significant reduction of overall energy consumption.

Having enormous power and rotating inertia, ball mills need in-depth research of their internal load dynamics by different tools. Using mathematical models, both analytical and DEM simulations can improve the understanding of the processed material flow, not available for direct observation. Nevertheless, 3D modeling of the grinding media and intra-mill material behavior is still being done with significant simplifications regarding particle shape, which are done due to the limited computational power, not sufficient to withstand industrial-scale simulations, and due to the fact that basic semi-empirical formulas do not account for the complex structure of raw material.

The most profitable mode of mill operation is when the central part oscillates up and down over the rest of the media treated, which corresponds to the synchronous mode of mill operation. This mode of parametric resonance can be achieved by regulation of input feeding rate, water supply (friction factor), or mill rotation speed (for variable speed motors).

The most reliable mill control is by electric motor current or power. These signals are easy for monitoring in the existing automation systems, but their capability is restricted to the diagnostics of the filling levels and wear rates of balls and protective liners.

Using vibration and acoustic signals from outside surfaces of the mill shell or other parts of the machine is a promising approach. However, its successful implementation for process control requires advanced signal processing methods and verification under non-stationary mill loading, and gradual wear of grinding bodies and internal protective liners, which affect the external measured signals. The neighboring mills' noise and vibrations can also interfere with measured signals. Therefore some methods of shielding or direct fitting of wireless sensors to the mill shell surface should be used. In the latter case, algorithms of data processing should account for the instantaneous sensor position depending on the mill rotation speed.

Optimal control of technological parameters in the ball mills should be combined with the simultaneous online monitoring and diagnostics of balls' and internal liners' wear, as their condition greatly affects the measured sound and vibration signals.

Further research is planned for both laboratory mills and industrial plants in order to achieve a resonance mode of operation and its control by the different signals available for measurement and control.

Author Contributions: Conceptualization, P.K. and R.Z.; methodology, M.G.; formal analysis, S.O.; investigation, M.G.; resources, R.Z.; data curation, S.O.; writing—original draft preparation, P.K., M.G. and S.O.; writing—review and editing, M.G.; supervision, P.K.; project administration, R.Z.; funding acquisition, R.Z. All authors have read and agreed to the published version of the manuscript.

Funding: This research has received funding from European Institute of Innovation and Technology (EIT), a body of the European Union, under the Horizon 2020, the EU Framework Programme for Research and Innovation. This work is supported by EIT RawMaterials GmbH under framework partnership agreement number 18253 (OPMO—Operational Monitoring of Mineral Crushing Machinery).

Conflicts of Interest: The authors declare no conflict of interest.

References

1. Walkiewicz, J.; Clark, A.; McGill, S. Microwave-assisted grinding. *IEEE Trans. Ind. Appl.* **1991**, *27*, 239–243. [CrossRef]
2. Musa, F.; Morrison, R. A more sustainable approach to assessing comminution efficiency. *Miner. Eng.* **2009**, *22*, 593–601. [CrossRef]

3. BCS Incoprorated for the U.S. Department of Energy: U.S. Mining Industry Energy Bandwidth Study. Available online: https://www.energy.gov/sites/prod/files/2013/11/f4/mining_bandwidth.pdf (accessed on 5 December 2020).

4. Jeswiet, J.; Szekeres, A. Energy consumption in mining comminution. *Procedia CIRP* **2016**, *48*, 140–145. [CrossRef]

5. Fuerstenau, D.; Abouzeid, A.Z. The energy efficiency of ball milling in comminution. *Int. J. Miner. Process.* **2002**, *67*, 161–185. [CrossRef]

6. Mular, A. *Mineral Processing Plant Design, Practice, and Control Proceedings*; Society for Mining, Metallurgy, and Exploration: New York, NY, USA, 2002.

7. Chelgani, S.C.; Parian, M.; Parapari, P.S.; Ghorbani, Y.; Rosenkranz, J. A comparative study on the effects of dry and wet grinding on mineral flotation separation—A review. *J. Mater. Res. Technol.* **2019**, *8*, 5004–5011. [CrossRef]

8. Wang, Y.; Forssberg, E. Enhancement of energy efficiency for mechanical production of fine and ultra-fine particles in comminution. *China Particuol.* **2007**, *5*, 193–201. [CrossRef]

9. Wills, B.A.; Napier-Munn, T. Comminution. In *Wills' Mineral Processing Technology*; Butterworth-Heinemann: Oxford, UK, 2005; Chapter 5, pp. 108–117. [CrossRef]

10. Sosa-Blanco, C.; Hodouin, D.; Bazin, C.; Lara-Valenzuela, C.; Salazar, J. Economic optimization of a flotation plant through grinding circuit tuning. *Miner. Eng.* **2000**, *13*, 999–1018. [CrossRef]

11. Gao, P.; Zhou, W.; Han, Y.; Li, Y.; Ren, W. Enhancing the capacity of large-scale ball mill through process and equipment optimization: An industrial test verification. *Adv. Powder Technol.* **2020**, *31*, 2079–2091. [CrossRef]

12. Nagar, N.; Garg, H.; Sharma, H.; Angadi, S.; Gahan, C. Influence of Grinding Time on The Bioleaching of Copper from Copper Slag. *Biosci. Biotechnol. Res. Commun.* **2019**, *12*, 512–522. [CrossRef]

13. Jonsen, P.; Palsson, B.; Haggblad, H.A.; Tano, K.; Berggren, A. Simulation of charge and structure behaviour in a tumbling mill. In Proceedings of the European LS-DYNA Users Conference, Strasbourg, France, 23–24 May 2011.

14. Astrom, K.; Wittenmark, B. *Computer-Controlled Systems: Theory and Design*; Prentice Hall: Upper Saddle River, NJ, USA, 1997.

15. Craig, I. Grinding mill modeling and control: Past, present and future. In Proceedings of the 31st Chinese Control Conference, Anchui, China, 25–27 July 2012; pp. 16–21.

16. Campbell, J.; Spencer, S.; Sutherland, D.; Rowlands, T.; Weller, K.; Cleary, P.; Hinde, A. SAG mill monitoring using surface vibrations. In Proceedings of the Third International Conference on Autogenous and Semiautogenous Grinding Technology (SAG2001), Vancouver, BC, Canada, 30 September–3 October 2001; Volume 2, pp. 373–385.

17. Su, Z.; Wang, P.; Yu, X.; Lv, Z. Experimental investigation of vibration signal of an industrial tubular ball mill: Monitoring and diagnosing. *Miner. Eng.* **2008**, *21*, 699–710. [CrossRef]

18. Bai, R.; Chai, T. Optimization control of ball mill load in blending process with data fusion and case-based reasoning. *Huagong Xuebao/CIESC J.* **2009**, *60*, 1746–1752.

19. Tang, J.; Chai, T.; Zhao, L.; Yu, W.; Yue, H. Soft sensor for parameters of mill load based on multi-spectral segments PLS sub-models and on-line adaptive weighted fusion algorithm. *Neurocomputing* **2012**, *78*, 38–47. [CrossRef]

20. Tang, J.; Zhao, L.; Yu, W.; Yue, H.; Chai, T. Soft Sensor Modeling of Ball Mill Load via Principal Component Analysis and Support Vector Machines. In *Lecture Notes in Electrical Engineering*; Springer: Berlin, Germany, 2010; pp. 803–810. [CrossRef]

21. Tang, J.; Zhao, L.-j.; Zhou, J.-w.; Yue, H.; Chai, T.-y. Experimental analysis of wet mill load based on vibration signals of laboratory-scale ball mill shell. *Miner. Eng.* **2010**, *23*, 720–730. [CrossRef]

22. Tang, J.; Zhao, L.; Yue, H.; Yu, W.; Chai, T. Vibration Analysis Based on Empirical Mode Decomposition and Partial Least Square. *Procedia Eng.* **2011**, *16*, 646–652. [CrossRef]

23. Das, S.P.; Das, D.P.; Behera, S.K.; Mishra, B.K. Interpretation of mill vibration signal via wireless sensing. *Miner. Eng.* **2011**, *24*, 245–251. [CrossRef]

24. Tang, J.; Chai, T.; Yu, W.; Zhao, L. Modeling Load Parameters of Ball Mill in Grinding Process Based on Selective Ensemble Multisensor Information. *IEEE Trans. Autom. Sci. Eng.* **2013**, *10*, 726–740. [CrossRef]

25. Mohanty, S.; Gupta, K.K.; Raju, K.S. Vibration Feature Extraction and Analysis of Industrial Ball Mill Using MEMS Accelerometer Sensor and Synchronized Data Analysis Technique. *Procedia Comput. Sci.* **2015**, *58*, 217–224. [CrossRef]

26. Nayak, D.K.; Das, D.P.; Behera, S.K.; Das, S.P. Monitoring the fill level of a ball mill using vibration sensing and artificial neural network. *Neural Comput. Appl.* **2019**, *32*, 1501–1511. [CrossRef]

27. Behera, B.; Mishra, B.; Murty, C. Experimental analysis of charge dynamics in tumbling mills by vibration signature technique. *Miner. Eng.* **2007**, *20*, 84–91. [CrossRef]

28. Cho, K.S.; Kim, S.H.; Lee, Y.H. Correlation between Acoustic Intensity and Ground Particle Size in Alumina Ball Mill Process. *J. Korean Ceram. Soc.* **2018**, *55*, 275–284. [CrossRef]

29. Aldrich, C.; Theron, D. Acoustic estimation of the particle size distributions of sulphide ores in a laboratory ball mill. *J. S. Afr. Inst. Min. Metall.* **2000**, *100*, 243–248.

30. Pax, R.; Djordjevic, N.; Hocking, R. Non contact acoustics measurement and validation of SAG mill operation. In Proceedings of the XXII International Mineral Processing Congress, Cape Town, South Africa, 29 September–3 October 2003; SAIMM: Cape Town, South Africa, 2003; Volume 1, pp. 371–377.

31. Wu, D.; Chen, W.; Yan, H.; Fischer, J.; Doolan, C. Identifying grinding mill dynamics using acoustic beamforming and numerical modelling. *Powder Technol.* **2020**, *371*, 231–243. [CrossRef]

32. Zeng, Y.; Forssberg, E. Monitoring grinding parameters by signal measurements for an industrial ball mill. *Int. J. Miner. Process.* **1993**, *40*, 1–16. [CrossRef]

33. Tang, J.; Liu, Z.; Wu, Y.J.; Zhao, L.J. Modeling Difficult-to-Measure Process Parameters Based on Intrinsic Mode Functions Frequency Spectral Features of Mechanical Vibration and Acoustical Signals. *Adv. Mater. Res.* **2014**, *989–994*, 3671–3674. [CrossRef]

34. Tang, J.; Yu, W.; Chai, T.; Liu, Z.; Zhou, X. Selective ensemble modeling load parameters of ball mill based on multi-scale frequency spectral features and sphere criterion. *Mech. Syst. Signal Process.* **2016**, *66–67*, 485–504. [CrossRef]

35. Tang, J.; Chai, T.; Yu, W.; Liu, Z.; Zhou, X. A Comparative Study That Measures Ball Mill Load Parameters Through Different Single-Scale and Multiscale Frequency Spectra-Based Approaches. *IEEE Trans. Ind. Inform.* **2016**, *12*, 2008–2019. [CrossRef]

36. Tang, J.; Qiao, J.; Liu, Z.; Zhou, X.; Yu, G.; Zhao, J. Mechanism characteristic analysis and soft measuring method review for ball mill load based on mechanical vibration and acoustic signals in the grinding process. *Miner. Eng.* **2018**, *128*, 294–311. [CrossRef]

37. Kolacz, J. Measurement system of the mill charge in grinding ball mill circuits. *Miner. Eng.* **1997**, *10*, 1329–1338. [CrossRef]

38. Morrell, S. Power Draw of Grinding Mills—Its measurements and Prediction. In Proceedings of the 5th Mill Operators Conference, Roxby Downs, Australia, 16–20 October 1994.

39. Melero, M.; Cano, J.; Norniella, J.; Pedrayes, F.; Cabanas, M.; Rojas, C.; Alonso, G.; Aguado, J.M.; Ardura, P. Electric motors monitoring: An alternative to increase the efficiency of ball mills. *Renew. Energy Power Qual. J* **2014**, 849–854. [CrossRef]

40. Pontt, J. MONSAG: A new monitoring system for measuring the load filling of a SAG mill. *Miner. Eng.* **2004**, *17*, 1143–1148. [CrossRef]

41. Cleary, P. Predicting charge motion, power draw, segregation and wear in ball mills using discrete element methods. *Miner. Eng.* **1998**, *11*, 1061–1080. [CrossRef]

42. Esteves, P.M.; Stopa, M.M.; Filho, B.J.C.; Galery, R. Charge behavior analysis in ball mill by using estimated torque. In Proceedings of the 2014 IEEE Industry Application Society Annual Meeting, Vancouver, BC, Canada, 5–9 October 2014. [CrossRef]

43. Campbell, J.; Holmes, R.; Spencer, S.; Sharp, V.; Davey, K.; Barker, D.; Phillips, P. The collection and analysis of single sensor surface vibration data to estimate operating conditions in pilot-scale and production-scale AG/SAG mills. In Proceedings of the XXII International Mineral Processing Congress, Cape Town, South Africa, 29 September–3 October 2003; pp. 280–288.

44. Martins, S.; Li, W.; Radziszewski, P.; Caron, S. Investigating the differences in charge dynamics due to a variation of the instrumented ball properties. *IFAC Proc. Vol.* **2007**, *12*, 463–468. [CrossRef]

45. Martins, S.; Li, W.; Radziszewski, P.; Caron, S.; Aguanno, M.; Bakhos, M.; Petch, E.L. Validating the instrumented ball outputs with simple trajectories. *Miner. Eng.* **2008**, *21*, 782–788. [CrossRef]

46. Martins, S.; Li, W.; Radziszewski, P.; Faucher, A.; Makni, S. Experimental and simulated instrumented ball in a tumbling mill—A comparison. *Miner. Eng.* **2013**, *43–44*, 79–84. [CrossRef]

47. Coetzee, L. Stabilising and optimising a primary closed-loop milling circuit feeding a flotation circuit using StarCS RNMPC. In Proceedings of the 19th IFAC World Congress on International Federation of Automatic Controll Congress, Cape Town, South Africa, 24–29 August 2014; Volume 47, pp. 9786–9791. [CrossRef]

48. Coetzee, L.; Ramonotsi, M. Applying StarCS RNMPC with Real-Time Optimiser to Pilanesberg Platinum Mines Primary UG2 Milling Circuit. *IFAC-PapersOnLine* **2016**, *49*, 78–83. [CrossRef]

49. Outotec. Outotec Automation Solutions for Grinding Optimization. Available online: https://www.outotec. com/globalassets/products/analyzers-and-automation/ote_act_grinding_optimization_eng.pdf (accessed on 5 November 2020).

50. FLSmidth. Maximize Grinding Efficiency with LoadIQ. Available online: https://www.kscape.com/loadiq (accessed on 5 November 2020).

51. Thwaites, P. Developments in Process Control—Grinding Controls. 2015. Available online: https://xps.ca/ files/31/Presentations/131/Developments-in-Process-Control---Grinding-Controls-by-P-Thwaites.pdf (accessed on 5 November 2020).

52. AMEplus. MillVis: System for the Optimization of the Grinding Process Control. Available online: https: //www.ameplus.pl/tag/millvis/ (accessed on 5 November 2020).

53. Chai, T.; Ding, J. Integrated automation system for hematite ores processing and its applications. *Meas. Control* **2006**, *39*, 140–146.

54. Kuzba, B.; Pawlos, W.; Konieczny, A.; Krzeminska, M. Optimisation Platform for copper ore processing at the Division of Concentrator of KGHM Polska Miedz SA. In *Proceedings of the E3S Web of Conferences*; EDP Sciences: Les Ulis, France, 2016; Volume 8, p. 01037.

55. Zhou, P.; Chai, T.; Wang, H. Intelligent Optimal-Setting Control for Grinding Circuits of Mineral Processing Process. *IEEE Trans. Autom. Sci. Eng.* **2009**, *6*, 730–743. [CrossRef]

56. Deineka, K.; Naumenko, Y. Revealing the effect of decreased energy intensity of grinding in a tumbling mill during selfexcitation of autooscillations of the intrachamber fill. *Eastern-Eur. J. Enterp. Technol.* **2019**, *1*, 6–15. [CrossRef]

57. Maryuta, A. *Frictional Oscillations in Mechanical Systems*; Nedra: Moscow, Russia, 1993; p. 240.

58. Naumenko, Y. *Fundamentals of the Theory of Working Processes of Drum Mills*; NUVGP: Rivne, Ukraine, 2014; p. 336.

59. Deineka, K.; Naumenko, Y. Establishing the effect of a decrease in power intensity of self-oscillating grinding in a tumbling mill with a reduction in an intrachamber fill. *Eastern-Eur. J. Enterp. Technol.* **2019**, *6*, 43–52. [CrossRef]

60. Alekseyev, M.; Alkhori, F. Automated control of ore-pebble mill charge by the signal of active power of magnetic separator electric motor. *Sci. Bull. Natl. Min. Univ. Ukraine* **2014**, *3*, 71–76.

61. Atutxa, I.; Legarra, I. Stepping forward: Using variable speed drives for optimizing the grinding process in SAG and ball mills. In Proceedings of the Sixth International Conference on Semi-Autogenous High Pressure Grinding Technology, Vancouver, BC, Canada, 20–24 September 2015.

62. Smotritskiy, A.; Smotritskiy, A.; Boriskov, F.; Chervyakov, S. Method for Grinding Materials in Rotating Drum Using Ferromagnetic Grinding Bodies. EU Patent EP3020483B1, 8 July 2013.

63. Vertimill. Metso Mining. Available online: https://www.metso.com/products/grinding-mills/stirred-mills/vertimill/ (accessed on 5 November 2020).

64. Isamill. Glencore Technology. Available online: https://www.isamill.com/en/downloads/Brochures/ IsaMillBrochure.pdf (accessed on 5 November 2020).

65. Krot, P. Telemetering systems for monitoring dynamic loads in drive lines of rolling mills. *Vib. Mach. Meas. Reduct. Prot.* **2008**, *1*, 46–53.

66. Krot, P. Methods and instrumentation for measuring wear in drivelines of rolling mills. *Metall. Process. Equip.* **2003**, *2*, 45–53.

67. Krot, P. Dynamics and diagnostics of the rolling mills drivelines with non-smooth stiffness characteristics. In Proceedings of the 3rd International Conference of Nonlinear Dynamic (ND-KhPI2010), Kharkov, Ukraine, 21–24 September 2012; pp. 115–120.

68. Tang, J.; Chai, T.Y.; Cong, Q.M.; Liu, Z.; Yu, W. Modeling mill load parameters based on selective fusion of multi-scale shell vibration frequency spectra. *Kongzhi Lilun Yu Yingyong/Control Theory Appl.* **2015**, *32*, 1582–1591. [CrossRef]

69. Meshcheryakov, L. Recognition of the operational state of the lining of drum mill. *Sci. Bull. Natl. Min. Univ. Ukraine* **2018**, *53*, 200–213.

70. Vinogradov, B. *Dynamics of Tumbling Mills*; Economic Herald of State Higher Educational Institution, USUCT: Dnipro, Ukraine, 2004.

71. Vinogradov, B.; Khristenko, A. The forced vibrations are in twin-engine synchronous drives of ball mills. *Sci. Bull. Natl. Min. Univ. Ukraine* **2012**, *6*, 72–76.

72. Krot, P.V. Dynamical processes in a multi-motor gear drive of heavy slabbing mill. *J. Vibroeng.* **2019**, *21*, 2064–2081. [CrossRef]

73. Borodai, V.; Borovyk, R.; Nesterova, O. Efficient transient modes of synchronous drive for mining and smelting mechanisms. *Mech. Mater. Sci. Eng. J.* **2017**, *8*. [CrossRef]

74. Putnoki, A.; Klevtsov, O.; Ermolenko, A.; Verenev, V.; Krot, P. Evaluation of operation of equipment at the rolling mill. *Stal* **2003**, *10*, 56–58.

75. Maryuta, A.; Kovalenko, A.; Meshcheryakov, L. Method for Determining Wear of Shaped Linings in Grinding Aggregate. Patent SU709173, 15 January 1980.

76. Beke, B. *The Process of Fine Grinding*, 1st ed.; Springer: Dordrecht, The Netherlands, 1981; Volume 1. [CrossRef]

77. Nussbaumer, H.J. *Fast Fourier Transform and Convolution Algorithms*; Springer: Berlin, Germany, 1982. [CrossRef]

78. Konieczny, A.; Ogonowski, S.; Kurzydlo, M.; Foszcz, D. Vision systems in O/ZWR as a support tool for production management and optimization. In Proceedings of the Mineral Engineering Conference (MEC2014), Istebna, Poland, 15–18 September 2014.

79. Thomas, A.; Filippov, L. Fractures, fractals and breakage energy of mineral particles. *Int. J. Miner. Process.* **1999**, *57*, 285–301. [CrossRef]

80. Naziemiec, Z.; Saramak, D. Analiza energochłonności procesów rozdrabniania kruszyw mineralnych. *Prace Naukowe Instytutu Górnictwa Politechniki Wrocławskiej. Studia i Materiały* **2012**, *134*, 209–220.

81. Weidenbaum, S.S. Mixing of solids. In *Advances in Chemical Engineering*; Elsevier: Amsterdam, The Netherlands, 1958; Volume 2, pp. 209–324.

82. Heim, A.; Gluba, T.; Obraniak, A. Bed dynamics during drum granulation. *Fizykochemiczne Problemy Mineralurgii/Physicochem. Probl. Miner. Process.* **2004**, *38*, 167–176.

83. Maryuta, A.N. Analysis of motion of mechanical systems with frictional interaction. *Sov. Appl. Mech.* **1989**, *25*, 1031–1040. [CrossRef]

84. Butikov, E.I. Analytical expressions for stability regions in the Ince–Strutt diagram of Mathieu equation. *Am. J. Phys.* **2018**, *86*, 257–267. [CrossRef]

85. Mardiansyah, Y.; Khotimah, S.; Viridi, S. Characterization of motion modes of pseudo-two dimensional granular materials in a vertical rotating drum. *J. Phys. Conf. Ser.* **2016**, *739*, 012148. [CrossRef]

86. Marjuta, A.; Krot, P. High frequency rolling mills chatter–mathematical identification and simulation. In Proceedings of the 1st International Symposium on "Multi-Body Dynamics Monitoring and Simulation Techniques", Bradford, UK, 25–27 March 1997; pp. 407–419.

87. Krot, P. Parametrical vibrations in the rolling mills. *Collect. Sci. Pap. Natl. Min. Acad. Ukraine* **2002**, *3*, 15–21.

88. Krot, P. Investigation of 'ribbing' defect and high-frequency oscillations of mills for strip cold rolling. *Proizvod. Prokata* **2002**, *3*, 21–23.

89. Kragelsky, I.; Dobychin, M.; Kombalov, V. *Friction and Wear: Calculation Methods*; Pergamon Press: Oxford, UK, 1982; p. 474.

90. Arbiter, N.; Harris, C. Scale-up and dynamics of large grinding mills-a case study. In *Design and Installation of Comminution Circuits*; AIME: New York, NY, USA, 1982; pp. 491–508.

91. Moys, M. A model of mill power as affected by mill speed, load volume, and liner design. *J. S. Afr. Inst. Min. Metall.* **1993**, *93*, 135–141.

92. Abou, S.C.; Tarasiewicz, S.; Remy, M. Mathematical modelling of ball mill charge slippage: Geometric approach. In Proceedings of the 4e Conférence International sur l'Automatisation Industrielle, Montréal, QC, Canada, 9–11 June 2003; CIAI: Montreal, QC, Canada, 2003. [CrossRef]

93. Wang, Z.H.; Han, Y.X.; Chen, B.C. A Mathematical Model for Predicting the Internal Parameters of Ball Mill. *Adv. Mater. Res.* **2012**, *454*, 151–156. [CrossRef]

94. Silva, M.; Casali, A. Modelling SAG milling power and specific energy consumption including the feed percentage of intermediate size particles. *Miner. Eng.* **2015**, *70*, 156–161. [CrossRef]

95. Kozicki, J.; Donze, F. YADE-OPEN DEM: An opensource software using a discrete element method to simulate granular material. *Eng. Comput.* **2008**, *26*. [CrossRef]

96. Cundall, P.A.; Strack, O.D.L. A discrete numerical model for granular assemblies. *Géotechnique* **1979**, *29*, 47–65. [CrossRef]

97. Mishra, B.; Rajamani, R.K. The discrete element method for the simulation of ball mills. *Appl. Math. Model.* **1992**, *16*, 598–604. [CrossRef]

98. Mishra, B.K.; Rajamani, R.K. Motion Analysis in Tumbling Mills by the Discrete Element Method. *KONA Powder Part. J.* **1990**, *8*, 92–98. [CrossRef]

99. Agrawala, S.; Rajamani, R.; Songfack, P.; Mishra, B. Mechanics of media motion in tumbling mills with 3d discrete element method. *Miner. Eng.* **1997**, *10*, 215–227. [CrossRef]

100. Radiszewski, P. Comparing three DEM charge motion models. *Miner. Eng.* **1999**, *12*, 1501–1520. [CrossRef]

101. Powell, M.; Nurick, G. A study of charge motion in rotary mills Part 1—Extension of the theory. *Miner. Eng.* **1996**, *9*, 259–268. [CrossRef]

102. Mishra, B.; Rajamai, R.K. Simulation of charge motion in ball mills. Part 2: Numerical simulations. *Int. J. Miner. Process.* **1994**, *40*, 187–197. [CrossRef]

103. Radziszewski, P.; Tarasiewicz, S. Ballmill simulation: Part II—Numerical solution to ballcharge model. *Trans. Soc. Comput. Simul.* **1989**, *6*, 75–88.

104. Powell, M.; McBride, A. A three-dimensional analysis of media motion and grinding regions in mills. *Miner. Eng.* **2004**, *17*, 1099–1109. [CrossRef]

105. Djordjevic, N. Influence of charge size distribution on net-power draw of tumbling mill based on DEM modelling. *Miner. Eng.* **2005**, *18*, 375–378. [CrossRef]

106. Cleary, P.W.; Morrison, R.D. Prediction of 3D slurry flow within the grinding chamber and discharge from a pilot scale SAG mill. *Miner. Eng.* **2012**, *39*, 184–195. [CrossRef]

107. Jahani Chegeni, M. Combined DEM and SPH simulation of ball milling. *J. Min. Environ.* **2019**, *10*, 151–161. [CrossRef]

108. Walton, O. Particle-Dynamics Calculations of Shear Flow. In *Mechanics of Granular Materials—New Models and Constitutive Relations*; Elsevier: Amsterdam, The Netherlands, 1983; pp. 327–338. [CrossRef]

109. Ting, J.M.; Khwaja, M.; Meachum, L.R.; Rowell, J.D. An ellipse-based discrete element model for granular materials. *Int. J. Numer. Anal. Methods Geomech.* **1993**, *17*, 603–623. [CrossRef]

110. Cundall, P. Formulation of a three-dimensional distinct element model—Part I. A scheme to detect and represent contacts in a system composed of many polyhedral blocks. *Int. J. Rock Mech. Min. Sci. Geomech. Abstr.* **1988**, *25*, 107–116. [CrossRef]

111. Hocking, G. The discrete element method for analysis of fragmentation of discontinua. *Eng. Comput.* **1992**, *9*, 145–155. [CrossRef]

112. Lin, X.; Ng, T.T. A three-dimensional discrete element model using arrays of ellipsoids. *Géotechnique* **1997**, *47*, 319–329. [CrossRef]

113. Williams, J.R.; Pentland, A.P. Superquadrics and modal dynamics for discrete elements in interactive design. *Eng. Comput.* **1992**, *9*, 115–127. [CrossRef]

114. Favier, J.; Fard, M.; Kremmer, M.; Raji, A. Shape representation of axi-symmetrical, non-spherical particles in discrete element simulation using multi-element model particles. *Eng. Comput. Int. J. Comput.-Aided Eng.* **1999**, *16*, 467–480. [CrossRef]

115. Zhong, W.; Yu, A.; Liu, X.; Tong, Z.; Zhang, H. DEM/CFD-DEM modelling of non-spherical particulate systems: Theoretical developments and applications. *Powder Technol.* **2016**, *302*, 108–152. [CrossRef]

Publisher's Note: MDPI stays neutral with regard to jurisdictional claims in published maps and institutional affiliations.

Article

Process Monitoring in Heavy Duty Drilling Rigs—Data Acquisition System and Cycle Identification Algorithms

Jacek Wodecki [1], Mateusz Góralczyk [1], Pavlo Krot [1], Bartłomiej Ziętek [1,*], Jaroslaw Szrek [2], Magdalena Worsa-Kozak [1], Radoslaw Zimroz [1], Paweł Śliwiński [3] and Andrzej Czajkowski [4]

1   Faculty of Geoengineering, Mining and Geology, Wroclaw University of Science and Technology, Na Grobli 15, 50-421 Wrocław, Poland; jacek.wodecki@pwr.edu.pl (J.W.); mateusz.goralczyk@pwr.edu.pl (M.G.); pavlo.krot@pwr.edu.pl (P.K.); magdalena.worsa-kozak@pwr.edu.pl (M.W.-K.); radoslaw.zimroz@pwr.edu.pl (R.Z.)
2   Faculty of Mechanical Engineering, Wroclaw University of Science and Technology, Łukasiewicza 5/7, 50-370 Wrocław, Poland; jaroslaw.szrek@pwr.edu.pl
3   KGHM Polska Miedź S.A., M. Skłodowskiej-Curie 48, 59-301 Lubin, Poland; pawel.sliwinski@kghm.com
4   Mine Master Spółka z o.o., Wilków, ul. Dworcowa 27, 59-500 Złotoryja, Poland; aczajkowski@minemaster.eu
*   Correspondence: bartlomiej.zietek@pwr.edu.pl

Received:13 November 2020; Accepted: 16 December 2020; Published: 21 December 2020

Abstract: The monitoring of drilling processes is a well-known topic in the mining industry. It is widely used for rock mass characterization, bit wear monitoring and drilling process assessment. However on-board monitoring systems used for this purpose are installed only on a limited number of machines, and breakdowns are possible. There is a need for a data acquisition system that can be used on different drilling rigs and for an automatic data analysis procedure. In this paper, we focused on the automatic detection of drilling cycles, presenting a simple yet reliable system to be universally installed on drilling rigs. The proposed solution covers hardware and software. It is based on the measurement of electric current and acoustic signals. The signal processing methods include threshold-based segmentation, a short-time envelope spectrum and a spectrum for the representation of results. The results of the research have been verified on a real drilling rig within the testing site of its manufacturer by comparing the results with the data of the on-board monitoring system installed on the machine. Novel aspects of our approach include the detection of the pre-boring stage, which has an intermediate amplitude that masks the real drilling cycles, and the use of the percussion instantaneous frequency, which is estimated by acoustic recordings.

Keywords: drilling rig; process monitoring; operational cycles; sound measurement; electric current acquisition; threshold-based segmentation; envelope spectrum

---

1. Introduction

All of the leading producers of drilling rigs in the recent three decades have devoted their attention to the development of process monitoring systems to enhance the control and quality of blast-hole drilling [1]. This initial step of ore extraction in mines using blasting technology has to be performed with an excellent, repeatable accuracy and high efficiency in order to ensure the economic feasibility of mining operations in the reality of the currently worsening mining–geological conditions, meaning decreasing thickness of the deposits, their depletion and deepening location. Since on-board monitoring systems have became a standard for this type of machine, to utilize their potential, it is crucial to develop accurate data processing algorithms that will enable the assessment of the efficiency of the blast-hole drilling process.

To date, there have been several research directions regarding monitoring and data analysis systems related to the drilling process. Equipment could be used for rock mass characterization, drill bit wear monitoring, the assessment of the performance of the drilling process and operator's skills, development of the design of a machine, minimization of the energy consumption for the drilling, etc. One of the most important parameters in blast-hole drilling is the service time of drill bits [2]. Their design and gradual deterioration are the principal components influencing overall productivity and machine down-times. Although some methods for the on-line condition monitoring and wear prediction [3] of drilling bits are currently under development, efficient diagnostics of these elements is still not a trivial task. The information about the number of drilling cycles as well as the quality of the drilling may be also used for maintenance policy.

Another issue arising in the real conditions of underground mines is the setting of appropriate drilling parameters for the usually not definitely known and varying rock media hardness, namely the drill bit rotation speed, feed pressure and percussion frequency. Although certain practical recommendations are given for each type of drilling tool by their producers, operators of machines need on-line data processing methods to adequately react to quickly changing geological conditions.

Regardless of the source of the measurement data, one can easily achieve very good results in the monitoring of the drilling process using basic analysis methods. In this work, it is shown that, with the help of very basic data acquisition systems, we obtain valuable measurements that allow us to avoid sophisticated and expensive measurement methods. The paper is organized as follows: first, we recall some work related to the monitoring of the drilling process; then, we define a research problem and propose a solution to it that includes experimental work, hardware and software solution. The validation of the proposed methods is conducted by comparison with an on-board embedded monitoring system installed on the machine as an option.

Monitoring of the Drilling Process—The State of the Art

The monitoring of blast-hole drilling performance, using both externally applied sensors and observations of the parameters of drilling tool-driving components, has been considered in numerous studies. Since the emergence of the concept of the so called "specific energy", described by R. Teale in 1965 [4], the main focus of research has aimed at upgrading the reliability, correctness and applicability of the rock mass characterization. The Measurement While Drilling (MWD) methods were used to evaluate roof strata based on feed pressure and acoustic data [5,6]. The detection of the abrupt changes in data patterns related to drilling, which was exhaustively presented in [7], has been applied to detect interfaces between different lithological types or to locate voids in the roof layers. A brief introduction to one of the techniques that allows for the detection of discontinuities, voids, cracks and the identification of different roof strata—a cumulative sum algorithm (CUSUM)-based program—has been shown in [8]. An accelerometer and acoustic sensor, independent of other sensors, have been successfully used to detect small apertures and voids in [9].

A simple and widely used support vector machine (SVM) pattern recognition algorithm has been successfully used to classify different soft and hard rocks in real-time, as presented in [10]. Input parameters such as the penetration rate, rotation pressure, feed pressure, vibration and acoustic signals have been used in combination with time series classification. Parameters such as the weight on bits or thrust and accelerometer data related with drilling—measured on the drill head as well as on other parts of the machine—have been used to estimate the mechanical properties of rock mass [11]. Another contribution to the evaluation of rock mechanical parameters with MWD measurements was presented in [12], in which an analytical model of a process that could be useful to evaluate the uni-axial compressive strength was described. The authors broke the process down into the repetitive cycles of feeding and cutting, and based on the fact that only cutting and indentation are effective, while friction on the flank surface and idle running do not contribute to the effective drilling work, an effective specific energy parameter that is independent of the penetration rate was derived.

MWD data, compared with geophysical loggings, were proven to be appropriate for determining rock properties in [13]. In [14], a method to derive operational parameters of the drilling jumbo (rate of penetration, rotary speed and torque and pulldown force) from voltages and currents was described. Data of this type have been applied to estimate rock-breakage characteristics, aimed at mine-to-mill optimization solutions in open-pit mining, as conceptually presented in [15]. Another example of the characterization of drilled material in open-cast mines, performed based on the drilling performance indicator linked with geomechanical parameters and in which the new measure of the Modulated Specific Energy (SEM) was introduced, can be found in [16]. Other interesting instance of the MWD technology application is the prediction of Excavation Damage Zones sizes, which are induced by blasting in underground excavations [17].

As can be seen, the focus of researchers was aimed at obtaining information about the rock mass, indirectly and automatically, leading to an increase in the understanding of the interactions between the machine and rock material.

In addition to the characterization of rock properties, process parameters can be used for the on-line diagnostics of drilling tool wear and other parts of machine equipment. By conducting laboratory tests on the wear resistance of drill bits inserts made of Cemented Tungsten Carbide [18] or their optimal selection and performance [19], it is difficult to approach the real contact conditions due to the possible dynamical effects related to elastic deformations of the drill-string and other structural elements [20]. For example, the effects of coupling modes on torsional, bending and axial vibrations can be observed in drilling units [21].

Therefore, the simultaneous observation of the acoustic data, voltages, currents and other signals available from the on-board data acquisition systems—e.g., bailing water pressure or temperature [22]—should help to better understand some hidden dynamical processes in the machine, leading to higher performance via its automation and control.

Acoustic telemetry and mud pulse telemetry are the communication methods used in deep-well drilling to provide valuable on-line information from the underground. However, acoustic signals have a significant attenuation and need repeaters for their transfer to on-surface monitoring systems [23]. The horizontal blast holes have a comparatively small depth (2–6 m); thus, acoustic signals can be directly registered by a properly installed microphone. This particular feature constitutes a physical basis for drilling process and tool condition monitoring by the acoustic waves generated in the rock deformation zone.

To correctly understand the events in the drilling process and equipment, the adequate segmentation of continuously recorded signals is absolutely necessary, as in any other types of underground mining machines. The task of the identification of operational cycles in the monitoring systems of underground vehicles is solved in [24].

Cycle extraction from process data has been discussed in [25]. Stefaniak et al. proposed several algorithms for the multidimensional analysis of the data from an on-board monitoring system used in the underground heavy duty load-haul-dump trucks (LHDs) [26]. The problem of drilling robotization has been discussed in [27]. Timusk et al. proposed a method for the automated operating mode classification of online monitoring systems [28]. Wodecki discussed long-term data analysis for condition monitoring purposes [29], Al-Chalabi et al. [30] discussed the problem of reliability in drilling. The problem of data acquisition, validation and analysis for LHD machines has been discussed in [31]. Acoustic emission and different acoustic signals have also been used for drilling monitoring in various contexts [32–37]. However, as mentioned in the discussed case, blast-hole drilling in the considered mine is a specific problem due to its enormous scale. Precise information about the number of holes, their lengths and the manner in which the drilling was performed is required; in that sense, the monitoring of the drilling process and drilling related-knowledge acquisition is very challenging.

2. Problem Definition

To properly assess the efficiency of the works performed by mobile machinery, the division of the generic process of blast-mesh preparation into sub-processes is required. Cyclically repetitive processes, which are completed by the drilling jumbo included in the preparation of the mining face before the injection of explosives in blast-holes, include the positioning of the drill, hole pre-boring, the actual drilling of the blast hole and blast hole flushing. The identification of the above-mentioned sub-processes and the definition of their duration is undoubtedly valuable from the perspective of efficiency and quality assessment. The problem can be approached by taking advantage of the electrical current signal, which varies when the load on the working unit changes consecutively with the succeeding sub-processes comprising the drilling cycle. Moreover, inspired by the operator's experience, an additional source of information regarding drilling performance can be taken into account, which is the acoustic signal. Being aware of the characteristic patterns in sound emission related to consecutive actions of a drilling rig, namely idle running (when the drill is positioned), pre-boring and the actual drilling, some informative features can be found in a pre-processed signal. The identification of such features can be used for the automatic distinguishing of the sub-processes and their duration and to count the number of drilled holes.

This paper outlines part of the fully automated evaluation of the operator's and machine's efficiency. The methods presented by the authors have potential for the monitoring of drilling in mining faces, with special attention being paid to the identification of the drilling of individual holes, based on three separate sources of information:

- The automatic identification of cycles and analysis of their features (electric parameters, time, amount, presence of pre-boring etc.) based on signals registered by the on-board data acquisition system;
- The automatic identification of cycles and analysis of their features based on the raw electric current consumption signal measured directly on the power line;
- The automatic identification of cycles and analysis of their features based on external noise recording.

3. Machine and Experiment Description

In underground mines that exploit deposits with the use of blasting technology, the drilling rig is the first machine in the technological cycle, the performance of which affects the general efficiency obtained in the exploitation area. Self-propelled drilling machinery for mining purposes can be decomposed into key elements, which are the operator's cockpit, electric cabinet, water hose and electric cable reeling units, diesel engine (or electric motor in case of battery-powered rigs), hydraulic system, leveling jacks, boom, arm and drill. In addition to drill carriages, there is sometimes a platform for a miner incorporated in the machine's structure. Currently, drilling jumbos are adjustable to local mining conditions, and therefore there are a multitude of available sizes—from the large sizes designed for large excavations and construction of tunnels to small, compact versions that are appropriate in cramped conditions (in ultra-low seams, for example). The rigs are equipped typically with up to four booms. It should also be noted that the other type of machinery—the bolting-rig—that is responsible for the reinforcement of the excavation after the blasting and loading of ore is very similar from a design point of view and also performs drilling to prepare anchor holes. Thus, not only can the performance of the drilling rig operating at the first step of the technological cycle be approached with the method described in this study, but the efficiency of rotary drilling and the preceding injection of the resin bolts can be evaluated as well.

3.1. Drill Rig Facemaster 1.7

All the measurements described in this article were taken on the FaceMaster 1.7 single boom drill jumbo, as shown in Figure 1). This drilling rig is an example of a diesel-powered machine designed for the preparation of a 3.2 m net length blast holes, with diameters from a threshold of 41–76 mm. The drilling rig is equipped with an HC95 LM hydraulic drifter that drives its rotary-percussive rock drill. After reaching the face and switching the combustion engine off, the main functionality of the machine is enabled by the supply of electrical energy from the mine's energy distribution network. The electric motor driving the hydraulic pump and supplying the machine's working system (the extension arm and feeder allow the positioning of the drill according to the established drilling-and-blasting mesh, together with the drifter providing rotation and percussion movements) is powered by an electric current.

Figure 1. The drilling machine used in experiments.

3.2. Geological Conditions

The experiment comprised the drilling of multiple blast holes with standard-diameter granitoid blocks at their full length (70–80 cm). The rock composition predominantly included feldspar, quartz and plagioclase, containing minor micas and amphiboles. Grains of the rock were relatively coarse and uniformly distributed. The lithology of the sample designed for the experiment was a rough approximation of the actual rock types encountered in the ore zone in the KGHM Polska Miedź mines, where the following lithological sub-types of rocks were present (see Figure 2): dolomites and sandstones; grey, streaky dolomite; dark-grey, clayey dolomite; dolomitic shale; pitchy, clay-organic shale; grey, dolomitic sandstone; grey, clayey sandstone; and red, clayey sandstone. Furthermore, drilling performed in a uniform rock mass that is devoid of discontinuities, voids and interbeddings does not reflect the conditions present in the mining faces. Such occurrences of non-uniformity are suspected to cause changes in the instantaneous performance of rock drill, and thus other parameters are related to it, including those analyzed in this study.

It should be noted that a granite rock was chosen because its average compressive strength was approximately the same as the average compressive strength in an ore deposit profile, at approximately 130 MPa [38,39]. The aforementioned rock was easily accessible on the surface and had similar mechanical properties to the ore raw material. This made the experiment a good simulation of the conditions in an underground copper mine.

3.3. Scenarios for Experiment

The underground working conditions in which a single boom drill jumbo tends to work are often unpredictable, and it is very demanding to extract signals only from the machine's drilling process. Based on ground experiments conducted at the MineMaster facility, measurements of vibrations, acoustics, and current consumption rates were taken. The principal approach was to implement many measurement sources for one process and distinguish the main features determining the drilling process phase between the different methods.

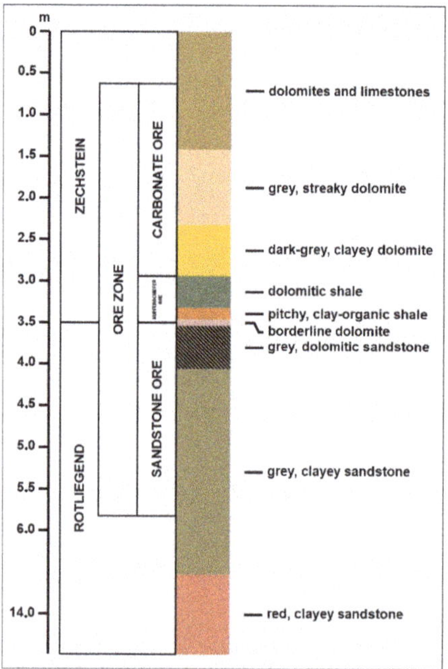

Figure 2. Geological profile of the Fore-Sudetic Monocline copper ore deposit (adapted from [40]).

Using the non-invasive method of the machine's workload assessment, a current measurement for instantaneous energy consumption was presented. An on-board diesel-engine powered only the steering system and a small hydraulic pump of the presented machine. As the power consumption of the drilling system was incomparably greater, the internal combustion engine was not used for this purpose. Thus, on these machines, the hydraulics of the drilling system were powered with an external AC 500 V or 1000 V three-phase voltage with a frequency of 50 Hz. An external source of electricity was connected to the machine's electric cabinet. The power supply was symmetrical, and consequently, a current clamp was mounted on any conductor of the supply line. It is important to underline that the level of energy consumption during the drilling process for the other machine subsystems, apart from the drilling system, was roughly constant. Owing to that fact and thanks to the current level analysis, we were able to gain insights into how the drilling process changed directly from current measurement.

Our approach allowed us to observe the movement of the drill bit and the number of holes, as well as the type and "style" of drilling. Moreover, the inrush current was measured and was found to be significantly above the rated current because of a cold start that occurred at the beginning of the work. Summarizing the above, the proposed current level measurement gave us a deep understanding of the drilling process from a new approach based on the level of current rather than, for instance, the pressure of drilling subsystems, which is currently widely used.

The experiment was divided into eight phases in which drilling points were deployed on the rock surface, as can be seen in Figure 3. The first experiment included the drilling of four holes in a straight line, and the second included five holes made in the same manner. The next stage consisted of five holes simulating an inexperienced machine operator, where the pre-drilling phase was slightly corrupted, as can be seen in the deeper analysis presented in the further Section 6.3. Following that, the fourth trial included five holes of blow/dry drilling, with a further reach of the drill into the rock. The fifth phase included another four holes made in a straight line, and the last one was made without the flushing of the hole. The sixth stage comprised one properly drilled hole, and later the operator quickly moved the drilling system of the machine back. The corresponding data obtained from current usage showed that our experiment differed significantly from the standard drilling process. In the seventh phase, one long-hole drilling process can be seen. The last step comprised two holes made when the rear support feet were retracted.

Figure 3. Granitoid block used in the experiment.

4. Data Sources

In the first stage of the extraction and processing of the excavated material, we used a single boom drill jumbo. To precisely control the process, it was necessary to acquire data from the drilling process of the aforementioned machine. Although the working parameters depended in large part on the physicochemical parameters of the rock, this process was repetitive, and it was possible to distinguish subsequent stages of the drilling process. From the perspective of a company in which tens or hundreds of these machines work continuously, it is important to send only key information about the machine's operation to the main supervisory control and data acquisition (SCADA) system. For this purpose, it is necesasry to determine the number and the quality of drilled holes. The system used for the diagnostics of working processes and technical conditions to support the management of the machinery in Polish rock mines is presented in [24]. This system, implemented for self-driving mining machines, consists of several types of gathered information, mainly regarding subsystem pressures or the mechanical parameters of engines.

In this research, we aimed to develop an accessible and effective way to gather information about drilling performance. Thus, we used data related to power supply parameters acquired from the on-board system installed on the tested machine. Moreover, we proposed our electric current monitoring system as well as acoustic measurement (a simple smartphone was used to record a video from which sound was extracted). All these data acquisition methods, together with the techniques of their processing, are described in the following subsections.

4.1. On-Board Data Acquisition System

The first source of data in the performed experiment was the on-board data acquisition system that registered various parameters related to the machine systems' power supply. The sampling frequency of the obtained dataset was equal to 10 Hz. For the purpose of the analysis described in this paper, the authors acquired the signals presented in Table 1.

Table 1. Signals obtained from the on-board system.

Variable Name	Description
Usr[V]	Voltage
Isr[V]	Current
rActivePowerTotal (kW)	Active power
rReactivePowerTotal (kVAr)	Reactive power
rApparentPowerTotal (kVA)	Apparent power
rPowerFactorTotal	Power factor (cosinus ϕ)
rActiveEnergyTotal (kWh)	Cumulative energy consumption

4.2. Induction Clamps for Current Measurement

As with many sensors used for current measurement in the automation industry, the majority of solutions are based on in-circuit measurement. It is necessary to apply this apparatus to the design level of machine development. On the other hand, current clamps are more convenient, as the current measurement system can be applied without any intervention in the machine's electric circuits. The center of the measured conductor should be installed in the current clamp jaw. Additionally, it is necessary to consider that the clamp is perpendicular to the conductor, which helps users to gather appropriate results.

For experimental purposes, Fluke's current clamp i400s (see Figure 4) was used. An output AC voltage signal of 0–400 mV was received. For data handling, the cDAQ-9171 USB chassis with an analog-to-digital converter (ADC) module was used. The frequency of measurements was 2000 Hz; as a result, we observed a quick-change pulse at the moment of the engine starting. The second approach relied on the microcontroller-based measurement system, which omitted the presence of additional software from the signal card producer. All data were saved in .txt format on an SD card.

Figure 4. Current measurement clamps mounted on the wire of interest.

In addition to the electrical parameters, the authors registered an audio–video recording of the drilling activity. The recording was taken using a smartphone camera with a video sampling frequency of 60 frames per second and an audio sampling frequency of 48 kHz. The acquired video data provided a reference for other data sources and allowed us to validate timings and actions. Additionally, the audio feed extracted from the video files provided an additional data source for process-related analysis (Figure 5).

Figure 5. The concept behind acoustic data acquisition.

5. Methods

The purpose of the methodology was to develop reliable raw data processing algorithms that are capable of extracting information about the drilling process. More precisely, it was expected that it would be possible to enable the identification of the number of cycles (the number of drilled holes) as well as information about the cycles' properties (duration of the cycle, pre-drilling phase, etc.) to be provided.

5.1. Data Processing Framework

The methodology proposed here is a multistep procedure. Below, we first provide a general step-by-step framework (see Figure 6); then, details of each presented step are further explained.

5.2. Density Distribution Estimation

In the presented application, density distribution was estimated using a kernel density estimator, which uses the estimated probability density function of a random variable [41,42]. For any real values of x, the kernel density estimator's formula is given by

$$\hat{f}_h(x) = \frac{1}{nh} \sum_{i=1}^{n} K\left(\frac{x - x_i}{h}\right),$$ (1)

where $x_1, x_2, \ldots, x_n$ are random samples from an unknown distribution, n is the sample size, $\mathbf{K}(\cdot)$ is the kernel smoothing function and h is the bandwidth. In this implementation, a standard Gaussian kernel is used.

The value of the bandwidth is obtained using the so-called Silverman's rule of thumb [42]. If Gaussian basis functions are used to approximate univariate data and the underlying density being estimated is Gaussian, the optimal choice for h (that is, the bandwidth that minimizes the mean integrated squared error) is

$$h = \left(\frac{4\hat{\sigma}^5}{3n}\right)^{\frac{1}{5}} \approx 1.06\hat{\sigma}n^{-1/5},$$ (2)

where $\hat{\sigma}$ is the estimator of the standard deviation of the samples and n is the number of samples.

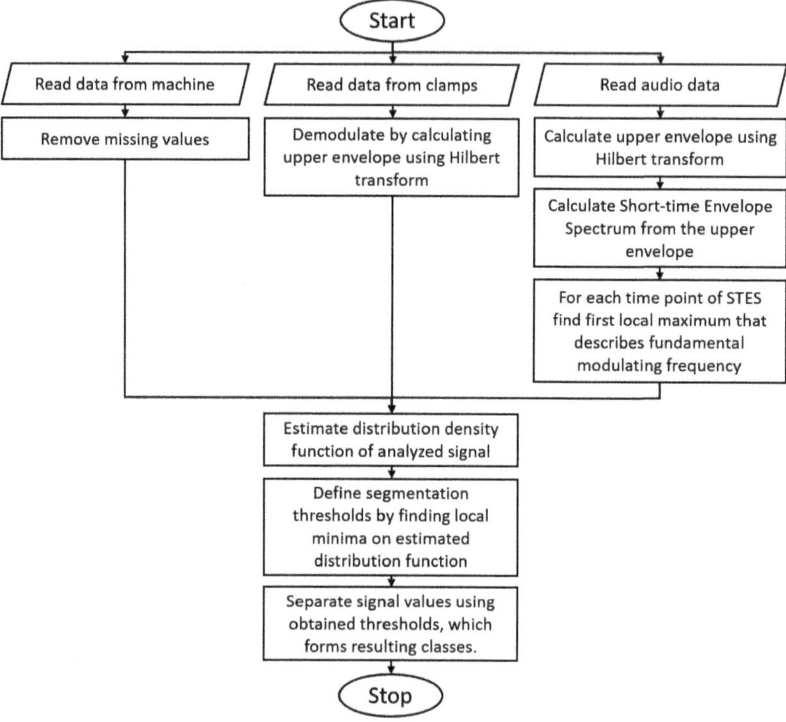

Figure 6. Flowchart of the performed analysis.

5.3. Threshold-Based Segmentation

The threshold-based segmentation of the signal is a fast and reliable method for the segmentation of data when individual states do not change over time. Thus, it can be useful for the processing of current signals, because current consumption levels (current consumption in the idle state, pre-boring state and drilling state) are typically well-defined.

For a given signal $X = \{x_1, \ldots, x_N\}$ of length N with a predefined set of thresholds T of size K, the class indicator c_n is assigned to each sample based on the regime (range of values between two consecutive thresholds) to which it belongs. Firstly, the vectors of thresholds need to be padded from both sides with the minimum and maximum value of the dataset, so that each regime can be described with lower and upper boundaries:

$$T = \{\min(X), t_1, \ldots, t_K, \max(X)\} \tag{3}$$

Thus, the new vector $C = \{c_1, \ldots, c_N\}$ of length N is created:

$$\forall_{n \in \{1:N\}} \exists_{k \in \{1:K+1\}} c_n = k \quad \text{for} \quad t_k \leq x_n < t_{k+1} \tag{4}$$

For ease of use, this formula has been presented in the form of pseudocode presented in Algorithm 1.

Algorithm 1: Threshold-based segmentation.

```
1  C = zeros(N,1)                          //declaration of N-element vector for cluster indicators
2  for n in 1:N do
3      /* for each signal sample...                                                              */
4      for k in 1:K+1 do
5          /* for each class between two thresholds...                                           */
6          if t_k ≤ x_n < t_{k+1} then
7              /* if a sample belongs between these two thresholds...                            */
8              c_n = k                       //Assign class number to a given sample
```

Thus, for each sample from the signal X, there is a value in vector C that holds its regime number. From this point, the easiest way to perform the actual segmentation is to calculate the numerical derivative of the vector C, where non-zero values will indicate the point of regime change, and the value for the next sample provides information about the class to which a given segment belongs.

5.4. Short-Time Envelope Spectrum

The short-time envelope spectrum (STES) applies the idea of envelope analysis, which is focused on cyclic modulations present in the signal, to the framework of time–frequency analysis. The authors decided to use the spectrogram as the base time–frequency representation. In the first step, the short-time Fourier transform (STFT) for discrete data $x[0], x[1], ..., x[N-1]$ is given by the formula [43]

$$\text{STFT}_{k,g} = \sum_{f=0}^{L-1} x[g+f]w[f]e^{-j2\pi kf/N},$$ (5)

where $0 \leq k \leq N-1$ is the frequency bin, g is the time point and $w[.]$ is a window of length L. One can observe that in STFT, for each time point, the Fourier transform is calculated using a fast Fourier transform (FFT). Furthermore, the spectrogram is the squared absolute value of the STFT:

$$\text{Spec}_{k,n} = |\text{STFT}_{k,g}|^2.$$ (6)

The idea behind this method is to use a spectrogram not for the raw signal but for its upper envelope, calculated as a modulus of the Hilbert transform:

$$Env(X) = \text{abs}\left(\mathcal{H}(X)\right)$$ (7)

where $\mathcal{H}(X)$ denotes the Hilbert transform of the signal X [44]. Applying the spectrogram algorithm to the envelope, one can obtain the STES:

$$\text{STES}(X) = \text{Spec}(Env)$$ (8)

6. Results

6.1. Process Monitoring Based on Electrical Data

For this analysis, the variable describing the electrical current was selected as an input signal due to its clarity and descriptive structure (see Figure 7). The main feature under consideration is the difference between levels describing the idle state, pre-boring and proper drilling.

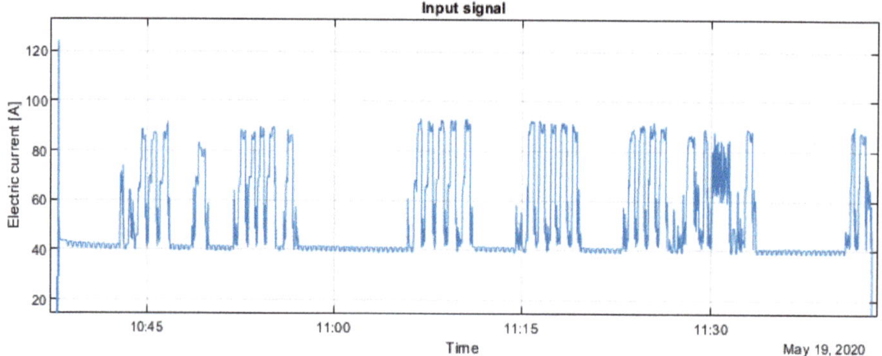

Figure 7. Input current signal.

In the first step, the distribution of data is estimated using the algorithm described in Section 5.2. According to the expectations, three main modes emerge that aggregate information about three states (idle, pre-boring and drilling), and the local minima between those modes indicate the optimal thresholds that will allow the separation of those states (see Figure 8).

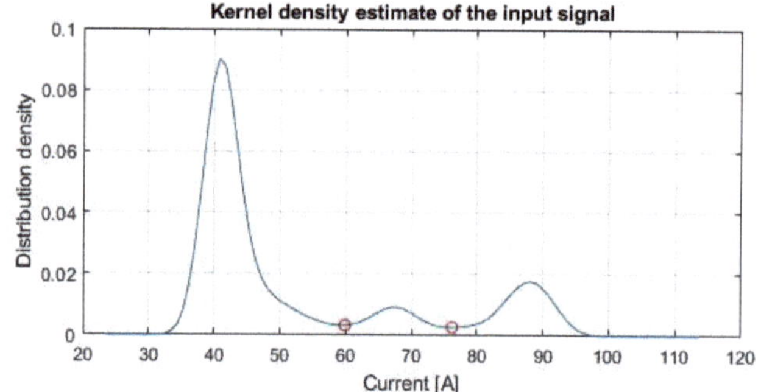

Figure 8. Kernel density estimate of the input signal. Detected thresholds are marked with red circles.

Finally, the signal is segmented according to the discovered thresholds (see Figure 9). Segments describing each type of activity are denoted with different colors. Separate segments denoting drilling are then counted, which allows the amount of drilled holes to be obtained; in this case, 28.

Additionally, assuming the presence of three states instead of two (one could simply perform segmentation into "drilling"/"no drilling" states) allows the assessment of the quality of the process. Namely, it allows us to detect if a particular drilling action has been performed, including the pre-boring phase. As a result, it was detected that the first 14 holes (drills before the 11:15 time mark) were drilled properly with pre-boring, while the next five holes (drills immediately after the 11:15 time mark) exhibited no detected pre-boring phase, according to the request of the machine operator (i.e., the pre-boring phase was omitted on purpose). The next four holes were drilled with a present but significantly reduced (barely noticeable) pre-boring stage, which may lead to difficulties in detection. The analysis showed that two of those four holes exhibited a pre-boring phase that was significantly shorter than that for holes made properly, and the remaining two holes did not exhibit pre-boring in the results, which was also an expected result. The rest of the activities (at around the 11:30 mark and later to the end of the studied time) were performed randomly by the operator: drills were supposed to be performed properly, improperly and with a longer or shorter drilling duration. From those, five drills were identified as proper but with varying drilling durations (some of them were incomplete, and thus shorter), and one drill showed a reduced feed power.

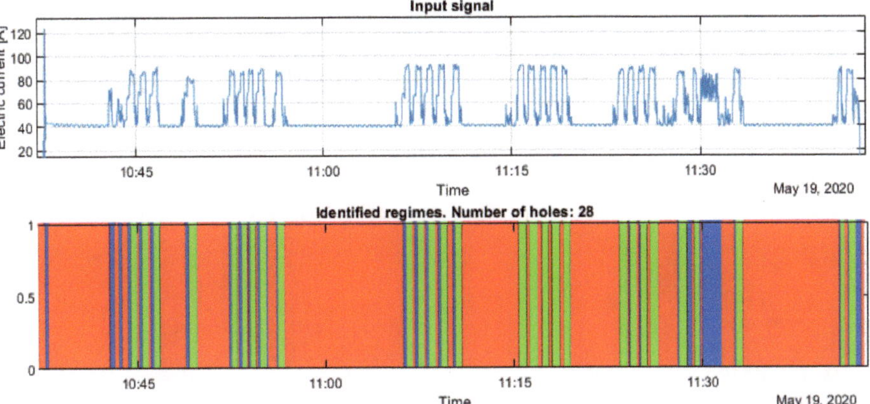

Figure 9. Segmentation results. Red: idle state, blue: pre-boring or ill-defined action, green: drilling.

Based on the obtained segmentation results presented by the authors, in Table 2, basic statistical parameters of the operational regimes are shown.

Table 2. Statistics of the identified regimes.

Regime	Mean	Standard Deviation
Idle	43.4	6.91
Pre-boring	67.8	5.15
Drilling	86.5	6.26

6.2. The Proposed Current Measurement Analysis

Data obtained from Fluke's i400s current clamp are presented in Figure 10 in blue color (after taking the absolute value to improve the temporal resolution of local maxima for demodulation, the raw signal is symmetrical around 0). The raw signal is a sine function that is amplitude-modulated with the actual signal amplitude, with a base (carrier) frequency of 50 Hz. The amplitude, calculated as the instantaneous current consumption and presented in Amperes [A], can be explicitly determined from the root mean square (RMS) signal value, converted directly from mV as a voltage output signal according to the amplification set on the current transformer built into Fluke's clamp. Data were taken with a 2 kHz frequency to enable further analysis.

After extracting the upper envelope using the Hilbert transform (as in Section 5.4), the obtained signal (the red plot in Figure 10) shows the values of the measured current consumption in Amperes. In this form, it is usable for analysis and, more importantly, it perfectly matches the shape and scale of the signal registered by the on-board system (see the Section 7.1). Due to this fact, it is possible to use those signals for cross-validation, and they can be analyzed using the same methods if desired. This is also the reason why the authors decided to limit the data analysis part of the current-related data to the machine-originated signals to avoid showing identical results twice.

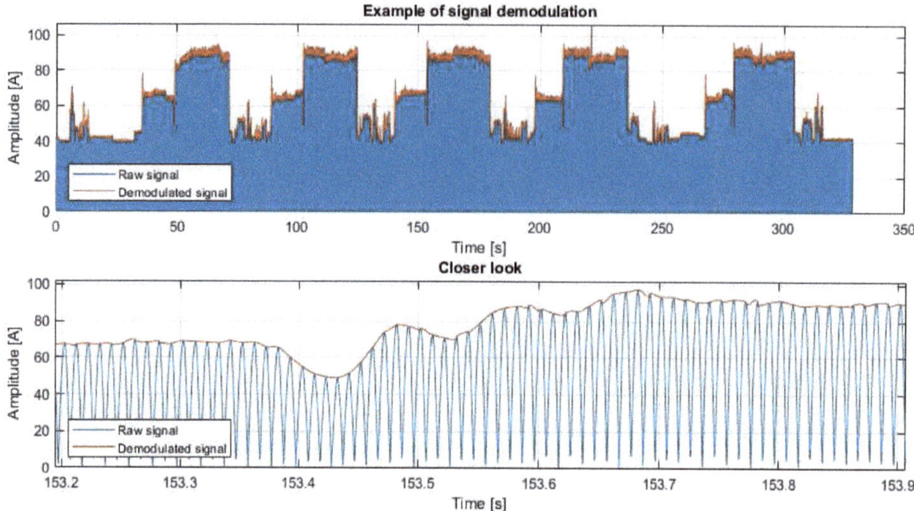

Figure 10. Raw data output from the current clamp sensor (blue), amplitude-demodulated to obtain proper values of the current (red).

6.3. Process Monitoring Based on Acoustic Data

In this example, acoustic data extracted from the ordinary video recording are analyzed (see Figure 11). The considered segment corresponds to the first four drilled holes from the previous example. During the experiment, video segments were recorded in shorter parts to avoid the necessity of working with large video files. Audio recordings were registered continuously because significantly less memory space per unit time was needed for their storage. As one can see, each individual drill is clearly visible. This signal could be processed very straightforwardly using the signal variance, which changed over time and was analyzed with the goal of finding segments of increased variance. However, considering that audio recordings can capture anything occurring in the neighborhood of the machine, the segments of increased acoustic energy could be caused by other sources. Thus, the authors decided to take advantage of the property visualized in Figure 12.

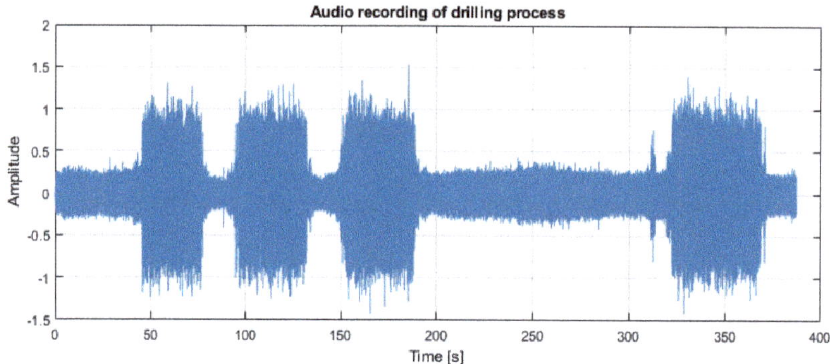

Figure 11. Example of the audio signal recorded during drilling. Sections of increased energy show individual drills.

As one can see, the frequency of drill percussion during pre-boring differs from that experienced during drilling. Since those impacts can be understood as an amplitude modulation, the envelope spectrum can reveal the spectral structure of this modulation and give access to the fundamental frequency of percussion in a given moment.

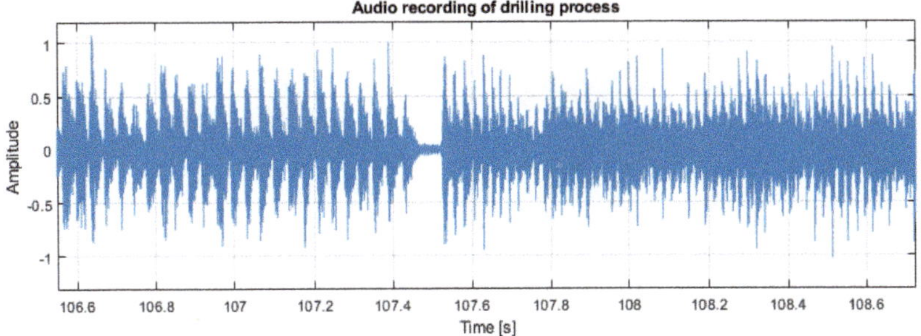

Figure 12. A closer look at the acoustic signal at the moment when pre-boring ended and proper drilling began. Visible portions of energy indicate individual percussion strokes. A difference in the percussion frequency between pre-boring (the left part of the signal) and proper drilling (the right side) is clearly observable.

To achieve this goal, a short-time envelope spectrum was constructed for this signal (see Figure 13 top panel), as described in Section 5.4. As can be seen, frequencies that are responsible for the modulations are presented and laid out over time. Sometimes, consecutive harmonic frequencies are also visible; however, for this methodology, we were interested only in the fundamental frequencies. Based on this representation, it is possible to detect fundamental frequencies over time by finding the first significantly energetic component above the high-energy mode centered around zero. This procedure is repeated for each timestamp, and the result is presented in the bottom panel of Figure 13.

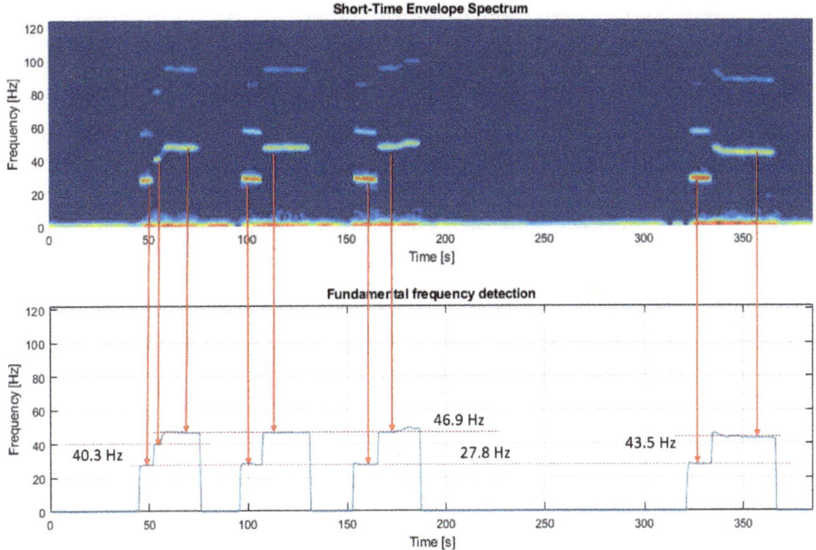

Figure 13. Results of audio signal analysis using the segment presented in Figure 11. The short-time envelope spectrum (in the top panel) allows us to observe the spectral structure of the modulation caused by the energy variation generated by percussion. Further processing allows us to extract and track the fundamental frequency of percussion (bottom panel).

This result describes the first four holes drilled during the experiment. Figure 14 presents the results obtained for the remaining parts with an indication of their location in time with respect to the continuous data obtained from the machine. The results are equally successful and, if desired, can be merged into one signal for further analysis. However, depending on the details of the solution deployed based on such measurements, it is possible that one can expect continuous audio measurement in the future.

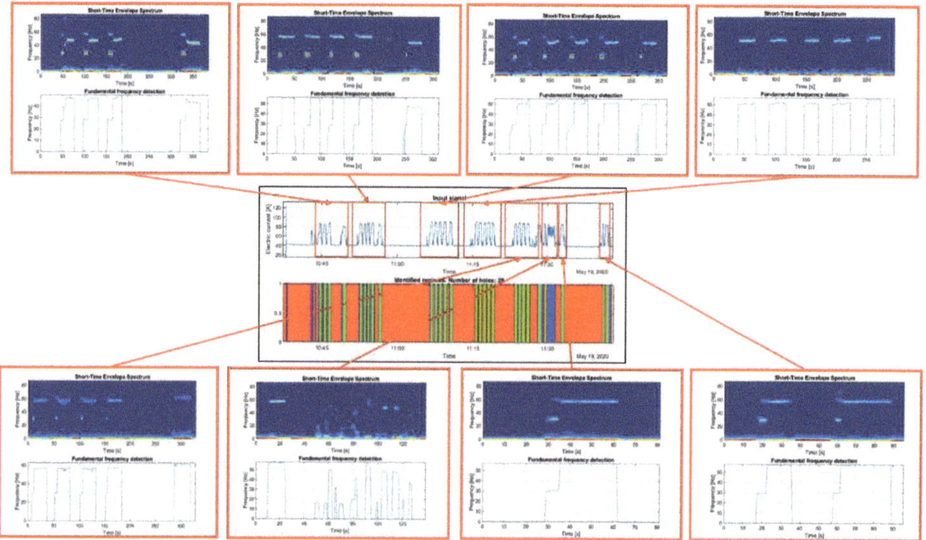

Figure 14. Results obtained for all of the sections of the experiment.

It is important to note three outcomes regarding this result. Firstly, this approach allows the verification of the proper operation of a percussion control system by assessing its frequency of operation during various activities. Secondly, the structure of the obtained output signal is similar to the structure of the electric current signal from the previous example; thus, this result can be further processed in the same way and provides a validation-related value in case there is any problem with the system registering electric parameters. Finally, this example shows that data acquisition for similar analysis can be performed very easily if it is desired to quickly assess the functioning of any drilling machine in operation by simply going near the machine and recording a short piece of audio data.

7. Validation

The monitoring system installed on the machine provides high-quality data that allows the precise analysis of the drilling process in terms of the number of holes drilled, the duration of a single cycle, etc. In this work, it was used as a reference system to validate our monitoring devices. The reason for preparing a new monitoring system is simple—the monitoring system proposed by the manufacturer is optional when purchasing the machine, and to the authors' best knowledge, there is no universal system for such a purpose that could be used in the case of the failure of the previous device. Moreover, in older machines, there are no monitoring systems for the drilling process.

The proposed system is based on electric current consumption analysis. It provides basic hardware as well as software to calculate the required parameters. The first version of the system is based on a standard Data Acquisition Card from National Instrument, while the second version is based on an Arduino micro-controller with other electronic components to support local data recording on the SD card of the processed signal (amplified and digitized). A schematic representation of different data-acquisition methods that can be used to assess the drilling rig's operation can be found in

Figure 15. Each monitoring system provides raw data in .csv format, which is then processed using the Matlab environment (signal processing toolbox). We used Matlab ver. R2020b and personal computers with an average office-class computational speed: an Intel I7 processor operating at 4 GHz and 8 GB of memory.

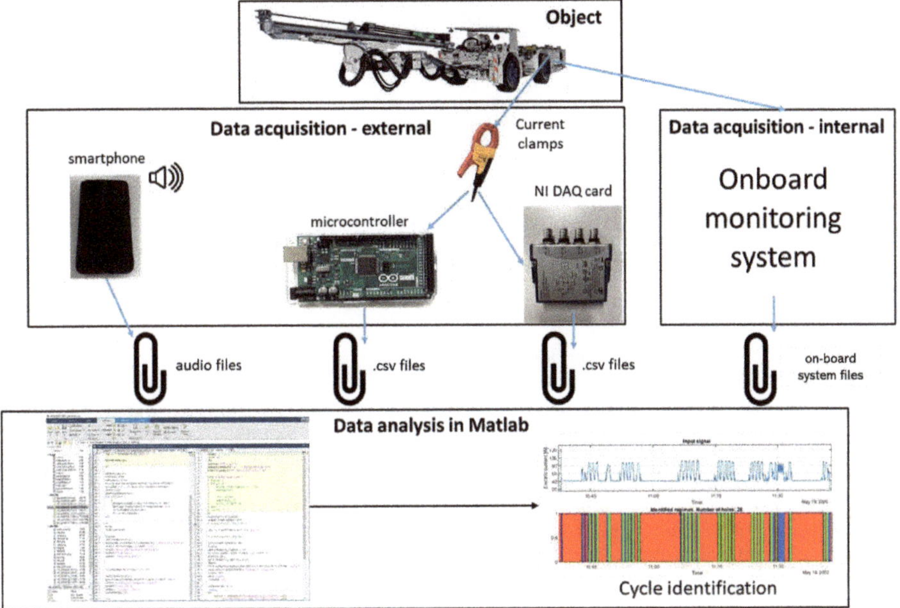

Figure 15. Schematic representation of different data-acquisition methods and the results obtained on the basis of various types of input data.

7.1. Audio–Video Recording

Both cases showed excellent technical parameters and allowed us to measure the current signal with a higher sampling frequency. Increasing the sampling frequency provides increased opportunities for transient events analysis in the signal. Current consumption requires special devices and integration with the electric circuits in the machine. It could be an interesting option to use acoustic signals for the monitoring of the drilling process. It should be mentioned that the diesel engine is switched off during drilling; the process is driven by an electric motor, meaning that background noise should be reduced.

In this section, as a reference, the signal from the on-board monitoring system was used. As presented in Figure 16, the signal acquired by the proposed system (blue curve) was very similar to the reference signal. As the sampling frequency was higher, more noise (that could be easily removed) could be observed, as well as some extra information related to transient events—low-frequency sampling was not able to capture this. The difference between the on-board and the proposed systems is related to the technologies used and sampling frequency.

Information extracted from acoustic signals has been discussed in detail in the previous section. It can be seen in Figure 14 that, thanks to instantaneous frequency monitoring, it was possible to identify the pre-drilling and actual drilling phase. We can therefore conclude that both electric and acoustic measurements are very useful for drilling monitoring.

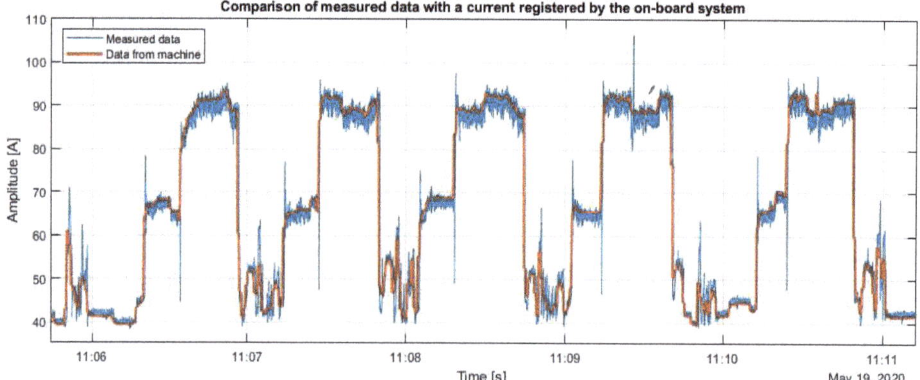

Figure 16. Comparison of the shape and scale of the measured signal (blue) to that registered by the machine system (red).

8. Discussion And Conclusions

In this paper, a data analysis methodology was proposed for three different data streams.

Case 1:Tthe data originating from the on-board monitoring system. The data, namely the current consumption, are a function of the load applied to the drilling tool. As the idle mode current is low, pre-drilling is associated with a medium load (60–70 A) and actual drilling means a heavily loaded system, so significant (>80 A) current consumption is visible. The signal is relatively clear, and so one can distinguish key elements of the cycle using simple statistical pattern recognition tools (threshold estimation based on probability density function). The statistical variability of real signal levels over the phases of drilling cycles is not very wide (up to ±5%); thus, the proposed method is robust and does not need subsequent tuning to account for changes of bits' parameters or rock properties within a certain region of mining.

Case 2: The data originating from the proposed electrical current monitoring system. These data should be pre-processed to a similar format as data from the on-board system; then, the processing is performed according to the procedure in case 1. Besides, the higher sampling frequency of a new monitoring system allows the development of algorithms for specific feature detection in the drilling process, which are currently under development.

Case 3: The acoustic signals from the microphones. These data require more advanced techniques as they are processed using the Hilbert-based envelope analyzed in the time–frequency domain in order to extract the instantaneous frequency. It appears that idle mode, pre-drilling and actual drilling have completely different acoustic signatures in terms of sound levels but also in terms of their frequency structure. Thus, instantaneous frequency (IF) may be used as a feature for drilling monitoring. Using the IF as a function of time, one may simplify the situation to case 1 and apply the segmentation procedure.

Finally, all three approaches were compared and a very good convergence of results was shown. A holistic solution was proposed that covers the data acquisition system (electric current measurement and acoustic signal acquisition) and data processing algorithms for the extraction of information about cycles related to drilling. The data acquisition system is relatively cheap and may be easily composed of accessible components (induction clamp, microphone and an Arduino microcontroller), but it requires low-level programming experience and knowledge related with embedded electronics design. The second version is even simpler and is based on a simple sound recorder—a standard smartphone was used to capture the video and sound.

Future work will focus on large-scale testing in an underground mine to estimate the influence of external noise and repetitive reverberations in the confined space of narrow tunnels.

Author Contributions: Conceptualization, P.K. and R.Z.; methodology, J.W. and A.C. and P.Ś.; formal analysis, J.W.; investigation, M.W.-K. and J.S.; resources, P.Ś. and A.C.; data curation, B.Z. and J.S.; writing—original draft preparation, M.G. and P.K. and J.W.; writing—review and editing, M.G., P.K. and B.Z.; supervision, R.Z. and A.C.; project administration, M.W.-K.; funding acquisition, M.W.-K. and R.Z. All authors have read and agreed to the published version of the manuscript.

Funding: This research has received funding from European Institute of Innovation and Technology (EIT), a body of the European Union, under the Horizon 2020, the EU Framework Programme for Research and Innovation. This work is supported by EIT RawMaterials GmbH under framework partnership agreement number 19036 (SAFEME4MINE. Preventive Maintenance system on safety devices of Mining Machinery). Moreover, this work has received funding from the European Institute of Innovation and Technology (EIT), a body of the European Union, under the Horizon 2020, the EU Framework Program for Research and Innovation. This work was supported by EIT Raw Materials GmbH under Framework Partnership Agreement No. 17031 (MaMMa-Maintained Mine & Machine).

Conflicts of Interest: The authors declare no conflict of interest.

References

1. Hoseinie, S.H.; Al-Chalabi, H.; Ghodrati, B. Comparison between simulation and analytical methods in reliability data analysis: A case study on face drilling rigs. *Data* **2018**, *3*, 12. [CrossRef]
2. Plinninger, R.; Ralf, J. Abrasiveness Assessment for Hard Rock Drilling. *Geomech. Tunn.* **2008**, *1*, 38–46. [CrossRef]
3. Thuro, K. Drillability prediction: Geological influences in hard rock drill and blast tunnelling. *Geol. Rundsch.* **1997**, *86*, 426–438. [CrossRef]
4. Teale, R. The concept of specific energy in rock drilling. *Int. J. Rock Mech. Min. Sci. Geomech. Abstr.* **1965**, *2*, 57–73. [CrossRef]
5. Bahrampour, S.; Rostami, J.; Ray, A.; Naeimipour, A.; Collins, C. Ground characterization and roof mapping: Online sensor signal-based change detection. *Int. J. Min. Sci. Technol.* **2015**, *25*, 905–913. [CrossRef]
6. Khanal, M.; Qin, J.; Shen, B.; Dlamini, B. Preliminary Investigation into Measurement While Drilling as a Means to Characterize the Coalmine Roof. *Resources* **2020**, *9*, 10. [CrossRef]
7. Basseville, M.; Nikiforov, I. *Detection of Abrupt Changes: Theory and Application*; Prentice Hall Englewood Cliffs; Prentice-Hall, Inc.: Upper Saddle River, NJ, USA, 1993; Volume 104.
8. Liu, W.; Rostami, J.; Keller, E. Application of new void detection algorithm for analysis of feed pressure and rotation pressure of roof bolters. *Int. J. Min. Sci. Technol.* **2017**, *27*, 77–81. [CrossRef]
9. Bahrampour, S. Instrumentation of a Roof Bolter Machine for Void Detection and Rock Characterization. In proceedings of the 32th International Conference on Ground Control in Mining, Morgantown, WV, USA, 30 July–1 August 2013.
10. Bahrampour, S.; Rostami, J.; Naeimipour, A.; Collins, G. Rock characterization using time-series classification algorithms. In Proceedings of the 33rd international conference on ground control in mining, Morgantown, WV, USA, 29–31 July 2014.
11. Lakshminarayana, C.; Tripathi, A.K.; Pal, S.K. Estimation of rock strength properties using selected mechanical parameters obtained during the rotary drilling. *J. Inst. Eng. Ser. D* **2019**, *100*, 177–186. [CrossRef]
12. Li, Z.; Itakura, K.I. An analytical drilling model of drag bits for evaluation of rock strength. *Soils Found.* **2012**, *52*, 216–227. [CrossRef]
13. Hatherly, P.; Leung, R.; Scheding, S.; Robinson, D. Drill monitoring results reveal geological conditions in blasthole drilling. *Int. J. Rock Mech. Min. Sci.* **2015**, *100*, 144–154. [CrossRef]
14. Khorzoughi, M.B.; Hall, R. Processing of measurement while drilling data for rock mass characterization. *Int. J. Min. Sci. Technol.* **2016**, *26*, 989–994. [CrossRef]
15. Park, J.; Kim, K. Use of drilling performance to improve rock-breakage efficiencies: A part of mine-to-mill optimization studies in a hard-rock mine. *Int. J. Min. Sci. Technol.* **2020**, *30*, 179–188. [CrossRef]
16. Leung, R.; Scheding, S. Automated coal seam detection using a modulated specific energy measure in a monitor-while-drilling context. *Int. J. Rock Mech. Min. Sci.* **2015**, *75*, 196–209. [CrossRef]
17. van Eldert, J.; Schunnesson, H.; Johansson, D.; Saiang, D. Measurement While Drilling (MWD) technology for blasting damage calculation. In Proceedings of the 12th International Symposium on Rock Fragmentation by Blasting, FragBlast12, Luleå, Sweden, 11–13 June 2018.

18. Tkalich, D.; Kane, A.; Saai, A.; Yastrebov, V.A.; Hokka, M.; Kuokkala, V.T.; Bengtsson, M.; From, A.; Oelgardt, C.; Li, C.C. Wear of cemented tungsten carbide percussive drill–bit inserts: Laboratory and field study. *Wear* **2017**, *386–387*, 106–117. [CrossRef]

19. Ergin, H.; Kuzu, C.; Balci, C.; Tunçdemir, H.; Bilgin, N. Optimum bit selection and operation for the rotary blasthole drilling using a Horizontal Drilling Rig (HDR)—A case study at KBI Murgul Copper Mine. *Int. J. Surf. Min. Reclam. Environ.* **2000**, *14*, 295–304. [CrossRef]

20. Sharma, A.; Srivastava, S.; Teodoriu, C. Experimental Design, Instrumentation, and Testing of a Laboratory-Scale Test Rig for Torsional Vibrations—The Next Generation. *Energies* **2020**, *13*, 4750. [CrossRef]

21. Songyong, L.; Xinxia, L.; Liu, X. Coupling vibration analysis of auger drilling system. *J. Vibroeng.* **2013**, *15*, 1442–1453.

22. Qiu, P.; Li, X.; Ning, J.; Wang, J.; Yang, S. Study on Thermal Energy Conversion Theory in Drilling Process of Coal and Rock Mass with Different Stresses. *Energies* **2019**, *12*, 4282. [CrossRef]

23. Shin, Y. Signal attenuation simulation of acoustic telemetry in directional drilling. *J. Mech. Sci. Technol.* **2019**, *33*, 5189–5197. [CrossRef]

24. Krot, P.; Sliwinski, P.; Zimroz, R.; Gomolla, N. The identification of operational cycles in the monitoring systems of underground vehicles. *Measurement* **2020**, *151*, 107111. [CrossRef]

25. Wyłomanska, A.; Zimroz, R. Signal segmentation for operational regimes detection of heavy duty mining mobile machines—A statistical approach. *Diagnostyka* **2014**, *15*, 33–42.

26. Stefaniak, P.; Zimroz, R.; Sliwinski, P.; Andrzejewski, M.; Wyłomanska, A. Multidimensional signal analysis for technical condition, operation and performance understanding of heavy duty mining machines. *Appl. Cond. Monit.* **2016**, *4*, 197–210. [CrossRef]

27. Stefaniak, P.; Wodecki, J.; Jakubiak, J.; Zimroz, R. Preliminary Research on Possibilities of Drilling Process Robotization. In *IOP Conference Series: Earth and Environmental Science*; IOP Publishing: Bristol, UK, 2017; Volume 95, p. 042027. [CrossRef]

28. Timusk, M.; Lipsett, M.; McBain, J.; Mechefske, C. Automated operating mode classification for online monitoring systems. *J. Vib. Acoust. Trans. ASME* **2009**, *131*, 041003. [CrossRef]

29. Wodecki, J.; Stefaniak, P.; Michalak, A.; Wyłomańska, A.; Zimroz, R. Technical condition change detection using Anderson–Darling statistic approach for LHD machines–engine overheating problem. *Int. J. Min. Reclam. Environ.* **2018**, *32*, 392–400. [CrossRef]

30. Al-Chalabi, H.; Lundberg, J.; Wijaya, A.; Ghodrati, B. Downtime analysis of drilling machines and suggestions for improvements. *J. Qual. Maint. Eng.* **2014**, *20*, 306–332. [CrossRef]

31. Zimroz, R.; Wodecki, J.; Król, R.; Andrzejewski, M.; Sliwinski, P.; Stefaniak, P. Self-propelled Mining Machine Monitoring System – Data Validation, Processing and Analysis. In *Mine Planning and Equipment Selection*; Drebenstedt, C., Singhal, R., Eds.; Springer International Publishing: Cham, Switzerland, 2014; pp. 1285–1294.

32. Dai, Y.; Xue, Y.; Zhang, J. Condition monitoring based on sound feature extraction during bone drilling process. In Proceedings of the 33rd Chinese Control Conference, Nanjing, China, 28–30 July 2014, pp. 7317–7322. [CrossRef]

33. Parsian, A.; Magnevall, M.; Beno, T.; Eynian, M. Sound Analysis in Drilling, Frequency and Time Domains. In Proceedings of the 16th CIRP Conference on Modelling of Machining Operations (16th CIRP CMMO), Cluny, Burgundy, France 15–16 June 2017. pp. 411–415, [CrossRef]

34. Du, S.; Feng, G.; Li, Z.; Sarkodie-Gyan, T.; Wang, J.; Ma, Z.; Li, W. Measurement and prediction of granite damage evolution in deep mine seams using acoustic emission. *Meas. Sci. Technol.* **2019**, *30*. [CrossRef]

35. Khoshouei, M.; Bagherpour, R. Predicting the Geomechanical Properties of Hard Rocks Using Analysis of the Acoustic and Vibration Signals During the Drilling Operation. *Geotech. Geol. Eng.* **2020**. [CrossRef]

36. Karakus, M.; Perez, S. Acoustic emission analysis for rock-bit interactions in impregnated diamond core drilling. *Int. J. Rock Mech. Min. Sci.* **2014**, *68*, 36–43. [CrossRef]

37. Seto, M.; Nag, D.; Vutukuri, V. In-situ rock stress measurement from rock cores using the acoustic emission method and deformation rate analysis. *Geotech. Geol. Eng.* **1999**, *17*, 241–266. [CrossRef]

38. Gogolewska, A.; Michalak, M. Classification parameters of roof rocks and seismic activity in "LUBIN" copper ore mine. *Min. Sci.* **2009**, *128*, 71.

39. Duszyński, A.; Jasiński, W.; Pryga-Szulc, A. Badanie wytrzymałości na ściskanie wybranych krajowych i zagranicznych surowców skalnych używanych do produkcji wyrobów galanterii drogowej. *Drogownictwo* **2014**, *3*, 88–91.

40. Fore-Sudetic Monocline Copper Ore Deposit. Available online: https://kghm.com/en/our-business/mining-and-enrichment (accessed on 14 October 2020.)

41. Peter D, H. Kernel estimation of a distribution function. *Commun. Stat. Theory Methods* **1985**, *14*, 605–620. [CrossRef]

42. Silverman, B.W. *Density Estimation for Statistics and Data Analysis*; CRC Press: Boca Raton, FL, USA, 1986; Volume 26.

43. Allen, J. Short term spectral analysis, synthesis, and modification by discrete Fourier transform. *IEEE Trans. Acoust. Speech Signal Process.* **1977**, *25*, 235–238. [CrossRef]

44. Marple, L. Computing the discrete-time "analytic" signal via FFT. *IEEE Trans. Signal Process.* **1999**, *47*, 2600–2603. [CrossRef]

Publisher's Note: MDPI stays neutral with regard to jurisdictional claims in published maps and institutional affiliations.

 *energies*

Article

A Portable Environmental Data-Monitoring System for Air Hazard Evaluation in Deep Underground Mines

Bartłomiej Ziętek [1,*], Aleksandra Banasiewicz [1], Radosław Zimroz [1], Jarosław Szrek [2] and Sebastian Gola [1,3]

[1] Faculty of Geoengineering, Mining and Geology, Wroclaw University of Science and Technology, Na Grobli 15, 50-421 Wroclaw, Poland; aleksandra.banasiewicz@pwr.edu.pl (A.B.); radoslaw.zimroz@pwr.edu.pl (R.Z.); sebastian.gola@pwr.edu.pl (S.G.)

[2] Faculty of Mechanical Engineering, Wroclaw University of Science and Technology, Łukasiewicza 5/7, 50-370 Wroclaw, Poland; jaroslaw.szrek@pwr.edu.pl

[3] KGHM Polska Miedź S.A O/ZG, Polkowice-Sieroszowice, 59-101 Kazimierzów, Poland

* Correspondence: bartlomiej.zietek@pwr.edu.pl

Received: 6 November 2020; Accepted: 27 November 2020; Published: 30 November 2020

Abstract: Air-quality measurements in a deep underground mine are a critical issue. The cost of ventilation, as well as the geometry of the considered mine, make this process very difficult, and local air quality may be a danger to miners. Thus, portable, personal devices are required to inform miners about gas hazards. There are available tools for that purpose; however, they do not allow the storage of data collected during a shift. Moreover, they do not allow the basic analysis of the acquired data cost-effectively. This paper aims to present a system using low-cost gas sensors and microcontrollers, and takes advantage of commonly used smartphones as a computing and visualization resource. Finally, we demonstrate monitoring system results from a test in an underground mine located in Poland.

Keywords: deep underground min; IoT devices; smartphone; gas hazards; Industry 4.0 in mining

1. Introduction

Although mining is often considered a "dirty" industry, its growth continues. This is due to the increasing demand for raw material, especially for rare earth material, which is needed for ICT, automotive, etc. industries. Unfortunately, raw materials are non-renewable resources. Intensive development of mining requires searching for new deposits in deep parts of the Earth. This leads to an increase in the level of hazard in an underground mine. In this article, we will selectively consider gas hazards, which are the subject of many studies [1–4]. In the considered mine, the most critical gases for miner health are hydrogen sulfide (H_2S), carbon monoxide (CO), and other NOx (gases in the nitric oxide family). They are present in mining voids because of the operation of LHDs (CO, NOx), blasting procedures (CO), and natural sources (H_2S) [5–8].

The cost of ventilation, the geometry of the considered mine and the mining technology used make the ventilation process very difficult. Moreover, the local air quality may fluctuate and pose a threat to miners. Thus, portable, personal devices are required, in order to inform miners about gas hazards. There are tools available on the market, meeting certain requirements for that demanding environment. However, although these allow the storage of data collected during a shift, they do not allow basic analysis of the acquired data in an online mode. In addition, these solutions are extremely expensive, and it is impossible to equip all underground workers.

The main aim of this article is to develop a basic monitoring system using low-cost gas sensors and microcontrollers (hardware modules) and take advantage of commonly used smartphones as computing and visualization resources. After the shift, when the miner leaves the mine and all collected data is in the phone memory, it may be automatically transferred to the cloud. Then, using advanced analytical tools, one may perform short-term and long-term analysis, including the spatial nature of data and its random character. During the shift, the miner may use a smartphone as a computer for simple visualization of past and current values on a plot, and some basic prediction (regression models) may be easily achieved to increase awareness to the miner.

Our solution is low-cost, intuitive, easily accessible (everybody has a smartphone these days), and can also be used as a mobile data-acquisition device for long-term analysis on the surface using advanced analytical tools. It should be highlighted that using mobile phones for measurements is not a new idea; however, we have not met such an approach in underground mining. Mobile phones have been used in tele-medicine [9], in condition monitoring [10,11], and environment monitoring [12,13]. See [14,15] for a review.

The rest of the paper is organized as follows: first, we will review the existing commercial products and some research activity in the area. Then, we will propose our system and in detail; each component will be described. To validate our measurement device, we performed two tests: one in laboratory conditions (to check the system's functionalities) and a second in a real underground mine with a gas hazard. The second test covers the comparison with results obtained by commercial measurement devices used in the mine regularly. Finally, we present a discussion and conclusions.

The places for underground measurements were selected by ventilation services. The selection of the measurement locations was based on the employee knowledge, supported by previous measurements in the given exploitation areas. The measurements were taken in mining divisions where increased concentrations of gases, such as hydrogen sulfide and carbon monoxide, occur. The measurements were carried out in fresh and used air streams.

This article presents a technique for a continuous measurement of gas concentrations, which has not yet been implemented in the underground mines of KGHM Polska Miedz S.A. The measurement system created is cheaper than other measurement systems available on the market. The use of this system will contribute to increased control over the safety of underground employees. The system is in line with the global trend of the Internet of Things, where microcontrollers are used and, by means of wireless communication, with low power consumption, give the possibility of measuring environmental conditions even for as demanding environment as a deep copper ore mine.

This article highlights the benefits from the implementation of a cost-effective environmental condition assessment system in line with Industry 4.0 fundaments. Presentation of the possibilities of additive manufacturing, wireless data transmission, and a set of sensors to be used as a new monitoring tool for the ventilation crew has been made. Presented results can be used for the better understanding of changes in the underground environment thanks to continuous measurement.

2. State of the Art—Mining Atmosphere Monitoring in a Deep Mine

Air pollution is now one of the most important problems all over the world. The problem of polluted air also applies to the mine environment, where workers are exposed to gas hazards every day, especially those providing services close to the mining face. Currently, global mining is moving towards deeper and deeper parts of the rock mass. It is a significant problem to transport adequate amounts of fresh air to the excavations. Malfunction of this process causes an increase in the gas hazard. That is why the ongoing activities in the world to develop a system for monitoring air quality and pollution in real time are crucial [16–18].

The article presents a measuring system for gases such as hydrogen sulfide and carbon monoxide. These gases are two of the most dangerous gases in the mine atmosphere. An employee who is in an environment with an increased concentration of H_2S is exposed to various health effects depending on the value of its concentration. At values up to 5 ppm, human beings suffer from eye irritation, headaches, and nausea. At concentrations of 5–50 ppm, conjunctivitis may occur, 100 ppm—olfactory disorders. At values of about 200–750 ppm, pulmonary edema and apnea appear, and at the concentration of about 1000 ppm, immediate respiratory paralysis and death [19–21]. The next gas analyzed is carbon monoxide. This is probably the most dangerous gas in the mine atmosphere. If an employee is exposed to this gas at a concentration of approximately 35 ppm, a slight headache may occur after 6–8 h. If the concentration is 100–200 ppm, a headache may occur after 2 or 3 h, and at 400 ppm, after 1 h. At a gas concentration of 1600 ppm, death occurs in less than 2 h, and at 12,800 ppm, a human passes away within 3 min [22,23].

Technologies of production of hazardous gas sensors differ from each other both based on materials detecting concentrations of elements and chemical compounds as well as by different methods of physical quantity transduction. The main groups of gas-sensing materials are metal-oxide semiconductors, conducting polymer composite, and other modern functional materials [24]. Most sensors rely on transduction methods such as chemoresistance, quartz crystal microbalance, and MOSFET technology [25]. Sensors with metal-oxide-sensing elements—ubiquitous for commercial purposes—use SnO_2. It absorbs gas molecules and activates reactions combined with carbon oxide detection [26]. There are many works concerning the increasing performance of SnO_2 as a gas sensor and to broaden the scope of use cases [27,28].

Another interesting approach for further hazardous gas sensor development is the usage of micromachining technology. Expansion of micro-electro-mechanical systems (MEMS), micro/nano-based structures, are going to play an active role in the detection industry. The measurements can be taken thanks to microcantilever deflection induced by an additional mass, attached to the end of the cantilever, which is generated by a mass gain on the sensitive element, and assigned to a specific gas concentration value. From this phenomenon, signals are taken and transformed by the Wheatstone bridge circuit. Calculation of the resonance frequency shift is made and directly the gas concentration is obtained [29–31].

Among the measuring devices, which monitor gas concentration values in excavations, the most popular are portable gas detectors. Those devices have the possibility of reading the parameters in real time. This experiment was conducted in underground conditions, and there are several accessories for measuring gas concentrations. Most often, PAC8500 and X-am 8000 gas detectors are used to measure the gas level of CO and H_2S. These devices, however, have only a discrete analysis function. Portable detectors are characterized by the possibility of instant assessment of the tested gas concentration [32]. Owing to the high cost of gas meters, only gas service employees and shift supervisors are equipped with those devices. However, more and more ideas are emerging, and new technologies are under constant development. It allows us to find new opportunities for ventilation measurements. The purpose of this article is to present the gas measurement system and applies it as personal equipment for every underground worker. Moreover, we are pursued to design a new approach of cost-efficient, easy-to-deploy, IoT-networked devices with smartphone usage for data presentation and acquisition.

One of the more interesting ideas is the use of the Arduino platform, which is also used for the gas concentration meter prototype in [33]. The popularity of smartphones creates an opportunity to design solutions with dedicated software. Among the various applications one may find one which could help to protect human health. An example is an application created for the Android operating system, which is used to assess air pollution. The application, connected to the appropriate gas sensors, sends information in real time about the state of the environment [34]. The development of modern industry, well known as Industry 4.0, puts an enormous impact on technological progress in the field of sensor devices. Smart metering, big data analytics, and cloud computing can be applied in the mining industry too. Unification of smart sensor networks and data handling for better understanding of the environmental properties in underground conditions is now widely taken into account [35,36].

Due to the geometry of the mine and room-and-pillar technology, online/real-time information about gas concentration/distribution in the whole mine is technically not feasible. It is well known that the vicinity of mining faces is the most critical part of the mine. Owing to dynamic changes because of blasting procedures, there is no infrastructure in place (sensors, monitoring systems, electricity, etc.).

Thus, to get information about gas concentration, the possible solution is by using an individual gas detector for each miner. Such devices are used in the mine presently; however, they do not register data for deeper, more comprehensive analysis. For testing purposes, and installation of the measurement device with several gas-type detectors and long-term data acquisition, has been considered in copper mining in Poland but is more a research project than everyday practice. The device is a prototype. It can work up to 50 °C (IP65 certificate) [37]. The structure of the monitoring system is presented in Figure 1.

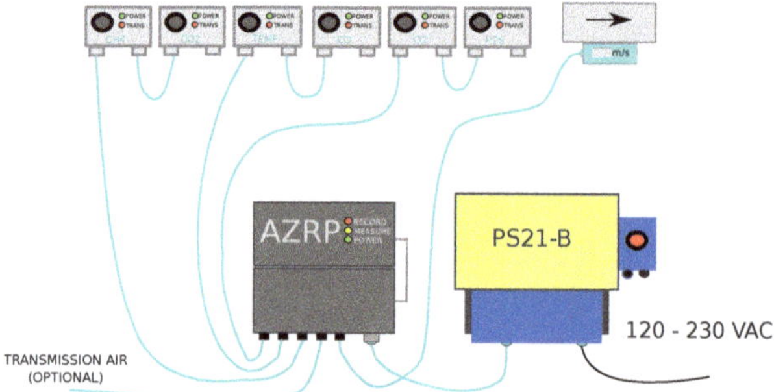

Figure 1. Measurement system-scheme [38].

More feasible solutions are associated with portable devices (Figure 2). They can measure gas concentration; however, they have a simple user interface and do not allow doing any analysis.

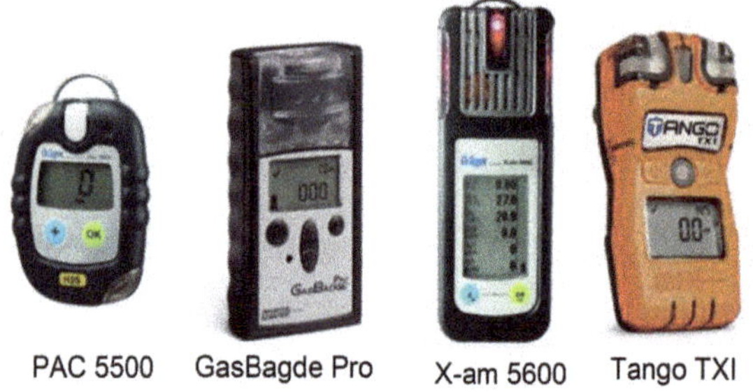

PAC 5500   GasBagde Pro   X-am 5600   Tango TXI

Figure 2. Examples of portable gas detectors used in the mine [32].

Thus, it creates an opportunity for research in this area. Air pollution was deeply studied in the literature. Also, the mining atmosphere quality measurement is an important topic for researchers [39]. Mine gas monitoring problems have been studied in [40]. Gas monitoring and testing in underground mines using wireless technology has been taken into consideration in [41]. Wireless sensor networks for the detection of explosive atmospheres in underground coal mines have been discussed in [42]. Portable gas sensors (including H_2S) have been discussed in [43–45]. A low-power, wearable, multi-sensor data-acquisition system for environmental monitoring has been studied in [46]. Wearable sensors have been proposed also in [47,48]. Air-quality monitoring systems based on microcontrollers as Raspberry Pi [49,50] or Arduino [51,52] have been also considered. The idea of the electronic nose concept has been proposed in [53]. CO detectors are based on a laser diode proposed in [54]. Statistical analysis of CO and H_2S in deep underground mines for long-term measurements have been studied in [3,4].

3. A General Concept of the System

The idea of the system is presented in Figure 3. It consists of 4 modules: sensor layers, data acquisition by the microcontroller, smartphone, and external IT infrastructure on the surface (optional). The main factor of portable environmental measurement unit construction is its low price and scalability attributes. It is based on an open-source platform and easy-to-use devices. Data visualization and additional storage functionality are satisfied thanks to the Android application (.apk format).

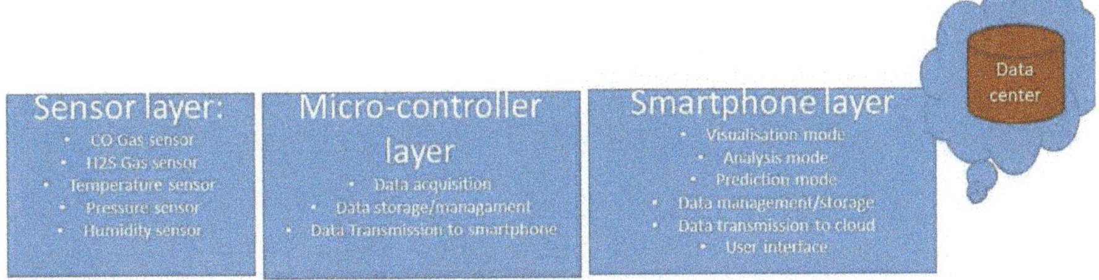

Figure 3. General structure of the system.

In this case, the ventilation crew at the underground mine Sieroszowice–Polkowice uses PAC850, among the variety of devices on the market, as a hazardous gas sensor. Table 1 shows the parameters of the device used.

The data presented in the table below presents the scope of the measuring device used in the underground mine of KGHM Polska Miedź S.A. Although the sensors used in PAC-type measuring devices have larger measuring ranges, in the calibration process, a range smaller than the measuring capabilities of the sensors is set as limit values. The measurement thresholds established during the calibration process are the values at which one should immediately withdraw from the excavations, as longer time spent breathing in such an atmosphere may have serious health consequences.

Table 1. Parameters of the measurement device in underground mine

	Drager PAC8500	
Measuring range	H_2S	0–100 ppm
	CO	0–2000 ppm
Ambient condition	Temperature	−30 to 50 °C
	Humidity	10 to 90% (without consolidation)
	Pressure	700 to 1300 hPa

The measurement data analyzed in the article was measured and generated by means of a constructed measurement system. The place of measurements was located in the underground mine in the areas where excavation was performed. The area of the survey was selected after consultation with the ventilation department employees, whose knowledge, supported by previous measurements and their analysis, allowed for the selection of gas hazardous zones. Concentrations of gases such as hydrogen sulfide and carbon monoxide were measured and then analyzed. The measurement was performed continuously, and the collected data displayed in the application on the phone were compared with the results of the measurements of the equipment used by the mine employees (PAC8000 and X-am 5600). The measurement results and their analysis are presented in Section 4.

3.1. A Hardware: Environmental Sensors

The main purpose of the presented prototype is to measure gas concentration in the mine environment. To achieve it, an MQ-9 sensor for CO concentration and MQ-136 and ZE03-H2S for H_2S concentration is used. These sensors use Tin Dioxide (SnO_2) as sensitive material, which changes conductivity depending on gas concentration. As an output, the voltage output is obtained. Using the formula from manufacturer datasheet [55,56] and calibration using certified devices from mine ventilation crew PPM concentration is calculated. Additionally, the DHT22 (temperature and humidity sensor) is used with temperature range: $-40\ °C$ to $80\ °C \pm 0.5\ °C$ and humidity from 0% to 100% $\pm$ 1% (see Table 2).

Table 2. System components.

Component	Role
Arduino M0 Pro	Board
HC06	Bluetooth controller
MQ9-Sensor	CO concentration
MQ136-Sensor	H_2S concentration
ZE03-H2S	H_2S concentration
DHT 22	Temperature
	Humidity
Polulu SD Card Reader	SD data acquisition

3.2. Hardware: Microcontroller

As the a main computer, Arduino M0 Pro with 32-bit ARM Cortex M0 core is used. The performance was satisfied with 48 MHz clocks and 32 Kb SRAM. Gas sensors are connected to analog pins. Other devices such as SD Card Reader, DHT22, and HC06 Bluetooth module [57] use UART, SPI, and (I^2C) communication protocols. Circuit voltage of sensors, especially the HC06 Bluetooth module, needs a stable 5 V voltage level, thus power bank RAXFLY 10,000 mAh was used. The system scheme is presented in Figure 4.

Environmental condition measurements are taken with 1 Hz frequency. Subsequently, PPM values are calculated, and the data frame goes further to the smartphone using the Bluetooth connection. With the aim of data protection against corruption and malfunction from a highly harsh and changeable mine environment, every 5 min the whole file with measurements is sent to a smartphone for data redundancy.

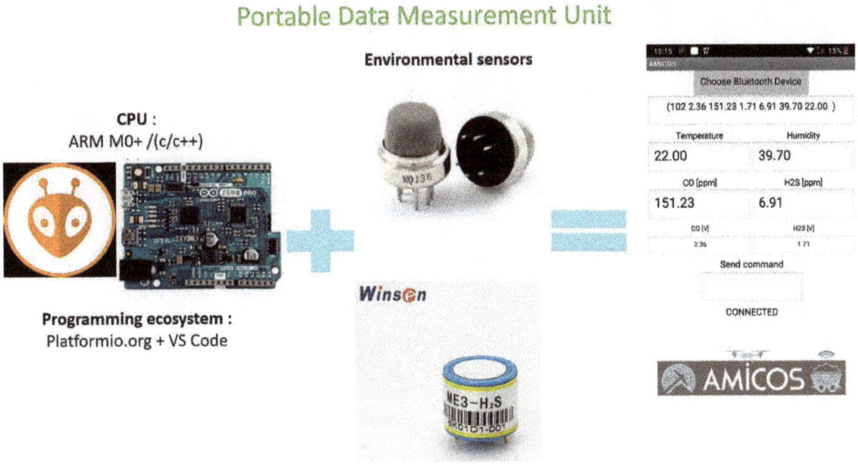

Figure 4. Portable Data Measurement Unit.

3.3. Smartphone Layer

First, the smartphone plays the role of visualization medium using data from the Bluetooth connection. The provided user interface can manage with a short command for the microcontroller layer to change the frequency of measurement (0.05–2 Hz). All data is stored in a folder on the smartphone with the option of auto-upload to a private cloud when a device is connected to the cellular/Wi-Fi network.

The second part of the used sensors come from a smartphone. Its embedded IMU sensor provides information about acceleration, velocity, etc. Frequency of 10 Hz helps to obtain additional information e.g., about several steps from smartphone sensors. In the case of the emergency state, when hazardous gas concentration increases, the application makes the phone vibrate and the background blinks red.

During the experiment, a standard smartphone with the Android operating system was used (Nokia 8). In future tests designed for the assessment of the entire mine, we will have to apply special devices approved by standards. In accordance with the Regulation of the Minister of Energy of 23 November 2016 (Chapter VI), the degree of protection of the device must not be less than IP 54 [37].

3.4. Rapid Manufacturing for Mining Purposes

As a part of rapid manufacturing, the most widespread and accessible method is 3D printing. Using a Prusa Mk3S printer a custom case for the onboard computer was created. A hand-sized $70 \times 75 \times 67$ mm case with holes for the gas sensors and control diodes protects the measurement unit from the harsh mining environment and simultaneously allows evaluation of ventilation condition correspondingly. The created case is presented in Figure 5. Polylactic acid (PLA) was chosen as a construction material. Its mechanical parameters are sufficient for use as a case in the mine area. Moreover, PLA is a biodegradable plastic, and its physical properties are excellent for a thermoplastic forming process [58,59].

Figure 5. Bottom and top side of 3D printed case for the experiment.

3.5. Software for Microcontroller

The open-source project has an enormous impact on beginners who want to start their adventure with microcontrollers. Arduino boards are the leader for education usage as well as IoT product development [60]. For this publication, integrated development environment (IDE) for the embedded project, platformio.org, is used to expand capabilities of used hardware [61]. This IDE, as an add-on for Visual Studio Code, enables rapid application development with a variety of boards and microcontroller models. Additionally, it maintains order in libraries that has an enormous impact on more complex project development.

3.6. Software for Smartphone

The HC06 Bluetooth module constitutes a bridge between the embedded ecosystem and eye-friendly data visualization [57]. Since more than 70% of mobile phones use Android OS, the dedicated application with data visualization and logging feature is created for that particular OS [62]. Data stored on smartphones enable unlimited possibilities to send information to the other applications, e.g., cloud provider storage, instantly (see Figure 6a,b). For that purpose, the MIT App Inventor, a block-based coding environment for mobile application development, was used. It ensures fast up-and-running mobile applications. Within the MIT App Inventor, block of code for Bluetooth connectivity and acquisition of data from IMU sensors was used. When a new data frame from a microcontroller occurs, data visualization is made instantly. In the background, every 100 milliseconds, the app saves information about the present values of acceleration and velocity. Other important functionalities of the application are measurement data management, simple visualization in the text field, and data plotting.

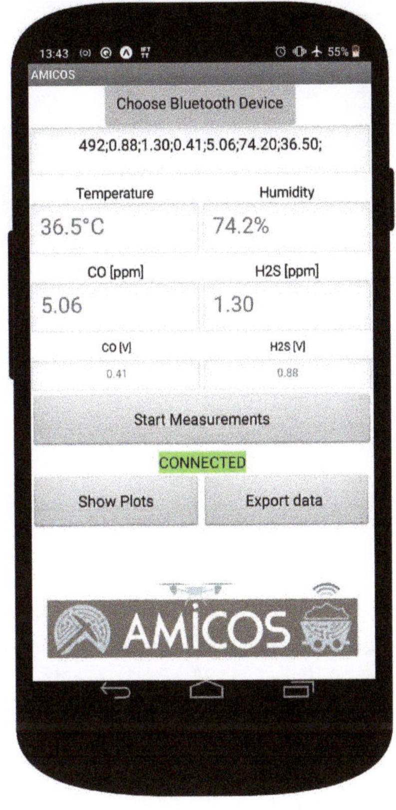

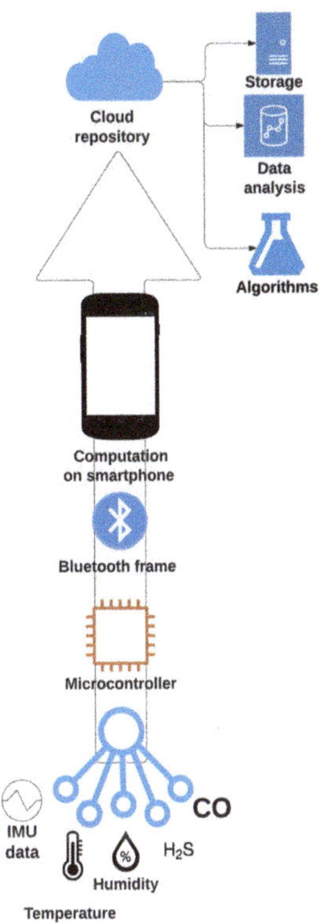

(**a**) Android application for data visualization.

(**b**) General scheme of the system.

Figure 6. Example of using the smartphone application for underground gas measuring system.

3.7. Cost Comparison and Repeatability Device Architecture

Hazardous gas sensors used in the aforementioned mine are devices from Dragger manufacturer shown in Figure 7. Devices, with basic functionality, which only covers the function of displaying the instantaneous gas concentration value, are PAC6000 and PAC6500—single gas meters. More convincing in usage are multiple gas meters—PAC8500—with cost ca. 850€. The most advanced sensor with additional functions such as data recording or complex data visualization is the PAC8500. The cost of the device is almost 3000€. The presented IoT device, together with the smartphone used as a component for data visualization, costs less than 100€. In addition, our prototype has options such as cloud data storage or drawing a plot for data visualization. Within the framework of Industry 4.0, the software developed is easily scalable, so there are no restrictions that these devices will be introduced to a greater extent to equip underground workers. This will significantly help to raise workers' awareness of the environmental conditions at work and allow them to react more quickly if a hazard arises.

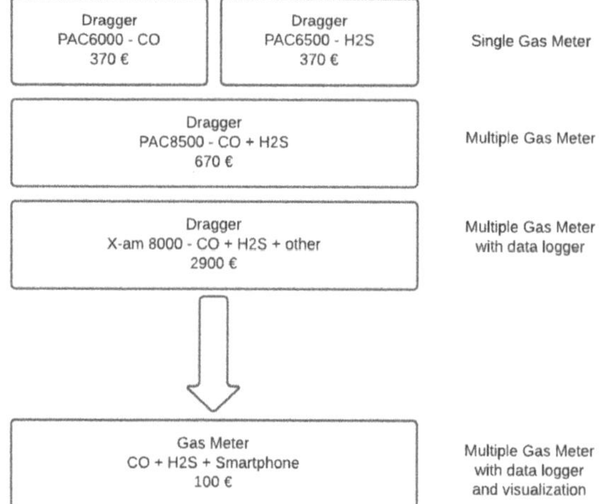

Figure 7. Price comparison of gas meters [63–66].

4. Experiments

The experience was divided into two parts. The first part focused on carrying out tests in laboratory conditions, while the second part concerned the performance of measurements in real conditions in an underground copper mine belonging to KGHM Polska Miedź S.A. The results of the tests of measurements of gases such as hydrogen sulfide and carbon monoxide, their interpretation, and analysis are included in this section.

4.1. Test in the Lab

Experiments in the mine are always difficult and may have a lot of limitations so first we decided to test our device in lab conditions. The experiment aimed to check measurement ability and preliminarily evaluate the sensitivity of the system. As a source of CO gas, we have used an unburned piece of wood. We produced a bit of smoke. By decreasing the distance between the source of the smoke and the sensor, we have increased gas concentration that could be seen in Figure 8, point A and B. We cannot imitate the source of H_2S in the lab, so for the second gas, we just tested the ability of measurement H_2S concentration in the air. For both cases, one may notice some minor fluctuations in gas concentration. For both gases, we were far from the limit value used in the mine (note red lines, 26 ppm for CO and 7 for H_2S, are based on mining law [37]). An example of environmental data acquired during the test, namely CO and H_2S, is presented in Figure 8. The conclusion from that simple test is that system is working, data can be acquired and stored in the device and at least the CO sensor is able to react to change in CO concentration.

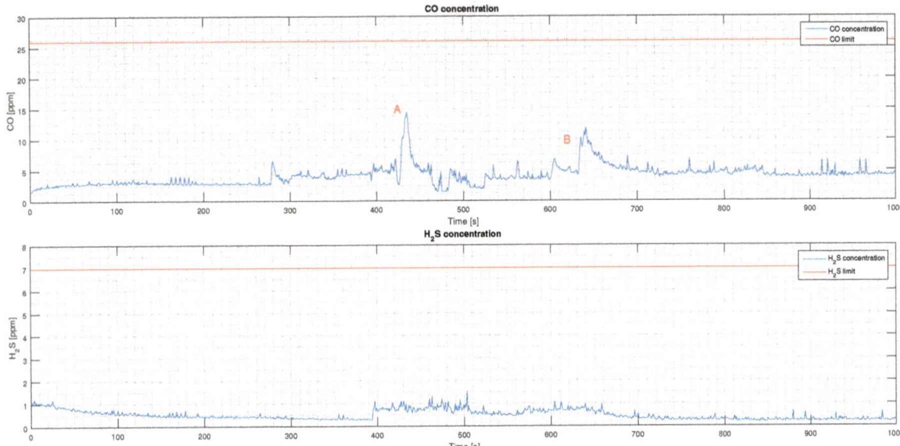

Figure 8. CO and H_2S sensors measurements in lab.

4.2. Mining Area Experiment

The examined mine is characterized by difficult operating conditions i.e., a large depth of exploitation, an extensive mine area, and harsh air atmosphere. The mine exploits minerals based on a room-and-pillar system. Excavations, where measurements were carried out, are marked in green circles on the map in Figure 9.

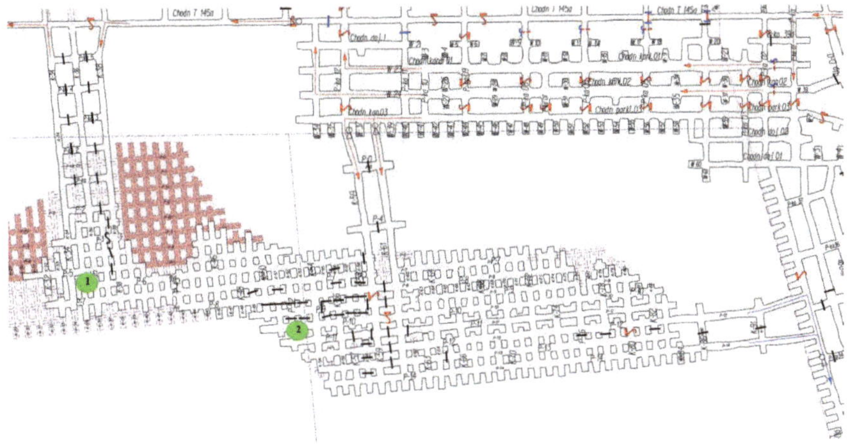

Figure 9. Map of investigated part of mine with fresh and used air stream flow and sensors location.

To achieve our goals, the plan of experiments has been discussed with the mining staff. This mine is in operation and as mentioned there might be potential serious hazards related to gas presence. The experiment was conducted in two places chosen by mine ventilation staff. Measurements were taken near the mining face, so higher concentrations of CO and H_2S gases, respectively, were expected. There are many factors influencing air quality in the mine: the presence of machines, the amount of time after blasting, distance to the shaft, and distance to fresh and used air streams. On the map, the red arrows symbolize the fresh air, while the blue arrows represent the return air. In addition, air regulators are marked on the map with red symbols—mainly air stopping. The expected progress of the mining face is marked in purple. A significant influence on micro-climate in the mining void (tunnel) is also geological conditions and the presence of machines (with diesel engines). To select an appropriate location for experiments deep knowledge about mining conditions is crucial.

4.3. Validation in the Mine

During underground measurements in the mine, difficult climatic conditions that prevail in underground excavations should be taken into account. First of all, there is the high temperature, which ranges from 35 °C to 39 °C in our experiment as well as high humidity 55–85%. The measurement time was about 60 min (3600 s). In addition, mining faces are characterized by a high level of dust and salinity.

The measurements allowed us to see that along with the changing concentration of gases in the excavation, the designed meter recorded concentration changes in a manner analogous to gas detectors equipped with ventilation services in copper mines (Drager PAC6500 and PAC8500). At moments of concentration increase or decrease on mine measuring instruments, identical variation in the shape of measured values on the meter prototype was noticed. Below are graphs of recorded CO concentration values and H_2S in the ppm unit. Other environmental parameter changes during the experiments also are presented as a variance of temperature and humidity.

In the graph above, showing the concentration distribution of the analyzed CO gas, it can be seen that none of the analyzed gases exceeded the limit values. The maximum value of the carbon-oxygen concentration in the mine excavation is 26 ppm [37]. The highest values of carbon monoxide were recorded at the very beginning of the measurement (see point A in Figure 10). The high concentration of this gas was caused by the work of machines in nearby excavations. As the distance from the working front had been moving away, the CO content decreased. Individual measuring peaks are probably caused by machines passing nearby the measuring route (point F). The maximum concentration of carbon monoxide was not close to the limit value.

The next graph shows the hydrogen sulfide concentration distribution. The maximum value of hydrogen sulfide concentration in the excavation is 7 ppm and during measurement. It was not exceeded [37]. The measurement was made in the excavation in the department, where the concentration parameter H_2S increased value significantly (point D). In other analyzed excavations, the hydrogen sulfide measurement sensor showed close to zero values.

This noticeable moment of measurement is at B and E points in Figure 10. It is crucial to consider that at point B and E we were in a fresh air inlet area that is notable on the humidity and temperature chart. A mixture of fresh air causes a decrease in temperature and humidity as well as hazard gas concentration. Other ventilation facilities, such as air stopping, play an important role in environmental parameter-shaping in the mine. In Figure 10 point C travel through several air dams is noticed.

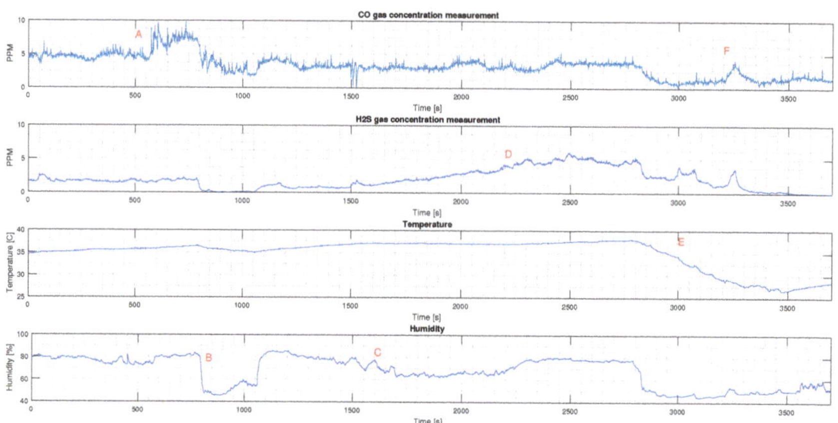

Figure 10. Mine environmental parameter measurement.

Using data from IMU sensors, built into the phone, distance calculation was performed with satisfactory accuracy. Signals are collected every 100 ms. When the main acceleration vector increases swiftly, then the application on the smartphone computes this event as a new step. Data is taken about acceleration and velocity in 3 axes. Based on that, when every considerable variation occurs, a step calculation function is triggered [67]. As can be seen in Figure 11 during the test, the length of distance is nearly 700 m. It could be helpful to estimate an overall distance made by the mine crew after the shift.

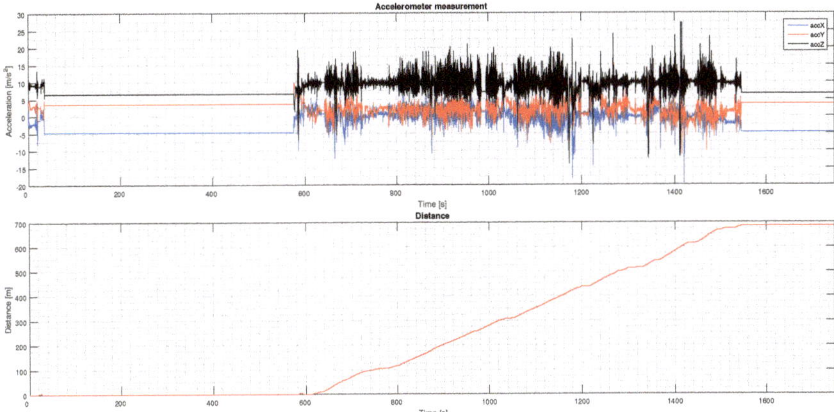

Figure 11. Distance calculation based on smartphone IMU sensors.

5. Validation

The standard process of sensor validation for industry usage involves special calibration equipment. These devices have required certificates and other approvals, which confirm their workability and operational reliability. It is how every industrial device is calibrated in an accredited laboratory. A described operation takes a long time, and it is expensive. In the case of a portable environmental measurement unit, proposed in this article, it is made in another way. On the ground of indications from devices widely used by mine ventilation crew, the calibration is conducted. After software-based calibration, data from certified portable sensors and prototypes are the same. This approach is, like the whole proposed solution, low-cost, not time-consuming, and has comparable accuracy.

For further analysis and scale-up, a prototype is necessary to confirm the correction of data, especially the calculation of PPM concentration. A device with a mixture of gas for calibration and validation measuring equipment such as proposed in [32] must be used. Valid data calculation needs additional information from the sensor's producers like cross-sensitivity. Thanks to that, we could separate the noise from the contribution of other gases to the sensor. Conducted calibration will allow us to bring the device into wider service in the mine environment.

6. Discussion

In an underground mine, detecting a gas hazard is essential. In this article, we focus on the most important and dangerous gases for underground workers—carbon monoxide and hydrogen sulfide. For KGHM Polska Miedz S.A. and other mining companies, this issue is critical from the underground employee safety point of view. As previous publications have shown, for example [19,22], which have precisely juxtaposed the appropriate gas concentrations with the corresponding symptoms, measurements of CO and H_2S are very important. Apart from the measurements, which are carried out in real time in the mine, there is also a need for continuous monitoring and data collection for subsequent analyses. Hebda-Sobkowicz et al. [3,4] show how important it is to analyze the data collected continuously and over a longer period of time in the process of gas hazard assessment. This work presents the approach that could be helpful to improve the work of ventilation services.

Thanks to the development of sensors, microcontrollers, and a wide usage of the smartphone, we proposed a system, which can be used for every underground worker. Our solutions enable real-time data collection and analysis to describe environmental conditions from the gas hazard point of view, in particular, H_2S and CO. Each mine employee could be equipped with a personal gas concentration sensor/analyzer and as a result of that, safety could rise easily. In the case of emergency, state workers evacuate themselves immediately. Furthermore, in accordance with the development of technology and the Internet of Things, our proprietary software and measuring system can be easily extended to test the conditions of other gases such as NOx, SOx, CH_4.

The main purpose of this work is to develop a low-cost, portable measurement system for the analysis of mine environment. To accomplish that, experiments under stable lab conditions and also in the underground copper mine were carried out. The created system allows us to conduct a basic analysis in situ as well as a long-term investigation of retrieved data. The environmental data-acquisition system, tested in the mine, can be successfully implemented.

Obtained data from each test shows that created system works well. Retrieved data is satisfying. In moments when there is a change in gas parameters such as measurements near mining machines or the place of natural hydrogen sulfide discharge, our prototype registers it. Collected data are the same as those gathered by the equipment of the mine's ventilation services crew.

7. Conclusions

There are detection systems on the market for the measurement of environmental parameters, namely concentrations of hazardous gases, which are both portable and stationary systems. However, it is difficult to obtain a stable source of power supply in underground conditions, and battery-based systems are difficult to maintain in practice. Moreover, due to the constantly moving mining front, in the mines using room-and-pillar with blasting technology, some underground mine areas have no infrastructure; in particular, there is no access to the electricity grid. The aforementioned blasting, possible seismic hazard etc., forces minimization of the use of stationary infrastructure near the mining face. Therefore, it is necessary to focus the efforts on the creation of battery-powered portable measurement systems, with long operating times, and the ability for wireless data transmission. This article presents an attempt at developing a solution according to the assumptions mentioned above.

The proposed measurement system is highly cost-effective in comparison with other existing solutions on the market. Moreover, implementation is less time-intensive and gives satisfying results simultaneously, compared with existing approaches used in the mining industry. The presented architecture of the system can easily extend for other sensors to measure various environmental parameters. Among CO and H_2S presented in this article, it is crucial to observe NOx or O_2 gases too. Subsequently, software for sensor handling, and mobile application and creation of a 3D-printed case was presented, describing the functionality of a device. Thanks to the proposed solution, it is possible to access environmental data in real time. In comparison to other available devices, the proposed solution can measure, visualize current gas concentration, and possess basic analytical functionality-historical data, trends, and basic prediction. It should be stated that the system is open, more functions may be added if required. Usage of an open-source platform for hardware systems and smartphone applications fits perfectly into today's software development trends, in line with the principles of Industry 4.0. Information can be used by ventilation crew both underground and above using a cloud solution for further data analysis. The application on a mobile phone stores messages from sensors and presents them immediately. When it is accessible, the connection to wireless communication, such as Wi-Fi or cellular data, are sent to the online repository. The results in this article confirm the correct work of the gas concentration meter prototype. Usage of the popular platform and efficient framework simplifies the process of code creation and management for embedded systems. The next step is to calibrate the prototype using a dedicated machine with the gas mixture. This mixture must be similar to the mine atmosphere. Thanks to that, the following experiment in the mine environment will be enrolled to ensure repeatability. After that, the proposed system could be used in a wider scope to empower the mine crew safety.

Considering all the above, the presented aspect may have enormous relevance for the mining industry. The problem of gas hazards in underground mines significantly affects the safety of employees. We try to solve this issue, generated by the industry, using the proposed system. Thanks to that, we could collect a lot of data and use it to determine underground ventilation changes such as hazardous gas concentration at a specific time. It gains knowledge about ventilation processes and generates new solutions. The designed measurement system is greatly flexible and allows us to expand measurements for other gases or specific mines' environment parameters in the future.

Author Contributions: Conceptualization, B.Z. and A.B.; Investigation, A.B. and R.Z.; Methodology, A.B. and S.G.; Resources, A.B., R.Z., J.S. and S.G.; Software, B.Z.; Supervision, R.Z., J.S. and S.G.; Validation, B.Z.; Visualization, B.Z.; Writing—original draft, B.Z. and A.B.; Writing—review & editing, R.Z. and J.S. All authors have read and agreed to the published version of the manuscript.

Funding: This work is supported by EIT Raw Materials GmbH under the Framework Partnership Agreement (Autonomous Monitoring and Control System for Mining Plants—AMICOS).

Acknowledgments: This activity has received funding from European Institute of Innovation and Technology (EIT), a body of the European Union, under the Horizon 2020, the EU Framework Programme for Research and Innovation. This work is supported by EIT RawMaterials GmbH under Framework Partnership Agreement No. 19018 (Autonomous Monitoring and Control System for Mining Plants—AMICOS).

Conflicts of Interest: The authors declare no conflict of interest.

References

1. Wierzbicki, M.; Skoczylas, N. The Outburst Risk as a Function of the Methane Capacity and Firmness of a Coal Seam. *Arch. Min. Sci.* **2014**, *59*, 1023–1031. [CrossRef]

2. Skoczylas, N. Estimating gas and rock outburst risk on the basis of knowledge and experience—The expert system based on fuzzy logic. *Arch. Min. Sci.* **2014**, *59*, 41–52. [CrossRef]

3. Hebda-Sobkowicz, J.; Gola, S.; Zimroz, R.; Wyłomańska, A. Pattern of H2S concentration in a deep copper mine and its correlation with ventilation schedule. *Measurement* **2019**, *140*, 373–381. [CrossRef]

4. Hebda-Sobkowicz, J.; Gola, S.; Zimroz, R.; Wyłomańska, A. Identification and Statistical Analysis of Impulse-Like Patterns of Carbon Monoxide Variation in Deep Underground Mines Associated with the Blasting Procedure. *Sensors* **2019**, *19*, 2757. [CrossRef]

5. Kurnia, J.C.; Sasmito, A.P.; Wong, W.Y.; Mujumdar, A.S. Prediction and innovative control strategies for oxygen and hazardous gases from diesel emission in underground mines. *Sci. Total Environ.* **2014**, *481*, 317–334. [CrossRef]

6. Pajdak, A.; Godyń, K.; Kudasik, M.; Murzyn, T. The use of selected research methods to describe the pore space of dolomite from copper ore mine, Poland. *Environ. Earth Sci.* **2017**, *76*, 389. [CrossRef]

7. Bian, Z.; Inyang, H.I.; Daniels, J.L.; Otto, F.; Struthers, S. Environmental issues from coal mining and their solutions. *Min. Sci. Technol.* **2010**, *20*, 215–223. [CrossRef]

8. Kudasik, M.; Skoczylas, N. Analyzer for measuring gas contained in the pore space of rocks. *Meas. Sci. Technol.* **2017**, *28*, 105901. [CrossRef]

9. Wen, C.; Yeh, M.F.; Chang, K.C.; Lee, R.G. Real-time ECG telemonitoring system design with mobile phone platform. *Measurement* **2008**, *41*, 463–470. [CrossRef]

10. Rzeszucinski, P.; Lewandowski, D.; Pinto, C.T. Mobile device-based shaft speed estimation. *Measurement* **2017**, *96*, 52–57. [CrossRef]

11. Cao, L.; Chen, J. Online investigation of vibration serviceability limitations using smartphones. *Measurement* **2020**, *162*, 107850. [CrossRef]

12. Corbellini, S.; Di Francia, E.; Grassini, S.; Iannucci, L.; Lombardo, L.; Parvis, M. Cloud based sensor network for environmental monitoring. *Measurement* **2018**, *118*, 354–361. [CrossRef]

13. Hatiboruah, D.; Das, T.; Chamuah, N.; Rabha, D.; Talukdar, B.; Bora, U.; Ahamad, K.U.; Nath, P. Estimation of trace-mercury concentration in water using a smartphone. *Measurement* **2020**, *154*, 107507. [CrossRef]

14. Daponte, P.; De Vito, L.; Picariello, F.; Riccio, M. State of the art and future developments of measurement applications on smartphones. *Measurement* **2013**, *46*, 3291–3307. [CrossRef]

15. Grossi, M. A sensor-centric survey on the development of smartphone measurement and sensing systems. *Measurement* **2019**, *135*, 572–592. [CrossRef]

16. Baris, K.; Aydin, Y. Atmospheric monitoring systems in underground coal mines revisited: A study on sensor accuracy and location. *Int. J. Oil Gas Coal Technol.* **2020**, *23*, 325. [CrossRef]

17. Fugiel, A.; Burchart-Korol, D.; Czaplicka-Kolarz, K.; Smoliński, A. Environmental impact and damage categories caused by air pollution emissions from mining and quarrying sectors of European countries. *J. Clean. Prod.* **2017**, *143*, 159–168. [CrossRef]

18. Dong, L.; Tong, X.; Li, X.; Zhou, J.; Wang, S.; Liu, B. Some developments and new insights of environmental problems and deep mining strategy for cleaner production in mines. *J. Clean. Prod.* **2019**, *210*, 1562–1578. [CrossRef]

19. Guidotti, T.L. Hydrogen sulfide intoxication. In *Handbook of Clinical Neurology*; Elsevier: Amsterdam, The Netherlands, 2015; pp. 111–133. [CrossRef]

20. Goyak, K.; Lewis, R.J. Application of AOP Networks in Human Health Risk Assessment: Increasing Confidence in Rodent Olfactory Nasal Lesions as the Point of Departure for Hydrogen Sulfide Exposure Limits. OSF Preprints, 2020. Available online: https://osf.io/kqszm/ (accessed on 6 November 2020). [CrossRef]

21. Tan, B.; Shao, Z.; Wei, H.; Yang, G.; Zhu, X.; Xu, B.; Zhang, F. Status of research on hydrogen sulphide gas in Chinese mines. *Environ. Sci. Pollut. Res.* **2019**, *27*, 2502–2521. [CrossRef]

22. Goldstein, M. Carbon Monoxide Poisoning. *J. Emerg. Nurs.* **2008**, *34*, 538–542. [CrossRef]

23. Ozmen, I.; Aksoy, E. Respiratory Emergencies and Management of Mining Accidents. *Turk. Thorac. J.* **2015**, *16*, 18–20. [CrossRef] [PubMed]

24. Wetchakun, K.; Samerjai, T.; Tamaekong, N.; Liewhiran, C.; Siriwong, C.; Kruefu, V.; Wisitsoraat, A.; Tuantranont, A.; Phanichphant, S. Semiconducting metal oxides as sensors for environmentally hazardous gases. *Sens. Actuators Chem.* **2011**, *160*, 580–591. [CrossRef]

25. Arshak, K.; Moore, E.; Lyons, G.; Harris, J.; Clifford, S. A review of gas sensors employed in electronic nose applications. *Sens. Rev.* **2004**, *24*, 181–198. [CrossRef]

26. Wang, C.T.; Chen, M.T. Vanadium-promoted tin oxide semiconductor carbon monoxide gas sensors. *Sens. Actuators Chem.* **2010**, *150*, 360–366. [CrossRef]

27. Ansari, Z.; Ansari, S.; Ko, T.; Oh, J.H. Effect of MoO3 doping and grain size on SnO2-enhancement of sensitivity and selectivity for CO and H2 gas sensing. *Sens. Actuators Chem.* **2002**, *87*, 105–114. [CrossRef]

28. Ansari, S.; Boroojerdian, P.; Sainkar, S.; Karekar, R.; Aiyer, R.; Kulkarni, S. Grain size effects on H2 gas sensitivity of thick film resistor using SnO2 nanoparticles. *Thin Solid Films* **1997**, *295*, 271–276. [CrossRef]

29. Wu Pan.; Ning Li. Micro-cantilever array and its application in gas sensor. In Proceedings of the 2008 International Conference on Microwave and Millimeter Wave Technology, Nanjing, China, 21–24 April 2008; Volume 3, pp. 1547–1550. [CrossRef]

30. Safavi, S.M.; rezazadeh, G.; Fathalilou, M.; Abazari, A.M. Couple Stress Effect on Micro/Nanocantilever-based Capacitive Gas Sensor. *Int. J. Eng.* **2016**, *29*, 852–861.

31. Aisah, N.; Aprilia, L.; Nuryadi, R. Piezoresistive microcantilever-based gas sensor using dynamic mode measurement. In Proceedings of the 2013 International Conference on QiR, Yogyakarta, Indonesia, 25–28 June 2013; pp. 5–8. [CrossRef]

32. Gajosinski, S.; Kulik, W.; Gola, S.; Turkiewicz, W. Uwarunkowania zastosowania metod pomiarowych w okreslaniu stezenia siarkowodoru w powietrzu kopalnianym. *CUPRUM—Czas. Nauk.-Tech. Gor.* **2018**, *88*, 43–53.

33. Rodriguez-Vasquez, K.A.; Cole, A.M.; Yordanova, D.; Smith, R.; Kidwell, N.M. AIRduino: On-Demand Atmospheric Secondary Organic Aerosol Measurements with a Mobile Arduino Multisensor. *J. Chem. Educ.* **2020**, *97*, 838–844. [CrossRef]

34. Kurubaran, A.; Varun Krishna, S.; Vignesh, K.; Bhaskaran, A. Air Pollution Monitoring System using Android Application. *Int. J. Curr. Res. Rev.* **2018**, *10*, 112–115. [CrossRef]

35. Li, X.; Li, D.; Wan, J.; Vasilakos, A.V.; Lai, C.F.; Wang, S. A review of industrial wireless networks in the context of Industry 4.0. *Wirel. Netw.* **2015**, *23*, 23–41. [CrossRef]

36. Lööw, J.; Abrahamsson, L.; Johansson, J. Mining 4.0—The Impact of New Technology from a Work Place Perspective. *Min. Metall. Explor.* **2019**, *36*, 701–707. [CrossRef]

37. Minister of Energy. Regulation of the Minister of Energy Related to Operations of Underground Mining (Available in Polish: Rozporzadzenie Ministra Energii z Dnia 23 Listopada 2016 r., w Sprawie Szczegółowych Wymagań Dotyczacych Prowadzenia Ruchu Podziemnych Zakładøw górniczych (Dz. U. z 2017r., poz. 1118). 2017. Available online: https://www.dziennikustaw.gov.pl/du/2017/1118 (accessed on 8 May 2020).

38. Sevitel sp. z o. o. Technical Documentation and User Manual of Measurement Device (Available in Polish: Dokumentacja Techniczno-Ruchowa. Instrukcja Użytkowania i Obsługi DTR SEV-256/2014 v3.2. Urzadzenie Pomiarowe AZRP). 2013. Available online: https://www.sevitel.pl/product,111, Iskrobezpieczny%20Czujnik%20St%C4%99%C5%BCenia%20Tlenku%20W%C4%99gla%20i.CO.html#tab1 (accessed on 9 September 2020).

39. Jo, B.; Khan, R.M.A. An Internet of Things System for Underground Mine Air Quality Pollutant Prediction Based on Azure Machine Learning. *Sensors* **2018**, *18*, 930. [CrossRef]

40. Vančura, V.; Otte, L.; Danel, R.; Létavková, D.; Šeliga, Z. Mine gas monitoring data analysis. In Proceedings of the 2014 15th International Carpathian Control Conference (ICCC), Velke Karlovice, Czech Republic, 28–30 May 2014; pp. 641–645. [CrossRef]

41. Anas, M.; Haider, S.M.B.; Sharma, P. Gas Monitoring and Testing in Underground Mines using Wireless Technology. *Int. J. Eng. Res.* **2017**, *6*. [CrossRef]

42. Romero Acero, A.; Marin Cano, A.; Jimenez Builes, J.A. SCADA System for Detection of Explosive Atmospheres in Underground Coal Mines Through Wireless Sensor Network. *IEEE Lat. Am. Trans.* **2014**, *12*, 1398–1403. [CrossRef]

43. Petrov, K.P.; Mine, V.; Curl, R.F.; Tittel, F.K. Fast sensitive trace gas detection with a portable solid-state mid-infrared laser sensor. In Proceedings of the 10th Annual Meeting IEEE Lasers and Electro-Optics Society 1997 Annual Meeting, San Francisco, CA, USA, 10–13 November 1997; Volume 1, pp. 182–183.

44. Guo, Y.; Qiu, X.; Li, N.; Feng, S.; Cheng, T.; Liu, Q.; He, Q.; Kan, R.; Yang, H.; Li, C. A portable laser-based sensor for detecting H2S in domestic natural gas. *Infrared Phys. Technol.* **2020**, *105*, 103153. [CrossRef]

45. Thai, N.X.; Duy, N.V.; Hung, C.M.; Nguyen, H.; Hung, T.M.; Hieu, N.V.; Hoa, N.D. Realization of a portable H2S sensing instrument based on SnO2 nanowires. *J. Sci. Adv. Mater. Devices* **2020**, *5*, 40–47. [CrossRef]

46. Serbanescu, M.; Placinta, V.M.; Hutanu, O.E.; Ravariu, C. Smart, low power, wearable multi-sensor data acquisition system for environmental monitoring. In Proceedings of the 2017 10th International Symposium on Advanced Topics in Electrical Engineering (ATEE), Bucharest, Romania, 23–25 March 2017; pp. 118–123.

47. Wilhelm, E.; Siby, S.; Zhou, Y.; Ashok, X.J.S.; Jayasuriya, M.; Foong, S.; Kee, J.; Wood, K.L.; Tippenhauer, N.O. Wearable Environmental Sensors and Infrastructure for Mobile Large-Scale Urban Deployment. *IEEE Sens. J.* **2016**, *16*, 8111–8123. [CrossRef]

48. Antolín, D.; Medrano, N.; Calvo, B.; Pérez, F. A Wearable Wireless Sensor Network for Indoor Smart Environment Monitoring in Safety Applications. *Sensors* **2017**, *17*, 365. [CrossRef]

49. Kumar, S.; Jasuja, A. Air quality monitoring system based on IoT using Raspberry Pi. In Proceedings of the 2017 International Conference on Computing, Communication and Automation (ICCCA), Greater Noida, India, 5–6 May 2017; pp. 1341–1346. [CrossRef]

50. Shete, R.; Agrawal, S. IoT based urban climate monitoring using Raspberry Pi. In Proceedings of the 2016 International Conference on Communication and Signal Processing (ICCSP), Melmaruvathur, India, 6–8 April 2016; pp. 2008–2012.

51. Sai, K.B.K.; Mukherjee, S.; Sultana, H.P. Low Cost IoT Based Air Quality Monitoring Setup Using Arduino and MQ Series Sensors With Dataset Analysis. *Procedia Comput. Sci.* **2019**, *165*, 322–327.

52. Rajalakshmi, R.; Vidhya, J. Toxic Environment Monitoring Using Sensors Based On Arduino. In Proceedings of the 2019 IEEE International Conference on System, Computation, Automation and Networking (ICSCAN), Pondicherry, India, 29–30 March 2019; pp. 1–6.

53. Jasinski, G.; Wozniak, L.; Kalinowski, P.; Jasinski, P. Evaluation of the Electronic Nose Used for Monitoring Environmental Pollution. In Proceedings of the 2018 XV International Scientific Conference on Optoelectronic and Electronic Sensors (COE), Warsaw, Poland, 17–20 June 2018; pp. 1–4.

54. Wang, Z.; Li, Y.; Zhang, T.; Hu, J.; Wang, Y.; Wei, Y.; Liu, T.; Sun, T.; Grattan, K.T.V. A Sensitive and Reliable Carbon Monoxide Monitor for Safety-Focused Applications in Coal Mine Using a 2.33-μm Laser Diode. *IEEE Sens. J.* **2020**, *20*, 171–177. [CrossRef]

55. Heavy Engineering Corporation Ltd. MQ9 Gas Sensor. 2020. Available online: http://www.datasheet.fr/parts/904644/MQ-9-pdf.html (accessed on 8 May 2020).
56. Heavy Engineering Corporation Ltd. MQ136 Gas Sensor. 2020. Available online: http://www.sensorica.ru/pdf/MQ-136.pdf (accessed on 8 May 2020).
57. Wavesen. HC06 Bluetooth Datasheet. 2020. Available online: https://www.olimex.com/Products/Components/RF/BLUETOOTH-SERIAL-HC-06/resources/hc06.pdf (accessed on 8 May 2020).
58. Wittbrodt, B.; Pearce, J.M. The effects of PLA color on material properties of 3-D printed components. *Addit. Manuf.* **2015**, *8*, 110–116. [CrossRef]
59. Mirón, V.; Ferrándiz, S.; Juárez, D.; Mengual, A. Manufacturing and characterization of 3D printer filament using tailoring materials. *Procedia Manuf.* **2017**, *13*, 888–894. [CrossRef]
60. Arduino.cc. Arduino Open-Source Platform. 2020. Available online: https://www.arduino.cc/en/Main/AboutUs (accessed on 8 May 2020).
61. Platformio.org. IDE for Embedded Development. 2020. Available online: https://platformio.org/ (accessed on 8 May 2020).
62. gs.statcounter.com. Mobile OS Market Share. 2020. Available online: https://gs.statcounter.com/os-market-share/mobile/worldwide/(accessed on 8 May 2020).
63. Amazon. Drager PAC 6000 CO. 2020. Available online: amazon.de (accessed on 4 November 2020).
64. GmbH, K. Drager PAC 6500 H_2S. 2020. Available online: https://www.kleinschmidtgmbh.com/Draeger-Pac-6500-H2S/en (accessed on 4 November 2020).
65. GmbH, K. Drager PAC 8500 H_2S and CO. 2020. Available online: https://www.kleinschmidtgmbh.com/Draeger-Pac-8500-H2S/CO/en (accessed on 4 November 2020).
66. GmbH, K. Drager X-am 8000 Multi-Gas Detector. 2020. Available online: https://www.kleinschmidtgmbh.com/epages/62984732.sf/de_DE/?ObjectID=183572492 (accessed on 4 November 2020).
67. Stefaniak, P.; Wodecki, J.; Michalak, A.; Wyłomańska, A.; Zimroz, R. Data Acquisition System for Position Tracking and Human-Selected Physiological and Environmental Parameters in Underground Mine. In *Proceedings of the 18th Symposium on Environmental Issues and Waste Management in Energy and Mineral Production Santiago (Chile)*; Springer: Cham, Switzerland 2018; pp. 241–248. [CrossRef]

Publisher's Note: MDPI stays neutral with regard to jurisdictional claims in published maps and institutional affiliations.

 *energies*

Article

Comminution of Copper Ores with the Use of a High-Pressure Water Jet

Przemyslaw J. Borkowski

Faculty of Geo Engineering Mining and Geology, Wroclaw University of Technology, 50-370 Wroclaw, Poland; przemyslaw.borkowski@pwr.edu.pl

Received: 4 November 2020; Accepted: 27 November 2020; Published: 28 November 2020

Abstract: The article presents research on the comminution of copper ore in a self-constructed mill using high-pressure water jet energy to investigate the usefulness of such a method for comminuting copper ore. As a result, ore particles are obtained that are characterized by appropriate comminution and a significant increase in their specific surface, in turn allowing for potential further processing of the mineral. A comparative analysis of the efficiency of copper ore comminution, primarily taking into account the unit energy consumption and the efficiency of the milling process, clearly indicates that the energy absorption of hydro-jet material comminuting is lower than during mechanical grinding, e.g., in a planetary ball mill. The applicability of the technique depends on the brittle nature of the host rock, e.g., it is especially appropriate for sandstone and shale ores.

Keywords: copper ore comminution; hydro-jet mill; high-pressure water jet

1. Introduction

Shredding processes are widely used in various fields of raw material processing. Such procedures are carried out in order to lower the costs of maintaining the cutter discs used in Tunnel Boring Machines (TBM) operating in hard rock formations [1]. They are also used to obtain fine-grained particles, sometimes even nanoparticles [2] or those that have an increased specific surface, which can be especially useful in pharmaceuticals [3].

Much attention is paid to the technical aspects of grinding specific mineral materials such as hard rock ores. For example, low-grade hematite ores generally only undergo crushing and screening, which is not energy intensive. Consequently, fine-grained magnetite ores require fine grinding, often to below 30 μm, to liberate the magnetite from the silica matrix, which usually involves traditional techniques. These processes generate greater costs and require higher energy consumption. The need to lower energy consumption is an issue stressed in the literature [4].

The discussion on the concept of dry and wet copper ore grinding in an electromagnetic mill (EMM) also focuses on the efficiency-related advantages of the former from the perspective of both different process parameters and also the energy costs [5,6]. However, a comparison of the results of EMM dry grinding with the results obtained using other standard solutions presented in the publication points to certain advantages of such a processing method. The authors of the above publications also indicate that the presence of sharp edges and the estimation of the processing time required to obtain a product of specific grain size remain a challenge and need further research.

Some other authors observe that reducing the size of minerals is the most critical step preceding their final separation. These authors refer to the fundamentals of the crack development processes and see them as being essential for optimizing both the comminution processes and the design of the comminution machines. They conclude that although many advances have been made in the design of comminution machines, the combined effect of crack development processes and ore properties such as ore texture is neither fully understood nor quantified [7].

An interesting and detailed comparative investigation of the behavior of ores processed in jaw, cone as well as hammer crushers, and in ball mills, was used to investigate the effects of comminution for different samples of chromite ores. The results served as the basis to analyze particle size distributions (PSD) for the products of grinding in order to evaluate the suitability of the Gates–Gaudin–Schuhmann (GGS) and Rosin–Rammler (RR) models. The results suggest that the particle size distributions depend mostly on the applied method of crushing. On the other hand, it was found that the particle size distributions of all chromites subjected to jaw and cone fragmenting were better characterized by the Gates–Gaudin–Schuhmann model than by the Rosin–Rammler model [8].

Another investigation focuses on determining the selection function parameters which describe the effect of the ball size on the milling rate. The investigation illustrates the issue with experiments conducted on coal. The identified balanced ball mixture was taken in order to analyze the effect of ball size distribution on the selection function, while an adequate mixture recommended by the manufacturer of the original equipment was taken to validate the model. The results show that the estimated parameters can be used with confidence for identification of the optimal distribution of balls for a series of operational restrictions [9].

In yet another study, the matrix model and the probability of breakage of the population balance model, were combined with the use of a MATLAB code. Such procedure let to predict the size distribution of the comminution products for different minerals. The resulting models were good well corelated with the particle size distributions achieved from a ball mill [10].

The literature also mentions research aimed at identifying the ball mill grinding characteristics for hematite products comminuted with the use of a high-pressure grinding roll (HPGR) and a traditional cone crusher (CC), focusing mostly on comminuting kinetics and technical efficiency [11]. Compared with CC products, HPGR products were comminuted faster and with an earlier occurrence of the phenomenon of over-grinding fine particles [11].

Analogical conclusions come from the analysis of high-pressure grinding conducted as part of another research project, in which copper ores were crushed by a jaw crusher and a high-pressure grinder, with the products being exposure by means of X-ray tomography. The analysis pointed out that the products from the high-pressure grinding have more particle damage and higher copper recoveries when compared with the jaw crusher [12].

Yet another research identified the mineralogical characteristics of comminuted Sn-Ta ore. The grinding process, carried out with the use of a HPGR in combination with a ball mill (BM), was confronted with a single ball mill process [13]. The combined method reduces the particle size distribution, and as a result helps to increase the flotation effect.

The method of dynamic comminution of brittle materials using a high-pressure water jet (HPWJ) is of particular importance here. Particles of material lifted by the water jet at high speed, during collision with a hard disc, are cracked and the resulting gaps are wedged due to water squeezing into them. The intensive fragmentation of particles, which occurs in such conditions, is caused by the fact that tensile stresses arise inside them, and also that brittle materials are much less resistant to these stresses [14,15].

The purpose of this article is to investigate the usefulness of such a high-pressure hydro-jetting method for comminuting specific brittle materials, e.g., copper ores, in order to analyze their susceptibility to micronization, which precedes the flotation process of copper recovery.

2. Mineralogical Characteristics of Copper Ores

The deposit of copper ore taken into consideration in this paper is owned by KGHM Polska Miedz mining company. It is situated in the south-west part of Poland and dips from 700 m up to 1500 m. Sandstone, dolomite as well as shale types of rock are characteristic for this deposit, which includes copper minerals.

After extraction, copper ore is enriched during mechanical processing composed of crushing, milling and after classifying, undergo flotation. Final procedure are thickening and drying of a concentrate. As a result, a concentrate with a 23% copper content is produced that can be processed further in a smelter.

Presently, the concentration of copper in ores ranges from 0.95 to 2.6%, depending on the KGHM mining region and the type of ore. As far as the mineralization of carbonate ores is concerned, sulphides can be found as fine, irregular inclusions occurring in the form of nodules and veinlets. The average content of the Cu in the carbonate rocks equals 0.8% [16].

The grade of the sandstone ore can range from 0.7% Cu up to 30% Cu but the average number is 1.8% of Cu. Sulphides occur in the form of fine anhedral and subhedral grains 20 up to 30 µm in size, and as aggregates coming as interstitial pore fillings [16,17].

The last shale ore type has more complex mineral characteristics. The content of Cu minerals usually exceeds 4% in weight. The chalcocite and bornite as well as chalcopyrite associations are dominant there. The minerals of the chalcocite group predominate at the Lubin area while the region of Sieroszowice mine is characterized mostly by digenite and chalcocite occurrence, and bornite or chalcopyrite being negligible [18].

3. Equipment and Research Methodology

An analysis of several applied design solutions of hydro-jetting mills [19–21], including vertical and horizontal designs, as well as different comminution mechanisms, allowed us to develop and build our own constructional solution in the form of a vertical hydro-jet mill (Figure 1). It provides the possibility to comminute materials in the pressure range of 50–300 MPa with a variable water flow in the range 0.2–0.5 dm^3/s. This device can be used to comminute brittle materials having a grain size of up to 2 mm, and with a significantly increased micronization efficiency, at the level of 50 g/s [22].

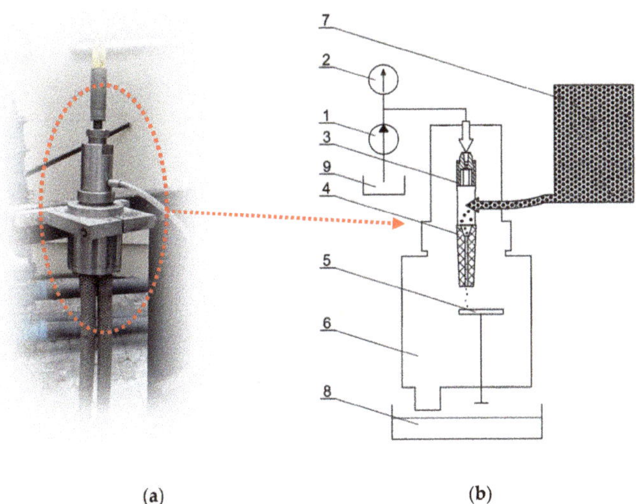

Figure 1. General view of experimental setup (**a**), and a diagram of a hydro-jet mill (**b**): 1—high-pressure pump, 2—manometer, 3—water nozzle, 4—comminuting nozzle, 5—comminuting disk, 6—shredding chamber, 7—feed tank, 8—product tank, 9—water tank.

The tests covered three types of copper ores found in the KGHM (Poland) mines (carbonate, sandstone and shale). These three types of ore were first mechanically ground in a jaw crusher to obtain granules of the required particle size # 0.5–2 mm, which were subsequently subjected to comminution with a high-pressure water jet, i.e., to the process being the focus of this article.

A FEI Quanta 200 Mark II scanning microscope was used to assess the shape and morphology of the copper ore particles.

Measurements of the fragmentation effects of the studied ores were carried out using the Analysette 22 MicroTec Plus particle size laser meter, which allows the analysis and preparation of particle size distributions of 0.08–2000 µm.

Tests of the efficiency of hydro-jetting comminution of copper ore were carried out:

- at a nominal water pressure of 100, 150, 200 and 250 MPa,
- with a diameter of the water nozzle $d_w = 0.7$ mm,
- using a comminuting nozzle diameter $d_h = 2.4$ mm,
- with a distance between the outlet of the homogenizing nozzle and the comminuting disc: s = 10 mm.

Depending on the needs, the number of parallel measurements ranged from 5 to a dozen or so trials, and their averaged results formed the basis for assessing the analyzed quantities.

4. Hydro-Jet Milling of Copper Ores

Tests of hydro-jetting comminution of three different types of copper ore yielded results in the form of particle distributions, an example of which is shown in Figure 2.

The tests were carried out at the nominal pressure of the water jet, and applied in the range of $p = 100$–250 MPa with an output of $Q_w = 0.2$–0.5 dm^3/s. The samples consisted of individual batches of m = 1 kg of copper ore, and the comminuting time (t) was also measured. For such conditions, the copper ores were found to show a relatively high resistance to comminution. However, different results were obtained for different ore types. The efficiency of the process, assessed on the basis of the reduction of the dimensions of particles obtained in the process of hydro-jet milling, is illustrated by the graphs presented in Figure 3. The influence of water jet pressure on the change in the passing size of the ore particles (a_{90}) is marked by the solid line, and the dotted line presents the efficiency (Q_c) of the milling process.

The research conducted on copper ores generally shows that as the nominal water pressure increases, the efficiency of micronization also increases, the phenomenon of which is assessed on the basis of the increasingly smaller particle sizes obtained from hydro-jet milling. Moreover, the efficiency of such milling also increases.

A detailed analysis of these charts shows that the courses for individual types of copper ore differ. In the case of the carbonate ore, for a water jet pressure of 100 MPa, the maximum characteristic grain size for 90% of the fragmented ore particles is $a_{90} = 130.7$ µm, for a pressure of 150 MPa it is $a_{90} = 75.2$ µm, and for a pressure of 200 MPa it is equal to $a_{90} = 63.8$ µm. For the highest applied jet pressure (250 MPa), the maximum particle size would be $a_{90} = 49.4$ µm. The efficiency of the hydro-jet comminuting process, however, is respectively: 15.1 g/s for the pressure of 100 MPa, 16.9 g/s for the pressure of 150 MPa, 18.4 g/s for 200 MPa, and 22.4 g/s for the maximum pressure of the water jet (250 MPa).

(**a**)

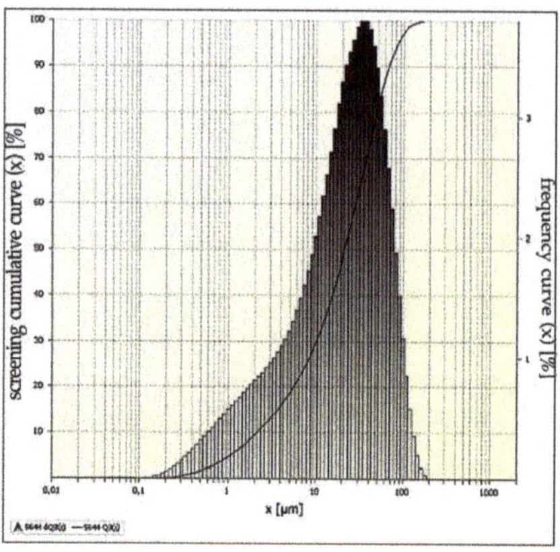

Figure 2. *Cont.*

(b)

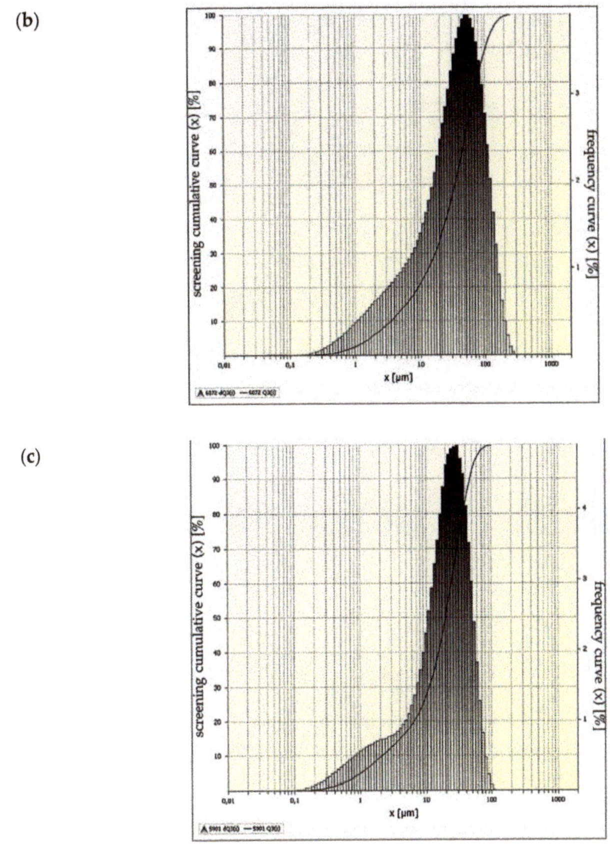

(c)

Figure 2. Example of the distribution of three types of copper ore comminuted in a hydro-jet mill at a water pressure of 150 MPa: (**a**)—carbonate, (**b**)—sandstone, (**c**)—shale ore, performed using the Analysette 22 MicroTec Plus particle laser meter.

(a)

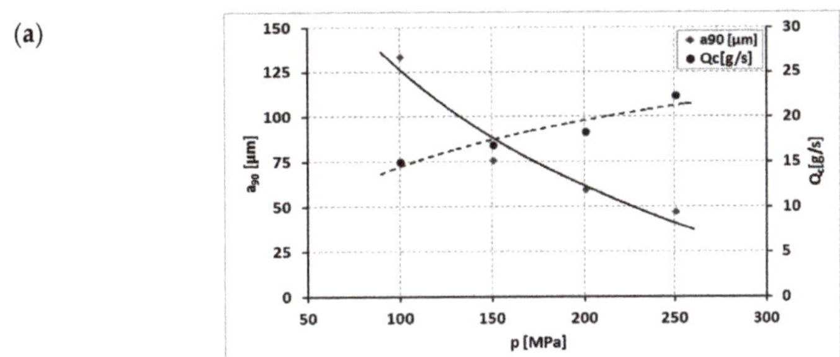

Figure 3. *Cont.*

(b)

(c)

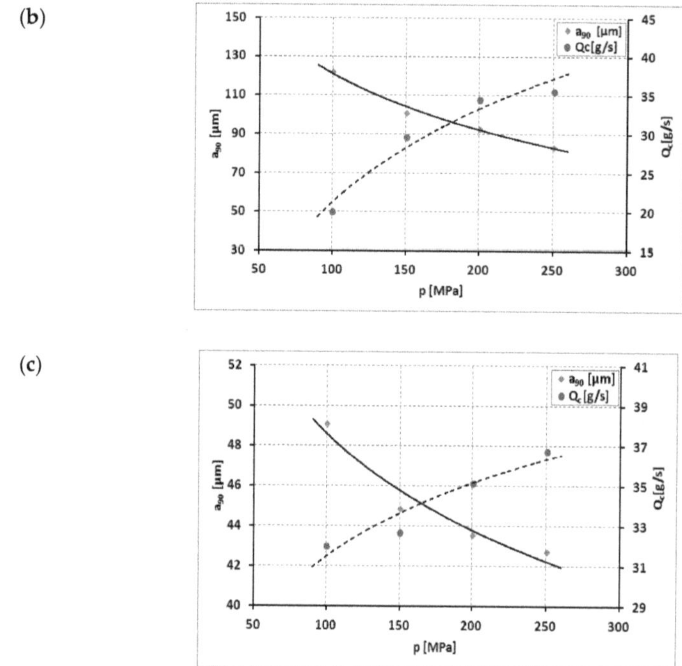

Figure 3. Influence of water jet pressure on the efficiency of the comminution (of characteristic dimensions) for 90% of ground copper ore particles (a_{90}), and the efficiency (Q_c) of its milling process for: (**a**)—carbonate, (**b**)—sandstone, (**c**)—shale ore.

The second type of sandstone copper ore has the following maximum grain sizes: a_{90} = 121.8 μm for a water jet pressure of 100 MPa, a_{90} = 102.5 μm for 150 MPa, a_{90} = 91.9 μm for 200 MPa, and a_{90} = 83.7 μm (the minimum size) for the highest jet pressure (250 MPa). The efficiency of the sandstone copper ore hydro-jet comminuting process can be described with the following values: 20.1 g/s for a pressure of 100 MPa, 29.7 g/s for a pressure of 150 MPa; 34.6 g/s for 200 MPa, and 35.6 g/s for the maximum pressure of the water jet (250 MPa).

The last studied type was shale copper ore. The experiments allowed the degree of its fragmentation to be estimated at a_{90} = 49.1 μm for 100 MPa. Higher values of water jet pressure, such as 150 MPa, cause the maximum particle size to decrease to a_{90} = 44.7 μm, and for the level of 200 MPa, it is even slightly lower, reaching a_{90} = 43.8 μm. The use of the highest water jet pressure of 250 MPa is uneconomical as it causes relatively insignificant changes in the maximum dimension a_{90} = 42.6 μm, which is characteristic for 90% of the fragmented ore particles. Considering the efficiency of the hydro-jet comminuting of copper shale ore, it should be noted that it increases with the increase of the nominal water pressure, reaching the efficiency of 32.0 g/s for a pressure of 100 MPa. In turn, for a pressure of 150 MPa it is 32.7 g/s, for 200 MPa it is 35.2 g/s, and for the maximum water jet pressure of (250 MPa) considered in the tests, the efficiency of this process reaches the value of 36.8 g/s.

Copper ore particles formed during hydro-jet comminution are usually characterized by an isometric shape with fairly expressive edges, while their rough surface usually takes the form of highly developed spatial micro-openings for a carbonate ore (a), a multiple trimming effect for sandstone (b), or a relatively smooth surface for shale ore. This is best characterized by the SEM images shown in Figure 4, which illustrate typical examples of their shape and morphology.

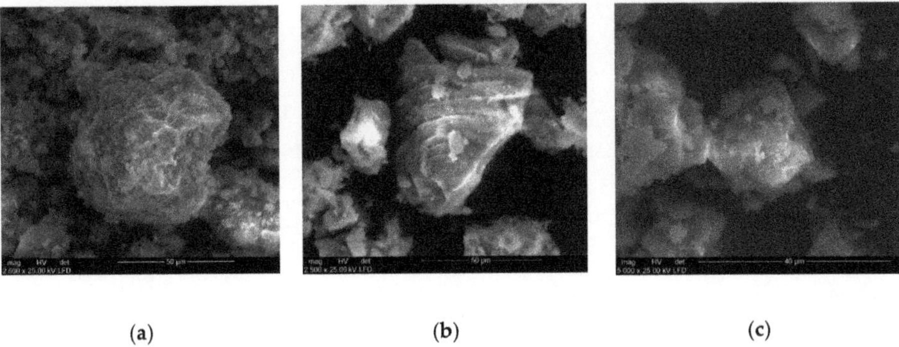

(a)   (b)   (c)

Figure 4. Examples of SEM images of copper ore particles comminuted in a hydro-jet mill at a water jet pressure of 150 MPa: (**a**)—carbonate, (**b**)—sandstone, (**c**)—shale ore.

The comparison of the above photos with analogous scan pictures of the same three types of copper ore comminuted in a PM 100 type laboratory planetary ball mill (Figure 5) is interesting. The surface quality of the copper ore particles obtained as a result of the mechanical grinding operations is significantly different. The surface of such particles is generally devoid of any cavities and is characterized by a relatively high smoothness, which probably causes certain problems in their effective enrichment in the flotation process. If other methods of mechanical grinding of copper ores, especially those used in actual production processes, provide powders of similar quality, the suitability of the industrial technologies used so far should be carefully analyzed.

(a)   (b)   (c)

Figure 5. Examples of SEM images of copper ore particles comminuted in a planetary ball mill type PM 100: (**a**)—carbonate, (**b**)—sandstone, (**c**)—shale ore.

5. Results Discussion

An essential feature of each production process is the energy expenditure necessary for its implementation. In order to determine the energy consumption of the process of comminuting copper ore, relevant tests were carried out in a hydro-jet mill and, for comparison purposes, also in a planetary ball mill type PM 100. The unit energy consumption of hydro-jet comminution is calculated according to formula (1), while the adequate consumption for a planetary ball mill is described by (2).

$$E = \frac{p\,Q_w\,t}{m} \tag{1}$$

where: p—water pressure, Q_w—water output, t—time of comminuting the batch of copper ore, m—weight of copper ore batch;

$$E = \frac{P\,t}{m} \tag{2}$$

where: P—power of the planetary ball mill.

Interesting effects resulting from the conducted tests are presented in Figure 6. In general, using the size of the carbonate copper ore particles obtained in the process of their comminution as a common criterion for the different methods of comminution analyzed in this article made it possible to compare the unit energy consumption of these processes.

The charts indicate the points for which joint tests with the use of both methods of comminution of such ore were carried out. As a result, it was possible to make unit comparisons of energy consumption in the two tested methods. In the case of carbonate copper ore (Figure 6a), the specific energy consumption of hydro-jet comminuting is two to five times more favorable than when using a mechanical planetary ball mill. The same characteristics are presented for the sandstone ore (Figure 6b) and show the highest advantage of hydro-jetting at a level approximately 40 times higher. Consequently, the comparison of the energy consumption of the two applied shale ore comminuting methods (Figure 6c) again points to the more favorable results of the hydro-jet method. This confrontation proves that the energy consumption of the proposed method of hydro-jet comminuting is generally three and a half times lower than in the case of the mechanical grinding method in a planetary ball mill.

The use of another criterion for comparing the energy consumption of material comminuting processes is not possible due to the use of completely different milling methods. When using a mechanical planetary ball mill, the grinding of the feed occurs as a result of the dynamic interaction of the ball grinding elements. Therefore, extending the operation time of such a mill contributes to an increased efficiency of the milling process by causing greater fragmentation of the particles (i.e., reducing their dimensions). In such circumstances, for the given conditions and for the given equipment of the mill, the impact of the effectiveness of the milling process on the energy consumption of the material grinding mainly depends on the duration of this process.

Completely different relations occur in the conditions of comminuting materials (including copper ore) using a high-pressure water jet. In a hydro-jet mill, comminuting of the feed occurs as a result of the synergistic interaction of a high-pressure water jet and the dynamic impact of the particles undergoing comminution. Therefore, in such circumstances, for the given conditions and mill equipment, the efficiency of the comminution process of the batch, assessed on the basis of the reduction of the dimensions of the obtained particles, is primarily determined by the level of the nominal water pressure.

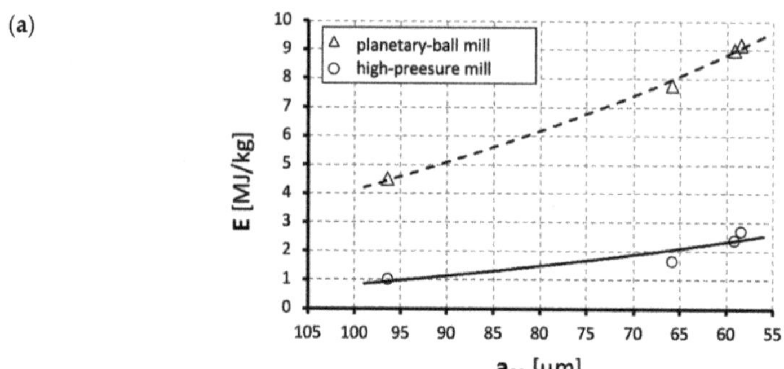

Figure 6. *Cont.*

(b)

(c)

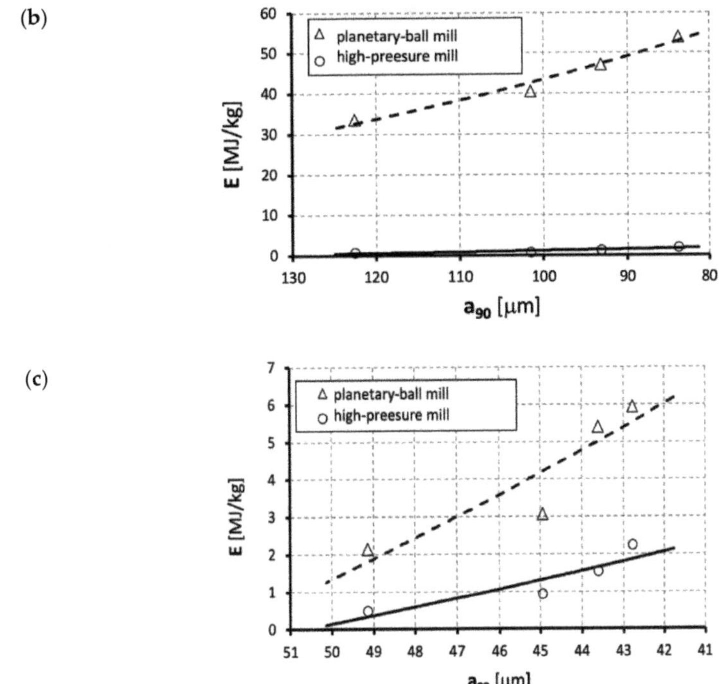

Figure 6. Dependence of the unit energy consumption of copper ore comminution on the efficiency of its milling process in a high-pressure hydro-jet mill (1) and in a planetary ball mill type (2) PM 100 ($n = 5$ s^{-1}) for carbonate copper ore (**a**), sandstone (**b**) and shale ore (**c**).

By properly confronting the energy consumption results of comminuting with both methods, presented in the form of the two curves in Figure 6, it was found that the energy consumption of hydro-jet comminuting of copper ore is distinctly smaller than during mechanical grinding in a planetary ball mill. The resulting benefits are determined by the following indicator:

$$W_k = \frac{E_M}{E_H}$$

(3)

where: E_M—energy consumption when crushing copper ore in a planetary ball mill, E_H—energy consumption when comminuting copper ore in a hydro-jet mill.

On the basis of the above dependence, an indicator of benefits occurring in relation to the tested conditions of the two compared methods was determined, ensuring in each case equal values of the maximum characteristic dimension for 90% of ground ore particles (a$_{90}$). The determined relationships allow finding the real values of the indicator of energy-consumption benefits, resulting from the comminuting of ore in a hydro-jet mill, as shown in Figure 7. With regard to milling in a PM 100 planetary ball mill, this comminution index in a high-pressure hydro-jet mill is usually very favorable. The analysis of the data in this graph indicates that the energy consumption benefit equals an average of 4 for carbonate copper ore (Figure 7a), 40 for sandstone copper ore (Figure 7b), and 3.5 for shale copper ore (Figure 7c).

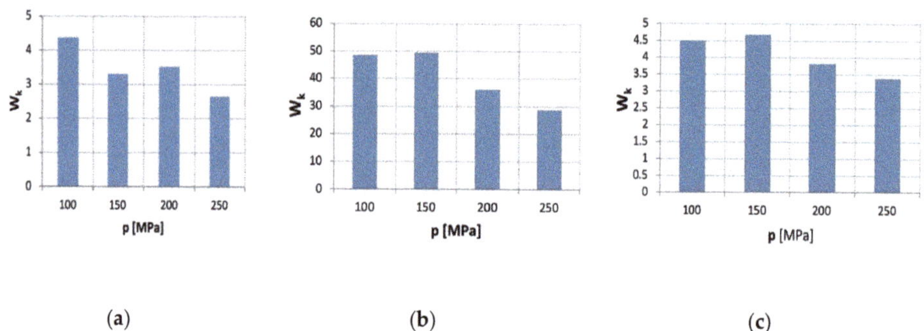

(a)   (b)   (c)

Figure 7. Values of the indicator of energy consumption benefits that result from the comminution of: (**a**)—carbonate, (**b**)—sandstone, (**c**)—shale copper ore in a hydro-jet mill.

6. Conclusions

Conducting detailed analyses of the research results made it possible to draw conclusions that are important for better understanding the process of the hydro-jet comminution. Increasing the water jet pressure has a beneficial effect on the hydro-jetting process of comminuting the discussed types of copper ore, as well as on the milling efficiency. A comparative analysis of the efficiency of copper ore comminution, primarily taking into account the unit energy consumption and the efficiency of the milling process, clearly indicates that the energy absorption of hydro-jet material comminuting is normally in the range of 4~40 times lower than during mechanical grinding, e.g., in a planetary ball mill. However, in terms of the energy consumption of the process, the increase in water pressure is unfavorable. During comminution of copper ore in a hydro-jet mill, particles with a fairly regular isometric shape are usually formed. Their most favorable morphology arises at a water pressure of 150 MPa for a carbonate ore, while a multiple trimming effect can be seen for sandstone ore or for a relatively smooth surface—the shale copper ore. Further increasing of the water pressure up to the maximum analyzed level of 250 MPa improves the micro-openings effect for sandstone copper and shale ore. This knowledge should be useful in the operations of the further processing of such copper ore. For the above reasons, the hydro-jet method, which uses a high-pressure water jet to comminute copper ore and other brittle materials, is very promising.

As the process in the current solution of the high-pressure hydro-jet mill (designed by the author) offers a significantly lower energy consumption, as well as other advantages, this prototype can be seen as a competition for standard milling devices that use conventional methods of mechanical disintegration of brittle materials. The current design solution of such a hydro-jet mill is the development base for the construction of derivative devices for comminuting various minerals in a wet environment.

Funding: This research received no external funding.

Conflicts of Interest: The authors declare no conflict of interest.

References

1. Cheng, J.L.; Jiang, Z.H.; Han, W.F.; Li, M.L.; Wang, Y.X. Breakage mechanism of hard-rock penetration by TBM disc cutter after highpressure water jet precutting. *Eng. Fract. Mech.* **2020**, *240*, 107320. [CrossRef]
2. Sitek, L.; Foldyna, J.; Martinec, P.; Klich, J.; Mašláň, M. On the preparation of precursors and carriers of nanoparticles by water jet technology. In Proceedings of the International Conference Water Jet 2011—Research, Development, Applications, Ostravice, Czech Republic, 3–5 October 2011; pp. 255–271.
3. Nakach, M.; Authelin, J.R.; Chamayou, A.; Dodds, J. Comparison of various milling technologies for grinding pharmaceutical powders. *Int. J. Miner. Process.* **2004**, *74*, S173–S181. [CrossRef]

4. Jankovic, A. Developments in iron ore comminution and classification technologies. In *Iron Ore. Mineralogy, Processing and Environmental Sustainability*; Woodhead Publishing: Cambridge, UK, 2015; pp. 251–282. [CrossRef]

5. Ogonowski, S.; Wołosiewicz-Głąb, M.; Ogonowski, Z.; Foszcz, D.; Pawełczyk, M. Comparison of Wet and Dry Grinding in Electromagnetic Mill. *Minerals* **2018**, *8*, 138. [CrossRef]

6. Wołosiewicz-Głąb, M.; Pięta, P.; Foszcz, D.; Ogonowski, S.; Niedoba, T. Grinding Kinetics Adjustment of Copper Ore Grinding in an Innovative Electromagnetic Mill. *Appl. Sci.* **2018**, *8*, 1322. [CrossRef]

7. Semsari, P.; Parapari, P.M.; Rosenkranz, J. Breakage process of mineral processing comminution machines—An approach to liberation. *Adv. Powder Technol.* **2020**, *31*, 3669–3685. [CrossRef]

8. Taşdemir, A.; Taşdemir, T. A comparative study on PSD models for chromite ores comminuted by different devices. *Part. Part. Syst. Charact.* **2009**, *26*, 69–79. [CrossRef]

9. Katubilwa, F.M. Effect of Ball Size Distribution on Milling Parameters. Master's Thesis, Faculty of Engineering and the Built Environment, University of the Witwatersrand, Johannesburg, South Africa, 2008.

10. Petrakis, E.; Komnitsas, K. Improved modeling of the grinding process through the combined use of matrix and population balance models. *Minerals* **2017**, *7*, 67. [CrossRef]

11. Liu, L.; Tan, Q.; Li, W.; Lv, L. Comparison of grinding characteristics in high-pressure grinding roller (HPGR) and cone crusher (CC). *Physicochem. Probl. Miner. Process.* **2017**, *53*, 1009–1022.

12. Kodali, P.; Dhawan, N.; Depci, T.; Lin, C.L.; Miller, J.D. Particle damage and exposure analysis in HPGR crushing of selected copper ores for column leaching. *Miner. Eng.* **2011**, *24*, 1478–1487. [CrossRef]

13. Hamid, S.A.; Alfonso, P.; Anticoi, H.; Guasch, E.; Oliva, J.; Dosbaba, M.; Garcia-Valles, M.; Chugunova, M. Quantitative Mineralogical Comparison between HPGR and Ball Mill Products of a Sn-Ta Ore. *Minerals* **2018**, *8*, 151. [CrossRef]

14. Averin, E. Universal Method for the Prediction of Abrasive Waterjet Performance in Mining. *Engineering* **2017**, *3*, 888–891. [CrossRef]

15. Yashima, S.; Kanda, Y.; San, S. Relationship between particle size and fracture as estimated from single particle crushing. *Powder Technol.* **1987**, *51*, 277–282. [CrossRef]

16. Borg, G.; Piestrzynski, A.; Bachmann, G.H.; Püttmann, W.; Walther, S.; Fiedler, M. An Overview of the European Kupferschiefer Deposits. *Econ. Geol. Spec. Publ.* **2016**, *16*, 455–486.

17. Oszczepalski, S. Origin of the Kupferschiefer polymetallic mineralization in Poland. *Miner. Depos.* **1999**, *34*, 599–613. [CrossRef]

18. Bartlett, S.C.; Burgess, H.; Damjanović, B.; Gowans, R.M.; Lattanzi, C.R. Technical Report on the Production of Copper and Silver by KGHM Polska Miedź S.A. In *the Legnica-Głogów Copper District in South-West Poland*; Micon International Co. Limited: Norwich, Norfolk, 2013.

19. Cui, L.; An, L.; Gong, W. Effects of process parameters on the comminution capability of high pressure water jet mill. *Int. J. Miner. Process.* **2006**, *81*, 113–121. [CrossRef]

20. Mazurkiewicz, M. Method of Creating Ultra-Fine Particles of Materials Using High Pressure Mill. U.S. Patent No. 6,318,649, 20 November 2001.

21. Zonghao, L.; Zhinan, S. Wet comminution of raw salt using high-pressure fluid jet technology. *Powder Technol.* **2005**, *160*, 194–197. [CrossRef]

22. Borkowski, P.; Borkowski, J.; Bielecki, M. Micronization of Carbonate Copper Ore With High-Pressure Water Jet Method. In Proceedings of the 22nd International Conference on Water Jetting 2014 Advances in Current and Emerging Markets, Haarlem, The Netherlands, 3–5 September 2014; Fairhurst, M., Ed.; BHR Group: Bedfordshire, UK, 2014; pp. 305–314, ISBN 978-1-85598-143-0.

Publisher's Note: MDPI stays neutral with regard to jurisdictional claims in published maps and institutional affiliations.

 *energies*

Article

The Impact of the Ventilation System on the Methane Release Hazard and Spontaneous Combustion of Coal in the Area of Exploitation—A Case Study

Magdalena Tutak [1,*], Jarosław Brodny [2,*], Dawid Szurgacz [3], Leszek Sobik [4] and Sergey Zhironkin [5,6,7]

1   Faculty of Mining, Safety Engineering and Industrial Automation, Silesian University of Technology, 44-100 Gliwice, Poland
2   Faculty of Organization and Management, Silesian University of Technology, 44-100 Gliwice, Poland
3   Center of Hydraulics DOH Ltd., 41-906 Bytom, Poland; dawidszurgacz@doh.com.pl
4   Polska Grupa Górnicza S.A. KWK ROW Ruch Chwałowice, 44-206 Rybnik, Poland; sobik55@poczta.fm
5   Department of Trade and Marketing, Siberian Federal University, 79 Svobodny av., Krasnoyarsk 660041, Russia; szhironkin@sfu-kras.ru
6   Department of Open Pit Mining, T.F. Gorbachev Kuzbass State Technical University, 28 Vesennya st., Kemerovo 650000, Russia
7   School of Core Engineering Education, National Research Tomsk Polytechnic University, 30 Lenina st., Tomsk 634050, Russia
*   Correspondence: magdalena.tutak@polsl.pl (M.T.); jaroslaw.brodny@polsl.pl (J.B.)

Received: 17 August 2020; Accepted: 17 September 2020; Published: 18 September 2020

Abstract: Various types of natural hazards are inextricably linked to the process of underground hard coal mining. Ventilation hazards—methane and spontaneous combustion of coal—are the most dangerous; they pose a major threat to the safety of the workers and decrease the effectiveness of the whole coal production process. One of the methods designed to limit the consequences of such hazards is based on the selection of a ventilation system that will be suitable for the given mining area. The article presents a case study of an active longwall area, where—due to increasing ventilation hazard (methane and spontaneous combusting of coal)—the whole system was rebuilt. The U-type ventilation system was used in the initial stage of the extraction process, however, it often generated methane in amounts that exceeded the allowable values. Consequently, such conditions forced the change of the ventilation system from a U–type to Y–type system. The new system was installed during the ongoing mining process, unlike the usual practice. The article presents the results of tests on mine gas concentrations and descriptive statistics for both types of ventilation system. The results clearly demonstrate that the U-type longwall ventilation system, in the case of high methane release hazard, prevents safe and effective operation. At the same time, the use of this system limits the carbon oxidation reactions in the goaf, leading to spontaneous heating and combustion, which is confirmed by the low concentrations of gases—by-products of these reactions. In turn, the use of the Y-type longwall ventilation system ensures safe and effective operation in areas with high methane release hazard, but at the same time deteriorates the safety associated with the spontaneous combusting of coal. The presented case—both from a scientific and practical perspective—is quite interesting and greatly broadens the knowledge in the scope of an efficient ventilation system for underground workings.

Keywords: underground mining; ventilation systems; methane release hazard; spontaneous combustion hazard; safety

1. Introduction

1.1. Preliminaries, Objectives and Outline of the Study

Despite ongoing changes in the global economy, hard coal is still one of the basic energy raw materials in the world. In many countries, it is a strategic raw material for energy independence and security [1,2]. Therefore, any work which results in an improvement in the safety and efficiency of exploitation of this raw material plays an enormous economic and social role. The environmental element is also important here. Improving the safety of mining operations reduces emissions of methane (a greenhouse gas) and other by-products of spontaneous combustion of coal. All these aspects shape the image of the energy industry and show that it is based on conventional energy sources and cares about the environment and occupational safety.

The underground exploitation of hard coal—as one of the basic stages of the mining process—is linked to the occurrence of several types of hazards, including ventilation hazards, making the whole process dangerous [3–9]. The natural hazards (methane, dust, rock burst, spontaneous combustion of coal and water) are particularly dangerous as they put the workers and the continuity of the underground exploitation process at risk. As indicated by the authors of the works [10–16], these hazards lead to mining accidents and disasters. The occurrence of such natural hazards strongly impacts on the efficiency of the whole process. When the permissible concentrations of gases defined in Polish regulations emitted in the process of coal exploitation are exceeded, they prevent the safe conduct of mining exploitation [17]. The electrical power supply must be cut off (voltage must be switched off) for all electrical equipment installed in the longwall (shearer, conveyor, etc.) and the main gate if the concentration of methane in the air flowing out of the longwall exceeds 2% or 1% in the air flowing into the longwall area (according Polish regulations). This results in a sudden stoppage of the entire exploitation process, disturbing its continuity and effectiveness.

The methane and spontaneous combustion of coal are both natural ventilation hazards [18–26]. When activated—either by the release and ignition of methane or the spontaneous combustion of coal—they lead to serious injuries to the workers, halts in the mining process, the need to isolate the exploitation areas, or many other disturbances; these all have a negative impact on the coal extraction process. For this reason, the paper focuses on the analysis of both hazards as they often occur simultaneously. In such cases, they are referred to as concurrent hazards [27–29]. A significant rise in the level of ventilation hazards has been observed in hard coal mines in recent years. It is caused by an increase in the amount of methane released from the rock mass and from the side of the cave-in and the spontaneous combustion of coal–mainly in the goafs.

The goafs act as activators of these hazards to a very large extent. The reasons for this include poor insulation of the goaf from the mining zone, coal that had not been extracted (e.g., some amount of coal left in the seam or the presence of an anticipated sub-economic coal seam), and the mining process conducted in the vicinity of old goafs. These factors increase the threat to the mining process. This applies both to the methane release hazard as well as to the spontaneous combustion of coal [30–35].

In the case of the extraction of coal beds that are rich in methane and, at the same time, in the areas impacted by the goafs, the methane release hazard caused by the flow of the gas increases. The probability of spontaneous combustion of coal is also higher when the extraction of deposits is extensive. Both are tightly linked to the ventilation of the longwall, including the flow of air (filtration) through the goafs. The above-mentioned facts support the conclusion that the goafs have a significant impact on the formation of methane release hazards and spontaneous combustion of coal in the area of active exploitation. Therefore, they must be treated as an integral part of the mining process.

In the case of the mining process conducted in areas rich in methane, the methane that is accumulated in the goafs and flows to the active workings can increase the probability of the ignition and combustion of methane. In turn, the air flows from the longwall to the goaf create conditions for spontaneous combustion of the coal left in these goafs. The air flowing through pores and crevices in the goafs can thermally activate the coal. This process may result in the emission of carbon combustion products into the atmosphere in the form of poisonous gases (CO) and smoke [36]. In addition, the spontaneous combustion of coal can be a factor that leads to the ignition of methane. Therefore, in order to ensure the safety of the workers and the continuity of the production process, various actions are designed to identify the processes related to the flow of gases through mining excavations and goafs that affect the formation of ventilation hazards in the whole area. One such process that has an impact on the shaping of the ventilation hazards—methane release and spontaneous combustion of coal—in the active mining area is the routine ventilation of longwalls.

The global mining industry uses several systems of routine ventilation of active longwalls. The most common and, at the same time, a popular system is the U-type ventilation system. In Poland, as much as 70% of active longwalls are based on this type of ventilation (Figure 1a). The second—applied relatively often—is the Y-type ventilation system (Figure 1b) [37].

Both systems are used in the areas where methane release and spontaneous combustion hazards occur simultaneously. However, they have a different impact on the formation of these threats in the area of exploitation.

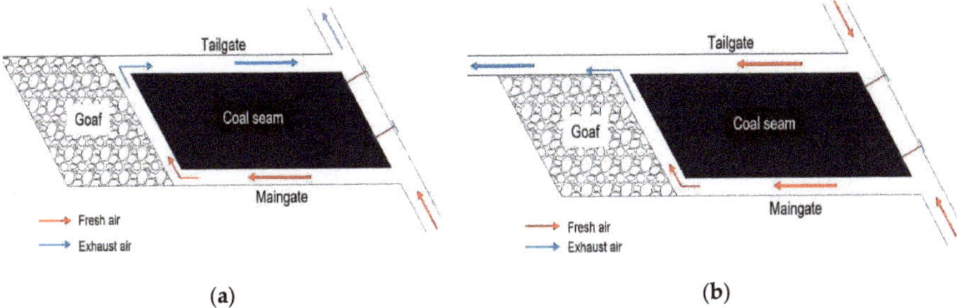

Figure 1. Diagram of the U-type ventilation system (**a**) and the Y-type system with routine ventilation of tailgate (**b**).

A common feature of both systems is the manner in which the air is supplied to the longwall through the tailgate; the difference is in the discharge of the used air. The U-type ventilation system discharges the air through the tailgate that runs along the body of coal; the Y-type ventilation system uses the tailgate that runs along the goafs. Additionally, the Y-type ventilation system involves the process of routine ventilation of the area of the longwall outlet (the intersection of the longwall and the tailgate). This involves a stream of fresh air fed through the tailgate located along the body of coal (Figure 1b) [38–40].

The manner of air distribution for both systems in the area of active mining works impacts on the development of the ventilation hazard related to the ongoing extraction process. This impact, however, is completely different.

The U-type ventilation system limits the probability of spontaneous combustion of coal that was left in the goafs. This is caused by the fact that this hazard is impacted by the flow of air rich in oxygen through the goafs—the air flows from the longwall. The air is a component required for the oxidation of coal that leads to spontaneous heating followed by spontaneous combustion of coal. The intensity of such a flow of air in the case of the U-type ventilation system is substantially smaller when compared to the flow in the Y-type ventilation system. In the case of the former, the air flows to the goafs only from the side of the tailgate and along the longwall. When using the Y-type system, the air flows also in the area where the longwall intersects with the tailgate. If there is a methane release hazard, especially in the area where the longwall intersects with the tailgate, its intensity will be greater for the U-type ventilation system than the Y-type. This intensity significantly depends on the physical parameters of the goafs, including the strength parameters of roof rocks (compressive and tensile strength), the porosity of the rocks the goafs, the total moisture content of rocks, the depth at which the exploitation is carried out and the thickness of the seam [37,41,42]. All these factors have a significant impact on porosity and, as a result, on the permeability of the goafs. The presence of hollow spaces in the goafs affects their permeability, which is the basic property characterising the porous medium and is fundamental for the process of airflow through these goafs and the emission of gases into the workings.

The highest amount of methane getting to the longwall working flows from the goafs. The highest methane accumulation when using the U-type ventilation system occurs in the area of the intersection of the longwall and the tailgate, whereas in the case of using the Y-type system, methane concentration is substantially smaller, as an additional air stream is supplied to this area [40,43].

Therefore, it can be concluded that the routine ventilation system activates, to a large extent, both hazards. The selection of the routine ventilation system—along with the physical parameters of the air supplied—should be preceded by a detailed analysis of the area where mining activities will be conducted.

Usually, the selection of an adequate longwall routine ventilation system in the case of concurrent hazards—related to methane and spontaneous combustion of coal—in the exploited deposit, impacts on the tendency of coal to spontaneously combust. The depth of exploitation in Polish coal mines constantly rises, which results in an increase in the initial temperature of the deposits, which in turn shortens the incubation time of spontaneous heating of coal [44]. This also increases the initial stress of the rock mass—later leading to the formation of bigger areas of cracks around the excavations. Such areas form porous centres that are permeable and can impact on the ventilation process [45].

If there is a high methane release hazard, the preventive measures used to reduce this hazard also have an adverse effect on the risk of coal spontaneous combustion, mainly due to the increased intensity of ventilation of the longwall and, consequently, increased airflow through the goafs.

The most preferable solution in this case is usually the system that is capable of limiting, as much as possible, the flow of the air through the goafs [37,40,46]. This reduces the possibility of spontaneous combustion of coal in the goafs. Hence, the most popular system in such cases is the U–type ventilation system. It, however, when applied in the areas with deposits rich in methane, prevents effective measures to minimize the methane release hazard. If methane release hazard prevents the effective conduct of the mining process, additional steps are necessary to limit the hazard. If they also fail, the type of the ventilation system must be changed from the U-type to the Y-type system. In practice, such change is not easy, but there have been cases when the system was changed successfully. The paper presents and discusses the case of such a change.

The issues that must be dealt with when changing the ventilation system have not been presented in any of the available papers to date. The area of the impact analysis of the ventilation system type on the concurrent hazards—methane and the spontaneous combustion of coal—have not been discussed sufficiently to date. In terms of safety of operation as well as its efficiency and efficacy, this research area plays a key role. Therefore, the authors hope that the case presented in the paper will fill the research and information gap and will point to a new approach to solving the ventilation problems. This proves that the solution presented in the paper is new and has not been applied to date.

The results of the analysis that are presented in the paper show that, in some cases, there is a need to undertake urgent and decisive measures in order to maintain the mining process. The objective of this paper is to present practical experience related to the change in the longwall ventilation system during ongoing mining and with concurrent hazards. This analysis also made it possible to determine the influence of the ventilation system on the formation of methane release hazard and the spontaneous combustion of coal (in Poland, it is referred to as endogenous fire) in the area of exploitation. As already mentioned, the analysis was conducted for the actual longwall. In terms of the formation of methane release hazard and spontaneous combustion of coal, the results of gas concentrations characterising this hazard were presented for two stages of longwall ventilation: for the initial stage of the U-type longwall ventilation system and for the Y-type ventilation system.

Undoubtedly, the method of solving the problem of exploitation in conditions of coexistence of ventilation hazards for the tested longwall is original and rarely used in practice. This is one of the reasons it deserves to be presented more extensively, all the more so because it enabled the safe operation of the longwall without additional obstacles. The results obtained clearly indicate a significant influence of the ventilation system on the formation of ventilation hazards for the two different ventilation systems tested.

Some of the features of the paper prove that it is original and innovative. First, to our knowledge, there are no studies in the scientific literature on the development of the hazard of methane release into the air of the longwalls with the Y-type ventilation system. Therefore, this study is an original scientific contribution devoted to the problem of the occurrence of methane release hazard for longwalls with this specific type of the ventilated system. This is especially important, since the Y-type ventilation system is often used in actual mines. However, the issue of the methane release hazard has, to date, mainly concerned only longwalls with the U–type system. Secondly, the testing ground is the actual longwall that, when active, had two types of ventilation system—the U-type and the Y-system. To date, to the authors' knowledge, there are no papers that present the results of research concerning the formation of the methane release hazard or spontaneous combustion of coal related to such a case. Thirdly, the authors used the measurements of the concentrations of gases emitted into the air of the excavations to demonstrate the impact of the applied longwall ventilation system on the formation of methane release hazard and the potential for the spontaneous combustion of coal, characterised by the concentrations of gases. Finally, there are no available publications that present and discuss the methane release hazard together with the spontaneous combustion of coal. Research activity within this area has been focused either on an analysis of the hazard of coal spontaneous combustion or methane release, whereas practice has shown that both hazards occur simultaneously.

All these factors prove that the research presented is original and, at the same time, show its broad range of application. The results extend the knowledge of the mining process.

1.2. Brief Literature Review of Ventilation of Longwalls and Methane and Coal Spontaneous Combustion Hazards

The issues concerning the occurrence of ventilation hazards in underground mine workings have been presented numerous times in the literature. Several researchers are engaged in intensive research on the recognition of processes related to the release of methane into the mine's air, e.g., [47–52] and the process of coal spontaneous combustion, e.g., [53–61].

The work on the release of methane from coal seams allowed, inter alia, for the construction of a simplified mathematical model of methane release from coal seams to mine workings [47] or the development of a mathematical model of methane release in coal seams [48], and provided a great extent of valuable knowledge related to the sorption capacity of coal and surrounding rocks, which affects the release of methane into mine workings [49–52]. In turn, the works [53–61] present several interesting results indicating that a side effect of the process of spontaneous combustion of coal is the release of gases such as carbon monoxide, ethylene, propylene, acetylene, and hydrogen into the air. The presence of these gases as indicator gases in mines' air has been used for many years to assess the process of spontaneous combustion of coal. The results are used for tests related to the ventilation of underground workings.

The subject of the selection of the longwall ventilation system and its influence on the co-occurrence of methane release hazard and spontaneous combustion of coal during mining is discussed in the world literature to a relatively limited extent. Most of the works focus on matters related to the issue of the independent occurrence of these hazards and the process of ventilating the workings in general [40,62–69].

One of the works devoted to the issue of ventilating the longwall is [62] and concerns the problem of designing a system for ventilating excavations at large depths. The authors of the work focused mainly on planning the ventilation parameters that are required during operation. The work represents a very general approach to this process. Gillies and Wu in their work [63] carried out tests of longwall ventilation systems in Australian mines. The authors presented the practices used to ventilate excavations during coal mining and drew attention to the basic problems of selecting a ventilation system for mining excavations. Mayes and Gillies [64] carried out an analysis of longwall ventilation methods in Australian mines and discussed popular solutions. The analysis covered sixteen large mining parcels, which allowed the authors of the paper to present the main problems occurring during the ventilation of excavations, as well as the best engineering practices to solve them. This work contains interesting practical tips for the underground ventilation process.

The paper [65] presents a critical analysis of the longwall ventilation systems in the context of methane release hazard. There, attention was drawn to the problem of a pressure drop (lasting at least one day) and its impact on the increase in methane release hazard. In turn, the work by Wang et al. [66] presents numerical research on the characteristics of ventilation airflow through the U-type longwall ventilation system. The results obtained confirm the general views on the use of this system, which were also confirmed in this work. The paper [67] presents the effect of ventilation on the spontaneous heating of coal.

Diamond and Garcia, in their work [68], evaluated the impact of mining practices on methane emissions and methane release hazard control systems. This work is a part of the presentation of practical experience in the field of longwall excavation ventilation. Wang et al., in their work [69], carried out research aimed at determining the characteristics of airflow through the longwall for different stages of the operation. This is another example of searching for methods to optimise the ventilation process. The paper [40] describes the characteristics of the most frequently used longwall ventilation systems in Polish hard coal mines. The material provided is valuable due to its practical approach to the problem of choosing the ventilation system.

The longwall ventilation systems are inextricably linked to the methane release hazard, which results in slightly less work being devoted to the fire hazard. There are also some valuable studies on this subject. The largest number of works in the area of the risk of coal spontaneous combustion concern goafs [70–73]. Furthermore, a large number of papers related to the methane release hazard are also devoted to goafs. This is due to the fact, as has already been mentioned before, that the goafs are of key importance for the formation of ventilation hazards in the area of mining operations.

In this respect, Karacan, in his work [70], pointed out that methane emissions from the longwall ventilation system are an important indicator of how much methane a mine produces and how much air needs to be provided to keep methane levels below the regulatory limits, including goafs. Zhang, in his paper [71], presented the results of the study on the distribution of air and gases in the sheared goafs of a longwall with the U-type ventilation system. Tutak and Brodny, in their work [72], studied the effect of the auxiliary devices used for the U-type ventilation system on the methane concentration in the upper corner of the longwall. They also presented the results of a study on the influence of the longwall ventilation system on the risk of spontaneous combustion of coal in the goafs [73]. In both cases, it was especially important to treat the goaf as a porous and permeable medium.

The problem of spontaneous combustion of coal left in the goafs and its influence on the ventilation process has also been discussed thoroughly in the works [74–77].

Lu and Qin, in their work [74], used numerical simulation to determine the distribution of gas concentration in the goaf, which resulted in the determination of a potential zone in which coal may spontaneously ignite. In turn, [75] presented the results of the analysis of coal spontaneous combustion in goafs using numerical modeling. The results obtained allowed to conclude that the risk associated with spontaneous combustion of coal is influenced by the concentration of oxygen in the goaf, the longwall ventilation system and the inclination of the seam. The paper [76] also presents the results of research on the prevention of spontaneous combustion of coal left in the goafs of one of the longwalls in the Haizi coal mine. Pan et al., in [77], used the measurement data on the concentration of carbon monoxide and oxygen in goafs and carried out numerical tests to determine three zones in goafs of one of the longwalls associated with the spontaneous combustion of coal.

Moreover, the authors of the works [78,79] have designated such zones in the goafs of the cave-in longwalls.

Taraba et al. [80,81] carried out numerical research in which they simulated the process of self-heating of coal in goafs. Yuan and Smith [82] used numerical research to predict the process of the heating and spontaneous combustion of coal in the goaf. The paper [83] also presents the results of similar studies.

In turn, the paper [84] presents the cutting roof and release pressure technique to determine the area of occurrence of three zones related to the process of spontaneous ignition of coal in the goaf.

Since spontaneous combustion of coal can lead to the ignition of methane, this subject has also been presented in several interesting and valuable works [85–87].

Brune and Saki, in their work [86], on the basis of research carried out with the use of Computational Fluid Dynamics (CFD), determined the places in goafs where methane occurs in explosive concentrations and presented a method of feeding nitrogen to goafs, which neutralizes the methane release hazard and, at the same time, spontaneous combustion of coal. The research referred to the longwall with the U-type ventilation system. Li et al., in turn, presented the results of the research consisting of determining the effect of spontaneous combustion of coal in goafs on methane decomposition [87]. The tests were also carried out for the U-type ventilation system. Zhang et al. presented the results of the numerical analysis of methane emission from goafs into the tailgate with the U-type ventilation system [88].

Zmarzły and Trzaskalik, in their work [89], presented the results of measurements of methane concentration in the area of the crossing of a longwall with a ventilation gallery for the U-type system.

The works presented here are only a part of the contribution to the broadly understood topic of ensuring ventilation safety in the process of the underground mining operation.

Therefore, the issues covered in this work should be a valuable addition to this material and indicate the practical possibilities of applying a completely new solution to improve the state of ventilation in case of combined hazards.

However, none of the works present the results of the influence of the ventilation system on the simultaneous formation of methane release hazard and spontaneous combustion of coal. Therefore, the issues covered in this work should be a valuable addition to this material and show a practical and completely new solution to improve the state of ventilation in case of combined hazards.

From both a scientific and utilitarian perspective, this issue seems to be extremely important and interesting.

2. Materials and Methods

The analysis of the impact of the longwall ventilation system on the formation of the methane release hazard and the risk of spontaneous combustion of coal was carried out for one of the longwalls in the mine in the ROW area (Rybnik Mining District) (Figure 2). The mine is located in the southern part of Poland. The size of its mining area is 21.76 km^2 and its operational resources amount to 144,643 thousand tons of coal.

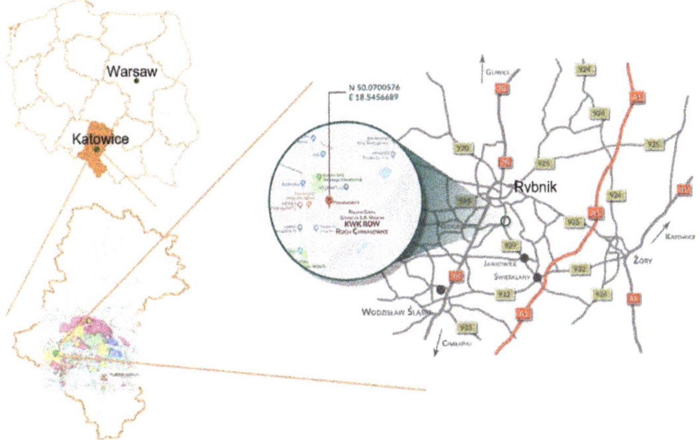

Figure 2. Location of the tested mine (own elaboration based on [90]).

The longwall was exploited with a longitudinal system with the roof rock cave-in in the direction from the boundaries of the field of exploitation. It was 240 m long, 870 m of coal was mined, and the thickness of the coal bed ranged from 1.6 to about 2.0 m.

The physical and chemical parameters of the rocks in the roof of the exploited seam are shown in Table 1 and the profile of the rocks in the direct roof is shown in Figure 3.

Table 1. Physical and chemical parameters of the rocks in the floor and roof of the exploited seam.

Type of Rock	Zone	R_c, MPa	Soakability	Separability	Density, g/cm³
sandstone	roof	41.6	0.8–1.0	block	2.65
shale	roof	26.6	0.6	block	2.50
coal (seam 408/1)	seam	17.5	-	-	1.28
shale	floor	31.7	0.6	block	2.56

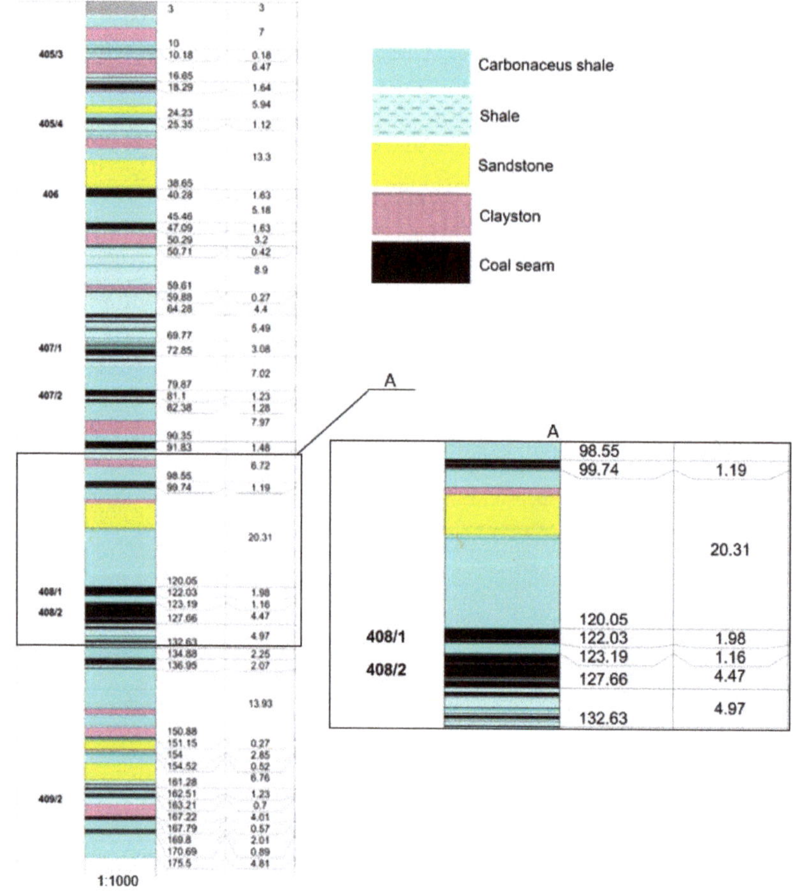

Figure 3. Profile of the seam and the direct roof of the longwall.

The hard coal deposited in the seam 408/1 is lustrous, rarely matte, with a strand-type structure and with occasional pyrite, and, less often, calcite coatings. The content of individual macerals in the coal is shown in Table 2.

Table 2. Petrographic composition of the coal deposited in the seam 408/1.

Petrographic	Content, %
Vitrinite	60–90
Exinite	5–20
Inertinite	5–25
Liptinite	5–8
Sulphides	1
Clay materials	1
Mineral matter	2–30

The susceptibility of the rocks in the roof and the floor to spontaneous sparking, which can ignite methane in the longwall due to the high variability of their quartz content, was determined as high. The maximum primary temperature of the rocks deposited at a depth of 780 m was 24.9 °C.

Ventilation parameters of the air, also known as the physical and chemical parameters (speed, pressure, and concentrations of gases in a mine) in the mine are measured by means of automatic gasometry sensors. These sensors include anemometers, automatic methane meters, oxygen meters, carbon monoxide, and carbon dioxide sensors, as well as absolute pressure, humidity, and temperature sensors. These sensors are part of the gasometric system of the mine (SMP system). They identify the composition and parameters of the underground air of the workings. This information is used to assess the hazard level in these workings. This can result, for example, in the automatic shutdown of the power supply to mining machines when the methane concentration exceeds the regulatory limit values [91]. The SMP system also provides a quasi-continuous communication with underground devices, such as sensors or voltage shutdown devices.

The applied automatic gasometry system distinguishes between the underground facility and the control stations located on the surface (Figure 4). The underground facility consists of sensors for measuring the ventilation parameters of air and underground stations, which perform the function of data concentrators and actuating systems. Surface stations, on the other hand, include data transmission systems and power supply systems. At the same time, they are an intermediate element for the computer part of the control system.

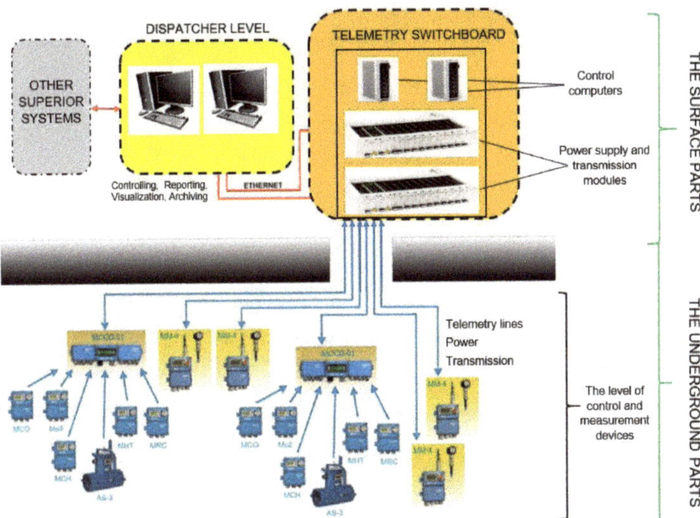

Figure 4. Structure of the system for monitoring the parameters of the air in the underground workings (own elaboration based on [92]).

Automatic methane meters type MM4 (manufactured by SEVITEL, Poland) with digital data transmission were used in the mine record methane concentrations with a resolution of 0.1 CH$_4$. The measurement is performed with a fixed sampling time of the recorded data. The measuring range ranges from 0 to 100%. These sensors are directly connected to the surface facility.

Anemometers, in turn, record air velocity with a resolution of up to 0.1 m/s, and barometric pressure gauges up to 1 hPa. An exemplary report of the results obtained from automatic methanometry is shown in Figure 5.

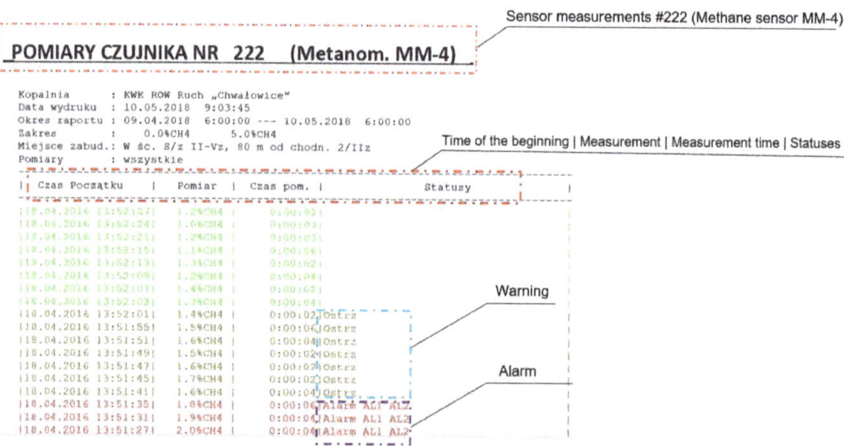

Figure 5. Report presenting methane measurement using an automatic methanometric sensor.

In the initial stage of exploitation, the longwall was ventilated with the U-type system. However, due to the level of methane release hazard and the impossibility of maintaining a permissible concentration of methane in the air (2%), it was necessary to change the ventilation system to the Y-type system. The diagrams of the longwall with the directions of airflow and the location of gasometric protections for the ventilation systems used are shown in Figure 6.

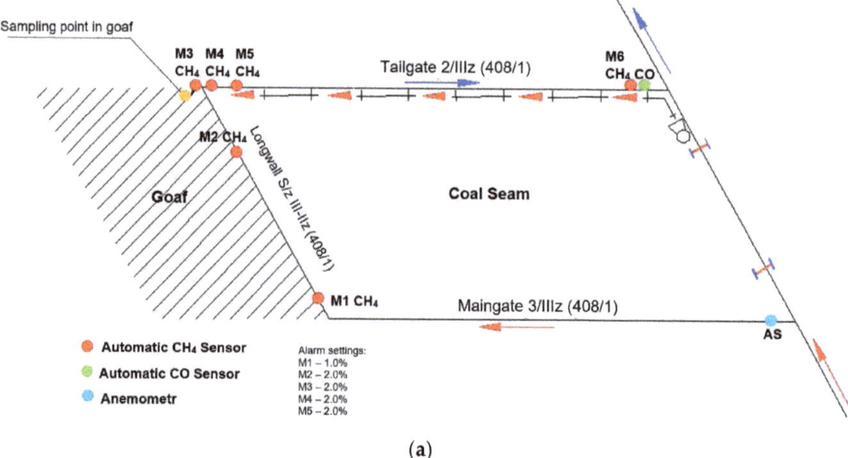

(a)

Figure 6. *Cont.*

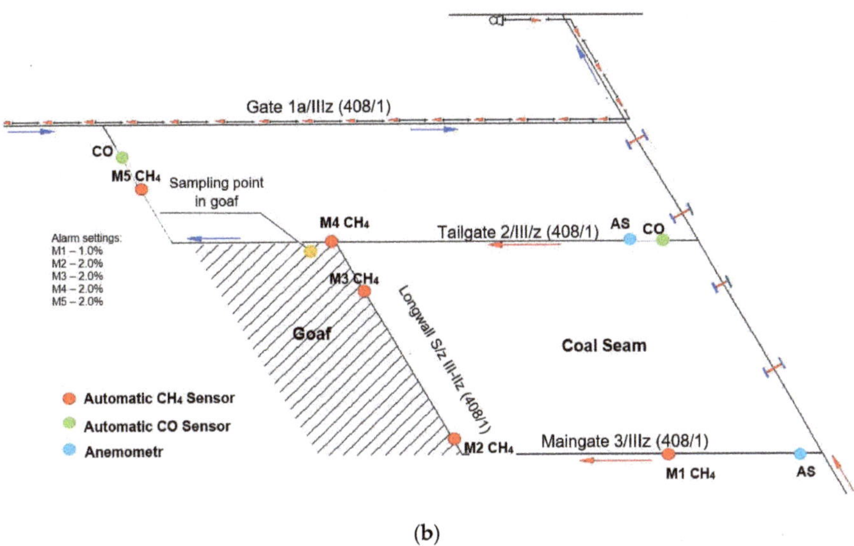

(b)

Figure 6. Diagrams of the tested longwall for the U-type (**a**) and Y-type (**b**) ventilation systems.

The measurement points for determining the actual concentration of methane released during the operation based on the U-type ventilation system were located in the longwall and the tailgate. Their detailed location is shown in Figure 6a. Sensor no. 1 (with the operating threshold of 1.0% CH_4), was installed under the roof in the longwall in the zone of about 10 m from the crossing with the tailgate. Sensor no. 2 (with the operating threshold of 2.0% CH_4) was installed under the roof in the longwall, at a distance of about 1/3 of the length of the longwall—counting from the outlet, in the place of the highest methane concentrations found. Sensor no. 3 (with the operating threshold of 2.0% CH_4) was installed in the tailgate at a distance of no more than 2 m from the excavation's liquidation line. Sensor no. 4 (with the operating threshold of 2.0% CH_4), was installed in the tailgate at the top of the sidewall, at the height of the longwall outlet. Sensor no. 5 (with the operating threshold of 2.0 CH_4) was installed in the tailgate at a distance of no more than 10 m from the outlet of the longwall.

The permissible concentration of methane, and consequently the operating thresholds of individual sensors, resulting in the exclusion of electricity supply to machines and equipment, depends on the place of their installation in excavations located in the area of the longwall on the basis of regulations applicable in Poland [91]. Methane sensors installed in the longwall area recorded methane concentrations with a resolution of 0.1% CH_4. The location of these sensors was consistent with the regulations valid in Polish hard coal mines (Figure 6a).

When using the U-type ventilation system, a volume flow rate of 1050 m³/min was supplied to the longwall by means of the main gate. Moreover, an extra airflow of 400 m³/min also supplied to the longwall.

Starting from the seventh month of its operation, the ventilation system was changed from the U-type to the Y-type. The method of arrangement of measurement sensors for this system is shown in Figure 6b.

Sensor no. 1 was installed under the roof of the tailgate at a distance of about 75 m before the inlet to the longwall. Sensor no. 2 (with the operating threshold of 1.0% CH_4), was installed under the roof in the longwall in the zone of about 10 m from the crossing with the tailgate. Sensor No. 3 (with the operating threshold of 2.0% CH_4), was installed under the roof in the longwall, in the zone from 2 m to the outlet from the longwall. Sensor No. 4 was installed in the tailgate maintained along the goafs, 10 m behind the longwall outlet (its operating threshold was 2.0% CH_4). Sensor No. 5 (with the operating threshold of 2.0 CH_4), was installed under the roof in the research cross heading 1a, between 10 and 15 m from the intersection with the tailgate 1a. When using the longwall ventilation system, volume flow of air of 1200 m^3/min (12.5% more than when using the U-system) was supplied to the longwall through the main gate; 450 m^3/min (11.11% more than when using the U–system) was supplied to the longwall through the tailgate. The need to increase the amount of air supplied to the longwall resulted from the fact that with less air it was impossible to maintain the permissible concentrations of methane.

The methane bearing capacity of the tested seam reached 5.3 m^3/Mg of pure coal and the gas desorption intensity index for coal from this seam reached 0.96 kPa. Determination of the seam's methane-bearing capacity is the foundation for determining the methane release hazard category under Polish regulations.

The tested longwall was located in the seam classified as category III methane hazard (as provided by Polish regulations). This means that the methane content of the accessible seam or its part is greater than 4.5 m^3/Mg but not greater than 8 m^3/Mg in terms of clean coal. The categories of methane release hazard in Poland are presented in Table 3.

Table 3. Category of methane release hazard in Poland (own elaboration based on [91]).

Category of Methane Release Hazard	Description
I	Category I methane release hazard includes the developed seam or its part if the quantity of methane of natural origin contained in a unit of weight in the depth of the body of coal, referred to as "methane carrying capacity", is between 0.1 and 2.5 m^3/Mg of dry ash-free coal.
II	Category II methane release hazard includes the developed seam or its part if the methane carrying capacity is greater than 2.5 m^3/Mg but not greater than 4.5 m^3/Mg of dry ash-free coal.
III	Category III methane release hazard includes the developed seam or its part if the methane carrying capacity is greater than 4.5 m^3/Mg but not greater than 8 m^3/Mg of dry ash-free coal.
IV	Category IV methane release hazard includes the developed seam or its part if the methane carrying capacity is greater than 8 m^3/Mg of dry ash-free coal or if there is a sudden methane release or methane and rock outburst.

Coal from the exploited seam was qualified to the IV group of spontaneous combustibility, i.e., with high susceptibility to spontaneous combustion. With a spontaneous combustibility index (*Sza*) of *Sza* = 116 °C/min and carbon oxidation activation energy (A) A = 47.9 kJ/mol, the spontaneous combustion incubation period determined under laboratory conditions was forty-seven days.

Due to the risk of coal spontaneous combustion, the area of the longwall was covered by an early detection system designed to report spontaneous combustion. Air samples were taken from chromatographic analysis on a weekly basis, which allowed determining the concentration of gases (ethane—C_2H_6, ethylene—C_2H_4, propane—C_3H_8, propylene—C_3H_6, acetylene—C_2H_6, carbon monoxide—CO, carbon dioxide—CO_2, oxygen—O_2, and hydrogen—H_2). Figure 6a shows the location of the goaf air sampling when using the U-type ventilation system and Figure 6b when using the Y-type system.

The exploited seam did not show a tendency to tremble: there were no records of frequent tremors. This seam and the excavations of the longwall area are classified as class *B* coal dust explosion hazards.

3. Results

3.1. Analysis of the Methane Release Hazard

It should also be noted that ventilation methane is the volume amount of methane released to the excavations associated with the longwall area and is expressed in m^3 per unit of time. On the other hand, total methane (absolute) should be understood as the sum of the amount of methane released into the mine workings (ventilation methane) and methane captured by methane drainage.

Based on the measurements of methane concentration, the courses were determined during ventilation of the longwall with the U-type system. The measurement points were put in the longwall and the top gate (Figure 6a). These courses represent weighted average values of methane concentration on subsequent days of the operation (Figure 7).

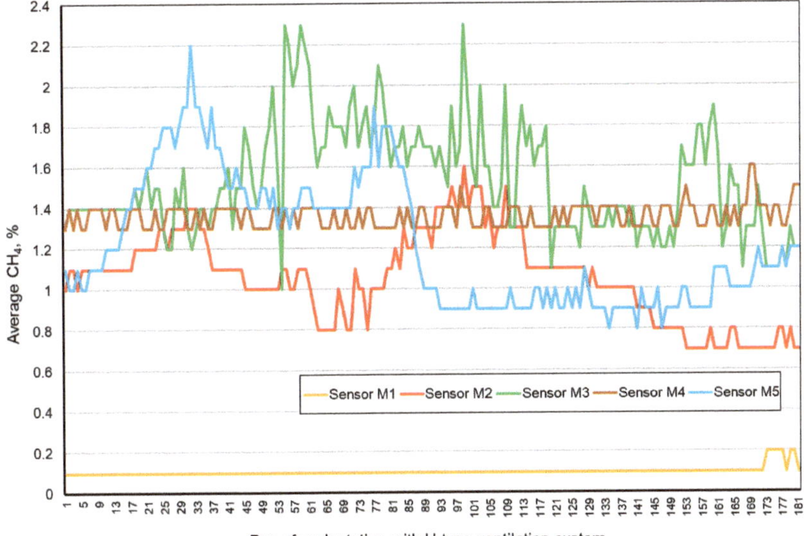

Figure 7. Registered concertation of methane for the U-type ventilation.

The analysis of the obtained values proves that the highest mean daily methane concentration of 1.6% was recorded on the M3 sensor, placed in the tailgate, at a distance of no more than 2 m from the line of liquidation the gate. A slightly lower value of average methane concentration, reaching 1.4%, was recorded on the M4 sensor in the tailgate placed on the upper sidewall at the height of the outlet of the tailgate. Methane concentration values of 1.2% were recorded on the M5 sensor located not more than 10 m from the outlet of the longwall. A quite high concentration of methane remained also in the longwall itself, about 160 m from the inlet to the longwall, and reached about 1.0% (M2 sensor). The lowest average daily methane concentration was found in the longwall that was 10 m long (M1 sensor), counting from the inlet to the longwall, and reached about 0.1% CH_4.

The analysis of the median values of average methane concentration on sensors located in the area of the longwall shows that for about half of the days of operation, on the M1 sensor, this value was about 0.1%. For the M2 sensor, the median value was 1.0%, whereas M3 and M4 sensors recorded (for half of the working days) an average value of methane higher than 1.4%. A slightly lower value occurred on the M5 sensor, of about 1.1%.

Taking into account the value of methane concentration in the case of 10% of measurements, the highest value of the 90th percentile occurred for the average concentrations recorded on the M3 sensor and was 1.8% and on the M5 sensor 1.6%. The 90th percentile value for the M4 sensor was 1.4%, and for the M2 sensor 1.3%. The lowest value was recorded on the M1 sensor and was 0.2%. The 10th percentile (mean concertation of methane in 90% of measurements) registered by the M1 sensor was 0.1%, M2—0.6%, M3—1.2%, M4—1.4%, and M5—0.9%.

Basic descriptive statistics for recorded measurements of methane concentrations at measurement points are presented in Table 4.

Table 4. Descriptive statistics of recorded methane concentrations for individual sensors for the U-type ventilation system.

Sensor	Mean	Median	Min	Max	10th Percentile	90th Percentile	Stand. Deviation	Variation Coefficient	Skewness	Kurtosis
M1	0.1	0.1	0.1	0.2	0.1	0.2	0.0	32.4	1.7	0.9
M2	0.9	1.0	0.6	1.6	0.6	1.3	0.3	29.0	0.3	−1.1
M3	1.6	1.5	1.0	2.3	1.2	1.8	0.3	17.7	0.7	0.0
M4	1.4	1.4	1.3	1.6	1.3	1.4	0.1	4.3	0.7	0.9
M5	1.2	1.1	0.8	2.2	0.9	1.6	0.3	23.7	1.0	0.3

During the operation of the longwall and the use of the U-ventilation system, already in the first months of operation, a significant number of exceedances of the permissible methane concentrations of at least 2% were recorded on sensors M3 and M5. The number of these exceedances may be treated as an indicator related to the magnitude of methane release hazard in the area of conducted exploitation. Exceeding the permitted concentration of methane in the excavation causes a risk of ignition of methane.

When the U-type ventilation system was used, the methane exceedances that occurred, therefore, posed a serious risk of methane ignition and simultaneously prevented operation. The sensors located in the area of operation (in the longwall and on the main gates) recorded a total of 1295 cases when the permissible methane concentrations were exceeded. The highest number was recorded by the M4 sensor and was 776. A significant number of exceedances of the methane concentration limit—423 cases—were also recorded on the M3 sensor (Table 5). On the remaining sensors (M2 and M5), a total of 96 exceedances of methane concentration limits were recorded. The total time for exceeding methane concentrations was about 25,900 min, giving almost 18 days.

Table 5. Summary of the number of exceedances of methane concentrations during longwall ventilation with the U-type ventilation system.

Sensor	No. of Exceedances, U-Type
M1	0
M2	40
M3	423
M4	776
M5	56

The main reason for exceeding the permissible concentration of methane on the M2 sensor was coal mining with a coal shearer or sudden gas outflows from relaxed rock mass (body of coal) during the operation. In turn, exceedances of the acceptable methane concentration recorded by the M4 and M3 sensors, located in the tailgate in the area of the intersection with the longwall, were most probably caused by methane outflows from the goafs. Taking into account the number of exceedances and the number of days during which the U-type ventilation system was applied, there were about 7.15 exceedances of the permitted concentration of methane per one day of extraction, which gives almost 2.5 h in total when the mining operation was halted.

When using the U-type ventilation system, the absolute methane and longwall ventilation values changed with the progress of the operation. In general, it can be concluded that these methane substances showed an upward trend (Figure 8).

The lowest mean absolute methane value occurred in the first month of the longwall mining and amounted to 15.39 m^3/min, and the highest in the sixth month and amounted to 44.50 m^3/min (when using the U-type system). In turn, the lowest mean value of the ventilation methane content occurred in the second month of operation and amounted to 14.11 m^3/min, and the highest in the fourth month and amounted to 17.72 m^3/min on average. The increase in absolute methane content was caused by the increase in the size of the relaxed zone in the rock mass and the gradual increase in extraction. The mean flow of methane volume captured by the system of methane drainage in the analysed period (from March 2018 to the end of August 2018), reached 19.13 m^3/min and increased from 1.18 m^3/min in the first month of exploitation to 28.09 m^3/min in the sixth month. It can be seen in Figure 6 that the changes in the amount of methane captured by methane drainage systems are similar to the changes in the ventilation methane content during this period.

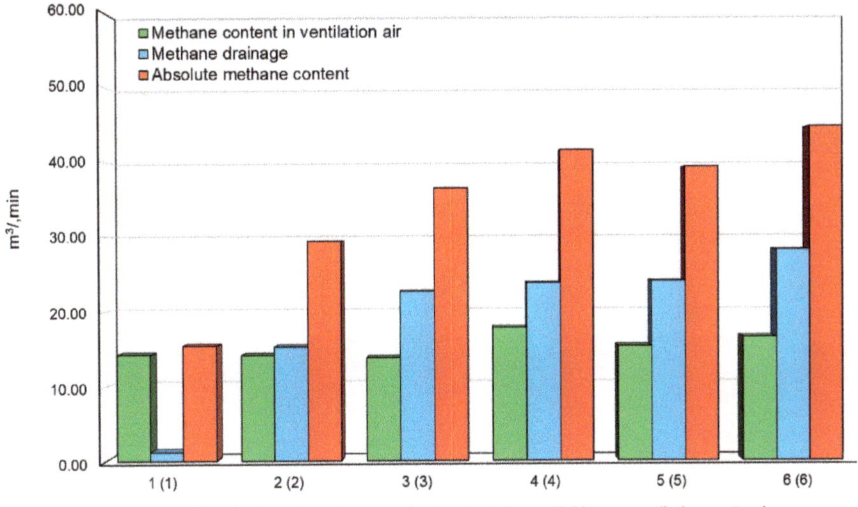

Figure 8. Absolute methane, ventilation and methane drainage rate when using the U-type ventilation system.

The analysis of the data presented in Figure 8 also shows that around the third month of operation there is some stabilisation in the range of the size of the methane flow captured by methane drainage systems and absolute and ventilation methane.

The number of exceedances of the permitted methane concentrations made it necessary to take measures to reduce the methane concentrations that persist in the air ventilation stream. To achieve this, it was decided to change the U-type ventilation system used and apply the Y-type system. For this purpose, it was necessary to carry out additional mining works in the form of excavations, which made it possible to change this ventilation system. The area of the longwall after the completion of the necessary mining works to enable the system to be changed is shown in Figure 6. The duration of mining work to change the ventilation system from the U-type to the Y-type was about one month.

Starting from the seventh month of its operation, the ventilation system has been changed from the U-type to the Y-type.

The weighted average daily methane concentrations when using the Y-type system are shown in Figure 9.

The analysis of the obtained average values of methane concentration during the application of the Y-type ventilation system (Table 6) shows that the highest values of this concentration, amounting to 1.30%, were recorded by the M3 sensor that was placed under the roof at a distance of about 2 m from the longwall outlet. Slightly lower values, up to 1.10%, were recorded by the M4 sensor in the tailgate along the goafs. Significantly lower methane concentrations were also recorded by sensors 2 and 5, which amounted to about 0.80% CH$_4$. The lowest concentration of methane was recorded in the air stream in the main gate (about 0.02% on average). The lowest concentrations of methane were recorded by the M1 sensor, due to the fact that this sensor is located at the inlet to the longwall, therefore it measures the methane content of the fresh air stream that is being fed to the longwall.

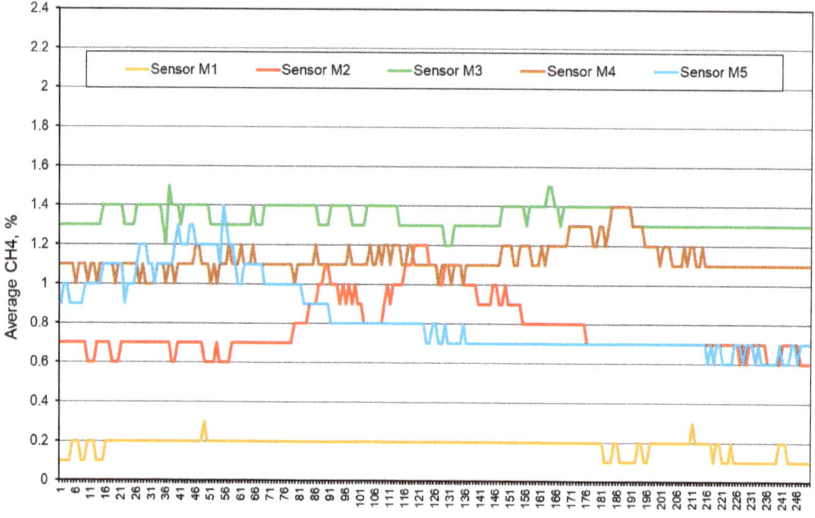

Figure 9. Concentrations of methane when using the Y-type ventilation system.

Table 6. Descriptive statistics of recorded methane concentrations for individual sensors for the Y-type ventilation system.

Sensor	Mean	Median	Min	Max	10th Percentile	90th Percentile	Stand. Deviation	Coefficient of Variation	Skewness	Kurtosis
M1	0.20	0.20	0.10	0.30	0.10	0.20	0.04	22.27	−1.30	0.83
M2	0.80	0.70	0.60	1.20	0.60	1.00	0.15	18.95	1.12	0.36
M3	1.30	1.30	1.20	1.50	1.30	1.40	0.05	4.09	0.23	−0.73
M4	1.10	1.10	1.00	1.40	1.10	1.20	0.08	7.10	1.36	2.31
M5	0.80	0.80	0.60	1.40	0.70	1.10	0.19	22.63	0.83	−0.50

The results from Table 5 show that for 50% of the days of the operation, the median value on M1 was 0.20%. For M2, the value was 0.70% and for M3, 1.30%. The median value for the M4 sensor was 1.10% and for the M1 sensor–0.80%.

The sensors distributed in the area of the longwall when using the Y-type ventilation system (251 days), only 47 exceedances of the allowed methane concentrations were recorded in total, the highest number of which was recorded by the M4 sensor and was 21. There are 17 exceeded methane concentrations on M5 and nine exceeded methane concentrations on M3 (Table 7).

Table 7. Summary of the number of exceedances of methane concentrations during longwall ventilation with the Y-type ventilation system.

Sensor	No. of Exceedances, Y-Type
M1	0
M2	0
M3	17
M4	9
M5	21

The total time for exceeding the permitted methane concentrations for this system was 940 min. Taking into account the number of exceedances and the number of days during which the Y-type ventilation system was used, there were 0.19 exceedances of the acceptable methane concentration per one day of extraction, which gives about 5 min of stoppage per day (for the U-type systems, it was

7.15 exceedances per working day, or about 2.5 h). This clearly means that the methane release hazard was many times lower when using the Y-type ventilation system than when using the U-type system.

The absolute and ventilation methane content and the amount of the methane drained were gradually decreasing during subsequent months when using the Y-type system (Figure 10).

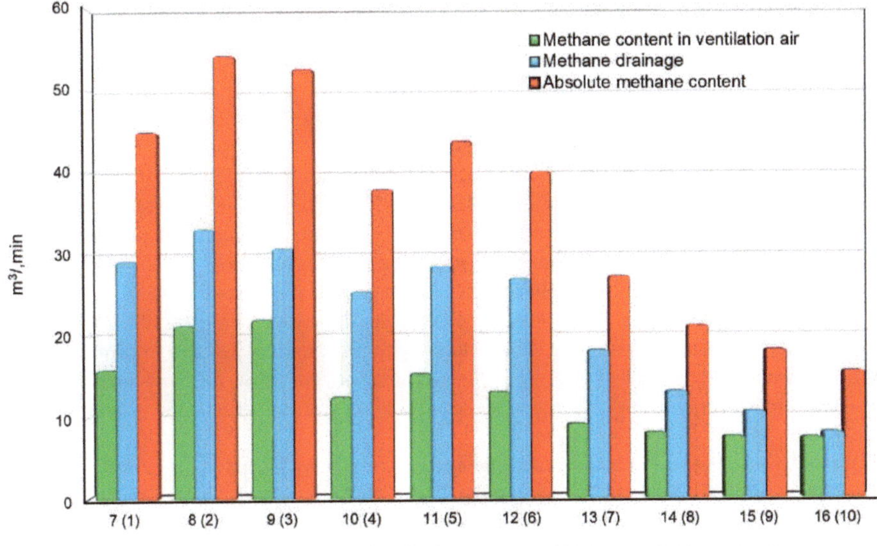

Figure 10. Absolute methane, ventilation and methane drainage rate when using the Y-type ventilation system.

The mean value of absolute methane in the first month when the Y-type system was used reached 44.89 m^3/min and was 0.39 m^3/min higher than in the last month when the ventilation was based on the U-type system. The highest absolute methane value was reached in the second month of the Y-type system (and in the eighth month of the mining in general) and was 54.18 m^3/min. The lowest mean daily absolute methane value occurred in the last month of the exploitation and amounted to 15.42 m^3/min and was similar to that occurring during the first month of exploitation and at the same time when the U-type ventilation system was used. The mean absolute methane value when using the Y-type ventilation system reached 35.46 m^3/min and was higher than the mean absolute methane value when using the U-type system by 1.1 m^3/min. For the Y-type ventilation, the methane flow rate recorded in the system for the methane drainage was 22.31 m^3/min on average and was higher than the flow rate recorded when using the U-type system by 3.18 m^3/min. Ventilation methane for the Y-type system amounted on average to 13.15 m^3/min and was lower than when using the U-type ventilation system by 2.1 m^3/min.

The mean values of absolute methane, ventilation and methane drainage rate for both ventilation systems used are listed in Table 8. The average absolute methane content reached 34.37 m^3/min in the case of the U-type ventilation system and 35.46 m^3/min for the Y-type ventilation system. The average ventilation methane content related to the volume content of methane in the ventilation air for the U-type system was 15.24 m^3/min and 13.15 m^3/min for the Y-type system. The methane drainage system drained 19.13 m^3/min of methane for the U-type ventilation system and 22.31 m^3/min in the case of the Y-type ventilation system.

Table 8. Mean absolute methane, ventilation and methane drainage values for the system on U and Y.

Variable	Ventilation System	
	U	Y
Absolute methane content, m^3/min	34.37	35.46
Ventilation methane content, m^3/min	15.24	13.15
Methane drainage m^3/min	19.13	22.31

Generally, it can be stated that during the application of the Y-type ventilation system, the average ventilation methane content of the longwall decreased by 13.72%, while the absolute methane content of the longwall increased by 16.63%, and the methane volume flow captured by methane drainage systems increased by 3.18%.

During the sixteen months of operation, the total absolute methane content of the longwall amounted to 807 465.6 m^3—of which 296,913.6 m^3 was for the U-type ventilation system and 510,552 m^3 for the Y-type ventilation system. The analysed ventilation methane content was 320,976 m^3, of which 131,659.2 m^3 of methane was released to the longwall area when using the U-type system, and 189,316.8 m^3 for the Y-type system. The methane drainage system captured 486,489.6 m^3 of methane.

However, it should be noted that period when the ventilation was based on the Y-type system was ten months, and the U-type system was used only for six months. The last four months of the operation when the Y-type system was applied were characterized by low absolute methane content (below 30 m^3/min). When using the U-type ventilation system, this phenomenon only occurred for two months.

During the application of both ventilation systems, the average daily output of the longwalls remained at a similar level—about 2300 tonnes per day. In the last month of the operation, the longwall was prepared for decommissioning, which results in a significant reduction in extraction. During this period, the reduction in absolute and ventilation methane and the value of the methane captured by drainage systems are visible.

A gradual decrease in absolute methane content for the Y-type ventilation system during the subsequent months of the exploitation was caused by the gradual decrease in the methane content of the deposit. Generally, two phases can be distinguished that are related to the formation of methane—the growth phase and the decline phase (Figure 11). There is no stabilisation phase.

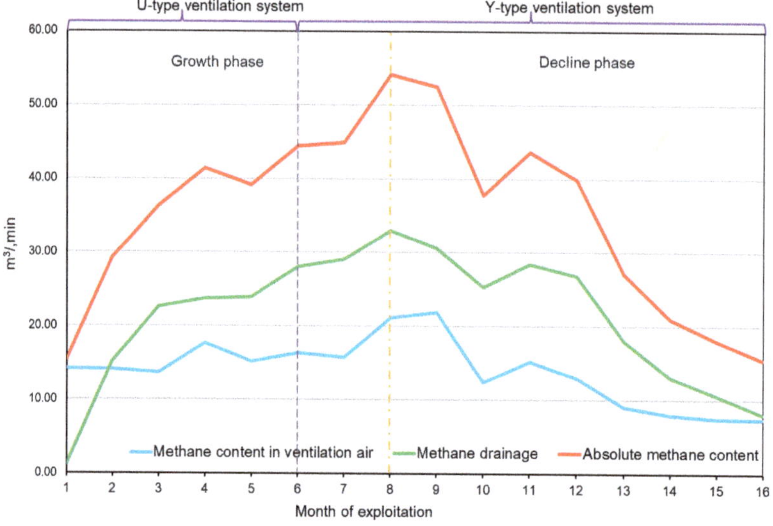

Figure 11. Phases related to the formation of absolute methane, ventilation and methane drainage.

The results show a strong positive correlation between absolute and ventilation methane, and absolute methane and methane drainage during the operation based either on the U-type or the Y-type ventilation system. The smallest positive correlation, although still significant, was between ventilation methane and methane drainage (Table 9).

Table 9. Correlations between absolute methane and methane drainage during the entire lifespan of the longwall.

Variable	Absolute Methane Content	Ventilation Methane Content	Methane Drainage
Absolute methane content	1.00	0.85	0.97
Ventilation methane content	0.85	1.00	0.69
Methane drainage	0.97	0.69	1.00

The analysis of the correlations between absolute methane, ventilation and methane drainage rate when using both ventilation systems prove that a stronger correlation between the variables occurred when using the Y-type system (Table 10).

Table 10. Correlations between absolute methane, ventilation and methane drainage for both longwall ventilation systems.

Variable	U-Type Ventilation System			Y-Type Ventilation System		
	Absolute Methane Content	Ventilation Methane Content	Methane Drainage	Absolute Methane Content	Ventilation Methane Content	Methane Drainage
Absolute methane content	1.00	0.62	0.99	1.00	0.96	0.99
Ventilation methane content	0.62	1.00	0.52	0.96	1.00	0.91
Methane drainage	0.99	0.52	1.00	0.99	0.91	1.00

3.2. Analysis of the Coal Spontaneous Combustion Hazard

The Figures 12–20 show the changes in gas concentration in the air samples taken from the goafs for both ventilation systems.

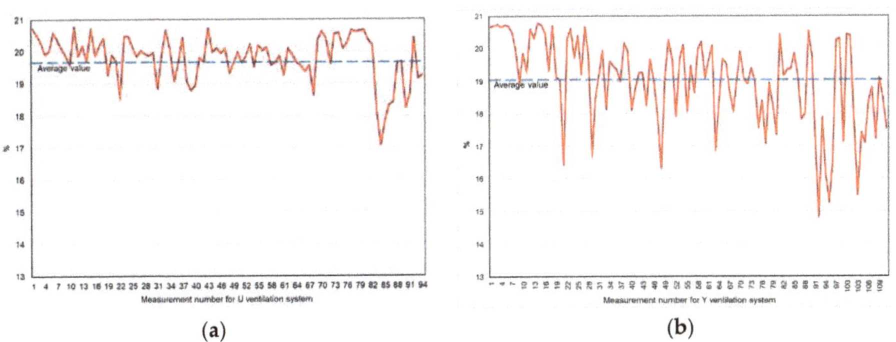

Figure 12. Changes in concentrations of oxygen in the goaf air—the U-type (**a**) and the Y-type ventilation system (**b**).

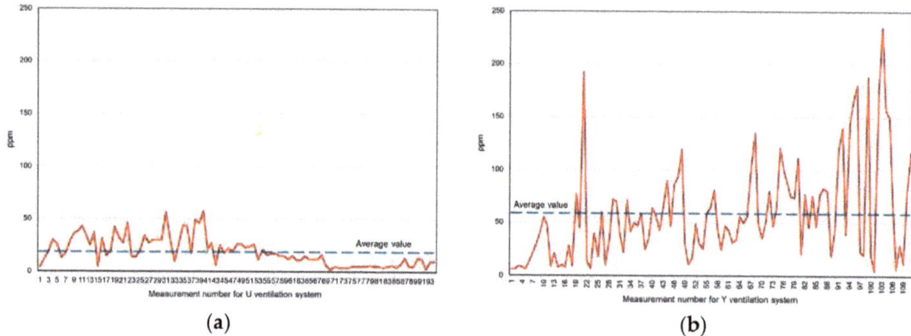

Figure 13. Changes in concentrations of carbon monoxide in the goaf air—the U-type (**a**) and the Y-type ventilation system (**b**).

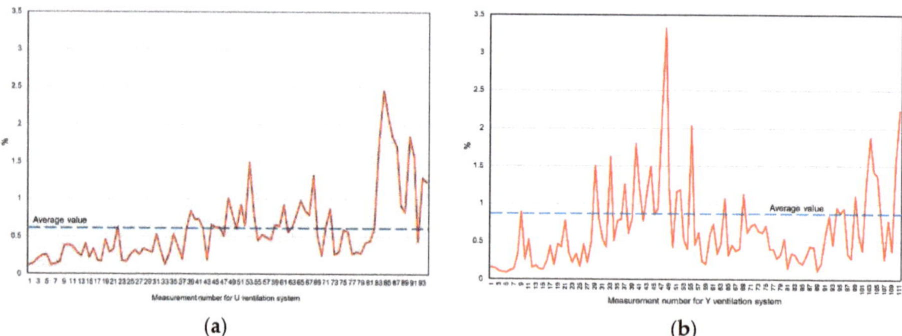

Figure 14. Changes in concentrations of carbon dioxide in the goaf air—the U-type (**a**) and the Y-type ventilation system (**b**).

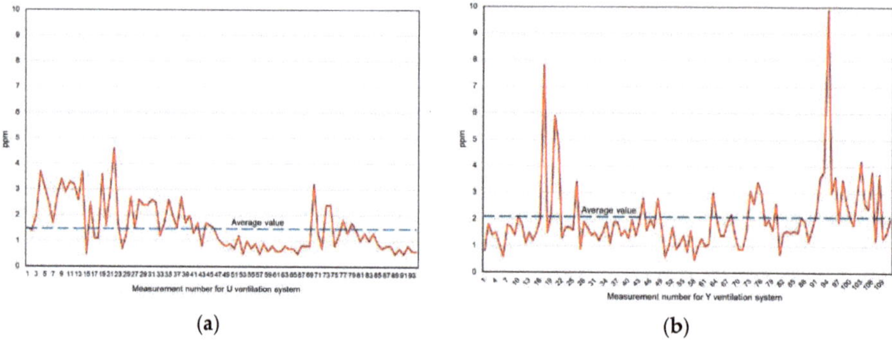

Figure 15. Changes in concentrations of hydrogen in the goaf air—the U-type (**a**) and the Y-type ventilation system (**b**).

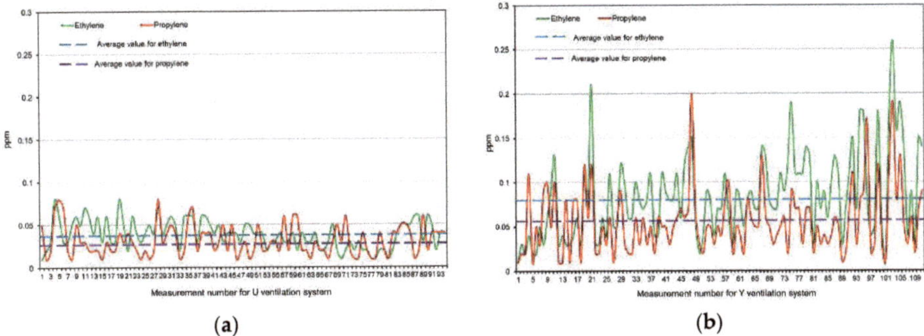

Figure 16. Changes in concentrations of ethylene and propylene in the goaf air—the U-type (**a**) and the Y-type ventilation system (**b**).

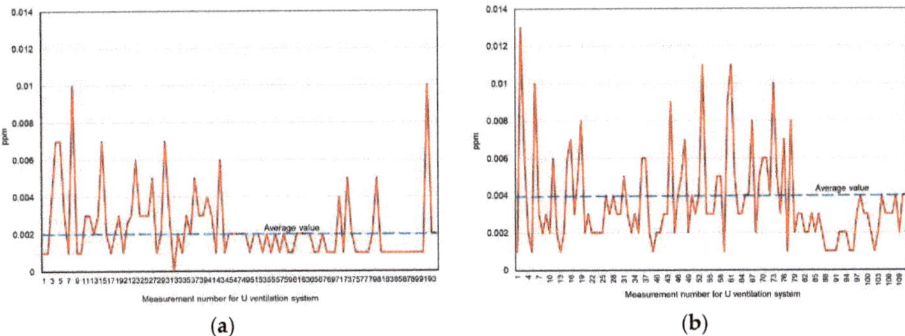

Figure 17. Changes in concentrations of acetylene in the goaf air—the U-type (**a**) and the Y-type ventilation system (**b**).

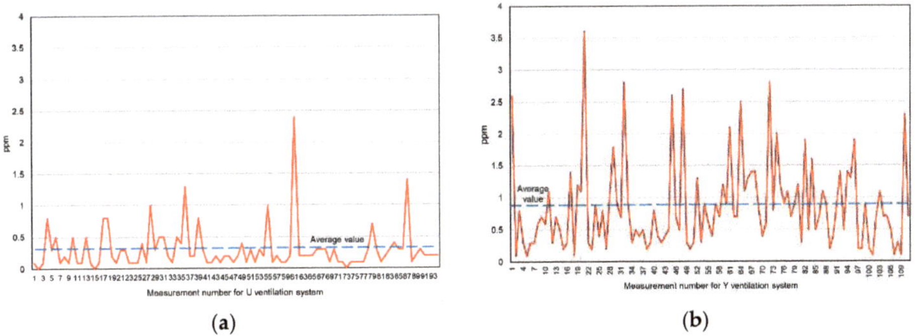

Figure 18. Changes in concentrations of ethane in the goaf air—the U-type (**a**) and the Y-type ventilation system (**b**).

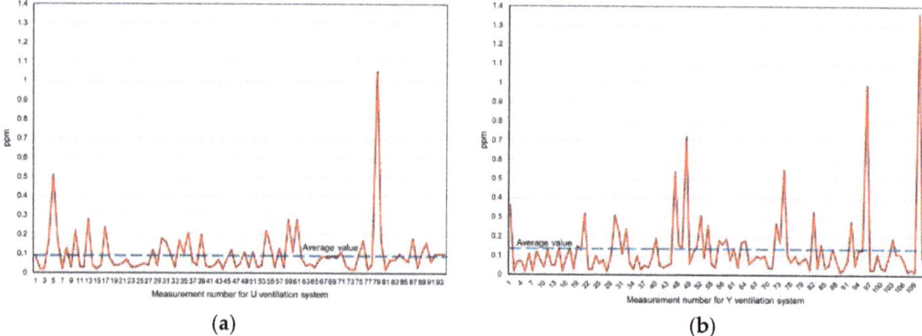

(a)　　　　　　　　　　　　　　　　　　　　(b)

Figure 19. Changes in concentrations of propane in the goaf air—the U-type (**a**) and the Y-type ventilation system (**b**).

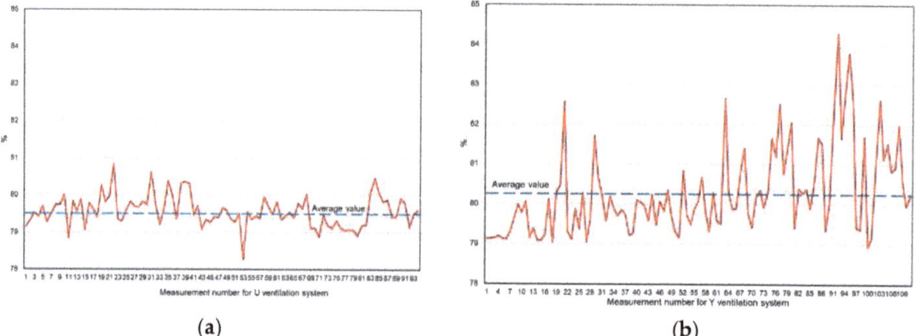

(a)　　　　　　　　　　　　　　　　　　　　(b)

Figure 20. Changes in concentrations of nitrogen in the goaf air—the U-type (**a**) and the Y-type ventilation system (**b**).

The analysis of the concentrations of gas changes in the air samples shows that the values depend on the type of ventilation and that the concentrations of oxygen (Figure 12) for the Y-type system were slightly lower than for the U-type system. This proves that the oxidation reactions of the coal in the goafs occurred. During such a reaction, the concentration of oxygen in the atmosphere in the goaf is reduced and the proportion of gases resulting from chemical reactions associated with the process of oxidation, spontaneous heating and spontaneous combustion is increased, which is confirmed by the concentrations of carbon monoxide, carbon dioxide, hydrogen and acetylene, and ethylene and propylene.

In hard coal mining, carbon monoxide is the basic indicator gas for the oxidation and spontaneous heating reactions that lead to spontaneous combustion of coal. The increase in its content (Figure 13) in the air for the Y-ventilation system in relation to the U-type system indicates the processes of the low-temperature oxidation of coal or spontaneous combustion of coal. This gas is present in the air only after these two processes occurred.

The characteristics of carbon dioxide concentrations (Figure 14) also indicate that for the Y-type system, the reactions of spontaneous combustion of coal occurred more intensively than for the U-type ventilation system. The process of spontaneous combustion increases the concentrations of carbon dioxide in the air (Figure 14) and the concentration of oxygen decreases (Figure 13). The dynamics of these changes are more or less the same. At the same time, it should be noted that in an underground hard coal mine, the presence of carbon dioxide is also connected with its secretion from the rock mass, so it should always be analysed, along with changes in concentrations of oxygen.

Analysis of changes in concentrations of hydrogen (Figure 15) ethylene and propylene (Figure 16), acetylene (Figure 17), propane (Figure 18) and ethane (Figure 20), were for the Y-type ventilation system compared to the U-type system. These gases are emitted as the temperature of coal increases, which means that the coal spontaneous heating when using the Y-type ventilation system is more intense.

Changes in nitrogen concentration for both the U-type and Y-type ventilation system (Figure 20), also indicate an increased risk of coal spontaneous heating for the Y-type system. During this process, the tendencies of changes in nitrogen content (Figure 20) and oxygen content (Figure 12) are divergent, because the nitrogen content in the air increases and the oxygen content decreases.

Table 11 presents basic descriptive statistics for the test gas concentrations for both ventilation systems. The results clearly show that for both systems, the level of probability of coal spontaneous combustion increases. This is evidenced by an increase in the concentrations of gases produced during chemical reactions associated with low-temperature oxidation, leading to coal spontaneous combustion with a simultaneous reduction in oxygen concentration.

Table 11. Basic descriptive statistics for tested gas concentrations in air samples taken from goafs when using the U and Y venting system.

Tested Gas	Mean		Median		Min		Max		90th Percentile		Stand. Deviation		Coefficient of Variation	
	U	Y	U	Y	U	Y	U	Y	U	Y	U	Y	U	Y
Ethane	0.30	0.86	0.20	0.70	0.00	0.10	2.40	3.60	0.70	1.90	0.34	0.70	112.30	80.61
Ethylene	0.04	0.08	0.04	0.08	0.01	0.01	0.12	0.26	0.06	0.15	0.02	0.05	45.98	57.04
Propane	0.10	0.13	0.07	0.09	0.02	0.02	1.05	1.36	0.20	0.27	0.13	0.18	123.32	135.04
Propylene	0.03	0.06	0.03	0.05	0.01	0.01	0.08	0.20	0.06	0.10	0.02	0.04	57.12	66.46
Acetylene	0.002	0.003	0.002	0.002	0.001	0.001	0.01	0.013	0.01	0.01	0.34	0.36	120.35	66.49
Carbon monoxide	19.18	59.65	16.00	47.00	2.00	5.00	57.00	235.00	38.00	121.00	13.18	47.64	68.74	79.86
Oxygen	19.81	19.00	19.92	19.21	17.09	14.81	20.78	20.77	20.62	20.60	0.70	1.34	3.56	7.06
Nitrogen	79.57	80.27	79.53	80.01	78.28	78.95	80.83	84.29	80.06	81.72	0.41	1.09	0.52	1.36
Carbon dioxide	0.61	0.68	0.51	0.49	0.12	0.09	2.45	3.33	1.29	1.43	0.47	0.56	76.67	82.38
Hydrogen	1.57	2.02	1.30	1.70	0.50	0.50	4.60	9.90	2.90	3.40	0.94	1.31	59.83	64.78

The table header spans "Ventilation System" across all U/Y column pairs.

4. General Discussion

The process of underground mining is continuously associated with the occurrence of natural hazards. Currently, the most burdensome and dangerous are ventilation hazards. The case study presented in the paper shows that different types of hazards occur simultaneously. This co-occurrence often significantly limits the methods designed to limit their impact. Measures to reduce one type of hazard can simultaneously activate another.

Such a case is presented in this article. Two hazards—methane and coal spontaneous combustion—accumulated in tested longwall. Both are extremely dangerous and can cause serious disturbances in the production process and pose a great danger to the workers.

One method to limit the consequences of these hazards is to choose a proper ventilation system. However, a system that has the capacity to reduce the methane release hazard simultaneously increases the coal spontaneous combustion hazard. Nevertheless, it is clear that a proper ventilation system selected as early as at the stage of designing the operation is one of the key factors that will ensure safe and effective underground operation. The longwall ventilation system must have the capacity to reduce the hazardous concentrations of methane and gaseous products of the processes associated with the spontaneous combustion of coal and to provide conditions that will allow for continuous mining. Unfortunately, it is extremely difficult to find a system that will meet both conditions.

The mining process of the case presented in the paper was initially based on the U-type ventilation system, which later proved to be very ineffective in reducing the methane release hazard. This system made it practically impossible to continue the mining process. Very frequent exceedances of the methane concentration limits caused interruptions, which, in turn, completely disrupted the entire production cycle. This resulted in the need to change the ventilation system to a more favourable and efficient one in terms of methane. The Y-type ventilation system managed to significantly reduce the concentrations of methane, but it led to an increase in the coal spontaneous combustion tendency. The Y-type system caused most of the air to enter the goaf, which can activate the combustion.

The results presented in Section 3 clearly indicate that the U-type longwall ventilation system, in the case of high methane release hazard, characterized by exceeding the permissible concentrations of methane in the air, prevents safe and effective operation. At the same time, the use of this system limits the carbon oxidation reactions in the goaf leading to coal spontaneous heating and combustion, which is confirmed by the low concentrations of gases that are the products of these reactions. In turn, the Y-type ventilation system enables safe and effective operation in conditions where there is a methane release hazard but deteriorates the safety conditions related to spontaneous combustion of coal.

It is evident that the presented case is remarkably interesting from a scientific and practical perspective. It clearly demonstrates the advantages and disadvantages of both ventilation systems used in the same longwall. Moreover, such a case when two different ventilation methods are used is extremely rare.

The authors believe that when methane concentration stays above 1.5% (median) for half of the working day, with simultaneous application of all possible preventive measures, in a longwall maintained in the mine air in the area of the intersection of a longwall with a tailgate, after the full launch of the longwall and the nominal mining capacity, the initial ventilation system must be changed—for instance, to the Y–type system. This involves specific financial outlays, but if the longwall is economically balanced, such a solution can help to successfully complete the exploitation of the longwall.

5. Conclusions

The paper presents the results of the analysis of the impact that the longwall ventilation system has on the formation of two basic ventilation hazards during mining: the methane release hazard and spontaneous combustion of coal.

The following statements and conclusions are based on the measurements, calculations, and analyses carried out:

- The average absolute methane content reached 34.37 m³/min in the case of the U-type ventilation system, and 35.46 m³/min for the Y-type ventilation system. The average ventilation methane content related to the volume content of methane in the ventilation air for the U-type system was 15.24 m³/min, and 13.15 m³/min for the Y-type system. The methane drainage system drained 19.13 m³/min of methane for the U-type ventilation system, and 22.31 m³/min in the case of the Y-type ventilation system;
- The ventilation methane content of the longwall when using the U-type system was almost 14% higher compared to the Y-type system. At the same time, the absolute methane content of the longwall when using the U-type ventilation system was lower than in the case of the Y-type system;
- The efficiency of the methane drainage system increased in the case of the Y-type ventilation system—on average, 14% more methane was drained compared to the U-type system;
- The number of times the permitted methane concentrations were exceeded when using the U-type system was 1295, and as "only" ventilation 47 for the U-type system. The highest number of exceedances of the permissible methane concentrations when using the U-type ventilation system was recorded by the M4 sensor (776 cases), and when using the Y-type ventilation system by the M5 sensor (21 cases). There is a clear reduction in the number of exceedances, which is very important for the continuity and efficiency of the mining process;
- The maximum average methane content in the ventilation air for the U-type ventilation system within the intersection of the longwall and the tailgate was 2.3%, and 1.5% for the Y-type ventilation system;
- The average concentration of gases (ethane, ethylene, propane, propylene, acetylene, carbon monoxide, carbon dioxide, hydrogen) produced during the process of oxidation and spontaneous heating of coal left in the goaf leading to its spontaneous combustion were higher for the Y-type ventilation system compared to the U-type ventilation system. At the same time, the average values of oxygen concentration in the goaf for the Y-type system were lower than for the U-type ventilation system, which indicated an intensified process of coal oxidation. The average content of carbon monoxide in the air samples taken from the goaf when using the Y-type longwall ventilation system was almost three times higher than when using the U-type system. The average content of propylene and ethylene was twice as high when using the Y-type longwall ventilation system, compared to when using the U-type system. The average oxygen content when using the Y-type ventilation system was lower than when using the U-type system; this indicates an intensified process of coal oxidation, leading to its coal spontaneous heating;
- In order to reliably assess the impact of the longwall ventilation system on the level of endogenous fire hazard in the goafs, it is necessary to conduct a full analysis of the change in the concentration of fire gases as a function of time treated as fire indicators. These gases include oxygen, nitrogen, carbon monoxide, and carbon dioxide, hydrogen, and unsaturated hydrocarbons. An analysis of the concentrations of individual gases does not provide a complete picture of the endogenous fire hazard and can lead to incorrect findings.

Following the tests carried out and the results obtained, it can be concluded that the methane release hazard is higher when using the U-type longwall ventilation system than when using the Y-type ventilation system. At the same time, the hazard of coal spontaneous combustion is lower when using the U-type system compared to the Y-type system. Such data are therefore very important, since it is necessary to ensure the safety of the coal extraction process.

The case presented by the authors is probably the first research of this kindconducted and presented. The authors hope that it will contribute to the discussion on the optimal selection and, if necessary, change of the ventilation system for active longwalls.

Author Contributions: Conceptualization, M.T. and J.B.; D.S.; L.S.; methodology, J.B. and M.T.; software, M.T. and J.B.; formal analysis, J.B. and M.T.; investigation, J.B. and M.T.; resources, M.T.; D.S.; J.B. and S.Z.; data curation,

M.T.; D.S.; J.B. and L.S.; writing of the original draft preparation, M.T.; writing of review and editing, J.B.; M.T.; visualization, M.T.; supervision, M.T. and J.B.; project administration, M.T. and J.B.; funding acquisition, M.T. All authors have read and agreed to the published version of the manuscript.

Funding: This publication was funded by the statutory research performed at Silesian University of Technology, Department of Safety Engineering, Faculty of Mining, Safety Engineering and Industrial Automation (06/030/BKM-20/0053).

Acknowledgments: This article is the result of statutory research performed at Silesian University of Technology, Department of Production Engineering, Faculty of Organization and Management (13/030/BK-20/0059) and the statutory research performed at Silesian University of Technology, Department of Safety Engineering, Faculty of Mining, Safety Engineering and Industrial Automation (06/030/BKM-20/0053). This article is the result of the research project No. PBS3/B6/25/2015, "Application of the Overall Equipment Effectiveness method to improve the effectiveness of the mechanized longwall system's work in the coal exploitation process", carried out in 2015–2018, financed by The National Centre for Research and Development.

Conflicts of Interest: The authors declare no conflict of interest.

References

1. Gui, C.; Geng, F.; Tang, J.; Niu, H.; Zhou, F.; Liu, C.; Hu, S.; Teng, H. Gas–solid two-phase flow in an underground mine with an optimized air-curtain system: A numerical study. *Process Saf. Environ. Prot.* **2020**, *20*, 137–150. [CrossRef]
2. Xiu, Z.; Nie, W.; Yan, J.; Chen, D.; Cai, P.; Liu, Q.; Du, T.; Yang, B. Numerical simulation study on dust pollution characteristics and optimal dust control air flow rates during coal mine production. *J. Clean. Prod.* **2020**, *248*, 119197. [CrossRef]
3. Aguado, M.; Nicieza, C. Control and prevention of gas outbursts in coal mines, Riosa–Olloniego coalfield, Spain. *Int. J. Coal Geol.* **2007**, *69*, 253–266. [CrossRef]
4. Cao, S.; Liu, Y.; Wang, Y. A forecasting and forewarning model for methane hazard in working face of coal mine based on LS-SVM. *J. China Univ. Min. Technol.* **2008**, *18*, 172–176. [CrossRef]
5. Kędzior, S.; Dreger, M. Methane occurrence, emissions and hazards in the Upper Silesian Coal Basin, Poland. *Int. J. Coal Geol.* **2019**, *211*, 103226. [CrossRef]
6. Shi, S.; Jiang, B.; Meng, X. Assessment of gas and dust explosion in coal mines by means of fuzzy fault tree analysis. *Int. J. Min. Sci. Technol.* **2018**, *28*, 991–998. [CrossRef]
7. Yang, S.; Hu, X.; Liu, W.V.; Cai, J.; Zhou, X. Spontaneous combustion influenced by surface methane drainage and its prediction by rescaled range analysis. *Int. J. Min. Sci. Technol.* **2018**, *28*, 215–221. [CrossRef]
8. Yu, Y.; Bai, J.; Wang, X.; Zhang, L. Control of the surrounding rock of a goaf-side entry driving heading mining face. *Sustainability* **2020**, *12*, 2623. [CrossRef]
9. Mark, C.; Gauna, M. Evaluating the risk of coal bursts in underground coal mines. *Int. J. Min. Sci. Technol.* **2016**, *26*, 47–52. [CrossRef]
10. Burtan, Z.; Stasica, Z.; Rąk, Z. The influence of natural hazards of disasters on the work safety conditions in Polish coal mining in the years 2000–2016. *Zesz. Nauk. Inst. Gospod. Surowcami Miner. I Energią PAN* **2017**, *101*, 7–18.
11. Burtan, Z.; Chlebowski, D.; Kapusta, M. The scale and conditions of disasters induced by the occurrence of natural hazards in the coal mining sector in Poland. *Bezpieczeństwo Pr. Ochr. Środowiska Górnictwie* **2018**, *7*, 3–11.
12. Liang, K.; Liu, J.; Wang, C. The coal mine accident causation model based on the hazard theory. *Procedia Eng.* **2011**, *26*, 2199. [CrossRef]
13. Ramani, R.V. Mining disasters caused and controlled by mankind: The case for coal mining and other minerals Part 1: Causes of mining disasters. *Nat. Resour. Forum* **1995**, *19*, 233–242. [CrossRef]
14. Saleh, J.H.; Cummings, A.M. Safety in the mining industry and the unfinished legacy of mining accidents: Safety levers and defense-in-depth for addressing mining hazards. *Saf. Sci.* **2011**, *49*, 764–777. [CrossRef]
15. Zhu, Y.; Wang, D.; Shao, Z.; Xu, C.; Zhu, X.; Qi, X.; Liu, F. A statistical analysis of coal mine fires and explosions in China. *Process Saf. Environ. Protect.* **2019**, *121*, 357–366. [CrossRef]
16. Harris, J.; Kirsch, P.A.; Shi, M.; Li, J.; Gagrani, A. Comparative analysis of coal fatalities in Australia, South Africa, India, China and USA, 2006–2010. *Coal Oper. Conf.* **2014**, 399–407.

17. Brodny, J.; Tutak, M. Analysing the utilisation effectiveness of mining machines using independent data acquisition systems: A case study. *Energies* **2019**, *12*, 2505. [CrossRef]
18. Karacan, C.Ö.; Olea, R.A.; Goodman, G. Geostatistical modeling of the gas emission zone and its in-place gas content for Pittsburgh-seam mines using sequential Gaussian simulation. *Int. J. Coal Geol.* **2012**, *90*, 50–71. [CrossRef]
19. Hu, S.; Yang, S.; Liu, W.V.; Zhou, X.; Sun, J.; Yu, H. A methane emission control strategy in the initial mining range at a spontaneous combustion-prone longwall face: A case study in coal 15, Shigang Mine, China. *J. Nat. Gas Sci. Eng.* **2017**, *38*, 504–515. [CrossRef]
20. Liang, Y.; Zhang, J.; Wang, L.; Luo, H.; Ren, T. Forecasting spontaneous combustion of coal in underground coal mines by index gases: A review. *J. Loss Prev. Proc.* **2019**, *57*, 208–222. [CrossRef]
21. Ma, D.; Qin, B.; Li, L.; Gao, A.; Gao, Y. Study on the methane explosion regions induced by spontaneous combustion of coal in longwall gobs using a scaled-down experiment set-up. *Fuel* **2019**, *254*, 115547. [CrossRef]
22. Ma, L.; Zou, L.; Ren, L.; Chung, Y.; Zhang, P.; Shu, C. Prediction indices and limiting parameters of coal spontaneous combustion in the Huainan mining area in China. *Fuel* **2020**, *264*, 116883. [CrossRef]
23. Wen, H.; Yu, Z.; Fan, S.; Zhai, X.; Liu, W. Prediction of spontaneous combustion potential of coal in the gob area using CO extreme concentration: A case study. *Combust. Sci. Technol.* **2017**, *189*, 1713–1727. [CrossRef]
24. Xia, T.; Zhou, F.; Wang, X.; Zhang, Y.; Li, Y.; Kang, J.; Liu, J. Controlling factors of symbiotic disaster between coal gas and spontaneous combustion in longwall mining gobs. *Fuel* **2016**, *182*, 886–896. [CrossRef]
25. Zhai, X.; Xu, Y.; Yu, Z. Numerical analysis on the evolution of CO concentration in return corner: A case study of steady U-type ventilation working face. *Numer. Heat Trans. Part A Appl.* **2018**, *74*, 1732–1746. [CrossRef]
26. Zhuo, H.; Qin, B.; Qin, Q.; Su, Z. Modeling and simulation of coal spontaneous combustion in a gob of shallow buried coal seams. *Process Saf. Environ. Protect.* **2019**, *131*, 246–254. [CrossRef]
27. Sanmiquel, L.; Rossell, J.M.; Vintró, C. Study of Spanish mining accidents using data Mining techniques. *Saf. Sci.* **2015**, *75*, 49–55. [CrossRef]
28. Gowrisankaran, G.; He, C.; Lutz, E.; Burgess, J. Productivity, Safety, and Regulation in Underground Coal Mining: Evidence from Disasters and Fatalities. Available online: https://www.nber.org/papers/w21129.pdf (accessed on 15 July 2020).
29. Xu, Y.; Li, Z.; Zhai, X.; Yu, Z. Potential dangerous zone of coal spontaneous combustion and gas coupled hazard in goaf under mining condition. *J. China Coal Soc.* **2019**. [CrossRef]
30. Ma, L.; Guo, R.Z.; Gao, Y.; Ren, L.F.; Wei, G.M.; Li, C.H. Study on coal spontaneous combustion characteristics under methane-containing atmosphere. *Combust. Sci. Technol.* **2019**, *191*, 1456–1472. [CrossRef]
31. Massanés, M.B.; Pera, L.S.; Moncunill, J.O. Ventilation management system for underground environments. *Tunn. Undergr. Space Technol.* **2015**, *205*, 516–522. [CrossRef]
32. Szurgacz, D.; Tutak, M.; Brodny, J.; Sobik, L.; Zhironkina, O. The Method of Combating Coal Spontaneous Combustion Hazard in Goafs—A Case Study. *Energies* **2020**, *13*, 4538. [CrossRef]
33. Su, H.T.; Zhou, F.B.; Song, X.L.; Qiang, Z.Y. Risk analysis of spontaneous coal combustion in steeply inclined longwall gobs using a scaled-down experimental set-up. *Process. Saf. Environ. Prot.* **2017**, *111*, 1–12. [CrossRef]
34. Tang, M.Y.; Jiang, B.Y.; Zhang, R.Q.; Yin, Z.Q.; Dai, G.L. Numerical analysis on the influence of gas extraction on air leakage in the gob. *J. Nat. Gas Sci. Eng.* **2016**, *33*, 278–286. [CrossRef]
35. Wu, J.J.; Liu, X.C. Risk assessment of underground coal fire development at regional scale. *Int. J. Coal Geol.* **2011**, *86*, 87–94. [CrossRef]
36. Fernánez-Alaiz, F.; Castañón, A.M.; Gómez-Fernández, F.; Bernardo-Sánchez, A.; Bascompta, M. Analysis of the fire propagation in a sublevel coal mine. *Energies* **2020**, *13*, 3754. [CrossRef]
37. Tutak, M.; Brodny, J. The impact of the strength of roof rocks on the extent of the zone with a high risk of spontaneous coal combustion for fully powered longwalls ventilated with the y-type system—A case study. *Appl. Sci.* **2019**, *9*, 5315. [CrossRef]
38. Mine Ventilation Systems. Available online: http://web.mst.edu/~{}tien/218/lab8-systems.pdf (accessed on 15 July 2020).
39. Kingery, D.S. *Introduction to Mine Ventilating Principles and Practices*; United States Government printing office: Washington, DC, USA, 1960.

40. Szlązak, J.; Szlązak, N. *Filtracja Powietrza Przez Zroby Ścian Zawałowych w Kopalniach Węgla Kamiennego*; AGH Uczelniane Wydawnictwa Naukowo-Dydaktyczne: Kraków, Poland, 2005.
41. Kłeczek, Z. *Geomechanika Górnicza*; Śląskie Wydawnictwo Techniczne: Katowice, Poland, 1994.
42. Palchik, V. Formation of fractured zones in overburden due to longwall mining. *J. Environ. Geol.* **2003**, *44*, 28. [CrossRef]
43. Brune, J.; Grubb, J.; Bogin, G.; Zipf, R.; Marts, J.; Gilmore, R.; Lolon, S.; Saki, S. A Critical Look at Longwall Bleeder Ventilation. Available online: https://www.semanticscholar.org/paper/A-Critical-Look-at-Longwall-Bleeder-Ventilation-Brune-Grubb/bb0d5819f17103a1e9faddb336930441510fc537 (accessed on 15 August 2020).
44. Chen, P.; Zhang, L.; Zou, D. Study of three-dimensional distribution of permeability in gob based on o-shape circle theory. *Min. Saf. Environ. Prot.* **2015**, *42*, 38–41.
45. Das, S.K. Observations and classification of roof strata behaviour over longwall coal mining panels in India. *Int. J. Rock Mech. Min. Sci.* **2000**, *37*, 585–597. [CrossRef]
46. Wang, K.; Tang, H.; Miao, Y.; Liu, D. Research on complex air leakage method to prevent coal spontaneous combustion in longwall goaf. *PLoS ONE* **2019**, *14*, e0213101. [CrossRef]
47. Gilman, A.; Beckie, R. Flow of coal-bed methane to a gallery. *Transp. Porous Media* **2000**, *41*, 1–16. [CrossRef]
48. Valliappan, S.; Wohua, Z. Numerical modelling of methane gas migration in dry coal seams. *Int. J. Numer. Anal. Methods Geomech.* **1996**, *20*, 571–593. [CrossRef]
49. Barker-Read, G.R.; Radchenko, S.A. Methane emission from coal and associated strata samples. *Int. Min. Ceol. Eng.* **1989**, *7*, 101–126. [CrossRef]
50. Bertrard, C.; Bruyet, B.; Gunther, J. Determination of desorb able gas concentration of coal (direct method). *Int. Rock Mech. Min. Sci.* **1970**, *7*, 43–65. [CrossRef]
51. Jolly, D.C.; Morris, H.; Hinsley, F.B. An investigation into the relationship between the methane sorption capacity of coal and gas pressure. *Min. Eng.* **1968**, *127*, 539–548.
52. Hildenbrand, A.; Krooss, B.M.; Busch, A.; Gaschnitz, R. Evolution of methane sorption capacity of coal seams as a function of burial history—A case study from the Campine Basin, NE Belgium. *Int. J. Coal Geol.* **2006**, *66*, 179–203. [CrossRef]
53. Kam, A.Y.; Hixson, A.N.; Perlmutter, D.D. The oxidation of bituminous coal—I Development of a mathematical model. *Chem. Eng. Sci.* **1976**, *31*, 815–819. [CrossRef]
54. Kam, A.Y.; Hixson, A.N.; Perlmutter, D.D. The oxidation of bituminous coal—II experimental kinetics and interpretation. *Chem. Eng. Sci.* **1976**, *31*, 821–834. [CrossRef]
55. Lu, P.; Liao, G.X.; Sun, J.H.; Li, P.D. Experimental research on index gas of the coal spontaneous at low-temperature stage. *J. Loss Prevent. Proc.* **2004**, *17*, 243–247. [CrossRef]
56. Xiao, Y.; Wang, Z.P.; Ma, L.; Ziai, X.W. Research on correspondence relationship between coal spontaneous combustion index gas and feature temperature. *Coal Sci. Technol.* **2008**, *36*, 47–51.
57. Adamus, A.; Sancer, J.; Guranova, P.; Zubicek, V. An investigation of the factors associated with interpretation of mine atmosphere for spontaneous combustion in coal mines. *Fuel Process. Technol.* **2011**, *92*, 663–670. [CrossRef]
58. Arisoy, A.; Beamish, B. Mutual effects of pyrite and moisture on coal self-heating rates and reaction rate data for pyrite oxidation. *Fuel* **2015**, *139*, 107–114. [CrossRef]
59. Dai, G.L. Study on the gaseous products in coal oxidation at low temperature. *Coal Mine Saf.* **2007**, *1*, 1–4.
60. Chamberlain, E.A.C.; Barrass, G.; Thirlaway, J.T. Gases evolved and possible reactions during low-temperature oxidation of coal. *Fuel* **1976**, *55*, 217–223. [CrossRef]
61. Chamberlin, E.C.A.; Hall, D.A.; Thirlway, J.T. The ambient temperature oxidation of coal in relation to early detection of spontaneous heating. *Min. Eng.* **1970**, *130*, 1–16.
62. Manohar Rao, A.; Ramalingeswarudu, S.; Venkateswarlu, G. Planning of ventilation requirements for deep mechanised long wall faces—A case study of Adriyala Longwall Project of the Singareni Collieries Company Limited (SCCL). *Procedia Earth Planet. Sci.* **2015**, *11*, 548–556.
63. Gilles, S.; Wu, H. Australian Longwall Panel Ventilation Practices. Available online: https://ro.uow.edu.au/cgi/viewcontent.cgi?referer=https://www.google.com/&httpsredir=1&article=2114&context=coal (accessed on 15 July 2020).

64. Mayes, T.; Gillies, A. An Analysis of Current Australian Longwall Ventilation Methods. In Proceedings of the Seventh International Mine Ventilation Congress, Cracow, Poland, 17–22 June 2001; Wasilewski, S., Ed.; Polish Academy of Sciences: Krakow, Pland, 2001; pp. 793–800. Available online: http://www.gwmt.com.au/Papers/2001/2001%20-%20June%20-%207IMVC%20LW%20ventilation.pdf (accessed on 15 July 2020).

65. Krog, R. Critical Analysis of Longwall Ventilation Systems and Removal of Methane. Available online: https://researchrepository.wvu.edu/cgi/viewcontent.cgi?article=7057&context=etd (accessed on 15 July 2020).

66. Wang, Z.; Ren, T.; Ma, L.; Zhang, J. Investigations of ventilation airflow characteristics on a longwall face—A computational approach. *Energies* **2018**, *11*, 1564. [CrossRef]

67. Yuan, L.; Smith, A. The effect of ventilation on spontaneous heating of coal. *J. Loss Prev. Process. Ind.* **2012**, *25*, 131–137. [CrossRef]

68. Diamond, W.P.; Garcia, F. *Prediction of Longwall Methane Emissions: An Evaluation of the Influence of Mining Practices on Gas Emissions and Methane Control Systems*; Report of Investigations No. 9649; National Institute for Occupational Safety and Health: Pittsburgh, PA, USA, 1999. Available online: https://www.cdc.gov/niosh/mining/UserFiles/works/pdfs/ri9649.pdf (accessed on 15 July 2020).

69. Wang, Z.; Ren, T.; Zhang, J. Numerical investigations of airflow patterns on a longwall face. *Int. J. Oil Gas. Coal Technol.* **2020**, *24*, 321–344. [CrossRef]

70. Karacan, C.O. Modeling and prediction of ventilation methane emissions of U.S. longwall mines using supervised artificial neural networks. *Int. J. Coal Geol.* **2008**, *73*, 371–387. [CrossRef]

71. Zhang, P. Numerical simulation of methane delivery law in U type ventilation work face gob. *Energy Proced.* **2012**, *14*, 632–636. [CrossRef]

72. Tutak, M.; Brodny, J. Analysis of the impact of auxiliary ventilation equipment on the distribution and concentration of methane in the Tailgate. *Energies* **2018**, *11*, 3076. [CrossRef]

73. Tutak, M.; Brodny, J. Impact of type of the roof rocks on location and range of endogenous fires particular hazard zone by in goaf with caving. *EPJ Web Conf.* **2018**, *29*, 00005. [CrossRef]

74. Lu, Y.; Qin, B. Identification and control of spontaneous combustion of coal pillars: A case study in the Qianyingzi Mine, China. *Nat. Hazards* **2015**, *75*, 2683–2697. [CrossRef]

75. Su, H.T.; Zhou, F.B.; Song, X.L.; Shi, B.B.; Sun, S.H. Risk analysis of coal self-ignition in longwall gob: A modeling study on three-dimensional hazard zones. *Fire Saf. J.* **2016**, *83*, 54–65.

76. Wang, Y.; Zhang, X.; Sugai, Y.; Sasaki, K. A study on preventing spontaneous combustion of residual coal in a coal mine goaf. *J. Geol. Res.* **2015**, *2015*, 712349. [CrossRef]

77. Pan, R.; Cheng, Y.; Yu, M.; Lu, C.; Yang, K. New technological partition for 'three zones' spontaneous coal combustion in goaf. *Int. J. Min. Sci. Technol.* **2013**, *23*, 489–493. [CrossRef]

78. He, X.; Zhang, R.; Pei, X.; Sun, Y.; Tong, B.; Huang, H. Numerical simulation for determining three zones in the goaf at a fully-mechanized coal face. *J. China Univ. Min. Technol.* **2008**, *18*, 199–203. [CrossRef]

79. Huang, Z.A.; Ma, Z.Z.; Song, S.Y.; Yang, R.; Gao, Y.K.; Zhang, Y.H. Study on the influence of periodic weighting on the spontaneous combustion "three-zone" in a gob. *J. Loss Prev. Process Ind.* **2018**, *5*, 480–491. [CrossRef]

80. Taraba, B.; Slovak, V.; Michalec, Z.; Chura, J.; Taufer, A. Development of oxidation heat of the coal left in the mined-out area of a longwall face: Modelling using the fluent software. *J. Min. Metall. B Metall.* **2008**, *44*, 73–81. [CrossRef]

81. Taraba, B.; Michalec, Z. Effect of longwall face advance rate on spontaneous heating process in the gob area—CFD modelling. *Fuel* **2011**, *90*, 2790–2797. [CrossRef]

82. Yuan, L.; Smith, A.C. Numerical study on effects of coal properties on spontaneous heating in longwall gob areas. *Fuel* **2008**, *87*, 3409–3419. [CrossRef]

83. Liu, W.; Qin, Y. Multi-physics coupling model of coal spontaneous combustion in longwall gob area based on moving coordinates. *Fuel* **2017**, *188*, 553–566. [CrossRef]

84. Chen, X.; Li, L.; Guo, Z.; Chang, T. Evolution characteristics of spontaneous combustion in three zones of the goaf when using the cutting roof and release pressure technique. *Energy Sci. Eng.* **2019**, *7*, 710–720. [CrossRef]

85. Xu, Y.; Li, Z.; Liu, H.; Zhai, X.; Li, R.; Song, P.; Jia, M. A model for assessing the compound risk represented by spontaneous coal combustion and methane emission in a gob. *J. Clean. Prod.* **2020**, *273*, 122925. [CrossRef]

86. Brune, J.; Saki, S. Prevention of gob ignitions and explosions in longwall mining using dynamic seals. *Int. J. Min. Sci. Technol.* **2017**, *27*, 999–1003. [CrossRef]

87. Li, L.; Qin, B.; Ma, D.; Zhuo, H.; Liang, H.; Gao, A. Unique spatial methane distribution caused by spontaneous coal combustion in coal mine goafs: An experimental study. *Process Saf. Environ. Prot.* **2018**, *116*, 199–207. [CrossRef]

88. Zhang, Y.; Zhang, X.; Li, C.; Liu, C.; Wang, Z. Methane moving law with long gas extraction holes in goaf. *Procedia Eng.* **2011**, *26*, 357–365. [CrossRef]

89. Zmarzły, M.; Trzaskalik, P. Comparative analysis of methane concentration near the junction of the longwall and top road. *Manag. Syst. Prod. Eng.* **2019**, *207*, 166–173. [CrossRef]

90. Polska Grupa Górnicza. Available online: https://www.pgg.pl/o-firmie/oddzialy/kz1 (accessed on 15 July 2020).

91. Ordinance of the Minister of Energy on Detailed Requirements for Conducting Underground Mining Operations of 23 August 2019 (Journal of Laws of 2016, No. 2017, Item 1118, as Amended). Available online: http://isap.sejm.gov.pl/isap.nsf/DocDetails.xsp?id=WDU20170001118 (accessed on 15 July 2020).

92. Bojko, B.; Mirek, G. Connecting mining monitoring system of environmental parameters with alarming system. *Pomiary Autom. Robot.* **2010**, *2*, 572–578.

MDPI

St. Alban-Anlage 66

4052 Basel

Switzerland

Tel. +41 61 683 77 34

Fax +41 61 302 89 18

www.mdpi.com

Energies Editorial Office

E-mail: energies@mdpi.com

www.mdpi.com/journal/energies